FARNBOROUGH AND THE FLEET AIR ARM

FARNBOROUGH AND THE FLEET AIR ARM

A HISTORY OF THE NAVAL
AIRCRAFT DEPARTMENT OF THE
ROYAL AIRCRAFT
ESTABLISHMENT
FARNBOROUGH, HAMPSHIRE

GEOFFREY G. J. COOPER

A record, comprising a personal and general account of
the author's involvement with some of the activities
undertaken by the Naval Aircraft Department –
previously the Catapult Section – of the Royal Aircraft
Establishment at Farnborough, Hampshire, during
World War 2, together with an overview of subsequent
developments in Flying Navy.

MIDLAND
An imprint of
Ian Allan Publishing

First published 2008

Copyright © Geoffrey G. J. Cooper 2008

ISBN 978 1 85780 306 8

Published by Midland Publishing
Midland Publishing is an imprint of Ian Allan Publishing Ltd
Riverdene Business Park, Hersham, Surrey KT12 4RG

Printed in England by Ian Allan Printing Ltd

North American trade distribution:
Specialty Press Publishers & Wholesalers Inc
39966 Grand Avenue, North Branch, MN 55056, USA
Telephone: 651 277 1400 Fax: 800 895 4585
Toll free telephone: 800 895 4585

Visit the Ian Allan Publishing website at:
www.ianallanpublishing.com

'There be three things which are too wonderful for me,
Yea, four which I know not:
the way of an eagle in the air;
the way of a serpent upon a rock;
the way of a ship in the midst of the sea;
and the way of a man with a maid.'

Proverbs 30, 18–19 (KJV)

To Agur, son of Jakeh, a Levantine sage of some 27 centuries past, these observations expressed his amazement and acknowledged his lack of understanding of them. The association of both first and third of these features into a combined activity – as practised in mid-20th century flying with the Fleet Air Arm of the Royal Navy – would have been absolute in its incomprehensibility. His thoughts may have reflected those of a contemporary:

'Such knowledge is too wonderful for me: it is high, I cannot attain unto it.'

Psalm 139, 6 (KJV)

A bows-on view of HMS *Ark Royal*[55] in her final commission. The proboscis-like projections are the van Zelm bridle retrieval units for the steam catapults. This ship was the final word in Flying Navy with the installation of the angled deck, the BS.4 steam catapults, the DA.2 spray-jet arrester gear and the projector landing sights – all NAD/RAE developments. (© Crown Copyright/MOD)

CONTENTS

FOREWORD

By Captain E. M. (Winkle) Brown CBE, DSC, AFC, MA, FRAeS, RN
Chief Naval Test Pilot at RAE Farnborough 1944–1949
C.O. Aerodynamics Flight 1947–1949

Geoffrey Cooper, who served as a 'boffin' in the Naval Aircraft Department (NAD) of the RAE Farnborough throughout World War 2 and the immediate post-war years, has chronicled the activities of that highly innovative group to give a unique insight into the design, construction and testing of the machinery necessary to operate aircraft from the flight decks of aircraft carriers.

Deck landing is a highly demanding task for naval pilots, but without the superb developments initiated by the NAD, the speed and weight of modern aircraft would have outstripped the abilities of the pilots to cope. Instead of arriving at such an impasse, there are now modern nuclear carriers with supersonic aircraft armed with powerful weapon loads being launched from high performance catapults and, on return, landing into high capacity arrester gear on angled flight decks – largely thanks to the ingenuity of NAD. This chronicle is a fine tribute to those very dedicated people involved in that work.

Eric Brown

Captain E. M. Brown (RN), as Commanding Officer of RNAS Lossiemouth. (Capt E. M. Brown)

PROLOGUE

In compiling notes for an autobiography, the intention was to include a chapter on my experiences within the Catapult Section/Naval Aircraft Department at the Royal Aircraft Establishment (latterly known as the Royal Aerospace Establishment) Farnborough. This period was during World War 2 when research into naval flying was running at a peak and, to a young engineer, involvement in such work was a deeply fascinating and most rewarding activity. The current, in-depth treatment of this fascinating subject was precipitated in remodelling the Establishment by the abandonment of the Farnborough 'Factory Site' in favour of a new location to the west of the airfield. This took effect in 1996.

The demise of the Royal Aircraft/Aerospace Establishment has meant that its well-known and respected acronym, 'RAE' (with its distinctive title, by Royal Command), is no longer valid. These initials will apply throughout this volume to my *alma mater*, The Royal Aircraft/Aerospace Establishment (1918–96) at Farnborough.

Similarly, the Air Branch of the Royal Navy has been identified under a variety of titles during its history. During my association with the Branch and subsequently, it has been known as the Fleet Air Arm, a title that is perpetuated for archival and historical purposes with the Fleet Air Arm Museum at Yeovilton. All FAA identities in this work relate to the Fleet Air Arm and not to the USA Federal Aviation Agency.

It is not my intention to add another volume to the plethora of works covering the history, art and practice of Flying Navy, nor to issue another schedule listing details of aircraft in operational service with the past or present Fleet Air Arm. These have been satisfactorily covered by others. The subject is fully documented in the numerous valuable and comprehensive biographies and historic treatises published during the past six decades. Many of these works are listed in the appended bibliography, all of which I have consulted and many of them forming my own library on the fascinating subject of naval aviation.

This work was compiled, and research undertaken, during the closing months of the twentieth century. Since then, a number of changes have occurred. The residual part of RAE has been absorbed (via DRA & DERA) into a new organisation known as QinetiQ and occupies a new 'greenfield' site. The archive, a vast array of records, housed at that time at RAE (DERA) Pyestock, was a veritable source of valuable material but, not being a revenue earner, it has been disbanded. Fortunately this occurred just after I had made good use of its facilities. My son, Peter, has referred to this in his latest volume (#*114 p.7*).

This book is written by a Chartered Engineer – it is about engineering achievements and the engineers who made it all happen. As a consequence, it contains a large amount of data, dimensions, terms and jargon associated with the mechanical and aeronautical engineering professions. Although most of the technical information relating to the activities of the Naval Aircraft Department (NAD) has been relegated to appendices, in a few cases such occurrences appear within the text itself. My intention has been to banish most of the specialist technology to these appendices for the reader to use at his or her discretion. A few lighter moments, mostly widely diverse from Flying Navy, occasionally appear in the text.

My aim has been to place on record the activities of the Government Department which helped to make it all possible, to set down the sometimes strange goings-on ashore designed to give those afloat the 'tools with which they were to do their job'. This work is dedicated to the memory of my colleagues in the Naval Aircraft Department (NAD) at the RAE, together with our associates within the Admiralty and Ministry of Aircraft Production. In doing this I echo the sentiments of Dennis, (#*26 p.10*), a fellow RAE ex-apprentice, who:

> 'makes no apologies for … diversions from the main theme, in common with many ex-RAE staff he feels that justice still remains to be done in recording the research work of the RAE.'

He repeats an observation by Tacitus of some eighteen centuries past, which is translated *ex* Penrose (#*68*) as:

> 'My purpose is not to relate everything at length, but only such things as were conspicuous for elegance: this I regard as the highest function of history – to let no worthy action be uncommemorated.'

I am honoured that Captain Eric M. Brown RN has agreed to provide a foreword to this book. For a while, we were young men together: he being at the 'sharp end' of much research and development in the practical side of Flying Navy while I was among the shore-based 'boffins' involved in providing material and facilities for the Fleet Air Arm. He was known as 'Winkle' throughout the Service and his many 'firsts' in his contributions to aviation science are recognised and acknowledged by the well-deserved honours accorded him. When a draft of my text was submitted to him he readily provided me with this commendation. I am also most grateful to him for putting me right on points with which he was personally familiar.

At this juncture I need to point out that all serving officers to whom reference is made in this record are given the rank they held at the time of the incident or occasion under consideration. Inevitably they enjoyed subsequent promotion and further attainments in their individual careers.

One cause for confusion is the apparent random presentation of either Roman or Arabic numerals for designating Mark numbers and the like. In the immediate post-war period (*c.*1948) a decision was made in the aviation industry to use exclusively the more familiar 'Arabic' numerals. However, in the numerous books and journals relating to aviation history, the familiar contemporary designations have persisted. For example, some of the Seafire marks remain as IIC, III, XV etc, even though later marks were identified as 45, 47 etc. I feel that, for the sake of historical accuracy, such nuances should be retained.

Similarly, in the narrower historical field where accelerators, catapults and arrester gear are involved, their designation generally was by Roman numerals. It may be thought that little cause remains to retain these. My successors in NAD appear to have accepted the change to Arabic but, as only five accelerators (two American) are involved from this period, I see no reason to change for change' sake. The use of Roman numerals, where appropriate, is retained, again on an historical basis.

A further point: the text deals with the technical differences between 'accelerators' and 'catapults'. This discrete identification was necessary during the years of World War 2 but, subsequently with the abandonment of catapults on the larger classes of warships (helicopters were about to be introduced to replace shipborne seaplanes), the term 'catapult' has become all-embracing. I retain the difference in my treatment of wartime events but for post-war descriptions the term 'catapult' is used exclusively.

Throughout this work reference is made to many aircraft carriers. There may be some ambiguity as, in a few cases, a famous name has been adopted by more than one carrier; examples of which are HMSs *Ark Royal*, *Eagle*, *Hermes*, *Illustrious* and *Ocean*. It may be evident from the context to which ship reference is being made but, to avoid any element of confusion, I have adopted the technique of adding the date of the original commissioning as a superscript to the name of the ship, for example, HMS *Ark Royal*[55]. I have not done this where the other ship was of a different type, for example a cruiser instead of an aircraft carrier.

As the events recorded on the following pages occurred long before the introduction of metrication, all calculations, dimensions, quantities etc in use then were the familiar Imperial units. I was tempted to ignore metric units completely when writing this book. I am also convinced that introducing both systems of measurement makes an untidy presentation and a distraction. It is, however, important to convey to the reader the physical sizes, parameters and properties in units which he/she is able to understand. I have therefore provided SI (*Système International d'Unités*) units alongside their Imperial antecedents.

Metric dimensions and values are inserted between braces as {***}. There is one exception: as the metric 'tonne' is approximately equal to the Imperial 'ton' (within a discrepancy of 1·6 percent), I have not provided conversions where these values occur. Again, to the sailor, be he above, upon or under the waves, his rate of progress is measured in knots. A knot is a unit of linear measurement equal to 6,080 feet, or ten cables-length (1·9km) but it also represents speed. The term 'knots' is preferred to 'knots per hour'. The metric equivalent (not recognised in *SI* units) to a speed of one knot is 0·514m/sec which conversion figure has been used throughout this work. Imperial values which appear in quotations abstracted from earlier works are not given their metric equivalents.

Where reference is made to buildings or locations of experimental equipment within the purlieu of the erstwhile RAE, and identified as building numbers, these are enclosed in square brackets [***] for ready reference to the relevant site plans. Such building/site identification numbers relate to the geographical location existing at the date of closure of RAE.

Having consulted 100 or so published works, and more than 200 technical reports, written by more than 100 authors, I am encouraged by a verse from an unknown source:

> Ye Engineer, his booke profound
> Him helpeth make ye wheels go round:
> Also, ye bookes in lighter vein
> They soothe, ye engineer, his brain.

A bibliography presents some of the works, published and unpublished, which bear relevance to the content of this book. This appears as Appendix 15. This bibliography is in two sections, the first being a list of published works, while the second contains RAE 'in-house' publications and technical reports of limited distribution. Both sections are presented alphabetically by author's name, sub-listed by date of issue. For the second section I am indebted to the courtesies of the staff of the (erstwhile) DERA Archive Unit at Pyestock in allowing me access to the original records in their custody. At that time, these documents were Crown Copyright. References quoted in the text are indicated by a 'hash' (#*) where '*' is the serial number, in the bibliographical list, of the work from which the quotation or reference is abstracted. Finally, a Glossary of terms and abbreviations is included at Appendix 16.

The illustrations have their origins in a variety of sources. The majority, by far, have come from the DERA Archive at Pyestock, the erstwhile repository of the RAE photographic records and a wide range of technical papers. A smaller number of illustrations, mostly relating to post-World War 2 activities, have been provided from records at DERA, Bedford. Other sources include Falcon Publications, the Fleet Air Arm Museum at Yeovilton, the Library of the Royal Aeronautical Society, engineering contractors, individual contributors and various personal records. All sources are acknowledged and numbered throughout, with a summary of credits as Appendix 17.

All images originating from the DERA archives are Crown Copyright, and I record here my gratitude to the Director of Corporate Affairs for permission to reproduce them in this work. The sources of all images are indicated in their captions.

References to post-war projects are outside my personal experience; for articles in this area I have depended heavily upon other writers. These sources are acknowledged fully in the text and bibliography.

By far, the greater amount of my information has been sourced from RAE contemporary records, both in text and photographs held at that time by the DERA Archive at Pyestock. Abstracts from text have mostly originated in the reports listed in the bibliography. This has been supported from the Research Centre in the Fleet Air Arm Museum at HMS Heron, Yeovilton. At the time of writing, the same DERA archive housed the vast collection of many thousands of photographic plates (stills) taken throughout the existence of the Establishment. While a few of the records were transferred to Bedford in the mid-1950s, the majority still remained at Pyestock. (The ciné records, unfortunately, are not to be found at Farnborough – some of these are thought to be in the possession of the Imperial War Museum.) Subsequent to the time of my research, this archive has been dismantled. Perhaps it is fortuitous that I was permitted to enjoy access to these records – a 'just-in-time' opportunity.

During my research I was privileged to have access also to the RAE Flight Books. These comprise the record of all RAE aircraft flight movements over the period 1922–52. From these I have abstracted all movements relating to flying operations involving the Naval Aircraft Department; these include the use of catapults, accelerators and arresting gear as well as rocket assisted take-off activities.

I am aware of other sources of information, particularly the Imperial War Museum and the National Archives, but my intention was to use as far as possible original material available from local sources – RAE itself. I apologise if any valuable information has been passed over through neglect of other records. I acknowledge any error in understanding and transcription to be mine. Where I have been bold enough to contribute a comment or two, that again is mine alone and does not necessarily reflect any 'official' opinion or attitude.

Finally, I am indebted to a host of people who have experienced some feature or another in the pursuit of excellence in the art and practice of Flying Navy. Some contributors were my erstwhile colleagues; others have followed the footsteps of us pioneers; many have operated the equipment in their service careers with the Royal Navy. To all I record my appreciation and thanks in Appendix 17, and submit my apologies for any contributor who, inadvertently and unintentionally, has been passed over.

INTRODUCTION

March 1940 was the month of my twentieth birthday and signalled the beginning of the final year of my four-and-a-half-years' training as an Engineering Apprentice with the Royal Aircraft Establishment (RAE) at Farnborough. For this terminal period each apprentice was invited to select two departments in which he would specialise and then, if satisfactory, be offered a post within one of them at the conclusion of his apprenticeship and the start of his professional career. The Catapult Section was not my choice – I was not even aware of its existence – but before I was required to make my selection I was invited by the head of the Section, Lewis Boddington, to an interview. I was accepted and thereby became the first apprentice to be appointed to the Catapult Section/Naval Aircraft Department. This started a trend which continued for decades. By this time much of the incidental activities of the place had become familiar to me. I now place on record some of these procedures before they disappear into oblivion.

In March 1940 the Royal Navy, Fleet Air Arm (FAA), had five aircraft carriers in commission. This number excludes HMS *Argus* (1918–46), which was mainly engaged upon training and ferrying activities, and HMS *Courageous* (1928–39), which had been lost through enemy action within the first month of World War 2. These 'service' carriers were:

HMS *Furious*	(1917–48)
HMS *Glorious*	(1918–40)
HMS *Eagle*	(1924–42)
HMS *Hermes*	(1924–42)
HMS *Ark Royal*	(1938–41)

(The first of the two dates given for each ship represents the year when the ship was completed as a carrier and when the first flights took place.)

Of these ships, four were vintage specimens, having been constructed upon the hulls and machinery of World War 1 ships. The sole modern ship was HMS *Ark Royal*[38]. Of the main fleet carriers only HMS *Furious* was to survive the war, the others by August 1942 all having been lost through enemy action.

My initiation into the intricacies of Flying Navy occurred at the time when aircraft in first line squadron service within the Fleet Air Arm (FAA) comprised the following types:

Blackburn	Skua (1937–41)
	Roc (1938–43)
Fairey	Swordfish (1934–45)
	Albacore (1938–43)
Gloster	Sea Gladiator (1938–41)
Hawker	Nimrod (1931–41)
	Osprey (1931–40)
Supermarine	Walrus (1933–45)
	Sea Otter (1938–46)

Apart from the first two, all were biplanes, each possessing, as one wag declared, 'a built-in head-wind and permanent induced drag'! These terms will be familiar to those whose studies included the subject of aerodynamics. The Skua was the victim of another trite comment – 'a predatory sea-bird which folds its wings for diving into the sea'.

The Fairey Seafox (1936–43), another biplane, was also in commission but operated as a floatplane and was not involved in operation from aircraft carriers. The Blackburn Shark (1933–8), Fairey IIIF (1926–36) and Fairey Seal (1933–8), although obsolescent, were still in use at RAE for experimental work. A few Shark aircraft were fitted with 'concrete engines' (ballast in lieu of engine) and used as dummies for unpowered, ballistic catapult trials – 'dead launches' – at RAE and on shipboard. One Fairey IIIF aircraft (S1317) was a venerable workhorse, and frequently featured in live launches from the catapults on the Jersey Brow site. This eleven-year-old aircraft had become obsolete by 1940. As a type it was superseded by the Fairey Seal which, in spite of itself having been replaced in FAA Squadrons by the Fairey Swordfish, was also in use at Farnborough. I was impressed by the size of the rear cockpit – it appeared to be larger than and as wide open as a bathtub!

Till (#*ibid. p.99*) writes:

'Only the Royal Navy went to war with large numbers of biplanes in its front-line forces. Its basic aircraft was the Fairey Swordfish, production of which was behind schedule and which did not perform as well as the types currently coming into service with the US and Japanese navies, such as the Douglas TBD Devastator and the Nakajima B5N2 Kate. The fighter situation offered little comfort either. Neither of the Navy's first two monoplanes, the Blackburn Skua fighter/dive bomber and the Roc, its turreted version, lived up to expectations. The FAA was accordingly forced to improvise with a few obsolescent Sea Gladiator biplanes for a few years more. Moreover, there were no designs in hand for the kind of single-seat high-performance fighter that was beginning to enter service with the US and Japanese navies. "There is," concluded the First Lord in January 1939, "a serious deficiency in fighter aircraft, of which the FAA has virtually none." By the beginning of the Second World War it was evident that something had gone seriously wrong with British naval aircraft design.'

As reported again by Till (#*ibid. p.85*), the Commander-in-Chief, Home Fleet, in July 1940 tersely summed up the situation:

'Our Fleet Air arm aircraft are hopelessly outclassed by everything that flies ...'

Yet within four months of this statement, and with the same Swordfish aircraft, the significant Fleet Air Arm victory of Taranto was achieved!

This was the environmental scene into which I was inducted in March 1940. I was a greenhorn with regard to knowledge of Navy flying but I was abounding in enthusiasm. Little could I foresee the outcome of this step, and the wonderful sense of satisfaction and achievement over the next five years and more that was to be mine.

SIFTING THE ARCHIVES

While this work is not intended to be an in-depth historical record dedicated to the science of naval aviation, a number of salient points with relevant dates will present a backcloth to the involvement of RAE with the Fleet Air Arm. Many of these dates have been abstracted from Caunter (#222). The page references are from this report.

S. Child and C. F. Caunter, both on the research staff of the Establishment, were commissioned to provide *A Historical Summary of the Royal Aircraft Factory and its Antecedents: 1878–1918* which was produced as an RAE internal Departmental Report, AERO 2150, in 1947. This was followed a couple of years later by a further report, authored by Caunter alone, *A Historical Summary of the Royal Aircraft Establishment: 1918–1948*, report AERO 2150A. The author adopted a chronological and departmental presentation of the highlights and essential facts of technical and scientific achievement and policy within RAE over this period of three decades. He admits to having researched a diversity of subjects and an abundance of data provided authoritatively by contemporaries. As a result of his painstaking work his report generally can be taken as a reliable record of events throughout that period. Cooper (#22) presents an up-to-date appraisal (1996) of some of these historic incidents.

The first Caunter reference to Farnborough's Naval activities (#ibid. p.23) is that:

'Percy Salmon began work in 1922 on the problem of launching aeroplanes by catapults, with Jersey Brow for a testing ground.'

Mr Salmon had been recruited in 1916, having received his engineering training with the London, Brighton and South Coast Railway workshops.

A footnote here relates:

'In the summer of 1909, Captain P. W. L. Broke-Smith, RE, erected a weight and pylon catapult on Jersey Brow for the launching of the Wright biplane which the Hon. C. S. Rolls had brought to the Factory for the training of Army officers.'

There appears to be no record extant of this equipment. However, on 13 July 1912 *The Aeroplane* published a proposal based upon similar principles for launching aeroplanes from capital ships: the driving weight was to operate within a hollow mast (see Lewis (#60 p.19)).

Continuing with Caunter, the first reference to an arrester hook (#ibid. p.43) states that in 1922,

'A restraining hook was designed and fitted to a Sopwith Snipe for deck landing.'

I am unable to trace any record of this but Penrose (#70 p.249) and Thetford (#97 p.307) both refer to a Sopwith Pup (N9497) as carrying an experimental arrester hook in 1918. Flying trials aboard HMS *Furious* (with the divided flight deck of that period) were deemed to be unsuccessful.

A later paragraph refers to pilotless aircraft, the RAE 1921 Target Plane and the Larynx, as being, in turn:

'launched by means of a falling water bag up an inclined runway; and later by a simple form of cordite catapult from the deck of a destroyer.'

In October 1922, Caunter continues (#ibid. p.44), with the Main Drawing Office initiating a development programme for aircraft launching catapults:

'The type of catapult evolved under the direction of P. Salmon was of the telescopic ram type actuated by compressed air. The Establishment pioneered the aircraft launching catapult in this country.'

Still in the Main Drawing Office, in September 1925 (#ibid. p.56, fig.42.1):

'The first compressed air telescopic ram type catapult for aeroplane launching was completed and installed on Jersey Brow and the first launch of an aircraft (SEAGULL amphibian) was successfully made with Flight Lieutenant R. de Haga Haig as pilot. A velocity of 66ft/sec (45mph) was reached in a total time of 1·04 seconds. With a maximum weight of aircraft of 7,000lb, a launching stroke of 34·5 feet and an acceleration of 2g were employed. This work was carried out under the direction of P. Salmon, and with the assistance of J. D. H. Pritchard, S. Child, C. Crowfoot and A. R. Crossfield.'

These names were still to be encountered a couple of decades later. In May 1926, Caunter continues (#ibid. p.61):

'The RAE compressed air catapult was converted to cordite operation and the first successful launch (Seagull) using cordite as propellant was made. The cordite was fired in the usual way by either a percussion or electric firing tube. Much development work was carried out by P. Salmon to determine a suitable charge.'

Also:

'A Parnell Peto light reconnaissance aircraft was test launched from the RAE catapult in order to prove it for launching in service from the Submarine M2.'

(Shortly after, this submarine was lost during diving trials in Weymouth Bay; it was thought that the ship foundered through the sea breaking into the hull via the hangar doors.)

In July 1927 (#ibid. p.62, fig.19.4) the first launch of a Larynx pilotless aircraft was made in the Bristol Channel from the destroyer HMS *Stronghold*. The paragraph continues with the recording of successful launches in the English Channel, culminating in final tests in Iraq. This catapulting programme ran for a couple of years.

In May 1928 (#ibid. p.67):

'Development work with a cordite-operated catapult enabled the first successful launch of a full size aircraft to be made at sea from a catapult from HMS *York*.'

The Main Drawing Office in January 1929 (#ibid. p.71, fig.42.2) records:

'A cordite catapult, subsequently known as the Farnborough IV.H, which resembled the Farnborough Mk.IV heavy type, was designed and built at the Establishment for the Admiralty and tested successfully prior to installation in HMS *Hood*. This had a considerably increased performance over the original unit of 1925, and incorporated new features, namely, the structure was capable of being folded for stowing on board and rotated for training on any bearing. Folding and opening was performed by compressed air engines. It was of the telescopic ram type, four concentric rams giving an aggregate stroke of 610 inches, a maximum velocity of 88ft/sec (60mph) and a capacity of 7,000lb with an acceleration of 2·36g.'

Years later, Alf Crossfield, (himself, an ex-naval stoker) who supervised the shipboard installation of this equipment, told me that his

The RAE, looking north-west, as it appeared until its recent closure and demolition. Jersey Brow is beyond 'A Shed', the white-roofed single-span hangar towards the top left corner. (Peter J. Cooper/Falcon Aviation)

changing into working overalls had to be done below decks; it was not 'correct' for officers to be compelled to witness such a menial activity occurring on the upper deck.

In a record against June 1930, Caunter (#*ibid. p.77, fig.42.3*) writes:

'catapult designed by P. Salmon to launch an aircraft of 18,000lb weight at 60mph in 100 yards with an acceleration of 1g. It consisted essentially of a windlass driven through gearing by two swashplate compressed air engines, by means of which the aircraft was accelerated on its own undercarriage wheels by a cable and hook device. The compressed air was at a pressure of 400lb/square inch, and the two power units were together capable of developing 4,000hp. This catapult was successfully demonstrated at Hendon during the Royal Air Force Pageant in 1931.'

Again, an entry against the Main Drawing Office in 1931 (#*ibid. p.81*):

'First controlled experiments to develop an arrester gear using athwartships wires were made on the airfield … initial landings were carried out with a Fairey IIIF aircraft. The arrester gear was supplied by the Admiralty and the aircraft was suitably modified by the makers. The Establishment initiated the 'snap gear' in these trials, which were followed in 1931 by the installation of an improved type of gear in HMS *Courageous*.'

Two years later (#*ibid. p.85, fig.19.5*) reference is made to catapult trials on Queen Bee pilotless aircraft.

The Main Drawing Office continued to be featured in November 1935 (#*ibid. p,92*) when:

'The design was commenced by P. Salmon of a large hydro-pneumatic catapult capable of launching the then projected heavy bomber aircraft which it was anticipated would weigh up to 65,000lb. During 1938/39, the work of installing this catapult proceeded at Harwell since there was no convenient site at the Establishment. Work proceeded during the early days of World War 2 but had to be dropped eventually because of pressure of other work . . .'

This is treated later in the text under the heading of 'Mark III Catapult'.

The Main Drawing Office features again in 1937 (#*ibid. p.99, fig.43.1*) with:

'A complete investigation of the problem of accelerator gear for launching aircraft from carriers was carried out by P. Salmon and L. Boddington, on the basis of Admiralty requirements of launching at the rate of an aircraft every 40 seconds, which was very high compared to previous achievements. Various schemes were devised and tested on Jersey Brow in addition to trials on carriers already in service. The design of the prototype equipment was started in March 1939; it was fitted to HMS *Illustrious* in May 1940 and was immediately taken into service. In the early part of 1939 it was possible for the Admiralty to crystallise their policy and a final scheme was produced by the Establishment which was accepted by the Joint Technical Committee for aviation arrangements in HM Ships.'

The development of this accelerator appears later in the text under 'Assisted Take-Off Gear (BH.III Accelerator)'.

The final relevant extract that I wish to take from Caunter (#*ibid. p.105*) occurs for September 1938 in which:

'The Catapult Section (L. Boddington) was formed within the Main Drawing Office under P Salmon.'

It was upon this scene that, eighteen months later, I made my appearance.

Caunter refers to other projects subsequent to the above 1938 entry. Those relating to Naval Aircraft Department and its forebears are treated in detail in my text, often as a result of my own personal involvement in their development. A historical summary of events in Farnborough's contribution to Flying Navy appears as Appendix 1.

As to be expected, a valuable source of historic material was, at the time of writing, the DERA Archive Department on the site of the National Gas Turbine Establishment (RAE, Pyestock) wherein were housed most of the documents issued by the Section. In addition. papers submitted to the contemporary Aeronautical Research Committee and previously held by the RAE library were stored in this archive. These records originally were official classified documents, distribution being closely restricted to those in the 'need-to-know' category within the defence constraints of the Official Secrets Act. The privilege accorded me by the archive personnel in the generous provision of making these papers available for research is appreciated and hereby acknowledged. The papers relating to Fleet Air Arm activities are listed in the Bibliography, Appendix 15.

One of the interesting features discovered in the reading of these papers is the record of the official attitude of the Admiralty to Navy flying in 1939. World War 2 was three weeks old, the FAA had already lost HMS *Courageous* to enemy action, yet ARC Paper 4211 of 25 September 1939, *Some Problems of the Fleet Air Arm* reports:

'The Fleet Air Arm exists solely as part of the Navy – it has no separate justification; it has no duties independent of the Navy. Its resemblances to the RAF are more obvious perhaps than its differences, but it is the latter which cause the Navy the most concern since these have necessarily received least attention in the past, and the necessity is increasingly felt for aircraft designed *ab initio* to meet these needs, dominated as they are by limitations peculiar to aircraft operation at sea. . . . The first duty of the FAA is to find the enemy fleet before it finds theirs, and the second is to destroy his carriers by immediate attack on them. The aircraft must then shadow the enemy fleet, and as the heavy ships approach within gun range, observations of enemy movements, and gunfire spotting, become the most important functions of the aircraft.

'The aim of the aircraft, when engaging the enemy, may be stated in three stages in the order of their occurrence:

Approach – requiring deep reconnaissance to find, and then shadow, the enemy fleet, and particularly his carriers.

Air Contact – requiring immediate attack on enemy carriers, or heavy ships.

Battle – requiring action, observation and spotting. Striking may be used as a subordinate function.

'In battle, the principal duties of the fighters are:- (1) defence of our own spotters; (2) attack on enemy spotters. Their use in protecting the fleet against air attack has hitherto been regarded as impracticable and although recent developments in RDF are making it a potent method of defence, considerable reliance for defence is still placed on AA gunnery.

'The characteristics, which an aircraft should possess in order to meet the needs, enumerated are:

I. Non-Attacking
    Reconnaissance – to find the enemy
    Observation – to report his movements
    Spotting – to direct gunfire, one aircraft for each firing ship
    Night Shadowing – to prevent escape during the night

II. Attacking
    Torpedo – the principal weapon
    Dive Bombing

III. Miscellaneous
    Fighters with fixed guns or turret guns
    Anti-Submarine patrol ahead of the Fleet
    Amphibian.

The paper continues with an appraisal of operating problems.

'Carriers:

Aircraft are limited in weight and dimensions by the width of the deck, size of lifts, hangar space, strength of lifts, and accelerators. The present absolute weight limit is 12,000lb and dimensions 50ft span, 13ft-6in height and 40ft length. Folded width is now 18ft, but it is hoped to reduce this to 13ft-6in and eventually to 11ft, giving respectively 3, 4 and 5 rows of aircraft in a 62ft hangar.

'Free take-off has been relaxed to 300ft against a 20 knots wind, in view of the impending introduction of Assisted Take-Off (a quick loading, long stroke accelerator giving 1⅛g). Accelerators give 3¼g as a maximum but are rather slow in operation. These permit take-off in cross winds up to 30 knots.

'Catapult Ships:

Though floatplanes are now carried they are likely to die out as regards sea operation, and need not be considered for the future, except possibly from the point of view of occasional conversion of the TBR type to operate in harbour only.

'Amphibians taking off from catapults have much the same limitations as aircraft launched from accelerators. Amphibians landing at sea must be extremely robust and must have a stalling speed of 45 knots or less. The landing run after touching down must be short to enable the boat to get down and stay down in the smooth patch produced by the turn of the parent ship. Good stability and control is essential when in the water, to enable the aircraft to come alongside and hook on to the crane while the parent ship is steaming up to 12 knots.'

This was the philosophy of Naval action with the FAA in 1939. It was hardly bettered when a subsequent ARC paper produced by another Admiralty department was dated February 1940, being a re-issue of a 1925 paper. This paper, over the name of L. W. Bryant, was titled *Note on Deck Landing – Methods of Reducing the Landing Run (#4,401)*. Bearing in mind that early experiments in deck arresting by mechanical means had been abandoned, the author intimates:

'The Aerodynamics Department of the National Physical Laboratory (NPL) has recently been asked to consider the possibility of reducing the landing run of an aeroplane on the deck of an aircraft carrier by means of an artificial blast or jet. In view of the fact that this suggestion has already been made several times in the course of the past six years the following notes have been prepared in order to state, as clearly as possible, the problem involved.'

The writer continues to declare the general problem of determining the shortest possible landing run to have been thoroughly examined in two reports to the ARC (1919–20) showing that:

1) mechanical brakes on wheels are advantageous,
2) wing incidence should be large,
3) air brakes are of little value,
4) a reversible pitch airscrew (propeller) being the most powerful form of brake that can be used,
5) the best way to reduce stalling speed was to provide variable-camber wings,
6) the effect of head wind.

Relating to the last feature, trials in HMS *Eagle*[24] had produced a comparison of landing runs on deck after touchdown against the relative head wind as below:

Head wind (knots)	Landing run (feet)
20	320
25	221
30	142
35	80
40	37

These related to a touchdown speed of 51 knots.

Having discussed the relative merits of the above points for reducing the length of the landing run, the writer continues his proposal for the production of an artificial current of air by the creation of a stream about 50ft wide and 15–20ft high along the deck. He acknowledged that this would need an elaborate and not very portable arrangement of motors and airscrews (propellers), about eight in number, each capable of producing a slipstream of at least 30 knots for operations with zero relative wind. Each of these airscrews would have required a power source of some 120hp. Whimsically, he continues with the conclusion that the suggested artificial current would be unlikely to find favour among the constructors and users of aircraft carriers, since they have to provide for the rapid adaptation of the deck arrangements to suit both launching and landing, and for stowing the current-producing apparatus during the very considerable periods when it would not be required. A later comment penned alongside the entry in the register is pointed and dismissive:

'None of the methods discussed are worth considering at present, 19–3–40'.

At least RAE can be credited with having produced something more feasible and reliable as effective, working solutions to the problems pertaining to the mechanical augmentation of flight deck operations.

ORIGINS

The Army Balloon Factory had been in existence on Farnborough Common for only a few years when, in 1909, the Viscount Haldane (Secretary of State for War) appointed Mervyn O'Gorman, under a seven-year contract, to the post of Superintendent of the Factory. O'Gorman was the third to hold this post and the first civilian to do so, following Col. J. M. B. Templer (1875–1906) and Col. J. E. Capper (1906–09). King (#56 p.71) writes that O'Gorman was to be responsible direct to Lord Haldane as his brief would incur hostilities from a number of sources, mostly military. Haldane had instructed the newly formed Advisory Committee of Aeronautics that its main task was to reform the Balloon Factory and put it on a more scientific basis. Haldane had seen that the practical, artisan approach of Capper and Cody, laudable as it was, would never lead to real progress. The Committee had visited the Balloon Factory on 5 July 1909 where it was clear that the workshops had been 'reduced almost to a shambles'. It probably never had been in any better shape.

King (#ibid. p.77) continues to state that O'Gorman was about ten years older than the aviation pioneers and was a man of commanding intellect and considerable verbal skill. He had worked as a consulting engineer in the automobile business in both Britain and France and could express his thoughts vividly in writing and colloquially. He also had the ability to choose good men to work with him and to retain their loyalty. He was convinced in those early days of the superiority of the aeroplane over all forms of lighter-than-air craft, revealing this position by changing the name of the set-up from the 'Army Balloon Factory' to the 'Army Aircraft Factory' and then, consequent upon a visit by His Majesty King George V in 1912, by Royal Command to the 'Royal Aircraft Factory'.

One man whose loyalty O'Gorman attracted was Frederick Michael Green, an engineer also in the automobile industry, who for five years had been Chief Engineer to the Daimler Car Company. He moved to Farnborough and immediately set about re-planning the workshops and creating the centre of research that Farnborough was to become. Child (#223) states that, at that time, the Factory consisted of one small machine shop, one shed [No.3 Building] for making balloons, and one airship shed [No.29 Building]; about 50 men and 50 women were employed. This was very much a 'basic' situation with a great potential for improvement. Green was appointed Engineer in Charge of Design and was responsible for the Beta, Gamma, and Delta airships, most of the aeroplanes from SE.1 and BE.1 and all of the engines (beginning from the RAF.1a) during a period of some seven years up to 1916. According to Child (#ibid. p.16), O'Gorman was responsible for introducing the nomenclature for the aircraft marques of that period, viz: BE – Bleriot Experimental, FE – Farman Experimental and so on. The events of that period already are well documented.

The Royal Flying Corps, formed by Royal Warrant on 13 April 1912, contained both a Naval Wing and a Military Wing. The intention was for them to operate as a single force rather than branches of two different services but, according to Rawlings (#81 p.11), neither the admirals nor the generals were sufficiently sold on aviation to see it as the potent force it might have become. (At this time the combined flying personnel of the Navy and Army totalled 19!) A month later the Central Flying School was formed at Upavon with Captain Godfrey Paine RN as Commandant and two pioneer Naval pilots, both ex-Eastchurch, as instructors.

Within a few weeks Farnborough came into the picture, but not to do with piloting of aircraft! The Admiralty appointed Lt L. R. Fitzmaurice

RN to full time development work on airborne W/T (Wireless Telegraphy), on detachment to Farnborough. His first job was to supervise the W/T equipment in the Army's airship, Gamma, for military manoeuvres. This was a significant appointment as, according to Child (#ibid. pp.14 & 17), it was only on 27 January 1911 that the first instance of 'wireless' communication between a ground station and an aircraft (Beta airship) was achieved. The first successful use of airborne W/T – with a BE.1 aircraft – for artillery spotting on Salisbury Plain occurred just a year later. Incidentally, a further ten years were to pass before the Wireless and Photographic Department was set up at Farnborough (Caunter – #222 p.44).

This was followed on 25 September 1912 by the Admiralty resurrecting the Naval Airship Section and putting Cdr E. A. Masterman RN in charge, with three lieutenants and some ratings in his command. This, again, was at Farnborough. In November 1912 the Air Department was formed within the Admiralty ostensibly to administer the Naval Wing of the Royal Flying Corps. In practice it proceeded with policy making for the Navy's own Air Service of the future, including an embryonic design for an aircraft carrier which proceeded no further. This is regarded as the 'real' date from which the Fleet Air Arm came into being and shortly after began to demonstrate its independence. It has to be remembered that, although the major part of the Royal Navy had no use for aviation, the strides made by the handful of naval aviators in the first two years of activity were formidable.

During World War 1 a certain aggressive tension was felt at home by many young men (some, including the author's father, with medical problems) who had not gone to the Western Front but were kept at home on 'Munitions'. In an attempt to counter this all the staff at the Royal Aircraft Factory were put into uniform and given military rank. O'Gorman, as Superintendent, was appointed Lieutenant-Colonel; his senior staff – including Green – were appointed Major, with corresponding ranks for the lesser fry. Not everyone was happy with this but it did serve a useful purpose. These commissioned ranks were sported by a few of their holders for the remainder of their days.

O'Gorman's contract as Superintendent terminated in 1916 and he was not offered an extension. His successor was H. W. Fowler, lately Chief Mechanical Engineer to the Midland Railway. King (#ibid. p.117) conservatively suggests that the latter was hardly the man for investigating the frontiers of technology in this new and exciting science. With Fowler's coming, Green and others were encouraged to leave Farnborough and relocate in private industry. Green went to Siddeley-Deasy Motor Company (later, Armstrong-Siddeley, one of the budding aero-engine manufacturers) as chief designer. His name reappears later in these pages.

Fowler himself, although receiving a knighthood, was replaced after two years as Superintendent. Apparently he returned to railway engineering, becoming Chief Mechanical Engineer to the London, Midland and Scottish Railway after the 1923 grouping, and became a Commissioner in the Boy Scout movement. Could he have been the instigator who decreed that the colour to be adopted universally for interior and exterior decoration of all RAE buildings was to be 'Midland Red' – the colour of LMS locos and rolling stock? On my entry to RAE, that was the predominating colour up to dado level, with broken white above, for laboratories, offices and workshops. Beta Shed was not exempt – the whole exterior was of this rusty shade.

In 1916 another railway engineer, Percy Salmon, was recruited from the London, Brighton and South Coast Railway to the post of Head of the Main Drawing Office. In the early days of Navy flying there seemed to be little liaison – on a regular basis – between the Royal Navy and the Royal Flying Corps. Brian Kervell, formerly Curator and Archivist of the RAE Museum before its closure in November 1994, writes (in a draft treatise on *The Officers' Mess on Farnborough Aerodrome*) that the RNAS (Royal Navy Air Service) seems to have had little to do with Farnborough during World War 1. While the Royal Flying Corps acquired its early aircraft from the Royal Aircraft Factory at Farnborough (the precursor of the RAE), the Royal Navy from the outset preferred to deal direct with independent manufacturers of aircraft. It was not until the formation of the Royal Air Force – which 'absorbed' Navy Flying into its 'Fleet Air Arm' – in 1918 that the newly entitled Royal Aircraft Establishment itself appeared to take any activity in naval matters.

Kervell continues:

'The coming of peace following the Armistice meant a very considerable contraction of military activity with, soon, much of the recently formed Royal Air Force deployed throughout the troublesome parts of the Empire. Thus Farnborough became aeronautically quiet, particularly so for over a year in the early nineteen-twenties, when flying moved almost in total to the ex-RNAS Testing Station on the Isle of Grain.'

From Salmon's appointment, RAE appeared to assume an increasingly prominent role in Flying Navy, presumably stemming from his interest in mechanical methods for launching aircraft.

FLYING NAVY

Flying Navy, at its inception, was completely devoid of any mechanical assistance. There was no flight deck, no catapult/accelerator, no arrester gear, no flying control – only the skill and initiative (dare-devilry?) of the pilot. Becoming airborne was a calculated risk. Recovery to the vicinity of the ship – certainly not an on-board landing – was unpredictable and as thrill-making as any young man could desire. Often it was a case of – using a parody of Cunningham:

> 'A wet (shirt) and a flowing sea,
> a wind that follows fast . . .
> The world of waters is our home,
> and merry men are we.'

The following paragraphs treat with early experimental work of shipboard flying-off (Getting Airborne) and landing-on (Returning to 'Mother') while under way.

GETTING AIRBORNE – UNASSISTED

Early feats of flying from the deck of a ship using platform structures are well documented elsewhere, for example Cronin (#23). These were known as 'Flying-Off Platforms'. In a few cases the aircraft, particularly a floatplane, was mounted on a wheeled trolley, sometimes running in rails fixed to these platforms. No external source of power was applied. Various schemes were tested. Among the first were flat and sloping platforms installed in the bows of the ship. Some platforms were erected over the larger, main armament gun turrets, demountable sections being supported by the gun barrels. Others were mounted as rearward-facing platforms, permitting the guns to be used without prior dismantling of the launching structure.

The first recorded take-off from a ship relates to an American civilian pilot, Eugene B. Ely, who flew from USS *Birmingham* in 1910. Late in 1911 work was put in hand in the UK to fit a timber stage ranging from – and over – the front turret to the fo'c'sle on two warships, HMSs *Africa* and *Hibernia*. On 2 May 1912, Lt. C. R. Samson took off along this sloping deck in a Short S.27 aircraft (No.38) from HMS *Hibernia* while under way. The plane, which was fitted with flotation bags in addition to its wheeled undercarriage, landed ashore.

Before long, platforms were fitted to one or another of the twin 12in {300mm} (or larger calibre) gun turrets on various capital ships. Firstly the decking ran from the top of the turret to the muzzles, supported by the gun barrels themselves. Wheeled aircraft were flown from these installations, one of the advantages being that a certain degree of orientation for take-off could be achieved by rotating the turret into a favourable wind direction. A disadvantage ensued from the need to unship the decking from the guns and to lash and stow it to the turret roof when the guns needed to be fired. A compromise was to reverse the direction of take-off from the turret roof, using a shortened platform. Here a trestle between the gun barrels could be used to support the tail of the aircraft. A few instances are recorded in which a special turntable was mounted on the deck of a ship on which an aircraft launching ramp was provided. Again the advantage claimed for this was that the turntable could be orientated (without the restriction of the travel of the gun turret) to suit the local wind direction and that the aircraft would be out of the way of flashback from gunfire.

Short modified S.27 aircraft, piloted by Lt Charles Rumney Samson RN, lifting off from HMS *Hibernia* in Weymouth Bay on 2 May 1912. This was the first flight of an aircraft from a ship while under way. (Royal Aeronautical Society Library)

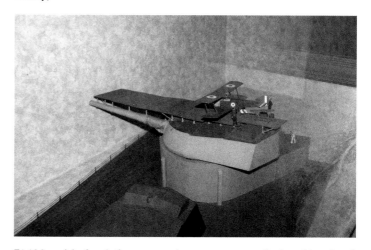

FAAM model of a platform mounted upon a gun turret for launching aircraft without mechanical assistance. This was standard procedure adopted during World War 1 and not abandoned until the introduction of catapults. (Author via FAAM)

Both the fixed fo'c'sle staging and the turret/turntable platforms were used for wheeled aircraft. A floatplane would be mounted on a simple trolley which ran along the staging. The tail wheel/skid would be mounted on a fixed trestle to support it in a tail-up position for the first yard or so of its take-off run. A simple holdback unit, comprising a quick-release link (Senhouse slip?) was employed to restrain the aircraft against propeller thrust until it was triggered to release the aircraft for take-off. The last turret-mounted platform installations were removed from HMSs *Emerald* and *Enterprise* in 1935/36.

Addressing the Royal Aeronautical Society in 1932, Salmon (#87) recalls (bearing in mind that the references to 'carriers' implies parent ships, not aircraft carriers as such):

'It was not until 1916 that any steps were taken in this country to develop the catapult. Up to that date aeroplanes operating with the Fleet were carried in vessels which had been obtained from the Mercantile Marine and converted for the purpose; these vessels being practically defenceless had to operate well in the rear of the Fleet. The seaplanes had to be hoisted overboard by a crane and

off a Sopwith Camel in July 1918. These lighters were fully decked, each carrying an aircraft handling crew. Hayward (*#42 p.31*) records that the deck of the lighter was camouflaged and carried a pair of aircraft identification roundels. Similar experiments were repeated during the early 1930s at Felixstowe with a lighter towed at high speed by a destroyer; this became known as the Porte system. The forward speed allowed the aircraft, with a short run, to take off, leaving no provision for its recovery, which had to be accomplished by flying ashore, or ditching. (Shades of the CAM-ship project a decade later!)

The earliest, purpose-built flying-off deck of any significance was that fitted in the bows of HMS *Furious* during her first commission as a cruiser, June 1917. This followed the adaptation of HMS *Campania* in 1916 until her loss in November 1918. At one period, HMS *Furious* was fitted with a guide slot in the forward deck in which a four-wheeled carriage supporting the aircraft floats/wheels would run.

A reference to HMS *Vindex* having been fitted with a catapult could be confusing. As the World War 2 ship, HMS *Vindex*[43], was not thus equipped, it is suggested that the fitting of a slotted tube launcher device to the seaplane carrier, HMS *Vindex*[15], is intended. This earlier *Vindex* was fitted with a flying deck of 64ft {20m} in length from which two Bristol Scout 'C' aircraft (1246 & 1255) were operated. Problems with the aircraft persistently veering off-course when flying from the deck had to be addressed. Experiments on a shore station (possibly Eastchurch) resulted in a section of pipe about twelve feet {3·7m} in length being supported about three feet {900mm} above ground on three trestles in alignment with the direction of take-off. A slot was cut along the top of this pipe to provide a guided runway trough for the tail skid of the aircraft, allegedly 'to maintain the tail at its most advantageous height for flying and getting off'. This elementary structure, given the name of launcher tube, was dismantled and transferred to the ship.

Penrose (*#70 p.248*) opens up the picture in explaining that during World War 1 the Navy flying establishment at Eastchurch (Isle of

flown off from the water, the carrier having to slow down in order to perform this operation. From then there was always the difficulty of getting the seaplane off the water and into the air in any but favourable weather conditions. In 1916, this difficulty was overcome by building a special deck on the carrier from which the aeroplane could fly off. This at the same time permitted the use of land aeroplanes having better performances compared with seaplanes. Other devices were later used in 1917, notably the fitting of revolving platforms on the turrets of light cruisers and capital ships from which the aeroplanes could take off head into wind without the ship having to alter course, use being made of the speed of the ship and the height of the turret above the water to gain the necessary speed for flight.'

In his paper, *Flying-Off Platforms, 1917–1934*, R. H. Nailer states that 33 battleships, nine battle-cruisers, two large light cruisers and 32 light cruisers were fitted with one type or another of these platforms. He also says that, in 1918, experiments with destroyers and mine-sweepers, fitted with similar platforms, were generally unsuccessful. Further experiments with versions of the Hein Mat, towed by HMS *Ark Royal*[14] met with only limited success. Rather more effective were lighters, towed by destroyers from which an aircraft could be flown and, hopefully, recovered on return from its sortie.

Thetford (*#97 p.17*) records the adoption of a towed lighter for launching aircraft at sea. Such lighters were 58ft long by 16ft beam {18m x 5m} and were towed behind a destroyer at a speed of 32 knots {16m/sec} without creating sheets of spray. The original intention was for this system to be used for launching floatplanes, thereby extending their operational range. A lighter towed by HMS *Truculent* was first used to fly

Short 184 floatplane exiting its guided, but unpowered, launching trolley on the forward flight deck in HMS *Furious*. July 1917 (Royal Aeronautical Society Library)

A rare picture of the Lee Richards annular biplane glider at an unidentified location but which could be Farnborough in 1912. (Royal Aeronautical Society Library)

Sopwith Pup, with wheel undercarriage, aboard HMS *Furious*. April 1918. (Fleet Air Arm Museum)

Sopwith Pup having been landed by Lt Acland on the after (landing-on) deck aboard HMS *Furious*. April 1918. (Fleet Air Arm Museum)

Sheppey) was augmented by the formation, in 1916, of the Marine Experimental Aircraft Depot at a new aerodrome in the nearby Isle of Grain. An experimental rig was developed the following year at the Marine Experimental Station at Grain using a railway truck as the launching carriage. Sq/Cdr Harry Busteed was appointed Commanding Officer and participated in a great many test flights. One of his reports records:

'Another experiment was launching seaplanes into the air from the land in order to eliminate risks and delays inseparable from operations off the water. I did this by putting them on railway trucks and charging down our siding at Grain under the aircraft's own power. It was entirely successful, gave one a thrill, and was good fun.'

Such procedures, although interesting in their development, were of little benefit and the adoption of catapults for launching floatplanes from capital ships was to become a regular feature of FAA operation.

GETTING AIRBORNE – POWER-ASSISTED

Although the foregoing installations were devoid of any form of mechanical assistance, there were forward-thinking people who could see the future need for some means of applying external forces to the take-off process. Cronin (#*ibid. pp.17&70*), who provides an interesting mass of detail on early days of Navy flying, records that a patent, No.3333, was taken out in February 1911 by M. F. Sueter (a prominent name in the history of naval aviation), F. L. M. Boothby and H. G. Paterson for a gravity-operated shipborne aircraft catapult. This scheme was for a sloping fo'c'sle staging on which a trolley was mounted. The trolley was rope-hauled towards the bows of the ship, the rope being routed back to a vertical shaft in the deck containing a multiple-reeved rope system on which was suspended a large weight – the prime mover for the system. A gravity-operated catapult, with the track tilted some 30° skywards, was installed in each of a couple of small ships. These were HMSs *Stronghold* and *Thanet*, destroyers of 900 tons, which, in 1921, were being used by the RAE for radio controlled 'target' trials, thought to have been an early stage of the 'Larynx' project. Progress on this project could be assumed to have been leisurely; one of these destroyers was used to obtain twelve launches between 1923 and 1924. Ten trials were undertaken in HMS *Stronghold* over the period 1922–5. A third ship, HMS *Castor*, was used in 1925 to launch a 'missile'. (Ref: 'Launching by Gravity' qv) Both HMSs *Stronghold* and *Thanet* were restored to their destroyer status and in 1942 were lost in the Pacific arena.

Following from what appears to have been a rather desultory start, the RAE began to take an increasing interest in the Fleet Air Arm, particularly in the catapulting of seaplanes from battleships and cruisers. The prototypes of all power-assisted take-off devices were two catapults built to a common specification issued in 1916. Both were completed by the end of the following year and became known as the 'Carey' and the 'Armstrong Whitworth' catapults respectively. Descriptions of these two machines are to be found in the chapter on catapults. Very quickly the production of aircraft launching trolleys – the catapult/aircraft interface – became a prominent feature of the RAE design offices and workshops. The provision of the catapult structures and propulsion systems remained an Admiralty province, centred upon the Royal Dockyard, Chatham.

Cronin (#*ibid.*) continues to write that an Admiralty invitation for tenders for hydraulic, electric and compressed-air catapults called for the launching of an aircraft 2½ tons AUW at 60mph {27m/s} with an acceleration not exceeding 2½g. Of these proposed systems only the

Prototype Sopwith Pup (N9497) fitted with a very early design of arrester hook for shore-based trials at the Isle of Grain in 1917. (Royal Aeronautical Society Library)

deck-arrester gear and the 504C (under the new designation 504H) became in 1917 one of the first aircraft to be launched by catapult gear: the pilot was F/Cdr R. E. Penny.'

As Caunter (#222) omits any reference to these events, it might be assumed that this work was undertaken before RAE became involved. On the other hand, the present writer believes it to have been the early results of the arrival of Percy Salmon at Farnborough in 1916. The catapult itself appears to have been that manufactured to the Admiralty specification by Waygood Otis (the Carey machine), which is discussed later.

The second catapult was built by Armstrong Whitworth and installed in HMS *Slinger*, a former steam hopper converted, and appropriately named, for shipborne catapult trials. According to Thetford (#97 p.21) a Fairey N.9 seaplane was used for these experiments. After the initial launching of a dummy Short 184 floatplane from HMS *Slinger* in October 1917 and the first live launch of a piloted aircraft, Fairey N.9/F127 in May 1918, the project was dropped. (These dates are from Cronin (#ibid. p.313).) The launching of naval aircraft from ships by

Blackburn Swift prototype aircraft (N139) with Bryer arrester units fitted to the undercarriage. 21 June. (© Crown Copyright/MOD)

Nieuport Nightjar (H8539), showing its Bryer arrester units attached to the undercarriage spreader bar. 22 June. (© Crown Copyright/MOD)

Sopwith Pup (N9497) seen at the end of a dummy deck arrestment. (Royal Aeronautical Society Library)

compressed-air option was taken up and by 1917 two catapults were ready for trials.

Of these two types, one designed by R. F. Carey was built by Waygood Otis (manufacturers of lift machinery) and was installed at Hendon. In May 1918 it was used to launch Avro 504 aircraft, three of which (N5261, N5269 & N5270) were adapted for this purpose. These originally were Avro 504C aircraft which, on modification, became 504H machines. (At the same time (1917) an Avro 504B was fitted with a rudimentary arrester hook for shipboard deck landing trials.)

Prior to this, when the Royal Naval Air Service was still in being, two interesting references are made by Halpenny (#40 p.136) and Thetford (#ibid. p.37) to events which occurred in 1917. Halpenny states that:

'An experiment of fundamental historical importance took place at Hendon, when an Avro 504 flown by Flight Commander R. E. Penny, RNAS, was successfully launched from a compressed air catapult on the airfield. This device was to revolutionise the operation of aircraft from ships at sea.'

Thetford amplifies this in stating:

'Two important experiments in naval flying ... were carried out by the Avro 504B and C. The 504B was used in pioneering work on mechanically assisted means was resurrected a few years later and became a subject for development in the early 1920s, generally coincident with the introduction of the RAE Mark I catapult.

Subsequently, the trials with these two original catapults were abandoned, as the installations were considered too cumbersome for installation in fighting ships. Flying operations from temporary or semi-permanent sloping platforms were considered to be adequate for the needs of the Royal Navy at that time.

There was a post-World War 1 lull in the experimental catapulting of aircraft. It was not until 1924 that, resulting from wind tunnel experiments, a modified Carey catapult was installed in HMS *Vindictive*. Trials with Fairey IIID aircraft in October 1925 were successful and installations were made in HMS *Resolution* in 1926 and the ill-fated HM Submarine M.2 the following year. These trials, together with those running concurrently at RAE, began to demonstrate that the day of the sloping platform method for shipborne launching of aircraft was drawing to a close.

Thetford (*#ibid. p.21*) states that the first catapult launch of a British aircraft at sea was from a post-World War 1 Cruiser, HMS *Vindictive*, in October 1925 using Fairey IIID and Flycatcher floatplanes at about 5,000lb {2,270kg} maximum all-up weight. Taylor (*#95 p.123*) identifies this Flycatcher as N9913 which had been adapted for catapult trials. HMS *Vindictive* was similar to her contemporary, HMS *Furious* (but with only half the displacement of the latter) in that she had separate take-off and landing-on decks. She saw little active flying and reverted to cruiser format in 1925, retaining a hangar and fitted with a catapult during this refit. Thetford (*#ibid. p.133*) adds that, in the early 1930s, a Fairey IIIF, Mk.III (S1317) was allocated to Farnborough.

With the advent of flight deck machinery for aircraft carriers, Farnborough became more deeply involved with both accelerator and arrester gear systems. It was in the light of this development that, in September 1938, the Catapult Section was set up as a unit within, but functionally separate from, the Main Drawing Office [top floor of Building 'Q.1'] at RAE.

With the progressively increasing weights of aircraft required to operate from warships the re-assessment of the catapult for launching became imperative. HMS *Ark Royal*[14] (the seaplane tender which replaced the World War 1 seaplane carrier and later to be renamed HMS *Pegasus* when the former name was required for the aircraft carrier commissioned in 1938) in 1930 was equipped with the first of a new design of launcher, to be followed by a sequence of others over the following two years or so. As a result of these trials, many of the Royal Navy capital ships were equipped with catapults for operating seaplanes (both floatplanes and hull-type machines), a process which was to continue until the early years of World War 2. By 1940, two types of catapults had been adopted: the 'light' equipment was capable of launching an aircraft of AUW 5,500lb {2,500kg} (eg Fairey Seafox) and the 'heavy' with a capacity of 8,000lb {3,600kg} (Fairey Swordfish, Supermarine Walrus and so on) at 60 knots {31m/sec}.

Pudney (*#80 p.55*) in 1948, probably obtaining his information from an early draft of Caunter (*#ibid. p.44ff*), writes:

'Catapults, originated by the RAE in 1922, had been in use for many years in *Ark Royal*[14] (Pegasus), *Courageous*, and *Glorious*, but the mechanism was of a type which made loading and launching a lengthy process. The superstructure, or trolley, which supported the aircraft was heavy and cumbersome. Aircraft in former days, being comparatively slow and light in weight, could more often than not be more easily flown off without this rather clumsy assistance. The modern technique of waging war at sea with aircraft carriers had not, moreover, been developed, and it was the advent of the new, more modern fleet carriers, together with the much heavier and higher-performance aircraft, that necessitated a new approach to the subject.

'This work was begun in 1937 at Farnborough. The design for the new *Illustrious* class carriers was already in hand, and the Admiralty required a catapult capable of launching an aircraft from the carrier at the rate of one in every forty seconds. The Establishment tried out various schemes from its runway (*sic*) on Jersey Brow, in addition to carriers already in service. A prototype equipment was started in March 1939. It was fitted to HMS *Illustrious* in May 1940, and taken immediately into service. The new gear required only five ratings to load the aircraft, and the new trolley was designed in such a way that it could be adjusted to suit all Fleet Air Arm aircraft in service at the time.'

(It is unfortunate that, in this statement, Pudney did not differentiate between the 1922 warship-type of 'catapult' and the 1935 type of 'accelerator' adopted for carrier use. This, perhaps, was due to the initial experimental work on this accelerator having been undertaken with the new flight deck machinery being mounted on the RAE Mark II catapult. The difference is explained later in this text. Also, there was no runway, as such, from Jersey Brow; it was a greenfield site until late 1940!)

In capital ships, the catapult often was located amidships, and mounted athwartships between a pair of the ship's funnels. On each side of the for'ardmost funnel a small hangar would be located, facing aft, each accommodating one seaplane with its wings folded. Variations to this practice did occur, one of which was the mounting of the catapult atop the roof of the after heavy gun turret. Another was mounted on a turntable on the quarter-deck – port side, of course! The placing of the aircraft onto the catapult trolley was a fairly involved exercise in which a special 'seaplane' crane and a 'shoe-horn' loading device was used. (See 'Launching Trolleys'.) After launching, and on completion of its sortie, the aircraft would alight on the sea alongside the ship (probably after the latter had turned in a complete circle, providing a smooth patch of water inside the wake) to be recovered by the crane while still under way.

This methodology was being overtaken by the greater use of the aircraft carrier, a concept different from what had become the accepted practice for gunnery spotting for the fleet. From this point the development of mechanically assisted launching of aircraft immediately prior to, and during, World War 2 is treated in greater detail below.

The development of catapult launching gear – and its scion, the flight deck accelerator – is recorded in greater detail later in this text.

Temporarily diverting from Fleet Air Arm related activities, and not for the last time, NAD became involved in many other facets of

Supermarine Seagull prototype (N158), showing the Bryer arrester hooks on the undercarriage legs. 22 August. [AS1](© Crown Copyright/ MOD)

Fairey Flycatcher (N9616) fitted with Bryer hooks on the undercarriage spreader bar for use with longitudinal (fore-and-aft) shipboard arrester gear. June 1925. (© Crown Copyright/MOD)

getting airborne. Some of these are treated in later chapters, highlighting such extra-mural excursions into the peripherals of boosting the take-off of aircraft by explosive, gravity, hydraulic, inertia and rocket procedures.

RETURNING TO 'MOTHER' – UNASSISTED

From the earliest days of navy flying, the question of shipboard retrieval of an aircraft, once it was airborne, did not arise. It either alighted ashore, on the nearest convenient piece of land, or, if it was a floatplane, on the sea itself. Difficulties began to appear when it was desired to recover the aircraft to its point of launch, particularly if a flight deck, however rudimentary, existed, forming an embryo aircraft carrier. In time, the situation became aggravated by the progressive increase in weight of aircraft adopted for naval service.

The early attempts of recovery onto the flying-off deck located forward in HMS *Furious* (August 1917) are well documented. The achievement of attempting to land a Sopwith Pup on this occasion bordered on the miraculous. It entailed the pilot drawing alongside the flight deck, side-slipping around the funnels and bridge structure in doing so and then being manually hauled aboard by those on deck grabbing a number of leather straps suspended beneath the lower wings. The death of the pilot, E. H. Dunning, in the third attempt was nothing short of a tragedy. At that time there simply was no other method of shipboard recovery.

Within a year of her commissioning in June 1917, and following the death of Squadron Commander Dunning, HMS *Furious* was relieved of her after gun turret and fitted with a landing-on deck abaft the ship's funnel and bridge structure. Chesneau (#21 p.85) quotes the size of this deck as 300 feet by 50 feet {91m x 15m} connected by a lift to a small hangar below. Deck-side palisades remained in evidence and the method of aircraft retrieval was still by flight deck crew manually grabbing the lower wings as the aircraft was stalled onto the deck. Before long, an installation of fore-and-aft wires was provided these ran the length of the landing-on deck for'ard of the lift up to a tall goal-post-like structure from which hemp ropes were suspended to form a basic safety barrier against an over-run of the aircraft into the ship's superstructure. Impact with these ropes was a frequent occurrence.

The prime reason for these deck wires, in conjunction with deck-edge palisades, was the prevention of aircraft from sliding sideways and disappearing over the side. In another instance, the fore-and-aft wires were supported edgewise on timber planks, which were knocked down as the aircraft trundled across them. This was not an unqualified success. The arrangement was heartily disliked by the pilots because of the air turbulence in the wake of the ship's top-hamper. Full-length flight decks did not appear in HMSs *Furious* and *Argus* until late 1918. For some time there was little need for flight deck arrester gear or safety barriers; most of the aircraft were capable of landing-on without mechanical braking.

Biggs (#7 p.14) states that HMS *Furious*, in its 1918 recommissioning, had a full-length flight deck in which the landing-on area again was fitted with fore-and-aft wires over a length of 350 feet {107m}. The wires were installed to a slight taper in spacing towards the forward direction and spanned a shallow well. These wires were supported at nine inches {225mm} above the deck, and aircraft, fitted with skids to their main undercarriages, landed along these wires. The intention was to apply a sideways pinching effect to the aircraft undercarriage, providing directional control and, hopefully, slowing it down. This was effective enough with the slow landing speeds of aircraft of that period, coupled with a high relative wind speed down the deck when the ship was steaming into wind. The fore-and-aft wires, however, presented a hazard to the flight deck personnel in restricting their movement about the deck.

A handicap was the time taken for the aircraft handling party to retrieve the aircraft from the landing pit, in spite of a ramp at each end. (In one instance a landing pit was formed through lowering the lift by a matter of a few inches.) Eventually, when full-length flight decks became the norm, this system was abandoned. History does not relate how, as the aircraft came to rest, it was retrieved from a maze of deck-level wires – it was no simple exercise. Biggs (#ibid. p.23) refers to the installation in HMS *Furious* as never having been popular with the pilots who, with the advent of full-length flight decks, were able to stop the aircraft on their own newly fitted undercarriage wheel brakes. He states that the longitudinal arrester gear was removed in the 1920s (perhaps during the 1922–25 refit?).

HMS *Argus*, commissioned shortly after HMS *Furious*, was intended to have been similar to her forebear with separate for'ard and aft flight decks. Adjustments to make her a through-deck carrier were implemented and she, too, was fitted with fore-and-aft arrester wires. Five years later,

Fairey Ferret Mark 1 prototype (N190/F538). This picture was taken with the aircraft on the apron fronting the RAE 'A' Shed [P.72]. June 1926. (© Crown Copyright/MOD)

in 1923, HMS *Hermes* was commissioned with longitudinal wires fitted as arrester gear. A year later she was followed by HMS *Eagle*[24] with her arrester gear also comprising longitudinal wires over a length of 190 feet. This ship was fitted with hinged metal flaps running across the deck at intervals and supporting the fore-and-aft wires. The flaps were knocked down on landing, helping to absorb some of the kinetic energy of the alighting aircraft.

These fore-and-aft wires, in common with the systems in her sister ships, were removed in 1926, with nearly a decade passing before the athwartships braking system was adopted. In spite of the lack of success of fore-and-aft arrester systems, they are believed to have been fitted in the subsequent commissioning of HMSs *Courageous* (1928) and *Glorious* (1930) as aircraft carriers. One stipulation required the wires to be no higher than 15 inches above the deck. About this time palisades were fitted in the carriers to prevent loss of life and aircraft over the deck edge. These were obliquely placed stanchions with wire nets stretched between them along the deck side. (In subsequent refits, the arrester systems in these ships were replaced by the pressurised hydraulic Mark III athwartships units.)

Penrose (#*ibid. p.248*) again quotes Busteed (who, as CO of the Marine Experimental Aircraft Depot at the Isle of Grain, was introduced a few pages back) as recording experimental flight investigations of deck landing systems, particularly arresting gear:

'We were given the task of trying to do this by flying an aeroplane into half a dozen or so steel-wire ropes suspended in a half catenary over a heavy wooden frame. Spring clips were fitted under the axle to grip the wire ropes. I had two or three shots at this, though I did not much fancy the job, but it worked all right in a clumsy sort of way. The system was quickly abandoned in favour of the experiments at Eastchurch using a hook arrester.'

Thetford (#*ibid. p.307*) recalls that shore-based experiments with a transverse cable across the deck to be engaged by a hook dangling from the rear fuselage were made on the Isle of Grain in 1918. A 'primitive' arrester gear comprising athwartships wires – 'light lines' – weighted with sandbags at each end, spanned the series of longitudinal 'guiding' wires. The aircraft used was a Sopwith Pup (N9497) fitted with a rudimentary type of arrester hook. As the hook picked up more and more of the transverse lines the aircraft was brought gently to a standstill. This system was fitted in HMS *Furious* in 1920. The idea proved not to be all that satisfactory and was soon abandoned. (The arrester-hook installation appeared to be more akin to the message-retrieving hooks fitted to the undercarriages of the Hawker Audaxes of the RAF No.4 Squadron, the Army Co-operation Unit as seen operating at Farnborough in 1937.) The Sopwith Pups embarked in HMS *Furious* at that time were fitted with rigid skid undercarriages, the 'arrester gear' on shipboard comprising the fore-and-aft installation of deck wires. Thetford (#*ibid.*) states that 'dog-lead' clips were fixed to the undercarriage which engaged these wires. Another aircraft fitted with devices to assist in landing was the Sopwith Camel 2F.I which, according to Thetford (#*ibid. p.314*), took part in some of the earliest deck arrester landings aboard HMS *Argus* in 1919. Rawlings (#*81 p.102*) refers to these devices as 'clothes pegs', probably early types of Bryer hooks.

During this interim, Caunter (#*ibid. p.45*), rather tantalisingly, carries a two-line reference to a restraining hook having been fitted to a Sopwith Snipe aircraft in 1922 for deck landing. This may be a misnomer, as there seems to be no record of this type ever having flown with Fleet Air Arm squadrons and there is no record of such an aircraft in the RAE Flight Books.

Within two years of the 1918 armistice Taylor (#*ibid. p.113*) records that naval aviation had been allowed to shrink to no more than a token force. In the 1920–2 period there was only one true aircraft carrier, HMS *Argus*, still in commission. In an attempt to effect financial economies in the immediate post-World War 1 period, the powers-that-

be ordained that a number of aircraft were to be built from surplus components of redundant DH9A aircraft. These machines were constructed by Westland and the marque was given the name Walrus (originally named Tadpole). According to Thetford (#*ibid. p.350*), it was one of the curious collection of ugly ducklings used in fleet spotting in the years immediately following World War 1. When brought into service in 1921, these machines were fitted with 'jaws' to the undercarriage spreader bar to engage fore-and-aft arrester wires.

It was not until 1922 that a specification was issued for a purpose-designed fleet aircraft suitable for use in all three versions as a landplane for carrier operation, a floatplane and an amphibian. This resulted in the Fairey Flycatcher fighter of which the first prototype went for deck-handling trials in HMS *Argus* during February 1923. According to Taylor (#*ibid. p.122*) further experiments were undertaken with Flycatcher aircraft with wheeled undercarriages. Flycatchers with float undercarriages were also landed-on without use of arrester gear. Flycatcher (N9616) appeared at Farnborough in 1925.

A general-purpose machine was the prototype Fairey IIID, an immediate post-war development of the maker's type III marque. Both the Flycatcher and the IIID were fitted with snatch-gear ('Bryer hooks', accredited to Fl/Lt G. M. (Peter) Bryer AFC) on the axle between the main undercarriage wheels, or skids, for landing into the fore-and-aft arrester wires. The DERA archive contains photographs of other aircraft appearing at Farnborough in 1921–2 fitted with Bryer hooks. These include a Blackburn Swift (N139), Nieuport Nightjar (H8539) and the contemporary Supermarine Seagull (N158). A later marque, the Fairey Ferret – generally considered to have been the missing 'E' variant in the company's marque III series – was fitted with a different pattern of snatch-gear. The first of three prototypes (N190/F538), with the Armstrong-Siddeley 'Jaguar' engine and displaying these 'hooks', was photographed at Farnborough in 1926.

Taylor (#*ibid. p.122*) states that the decision finally to abandon the use of longitudinal arrester wires was made following a fatal accident to a Fairey IIID operating from HMS *Hermes*[24] in 1926. In fact, questions were already being asked in high places why flight deck machinery was required at all, as aircraft by then were equipped with brakes as standard. These arrester systems failed through their inability to prevent yawing of the aircraft on landing and, resulting from the fatality in HMS *Hermes*[24], the principle was abandoned, the fore-and-aft systems being removed from all ships. For almost a decade there were no mechanical aids for recovering aircraft operating from a carrier flight deck; the performance of the aircraft in use at that time was within the aerodynamic parameters obtained naturally through the ship's own forward movement into the available natural wind. By the early 1930s, mechanical assistance to shipboard landing was investigated anew and found to be essential as the operational weight of naval aircraft approached the 7,000lb {3,170kg} mark.

Rebecca John (#*52 p.70*), in the biography of her father, Admiral of the Fleet Caspar John, quotes his referring to the Fleet Air Arm as:

'... "an awkward child for which no legal birth certificate exists". The Royal Naval Air Service had been formed in 1914 but had been amalgamated with the Royal Flying Corps on 1st April 1918 to form the Royal Air Force. ... The Navy surrendered its expertise in aeronautical matters, and it was not until 1924 that Naval flying was restored. Even then the (Naval Officers' Flying) Course was run by the RAF until the day of liberation in 1937.'

In the interim, Rebecca John (#*ibid. p.78*) again writes of her father:

'When Caspar made his first deck landings (*1926*), not a single facility existed on the ship's deck – or on the aeroplane – to aid the notoriously difficult and frightening process of "landing-on". There were neither arrester wires nor crash barriers, no angled deck or mirror landing sight on the carrier, and no hook, wheelbrakes or tricycle undercarriage on the aircraft. There was nothing, save the wind, an empty deck and the skill of the pilot. "Those men who pioneered deck landing had to *feel* their way back down to the carrier," Caspar wrote, "added to which the aircraft we were flying were out of date and weren't designed for the accuracy needed for deck landing."'

Apropos the foregoing, John (*#ibid. p.80*) portrays a 1926 photograph of her father – then a Lieutenant – standing, rather disconsolately, by his aircraft which, following an engine failure, is lying inverted in a knee-high Scottish cornfield. The aircraft was a Fairey IIID and clearly shows a set of three Bryer hooks affixed to the undercarriage. The caption states that these mechanisms were designed for aircraft carrier wires, not for cornfields.

Taylor (*#ibid. p.152*) tells of a Fairey IIIF floatplane (S1148/F889) equipped with strengthened floats used for landing experiments onto HMS *Furious* during 1930. Without arrester gear the landing was somewhat perilous through lack of directional control. The aircraft usually ended its landing run near the forward edge of the flight deck after touching down in a shower of sparks from the reinforced floats scraping along the steel deck.

The first mechanical arrester gear of the transverse-wire type was based upon a dry friction brake-drum system (Mark 2). This was not very effective and before long the multiple-reeved Mark 3 hydraulic system was introduced.

There was one experiment carried out with HMS *Pegasus* (ex *Ark Royal*[14]) in the early 1930s which was known as the Hein Mat. This was a floating mat, towed by the ship, onto which an aircraft (originally a floatplane) was expected to taxi after alighting on the water. From the outset this was a non-starter, the forward motion of the aircraft on contact rumpled the mat, pushing the mat ahead of itself in the process.

RETURNING TO 'MOTHER' – MECHANICALLY ASSISTED

First attempts at mechanical assistance in the landing-on process were undertaken in the early 1920s with an across-deck wire supplementing the longitudinal wires. The ineffectiveness of the earlier sandbag braking units (the Mark I Arrester Gear) prompted investigation into the adoption of mechanical devices. The next step was to replace the previous fore-and-aft system of wires with an across-deck wire, the ends of which were taken below deck and terminated in a set of friction brake drums as the arrester unit. The drums were reset by electric motors. This Mark 2 system was not a success as the aircraft on landing tended to yaw violently due to variances in the performance of the individual brake drums.

According to Taylor (*#ibid.*) and Thetford (*#ibid.*) at the period when the fore-and-aft arrester-wire technique was out of favour, a Fairey IIIF Mark III aircraft (S1781/F1518) was fitted with a deck hook, later to become a standard fitment on other biplanes of the 1930s. The illustration shows a Mark I squadron aircraft (S1182) fitted with an elementary

hook of this type. This appears to have been the prototype for the standard pattern of a swinging 'V'-frame, with the hook at the apex, suspended below the rear fuselage and under the control of the pilot for both deployment and (sometimes) retraction. Interesting entries appear in the RAE Flight Book for September/October 1930 in which a Fairey IIIF aircraft (S1351) undertakes Deck Landing Arrester Gear trials at Farnborough. The author has not been able to trace any reference to the equipment itself or to identify its location on the Farnborough airfield. It may be surmised that the equipment was a version of the Mark 2 arrester gear employing brake-drum energy absorbers. On completion of the trials, the rig may have been uplifted for transfer to one of the carriers (HMS *Courageous*?). Initial shipboard trials took place aboard HMS *Courageous* in 1931, aircraft landing into the first set of transverse arrester wires to be connected to an under-deck braking system.

From this simple beginning the basic technique of arrested landings has continued without variation to the present day. A trailing hook attached to the aircraft fuselage is made to engage a transverse steel wire rope spanning the carrier flight deck from a pair of pulleys situated at the deck edges and continuing below deck to a mechanical braking system. On engagement by an aircraft, the arrester wire pays out in a 'V' plan configuration against the resistance imposed by the brake system and brings the aircraft to rest.

With the hydraulic 'jigger' system, two sets of pulleys are fixed to the cylinder/ram combination, one set being mounted on the fixed end of the cylinder with the other at the extremity of the moving ram. The cylinder is filled with a water/glycerine mixture. The deck wires are reeved around these pulleys to give a ratio (generally 6:1) of pull-out to ram movement. When set, the ram is fully extended and, on the engagement of an aircraft, the pull-out of the arrester wire draws the ram into the cylinder against hydraulic pressure. The displacement of the hydraulic fluid at the back of the ram is restricted through a series of control orifices and valves against a pneumatic pressure in an accumulator. At the end of an arrestment, the wire is disengaged from the aircraft and the system is reset by the air pressure in the accumulator being employed to return the ram to its original position, ready for the next engagement.

The first mechanically assisted deck landing, using a complete system with multiple arrester wires rigged athwartships and connected to an under-deck retarding mechanism was undertaken in 1933, again aboard HMS *Courageous*, followed shortly after with HMS *Glorious*. This equipment became known as the Mark 3 arrester system. The first aircraft to be fitted with a swinging 'V'-frame arrester hook was a Fairey Seal, a navalised development of the Fairey IIID aircraft. When the Fairey Flycatcher went out of squadron service in 1935, its place was taken by the Gloster Sea Gladiator. With delivery beginning in 1936, the first of these marques were direct adaptations of aircraft ordered for the RAF, the most salient modification in navalisation being the fitting of arrester equipment to the rear fuselage. The new 'V'-frame assembly with a single hook was progressively fitted to these aircraft, one of which was N5517 used in arrester trials aboard HMS *Courageous* in 1939.

The subsequent development of mechanically assisted devices and machinery for shipboard recovery of aircraft is treated in detail later in this text.

THE CATAPULT SECTION – ITS HERITAGE

Consequent upon Salmon's submission of his proposals to the Superintendent of RAE, the Catapult Section was formed within his Department in September 1938 under the supervision of Lewis Boddington, the newly recruited engineer assistant to Salmon. A few months later, by May 1939, Salmon had produced a paper, *A Review on Launching Schemes for Aircraft (#283)*. In effect, this document of some 50-plus pages proved to be a virtual 'Job Specification' for the new Section. It presented a complete overview of ideas, both practical and naïve, for getting an aircraft airborne, together with the author's impressions on the validity of each scheme. Very little of the content of Salmon's paper is thought to have appeared elsewhere. Most of the material in the following paragraphs has been abstracted from his original. Because of this the present tense, as used by Salmon in his description of 1939, is retained, and concessions to SI (metric) units have not been made. In this abstract the current author's personal comments appear as (*bracketed italics*).

NAVAL DEVELOPMENT

In considering means to assist aircraft to take off, it is necessary to have a clear idea of the conditions under which the aircraft will have to operate. The conditions under which the Fleet Air Arm operates, which have built up since 1924, are worth consideration. Broadly speaking there are two sets of conditions:

a) Those required to suit capital ships, ie battleships and cruisers
b) Aircraft carriers.

In the case of (a), seaplanes and ship planes have to be dealt with; in the case of (b), ship planes only.

In capital ships the space available for a catapult is only a strip of deck amidships, the length being the beam of the ship – about 90ft in a battleship and 60ft in a cruiser. The aircraft is mounted on the launching trolley and can be released, as soon as the engine is warmed up, under all weather conditions suitable for flight. Launches can be made within the following wind conditions, either to port or starboard providing the rolling of the ship does not exceed 10°:

a) Combined catapult and wind speeds down the catapult 100kts with wind across the catapult limited to 26kts – one extreme limit.
b) No wind down catapult, and 40kts across being the other extreme limit.
c) Any combination of winds down and across causing stresses in the aircraft within these limits – charts are prepared for each aircraft.

The attitude of the aircraft on the catapult being with the angle of incidence such as to give 33% lift at 50kts, and the axis of the aircraft along the axis of discharge. Generally speaking the effects of these conditions are that, under a heavy side wind, the aircraft yaws and side-slips into the resultant wind; and, when there is no wind down the catapult, the aircraft may drop as much as 15ft. Severe as these conditions may appear, eight years' experience has proved them satisfactory.

In aircraft carriers the normal take-off along the deck is not always possible. Accelerators at the forward end of the flying deck are provided (the usual number being two). The principle of accelerators is the same as that of catapults, with different facilities for loading the aircraft, being so arranged to allow the aircraft to be taxied into position ready for attaching to the launching truck. The rate at which aircraft can be discharged from the present type of accelerator is approximately one per minute per accelerator, but in the ships now under construction it is anticipated that the rate will be increased to approximately one every 40 seconds. Two accelerators will give a rate of launch equivalent to that of normal take-off along the deck. In other respects the conditions under which accelerators operate are similar to those of catapults. Other interesting points worth mentioning are:

a) All the Fleet Air Arm aircraft are fitted with standard catapult 'points', and the cradles carrying the aircraft attached to the launching trucks are 'universal' and can be adjusted to suit any aircraft.
b) At the end of the acceleration run and the catapult launch these cradles collapse, leaving the aircraft free.
c) All the catapults in capital ships are cordite operated, and the accelerators on the carrier ships are operated by compressed air.

It will be seen that the policy adopted by the Fleet Air Arm is now definite and all the specifications for new FAA aircraft embody all the requirements for these conditions.

DEVELOPMENT IN THE ROYAL AIR FORCE

Until now, RAF aircraft have not needed assistance in take-off (with the exception of target aircraft) as the size and surface of aerodromes have satisfied their requirements. With the introduction of heavy wing loadings and their higher take-off and landing speeds, the RAF is faced with a similar problem as confronted the FAA fifteen years ago. The position, however, has not become urgent at present, as none of these aircraft are yet in service, particularly since many of the designing firms are very optimistic as to the ability of their heavy wing loaded craft to take off from existing aerodromes.

Four years ago (*1935*) the Air Ministry, realising that assisted take-off might become essential, directed the RAE to review the work done on catapults and put forward suggestions. RAE responded by pointing out the advantages to be gained by using large bombers of weight 40,000lb with a wing loading of 51.lb/sq.ft requiring a take-off speed of 110mph. A design for a multiple ram type catapult to deal with this weight and speed that could be installed in the centre of an aerodrome was submitted. The design provided for launching in four directions. The rams and launching track are mounted in a turret, with the top of the turret flush with the ground. The four runways, when not in use, are covered with plates, thus presenting no obstruction to aircraft using the aerodrome. With such a catapult it should be possible to launch twenty aircraft an hour under all weather conditions. In November 1935 it was decided to proceed with the design and construction of this catapult and install it in the first instance at Farnborough, but subsequently it was decided to install it near the eastern boundary of Harwell aerodrome with two runways only. (*This became the RAE Mark III catapult. qv.*)

Fourteen other launching schemes have been examined over the past four years (*1935–9*). A list of these is given below. An appendix by Lewis Boddington is devoted to a mathematical survey of assisted take-off schemes varying from the catapult at one end of the scale to the normal take-off at the other. Curves give a very clear conception of the magnitude of additional energy required and the factors which limit their application, particularly in assister trucks. Nearly all these are of the power-driven truck variety, which are not really practical propositions. As most of the ideas submitted for consideration show a lack of appreciation of the magnitude of the energy required, this way of presenting our criticisms is more effective and does not require the consideration of the merits of the various inventions.

(*The argument continues with details of the various airfield systems including overall schemes coupled with airfield arrester gear making a composite whole. The problems of installing such airfields with mechanical equipment without disturbing flying activities were also addressed.*)

The main sources of power used on catapulting schemes at present are:

a) Cordite
b) Compressed air
c) Hydraulic
d) Flywheels or inertia systems
e) Electric."

(*In the original paper these are discussed in detail extolling their various merits and pointing to their disadvantages.*)

Aircraft designed to be catapulted with acceleration of between 2g & 3g have to withstand additional stresses and therefore will be slightly heavier. This extra weight, expressed in terms of its all-up weight, is between 0·5% & 1%. In the case of the Manchester it is about 300lb. Although much is made of this increase in weight, when considered against the advantages to be obtained by making full use of heavy wing loading and the easy means of getting safely into the air, one is inclined to take the opposite view, that a great deal is obtained for very little.

Other means to enable these heavily loaded planes to take off in the normal manner such as bi-fuel systems, two-speed gear, and variable pitch airscrews, not only add weight to the aircraft but call for more elaborate and complicated mechanisms, increasing liability to engine failure, as well as increased time and cost of production. Additional strain is also imposed on the pilots. Catapults, although in the first place, are costly, their life and useful service is a matter of years, whereas the increased cost incurred on every plane will soon exceed the entire outlay on catapults.

LAUNCHING SCHEMES FOR AIRCRAFT

Catapults
1) The Carey Catapult
2) RAE – Ram Type

3) RAE – Portable (Haulage)
4) Vickers Aviation – Electric haulage
5) Margeram – 'Escalator'
6) Mitchell – Slotted Cylinder
7) Merz – Slotted Cylinder
8) Sandor – Inertia Catapult

Assisters
9) Metropolitan Vickers – Electric Locomotive
10) Armstrong Whitworth Aircraft – I.C.E. Locomotive
11) Muir & Braeme – Locomotive
12) Courtney – Airscrew-driven carriage
13) Whittle & Kirk – Turbine form
14) Rocket
15) Flywheel locomotive
16) Rowley – Powerless carriage
17) Direction Controlled Take-Off – RAE
18) Walsey."

(*Of the above, the projects at 1, 2, 3, 6, 8, 14, 15 & 17 are of deeper interest and are treated in greater detail in this work under their respective chapter headings.*)

CATAPULTS

1) THE CAREY CATAPULT
The present (*1939*) type of catapult in use with the Admiralty is a development of the Carey catapult. It is a rope type operated by a pneumatic or hydraulic ram through a multiple reeve and is particularly adaptable for ship operation. Its use is confined to accelerations of 2–3g for machines of 12,000–14,000lb launched at 60 knots. The application to larger aircraft and higher speeds is feasible although each launching rope used at present would need to be duplicated or even trebled to carry the loads. For lower accelerations in the region of 1g a very much greater reeve than the 8:1 used at present would be necessary to produce a reasonable length of ram. The equivalent weight of the apparatus would not be much less than the ram or direct haulage (drum) type and would therefore develop practically the same horsepower.

2) RAE RAM TYPE CATAPULT – MARK III
The catapult now being constructed (*at RAF Harwell*) will launch an aircraft of 60,000lb at 110mph with a mean acceleration of 2g (ie with a launching run of 202ft) every three minutes. The complete catapult is housed in a 100ft turret that is flush with the ground and can be fixed in particular directions in prepared slots in the ground arranged to radiate to cover all wind conditions. In the present equipment, two such runways are provided.

The catapult is based on the ram system developed at the RAE and is pneumatic-hydraulic in operation. The power plant is housed in the turret structure, the scheme is compact and suitable for high rates of launching. (*The scheme was proposed as a composite installation for use in association with airfield arrester gear.*)

3) RAE PORTABLE CATAPULT (HAULAGE)
The portable catapult, or field accelerator, was developed to launch at 60mph a 20,000lb machine at 1g acceleration. The aircraft is towed along the aerodrome, away from the winding gear, on its own undercarriage with the tail supported on a special carriage. The winding rope passes over a return pulley suitably anchored in the ground and back to the winding drum. A tail rope is also carried on the drum which winds off during acceleration and is used to retard

the towing gear and tail carriage. Such a catapult could be set down in any desired direction by provision of suitable anchorages. To apply such a scheme to 120mph launching would require a new definition of 'portable'.

A proposal, however, involving flywheels looks promising although a high rate of launching could not be attained. The behaviour of long lengths of rope is unknown and the practical limits can only be investigated in a full-sized apparatus. This could limit the possibilities of the scheme, since low accelerations are essential. Such a scheme has been investigated fundamentally. To launch a 30,000lb aircraft at 120mph would require a flywheel of approximately 6·5 tons. Such a flywheel would have to be specially constructed. Drive of the drum, which may be conical or parallel, would be through hydraulic gearing and a flexible coupling so that the apparatus could be split into two convenient units for transport.

4) VICKERS AVIATION LTD
Four schemes were submitted for launching at 110mph a 25,000lb and 40,000lb machine at 0·5g and 1·5g with a launching rate of one every three minutes. In the aerodrome scheme two haulage tracks are provided at 90° to operate in both directions with the winding house at the intersection of the tracks and everything, except the cradle, is below ground. The Whiting system of winding is employed using a continuous rope lapped eight times round the driving pulleys arranged in tandem. Each driving pulley is coupled to a reversible DC motor which takes its supply from a flywheel motor-generator (Ward-Leonard-Ilgner system). (*Salmon estimates the demand for mains electricity to be almost 4MVA.*)

5) MARGERAM
This scheme proposed a moving mat, similar to an escalator. Accelerations would necessarily be low dependent on the coefficient of friction between aircraft wheels (which would have to be locked) and the nature of the surface. In the simple conception, it is very cumbersome and offers no practical solution to a scheme which has been brought up and discussed from time to time.

6) MITCHELL
The scheme was proposed when the Atlantic Flying Boat scheme was being investigated. (*This flying boat project proceeded no further than a paper scheme prepared by MacTaggart Scott of Midlothian and is not included in my text.*) It comprises a piston and cylinder which has a longitudinal slot through which a fin plate projects from the piston. This fin is attached directly to the carriage carrying the aircraft. The cylinder extends the full length of the take-off run with the piston being forced through the cylinder by pneumatic pressure. The slot in the cylinder is sealed by a flexible belt located in the cylinder, the belt being pressed into position so that the seal is effected before and after the piston but passes through the piston under the fin. The aim of this proposal was to overcome the disadvantages of using wire ropes. Whether the difficulties of providing the seal to the slot are to be preferred to the known deficiencies of the ropes can only be decided by practice. Certainly the scheme in its conception, is a very simple form of catapult.

(*This idea was patented by C. C. Mitchell in 1938. His name occurs again in this work, particularly in connection with the Steam Catapult – a development of his idea – some two decades after the date of Salmon's paper. In the 1950s, the prototype RAE slotted tube catapult was constructed from a German V1 launcher,*

[Did the Wehrmacht poach the original British patent in developing this weapon?])

7) MERZ
This scheme is identical with the Mitchell proposal except that the practical difficulties are trebled by the introduction of three cylinders. This development is obvious if the distortion of the cylinders is to be reduced, but construction is practically impossible.

(*A thumbnail sketch appears on the original. This shows the three launcher tubes arranged within a larger tube with the slots occurring at 4, 8 & 12 o'clock cross-sectional orientation. The production Steam Catapults employed a pair of cylinders in parallel, with the pistons spanned by a bridge-piece which, in turn, carried the launch spigot [shuttle]. This is described more fully in the chapter on the Steam Catapult.*)

8) SANDOR AIRCRAFT ACCELERATORS LTD
The scheme presents a catapult which is mobile and which utilises the linear momentum of a large mass to launch the aircraft. A 60,000lb machine is launched at 120mph at a resulting acceleration of 0·5g. A normal railway construction of a closed circuit round an aerodrome is laid in curved and straight runs. The equipment consists of a train of carriages to form a platform for the launching cradle to run on. The aircraft is mounted on the cradle at the rear of the train and locked thereto. Under the thrust of the airscrews, and/or the assistance of a locomotive, a speed of 54mph is attained. The total weight of the train is approximately ten times the weight of the aircraft, viz 240 tons. The aircraft cradle is attached to a rope gear which is reeved over a pulley on the forward end of the train through a 2:1 system to what is described as a 'register bogie'. This is arranged to grip a special rail when the desired speed is attained and the aircraft (the cradle is automatically unlocked) is launched along the top of the train at a relative velocity of 80mph, the train meanwhile dropping to 40mph. The velocity of launch relative to the ground is therefore 120mph. The energy is drawn from the kinetic energy of the train.

The real development of this scheme may be as a portable launching gear to run on long straight portions of existing railways. It would be as portable as any railway rolling stock. No insurmountable engineering difficulties are foreseen, although many new problems would have to be solved. Using the aircraft itself as a means of accelerating the whole train is undesirable.

ASSISTERS

9) METROPOLITAN VICKERS LOCOMOTIVE
The scheme provides a carriage for the aircraft driven by an electrically operated locomotive. It was agreed by the firm at the time of the proposal that the difficulties confronting the scheme were probably too great. The main problem was the adhesion of the wheels at the high speeds with a progressive reduction in load as the aircraft became more airborne. The production of a locomotive with a low weight/power ratio is also a serious factor. No possible development of the scheme can be foreseen.

10) ARMSTRONG WHITWORTH AIRCRAFT – I.C.E. LOCOMOTIVE
To increase the tractive effort, it was proposed to run the locomotive on pneumatic tyred wheels and it was claimed that a coefficient of friction as high as 0·8 could be obtained. The scheme, which was developed to launch a 32,000lb aircraft at 90mph, had a locomotive

carriage of 24,000lb powered by six Rolls-Royce Kestrel aero engines (equivalent to 3,900bhp). In a low gear, a speed of 45mph is obtained in 108ft at a mean acceleration of 0·61g. In second gear a speed of 90mph is obtained at a mean acceleration of 0·38g. Such accelerations may be obtained in ideal conditions; in wet weather, ice or snow the values must be lower, probably as low as 0·2g.

11) MUIR & BRAEME

The scheme proposed is similar to those by Metro-Vickers, A.W.A. and Courtney and covers such schemes only in a general way. No new ideas or ground is covered. (*Both Muir and Braeme at that time were senior members of the Engine Department research staff at RAE.*)

12) COURTNEY

This scheme is similar to both the Vickers and A.W.A. locomotive schemes but with the aircraft mounted on a swivel carriage so that it may take off into the resultant wind. A new feature is introduced by the use of airscrews to drive the power carriage, but it must suffer from the same limitations as the tractive type, ie the acceleration at the end of the run is low. To use ordinary gauge railway would mean a correspondingly heavier truck to provide stability under side loads etc, and at high take-off speeds check rails to prevent 'jumping' would be necessary.

13) WHITTLE & KIRK

The carriage and the track are related to each other in the same way as the stator and rotor of a turbine. The propelling forces are caused by the dynamic effects of swiftly moving fluids engaging suitable shaped passages on the track and carriage respectively. The scheme is beyond the bounds of practical possibility on any large scale.

14) ROCKET

The possibility of assisting flying boats in take-off by rockets has been considered, using cordite. The mass of cordite to be carried and the control of its burning makes the proposition extremely dangerous.

15) FLYWHEEL

Energising locomotives by spinning flywheels has been investigated and whilst the flywheel represents a simple form of storing and giving out energy, limitations due to weight and friction coefficient apply as in other forms of powered carriages.

16) ROWLEY

The scheme proposed a closed circuit on which a carriage carrying the aircraft runs. Both carriage and aircraft were accelerated by the power plant of the latter and take-off effected under the pilot's control when the desired speed is obtained. A track 1,600ft diameter was proposed and the railway super-elevated. As the velocity was not constant, super-elevation of the track is not effective.

17) DIRECTION CONTROLLED TAKE-OFF SCHEME

An experimental track is being constructed at present to investigate the railway system of take-off. No one can claim the monopoly on a rail-track system as it was at the RAE between 1917 and 1924 for the take-off of aerial targets. The present scheme, which introduces some new features, consists of a rail track which permits of a special anti-friction truck being used with no additional power plant. By making the gauge of the track to suit the pitch of the undercarriage, a light form of truck is possible. The aircraft and truck are accelerated by the aircraft itself and the overall length of the track including the braking portion is 1,150 yards. For a 60,000lb machine the take-off run is estimated at 880 yards for 120mph.

18) WALSEY

The scheme provides for high towers and literally pushing the aircraft off the edge of a platform at the top, thus providing them with height to attain speed! The scheme is too fantastic to be dealt with seriously.

THE CATAPULT SECTION – A NEW VENTURE

So much for Salmon and the position as it appeared to him in 1939. From the viewpoint of six decades later, many of the schemes enumerated by Salmon have been addressed and tackled successfully. Others have become lost because the problems either were incapable of ready solution or they just went away. It is easy to be wise from hindsight and, perhaps, to give a wry smile to some of these matters, but just prior to World War 2 they were very real challenges, requiring attention and exploratory work to be undertaken with positive results as the outcome. This was the remit for the next four decades for the new Catapult Section.

With the development of the aircraft carrier and the expansion of naval flying within the Fleet Air Arm, in addition to the development of deck arresting, a fresh approach to the methodology associated with the mechanical launching of aircraft was needed with a view to speeding up the operational procedures under service conditions. At around the same time that HMS *Courageous* had been fitted with an arrester-gear installation she was equipped with the prototype accelerator gear for launching aircraft from the deck, as distinct from being 'projected' into the air from a catapult.

In 1938, only five years after the initial concept of flight deck machinery (to the design of which RAE already had made a significant contribution), it was felt that a new specialist design and research team should be assembled to work with the Admiralty on the development of these features. This unit would have as its objective the concentration and co-ordination of study on the operational interfaces between ship and aircraft. On 9 March 1938, Percy Salmon submitted his case to the Chief Superintendent of RAE for forming such a unit. (See Appendix 2.) He identified requirements for a team of seven technical staff, two of whom were to be promoted from the workshops supervisory staff. Hence the Catapult Section – the precursor of Naval Aircraft Department – was set up at Farnborough. This Section assimilated all research and development into the complete field of mechanically assisted take-off and landing of aircraft of all types but with priority for the Royal Navy.

Percy Salmon received his engineering training in the workshops of the London, Brighton and South Coast Railway. He joined the Royal Aircraft Factory at Farnborough in 1916, originally working on the design of aero engines. He, with Mr J. D. H. Pritchard, became involved in the development of catapults in 1922. Subsequently, as Chief Draughtsman and head of the Main Drawing Office [04 Department], Salmon held the oversight of this new Catapult Section, with the field of mechanical assistance to take-off and landing procedures of aircraft becoming his primary interest.

Lewis Boddington was appointed Senior Technical Officer in charge of the actual running of the new Catapult Section. Boddington studied engineering at the University College and Technical College at Cardiff, obtaining the joint diploma of both colleges and the Sir Edward Nicholl Scholarship. He joined RAE in 1936 as a technical assistant to Salmon.

Sam Child, whose name appears in Caunter's record, was the principal draughtsman in developing Salmon's schemes on the drawing board. Although he did not become an integral part of the new Catapult Section, his signature features on many manufacturing drawings of catapult trolleys. (Child's daughter, Barbara, was a schoolmate of the author and, through marriage, is distantly related to him.)

There were three key people at RAE at that time who had been working on the practical side of the manufacture and operation of the RAE Mark I and Mark II catapults and were engaged on a further major project known as the Mark III Catapult. Charles Crowfoot was the first to be appointed. Previously he was head of the RAE Fitting Workshop [27 Department, Building R.21] where all the manufacture of components and experimental rigs was undertaken; he was appointed Technical Officer in the new Section.

Two of Crowfoot's craftsmen-managers, Alf Crossfield and Norman Skelton, were promoted to be Technical Assistants to work with him in his new appointment. The craftsman-managers who replaced these in the Fitting Shop – which carried out most of the local engineering work on the catapults – were Jack Marsh and Bob Pawson, with occasional assistance from 'Big' Jim Howie – whose son later joined the Section. Recruitment of additional technical and professional staff into the Catapult Section continued apace until by 1940 – when the author joined the outfit – there was a staff of 11 accommodated at the eastern end of the Main Drawing Office [Building Q.1]. (See Appendix 2.) This pattern of recruitment from wide-ranging sources accented the requirements for a good, practical, 'hands-on' grounding in addition to professionally qualified engineering designers to be the core of the Section. This is a feature which has had the author's support throughout his subsequent career in engineering design and management within the field of research and development!

From the archival material available to the author it would seem that Salmon, with his small band of designers and draughtsmen, worked as a unit either oblivious to, or ignorant of, experiments in other countries. Catapults were in being, developed in the USA, Germany (Heinkel K.2 in SS *Bremen* and K.4 in SS *Europa* for mail planes), France (SS *Ile de France*) and Italy (abandoned). Of these, no reference appears in Salmon's reports. The submarine aircraft carrier also featured in some foreign navies, particularly the Japanese IJN Class I-400 boats. According to Treadwell (*#100*) each of these 3,000 ton boats was equipped with a catapult and a large hangar containing three 'Jake' aircraft. These are reported to have been in service during World War 2.

Salmon retired in September 1942. His place as Head of the Main Drawing Office (later, the Designs Department) was taken over by Lewis Boddington, who for six years had acted as personal and technical assistant to Salmon.

The Catapult Section was originally housed adjacent to Salmon's office on the newly completed second floor extension of the Main Administration Office block, later known as Building Q.1. With the expansion of the Section, further accommodation was sought and eventually it was 'out-stationed' to a couple of prefabricated buildings on Jersey Brow, just to the west of the RAE catapults, and the location of its main centre of activities. Before long, a substantial building was erected as a Design Office, and a small hangar became the Section's laboratory. This last building provided research facilities on a model scale for a number of ensuing major projects, the flexi-deck and safety barrier programmes being two instances. A simple launcher was devised for use with a model flight deck on which various configurations of arrester wires and safety barriers were mounted. Model aircraft were launched onto the deck by a gravity-operated catapult with its trolley

being brought to rest by a basic 'bungee' system. On occasion, the lab also became an exhibition area for the display of work in progress and new projects.

Subsequently, in order more accurately to identify the specific sphere in which the original Catapult Section was functioning, the new title of 'Naval Aircraft Department' was adopted on 17 April 1945 by which time the original staff complement had tripled to 33. (Details of this growth appear as Appendix 2.) It was under this later title that the Department became independent of its parent, by then the 'Central Designs Department', and it is within this context that this record is compiled.

In September 1946, the Naval Aircraft Department was augmented by the creation of the Carrier Equipment Division (CED), its complement of various Admiralty personnel operating in parallel with NAD within the RAE campus. The two units, on removal to RAE Bedford in the mid-1950s, merged to form the Naval Air Department. Boddington was transferred on promotion to a headquarters post as AD/RDN in November 1951 with A. W. Hotson appointed as his replacement. L. H. G. Sterne was next to assume the post as head of NAD in 1957, under whose command the Department completed its transfer to Bedford.

An early member of the Carrier Equipment Division, Lt/Cdr Garstin, stated in his paper (*#35*) that, at that time, the major items of flight deck machinery:

'are the catapult for launching the aircraft and the arresting gear for recovery. In addition there are: aircraft-handling equipment, particularly for loading the aircraft on to the catapult; crash barriers for last-ditch recovery of aircraft that cannot be recovered normally into the arresting gear; lifts of exceptionally large size to transfer aircraft from the flight deck to the hangar beneath, and vice versa; operating gear to lower obstructions out of the way when flying, such as wireless masts and the like; hangar doors, and a variety of chocks let into the deck and designed to be raised to prevent aircraft being inadvertently manoeuvred over the side. As a recent development, hydraulically operated deflection plates of large size have been installed on flight decks to catch and deflect upwards the blast from jet engines.'

Even this list is not complete. Other top-side features were the collapsible wind shields fitted across the for'ard end of the flight deck in some of the *Illustrious* class ships; the bomb/torpedo lifts for arming the aircraft on the flight deck; the bow-spring depressers supporting the centre spans of the arrester wires; and, latterly, the provision of visual landing aid equipment. Of all this machinery, NAD was mostly concerned with the development and testing of launching and arresting devices (which included safety barriers), their peripherals and the aircraft/machinery interfaces. In addition, the Department adopted a more than passing interest in the development of the early mirror landing sight. The remaining flight deck machinery was the remit of the Admiralty E-in-C Department, but NAD was able to make its contribution to these features, alongside CED, when required to do so.

CATAPULT *VIS-À-VIS* ACCELERATOR

Over a period of years, within naval parlance, the terms 'catapult' and 'accelerator' have become synonymous although, technically, it is possible to define either title in specific terms. The titles were originally related to the launching of aircraft from battleships and cruisers by 'catapult', and that from aircraft carriers by 'accelerator'.

For strict adherence to formal definitions, a catapult is a device for launching an aircraft into flight from a system that completely supports the aircraft and, virtually, throws it into the air. Much of the early history of assisted take-off of sea-going aircraft was the installation of catapults on the larger warships of the Royal Navy for launching spotter seaplanes. Such floatplanes were recovered from flight by alighting on the sea alongside the parent ship, being hoisted aboard by crane and reloaded on to the catapult for the next sortie, or stowed in the hangar with its wings folded. This subject is treated in greater detail, *vide* 'Launching Trolleys' and 'Mark I and Mark II Catapults'.

In contradistinction, an accelerator specifically is a device that propels an aircraft along the ground, or the flight deck, assisting its launch by shortening the take-off path. Apart from early airfield launching experiments with a Vickers Virginia aircraft (1930), it was not until naval aircraft began to increase in weight – passing 5,000lb {2,300kg} – that assisted take-off in the 'accelerator mode' for carrier-borne aircraft was to become important. This type of installation is discussed in greater detail, *vide* 'Assisted Take-Off Gear – (BH.III Accelerator)'. Basically, with a catapult the aircraft is loaded into the trolley by a crane while with the accelerator the loading is achieved by the aircraft taxiing to the loading position. The trolley is then inched to the aircraft while the latter remains standing on its own undercarriage. With the BH.I and the early days of the BH.III accelerators the aircraft was fully supported in the trolley with the undercarriage just above the deck. It was early in World War 2 that it was found the system worked just as well with the wheels running along the deck and, thereby, the process of loading was made easier and quicker.

Throughout the years of World War 2 the Royal Navy retained the term 'accelerator' for the launching devices to be found as part of the flight deck machinery on aircraft carriers. With the arrival of US Navy escort carriers upon the scene, American terminology began to insinuate itself into Navy talk and the specific definitions tended to become less clinical. On some ships the accelerator unofficially became the 'launcher' or even the 'booster'. Perhaps the pronunciation of 'catapult' was clearer when piped over the ship's tannoy (public address) system. With the 1943 phasing-out of catapult installations in capital ships, accelerators then adopted the old title and themselves became 'catapults'.

The type of accelerator under development at RAE for HMS *Illustrious*[40] and her sister ships, initiated in March 1939, was termed the Assisted Take-Off Gear (ATOG), making an early distinction between this system of launching and the basic catapult principle. A development of this system, using a cordite charge (instead of high pressure air) as the primary propulsion element, became the Cordite Assisted Take-Off Gear (CATOG) and the rocket-propelled catapult became the Rocket Assisted Take-Off Gear (RATOG). Before long, confusion ensued through the introduction of direct-mounted rocket motors on the aircraft itself, also using the identical title and initials RATOG. Honour was satisfied when ATOG became the BH.III accelerator (*vide* 'Assisted Take-Off Gear – [BH.III Accelerator]'), CATOG became the BC.I Accelerator, and the original RATOG became the P.I Catapult. Details of all these projects are given in their respective sections 'Rocket Assisted Take-Off Gear – RATOG', 'The 'P' (Projectile/Pyrotechnic) Catapult', 'The 'C' (Cordite) Accelerator – (BC.I/II)' and 'The 'K' (Inertia) Accelerator – (BK.I/II)'.

It may be an advantage here to attempt to unravel the method of specifying the various accelerator types used during World War 2. The first batch of American Lend/Lease escort carriers (CVEs) were fitted with what was designated by the US Navy as the H.2 gear for operating aircraft of AUW of 7,000lb {3,200kg} in tail-down mode. Following this,

the Royal Navy decided that the new 'Illustrious' type of accelerator would be known as the H.III, with the prefix 'B' for British, hence the BH.III, and the American World War 2 upgrade being designated AH.IV.

The BH.III installation was used throughout World War 2, undergoing a number of developments in the process, mostly the upgrading of its performance. Also, the initial single-track system was adapted for bridle (tail-down) launching but its limitations were soon realised and the transition into a twin-track system was introduced in HMSs *Implacable* and *Indefatigable* and subsequent ships until replaced by the BH.V single-track accelerator (BH.5 Catapult).

The initial installation – later, designated (and conveniently back-dated) as the BH.I Accelerator – was operational aboard HMS *Ark Royal*[38] (1938) and her contemporaries HMS *Glorious* (1935), HMS *Courageous* (1936) and, later, HMS *Argus* (1938). The prototype installation initially possessed the capacity for launching aircraft of 7,000lb {3,200kg} AUW at 53kts {27m/sec}, progressively being upgraded in its development. Friedman (*#34 p.139*) states that the specification for this equipment required a launching stroke of 95ft {29m} with a 40 seconds launching interval. It goes without saying that this extremely restricted performance cycle was impractical and was never achieved.

In comparison, the accelerators aboard the American CVE Escort Carriers remained virtually unchanged throughout the Lend/Lease programme. The ships were of four classes, the earlier ones of the 'Archer', 'Attacker' and 'Avenger' Classes carrying the AH.II installation with upgraded capacity for launching an aircraft of 7,000lb AUW at 70kts {3,200kg, 36m/sec}. Later ships of the 'Ruler' Class carried the AH.IV installation upgraded, under Mark C format, to 16,000lb AUW at 74kts {7,200kg, 38m/sec}. These operated only with the shuttle powered, tail-down system of launching. The units were thought to have been able to generate an acceleration rate of up to 4¼g, a figure considered at the time to have been excessive for its purpose.

Before the end of World War 2 the need was felt for an accelerator with the capacity for launching an aircraft of 30,000lb {13,600kg} AUW. While this was being designed and manufactured it was thought that RAE could produce economically an interim system from redundant material already available. The source of the major components was the remains of the uncompleted Mark III Catapult originally at RAF Harwell. Some of the telescopic tubes and the hydraulic gear were brought to Jersey Brow. This installation was given the prosaic title 'The 30,000 Accelerator'. Problems were encountered with the development of the control system and the new 'jury' installation was never successfully operated. By the time the bugs were being eliminated, the 'real' 30,000 accelerator was ready and became the BH.V Accelerator – BH.5 Catapult – with a successful life on Jersey Brow.

The BH.V was fitted to carriers in the immediate post-war period. Some of the older ships were equipped during major refits, while the later ships were equipped for their initial commissions. This machinery, too, was progressively updated in performance. The ultimate post-war achievement was the Steam Catapult; in reality, still an accelerator! This was developed from the experimental prototype unit (BXS.I), becoming the BS.IV Accelerator, eventually recorded as the BS.4 Catapult.

The performance specifications for all shipboard accelerators, and their host carriers, are listed in Appendix 6.

CATAPULTS – PRE-1939

No attempt has been made by the author to expand the treatment of this subject; however, use has been made of whatever information is readily accessible. According to Salmon (#87) in 1932, the aeroplane-launching catapult was a means whereby, in a very short distance, the speed of an aeroplane could be accelerated from rest to that at which it became airborne. He writes:

'Fundamentally the aeroplane launching catapult comprises a carriage on which is mounted the aeroplane to be launched, a motive plant which causes the carriage to travel at an increasing speed along a prepared track or structure and a means of arresting the carriage after the aeroplane has attained the necessary flying speed. The energy of the motive plant is usually derived from compressed air or some form of slow burning powder such as melinite or cordite.'

Percy Salmon, who was recruited to the Royal Aircraft Factory in 1916, was instrumental in issuing a performance specification for aircraft catapults. The specification was open-ended in that firms tendering were free to specify the method of powering the catapult. The working requirements were for launching an aeroplane weight of 2½ tons at a speed of 90ft/sec {27m/sec} with an acceleration not exceeding 2½g in a distance of 60ft {18m}, and that, since it was intended for use on board ship, it was to be as light and compact as possible. Offers submitted by Armstrong Whitworth & Co and Waygood Otis, the latter to a design by R. Falkland Carey, were accepted.

THE FIRST CAREY CATAPULT

This catapult, designed by Carey and built by Waygood Otis, was installed in a shallow pit at RAF Hendon and was in use by October 1917. Few engineering details are available relating to this machine. It is known to have been powered by compressed air and embodied the jigger type power unit with, possibly 2:1 or at the most 3:1 rope reeving driving a trolley along a beam. It was used for launching an Avro 504H aircraft, a strengthened Avro 504B with catapult connecting points and headrest for the pilot, Fl/Cdr R. E. Penny. Cronin (#23 pp. 72/73) states that three converted aircraft were flying at Hendon in May 1918, these being N5261, N5269 & N5270.

THE ARMSTRONG WHITWORTH CATAPULT

The catapult designed and built by Armstrong Whitworth was installed (Cronin, #ibid. p.313) in HMS *Slinger*, an appropriately named former steam hopper of 875 gross tonnage, acquired by the Isle of Grain Experimental Aircraft Depot, and converted for shipborne catapult trials. The dummy aircraft (Short 184 floatplane) was mounted on a trolley which was propelled along a three-railed track by a compressed-air power unit driving a rope haulage system. Dummy launches of

Fairey F127 (N9) floatplane being prepared for launch from HMS *Slinger's* catapult in May 1918. (Fleet Air Arm Museum)

Fairey F127 (N9) floatplane exiting the Armstrong Whitworth catapult in HMS *Slinger*, May 1918. (Author)

HMS *Slinger* returning to a Tyneside quay with the wreckage of a Short 184 catapult dummy, 1 October 1917. (Author)

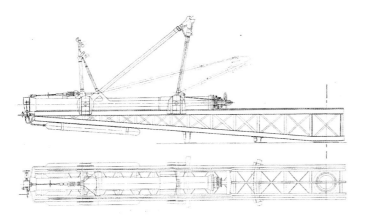

An arrangement drawing of the Farnborough Mark I catapult, one of the 'F' series with a pneumatic-operated, telescopic ram unit propelling the aircraft trolley. The dotted outlines indicate the trolley in collapsed mode. (Author)

aircraft loaded to 5,000lb at 30mph (26 knots) {2,300kg, 13m/s} and 4,500lb at 40 mph (35 knots) {2,000kg, 18m/s} were made on Tyneside in October 1917. Flying trials were conducted off the Isle of Grain with the first 'live' launch made by H. R. Busteed in May 1918 flying a Fairey N.9 floatplane (F.127), a precursor of the renowned Fairey III series. Their Lordships were not impressed with the system of launching and the experiments were terminated, with the ship being taken out of commission shortly afterwards and sold by October 1919.

An interesting feature was the design of the launching trolley. This took the form of an arched bridge spanning three rails of track. The two outer rails supported the weight of the launching trolley and aircraft, the float undercarriage sitting on an open-slatted platform on each side. A cross-beam with a central arch spanned the middle rail but was attached to the main thrust slider which ran on this rail. To the rear of this beam was attached a pair of arms sloping forward and upward with a hook unit at each forward extremity. These hooks were matched to fitments on the aircraft fuselage (rudimentary catapult spools?) to transmit the forward propulsive thrust.

INTERLUDE

Although both Carey and Armstrong Whitworth catapults successfully launched aeroplanes, they never attained the specified performance. With the declaration of the 1918 Armistice the catapult project was abandoned. Interest was resumed a few years later when RAE began its deeper involvement in the field of mechanical launching of aircraft.

In the autumn of 1922 the project was re-started with a revised specification to launch aeroplanes up to 7,000lb {3,200kg} AUW at a speed of 45mph {20m/sec} in a distance of 34ft {10m} with a mean acceleration of 2g. Two designs were put in train, the first to a development of Carey's prototype, while the second embodied a new principle proposed by Salmon. These are discussed separately below.

The development of the ship's catapult, as distinct from the flight deck accelerator, had probably reached its peak by the outbreak of World War 2. The following information has been gleaned from a variety of sources in addition to RAE records, some of them being sparse in the extreme. One useful, though rather limited, source was the small book among the FAA Museum records titled 'General History of Catapults, 1930–1940' carrying the name Barclay Jenkins and presented to the Museum in 1971.

From the mid-1920s, a variety of catapults came into existence. These carried individual type references for which those able to be identified are listed as Appendix 6. Both the Carey and the Armstrong Whitworth machines together were the forebears of subsequent generations of catapults fitted in warships. Both designs were similar inasmuch as they each employed a trolley of sorts running along a beam. The beam was an open structure like a Warren girder with a track along the top. The trolley was hauled by a wire rope along the track, the wire being multiple-reeved through a jigger which, in general, was housed horizontally within the

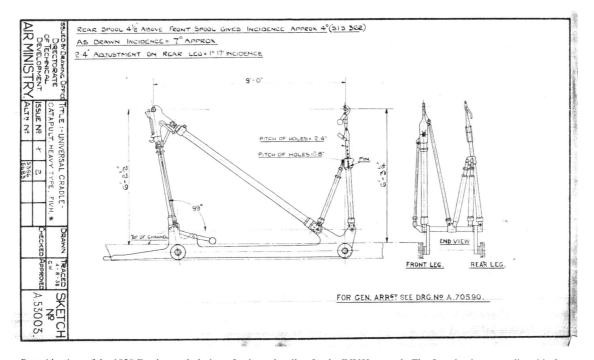

Port side view of the 1938 Farnborough design of universal trolley for the F.IV.H catapult. The front hooks were adjustable for width while the rear hooks provided the universal capability of adjustment in all directions. 1938. (© Crown Copyright/MOD)

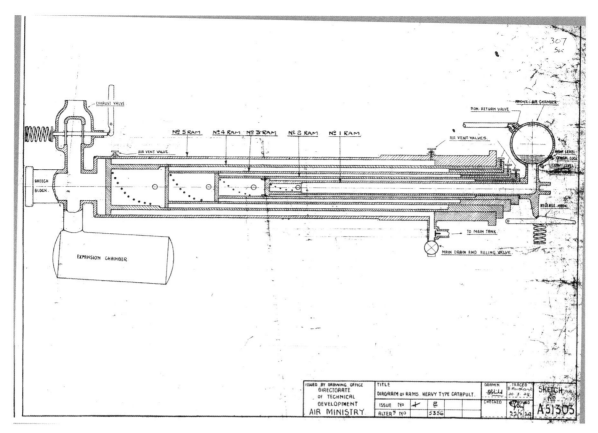

Diagrammatic presentation (1929) of the principles of the telescopic tube construction of the 'F' series – Farnborough design – catapults. 1929. (© Crown Copyright/MOD)

beam. At the outset the working fluid was an oil/water mixture, pressurised by compressed air but latterly most ships' catapults were cordite operated. The detail differences were mostly in the design of the trolley and its attachment to the aircraft and the reeving of the hauling wires. Details of some of these trolleys are given in the chapter dedicated to the subject.

The designs comprised a single hydraulic ram and cylinder with a cluster of pulleys at each end hauling on a launching cable of steel wire to which the trolley was attached. (This type of installation is known in hydraulic engineering parlance as a 'jigger' and was often employed in dockside hydraulic cranes.) With multiple turns in the reeving of the

system, the motion of the trolley along its track was the extension of the jigger multiplied by the number of turns in the reeving. In other words, an extension of ten feet in the jigger with a 6:1 reeve would move the trolley a distance of sixty feet. With this system a second similar reeve of wire was attached to the tail of the trolley for braking purposes in that, as the towing cable was hauled in, so the trailing cable was paid out. This required a second set of sheaves to be mounted on the moving crosshead of the jigger. A hydro-pneumatic braking system was incorporated in the jigger, and the prime mover again was compressed air. The double-reeving principle was adopted for all similar catapults built subsequently.

Fairey IIID (N9463) being loaded into the 'E' type catapult trolley aboard HMS *Vindictive*, 30 October 1925. This was to become the first standard seaplane of the FAA to be catapulted from a ship at sea. The pilot was Sqn Ldr E. J. Burling, RAF. (Fleet Air Arm Museum)

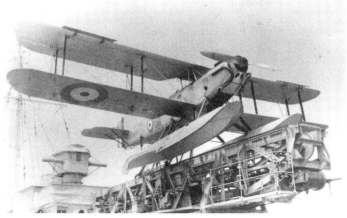

Fairey IIIF Mark IIIB (S1837) floatplane of 447 Flight sitting in the trolley atop an 'E' type catapult in HMS *Shropshire*. (Fleet Air Arm Museum)

Fairey IIIF, Mark IIIB (S1816, Code 716 of 444 Flight) making a clean exit from the trolley of an 'E' type, extending catapult on the quarter-deck of an unidentified ship – possibly HMS *Valiant*. (MacTaggart Scott)

Fairey IIIF Mark IIIB (S1782) seen stowed and lashed to the trolley of an 'E' type catapult, aboard HMS *Exeter*. (Royal Aeronautical Society Library)

Salmon designed and produced a machine which embodied a different principle. He employed a system of telescopic tubes which progressively extended on the application of fluid under pressure at the back end of the largest tube. A fully collapsible trolley embodying, generally, a four-point support for the aircraft, was attached to the outer end of the smallest tube. All moving parts of this system were contained within a beam structure, similar to those for the cable-operated machines. One main advantage with Salmon's system was that the catapult beam – with the tubes retracted – could be stowed in a folded position by rotating the forward end around a vertical hinge located in mid-position. This facility was adopted by a number of shipboard catapults.

At first, the primary source of power was compressed air but as early as October 1924 Salmon was in discussion with the Royal Arsenal at Woolwich and HMS *Vernon* (Portsmouth) on substituting cordite for this purpose. With a gas (air or cordite combustion products) as the prime mover the installation could be identified as a 'pneumo-hydraulic' system. As this is rather a mouthful, the term 'hydro-pneumatic' is preferred, even though, to the pedant, this designation would be incorrect.

The descriptions following present only a general outline of these early machines and their working. The reality was a greater sophistication in content, particularly in the control systems, safety interlocks and hydraulic/pneumatic circuitry. Such auxiliary features are not discussed in this work.

DEVELOPMENT OF THE CAREY CATAPULT

This catapult was designed, to the new 1922 specification, by Carey but built by HM Dockyard, Chatham. It used a system of ropes and pulleys driven by a multiple-reeved jigger powered by compressed air. The structure was some 50ft {15m} long within which was mounted the power unit and the compressed air storage cylinders. The top members of the girder structure served as a runway for the launching carriage. The whole structure was mounted on a vertical spigot allowing it to be slewed into any direction to permit the aircraft to be launched into wind.

The power unit comprised a steel cylinder mounted horizontally within the structure along its longitudinal axis. A piston travelled the length of the bore with rams extending through packed glands at each

end. The outboard ends of these rams carried crossheads each with rope sheaves. Between the extremities of travel, athwart the cylinder were two sets of fixed rope sheaves. Two rope systems were reeved through these sheaves: one pair of ropes was used for hauling the carriage during the accelerating portion of the launching stroke, while the other pair was used for braking the carriage after launch and for recovery of the system to its launch position. In this first installation the rope reeving was 3:1, in other words, for every foot of travel of the jigger, a movement of three feet was attained by the carriage. For the launching stroke, compressed air was applied to the back of the piston which drove it forward, hauling in the accelerating head-ropes and paying out the retarding tail-ropes in the process.

For bringing the system to rest at the end of the launching stroke, pressure was applied at the front of the piston to counter the driving pressure. For most of its travel, the downstream side of the piston was vented to atmosphere. A small, controlled quantity of hydraulic fluid (a water and glycerine mix) was retained in this downstream side. The braking system came into effect in the final two feet of the travel of the piston. A vent in the cylinder at the braking end was progressively closed by the travel of the piston until it was fully blanked off. At this stage the hydraulic fluid would have taken over and the residue at the front of the piston was forced through a number of smaller, graduated orifices in which further blanking-off was progressively introduced.

Fairey Seal (K3479, Code 719 of 444/701 Flight) exiting the trolley of an 'E' type catapult in HMS *Valiant*. (Royal Aeronautical Society Library)

Supermarine Seagull V/Walrus ready for launch to starboard on a midships, E type, extendible track catapult. The cruiser has not been identified but the aircraft identification A2–2, Code 076, would indicate it to be HMAS *Sydney*. May 1937. (© Crown Copyright/MOD)

The fluid resistance thereby imposed the braking effort, bringing the system to rest.

This machine was built at HM Dockyard, Chatham, and underwent its commissioning trials locally in March 1925, using dead-loads. Subsequently it was mounted in HMS *Vindictive* and sea-going trials were conducted at Chatham, again with dead-loads, and a dummy Fairey IIID aircraft (N9481) without engine and ballasted to 5,043lb {2,300kg}. Salmon (#87) continues:

'The forward guns had been removed from this cruiser and a hangar, capable of housing three or four aeroplanes built in front of the bridge, the catapult being installed on top of this hangar. The aeroplanes were hoisted on board by a crane which also served to hoist them from the hangar and mount them on the catapult.'

Salmon, in his report 'Tests of Carey on Vindictive', dated August 1925, records that the installation was designed for launching 7,000lb at 39kts {3,200kg, 20m/sec} with maximum acceleration 2g. The launching run of the trolley was 39ft {12m}, powered by a hydro-pneumatic jigger with a 3:1 rope reeving. The first successful live launch followed some months later, on 30 October 1925, with a Fairey IIID aircraft (N9740) piloted by Sq/Ldr Burling.

The log book, for this catapult in HMS *Vindictive*, is held by the FAA Museum at Yeovilton. It is a meticulously kept record of the life of this machine, including launches and details of maintenance. It was written in the readily legible hand of the catapult engineer officer, Fl/Lt L. G. Craxton, Flight Commander of 444 Flight. The following are extracts from the opening pages of this record:

'The catapult, designed by Mr F. Carey of Lea-on-Sea (*sic*), was constructed at Chatham Dockyard in 1925 and underwent trials there, and at Portsmouth the same year. The advantage of launching aircraft by means of a catapult is that machines may be launched independent of the ship's movements. With no catapult the ship has to steam at a speed and in a direction suitable to the prevailing wind, so that the machine can take off 'head to wind' and into a suitable air stream. This would be a serious disadvantage to an aircraft carrier going into action.

'Late in 1925 the first catapult to be constructed in England was fitted in HMS *Vindictive* and trials were commenced. This consisted of launching various weights of sand and cast iron at various pressures and the acceleration of the trolley on which the machine rides was measured.

'When these trials had proved satisfactory an old Fairey 3D machine with no engine but weighing approximately 6,000lb was fired off the Catapult into the basin at Chatham dockyard on September 20th 1925. The machine rose on leaving the catapult and proved the catapult to be a working proposition.

'The catapult was now dismantled for an internal inspection. All alterations made are noted in the log book. Before dismantling the catapult it was decided to give it several runs with no weights on the trolley, for the purposes of working the leathers and other working parts. On September 16th 1925 the Catapult was given twenty-one runs for this purpose. [*Some technical details appear in the original, but are omitted here.*] On the following day the catapult was given another 44 runs.

'The 'Vindictive' now proceeded to Portsmouth where similar trials with weights were carried out and another Fairey 3D with no engine was launched on October 10th 1925 with excellent results. On October 21st and 28th two Fairey 3D machines, with Rolls-Royce engines running at 1,480rpm, were launched with no pilots. The results on October 21st were very good, the machine flying for some distance before crashing into the sea. On the 28th the results were not so good owing to the rigging of the aircraft being at fault.

'On October 30th 1925, Squadron Leader Burling DSC made the first flight off an English Catapult in Fairey 3D 9740. This flight fully demonstrated the utility of the Catapult.'

Parnall Peto (N181?) stowed in its hangar aboard HM Submarine M.2. (Royal Aeronautical Society Library)

Parnall Peto launched from the catapult aboard HM Submarine M.2. (Author)

Thus a milestone in aviation history was attained – and passed.

Over a two-and-a-half-year period, up to April 1928, this installation in HMS *Vindictive* performed some 189 dummy launches and 95 live launches of aircraft.

The Carey system was improved initially by Chatham Dockyard where two or three machines were produced. In 1927 Messrs MacTaggart Scott, marine engineers of Loanhead, Edinburgh, were invited to co-operate with the Admiralty and Farnborough in the further development of this machinery for sea-borne operation. This company quickly became involved in the manufacture of a number of these machines but the complete output was shared by Messrs Brown Brothers, also of Edinburgh, and Messrs Ransomes & Rapier of Ipswich. (MacTaggart Scott also held the virtual monopoly in the manufacture of arrester gear and safety barrier units for aircraft carriers. Brown Brothers later became more closely involved with producing the accelerators for aircraft carriers.)

In 1930 MacTaggart Scott prepared a detailed engineering report on one series of these machines – the 'E' type, double-extension catapult – which was used as a basis for illustrated articles published in *The Aeroplane* for 25 February 1931 and in *Engineering* for 15 and 22 April 1932.

This catapult consisted of a girder-like structure along the top of which a trolley, which supported the aircraft during its launch, was propelled by compressed air or a cordite charge. Motion was transmitted to the trolley by a multiple-reeved rope system powered by a hydro-pneumatic jigger. The basic catapult beam was 45 feet {14m} long, some 4 feet {1·2m} wide and 6 feet {1·8m} high with retractable extensions at both ends to provide an overall length of 76 feet {23m}. It was designed to launch an aircraft of 8,000lb {3,600kg} weight with a launching stroke of 53 feet {16m} and a trolley retarding length of 13 feet {4m}. This specification and date would suggest this machine to be the E.I.H. catapult destined for HMS *Valiant*.

The actuating mechanism consisted of a power cylinder containing a piston and ram. The ram was connected to the trolley by a series of pulleys and wire ropes reeved so that a four-to-one multiplication of the ram speed and movement was transmitted to the trolley. Two pairs of ropes were used, the accelerating pair conveyed the power to the trolley for launching while the arresting pair produced the braking effort to retard the trolley at the end of the launch. The power cylinder was rigidly fixed to the structure. The piston/ram unit carried a double crosshead at the outer end which carried eight pulleys, four each for towing and retarding ropes. Throughout its launch the extension of the crosshead was supported by guide rails within the structure.

The piston was fitted with a tapered sleeve, equal in length to the retarding stroke of the piston. The largest diameter was slightly less than

Hawker Nimrod II (K3654) about to be launched from a V.1.H type, telescopic tube catapult at RAF Leuchars. (Royal Aeronautical Society Library)

the bore of a neck ring through which it passed. The annular space in front of the piston was connected to a pressure vessel (hydraulic accumulator) and was filled with a 50/50 mix of water and glycerine. (MacTaggart Scott applied the name 'oleo' to this mixture but this did not catch on. The term 'oleo' soon became associated with sophisticated shock absorber elements in aircraft undercarriages.)

When a launch was made, pressure was generated on the back of the piston in the power unit, either by the admission of high pressure air or by the combustion of cordite. This pressure drove the piston forward at an ever increasing speed. This, in turn, forced the fluid in front of the piston through the annulus between the neck ring and the ram into the receiver against a pneumatic back-pressure. At the end of the accelerating stroke, the tapered sleeve would enter the choke ring, partially restricting the passage of the fluid through the annulus and thereby increasing the hydraulic back pressure downstream of the piston. This pressure rise checked the speed of the ram and, via the retarding ropes, the movement of the trolley. At this point, the aircraft left the trolley in free flight.

The taper of the retarding sleeve was profiled to enable the clearance between it and the choke ring to provide a variable restriction (controlled orifice) so that a constant retarding pressure was maintained in front of the piston, allowing the trolley to come smoothly to rest at the end of the stroke.

The catapult had many other features: its installation often incorporated a pivot mounting to allow the beam to be slewed relative to wind direction together with facilities for deploying and stowing the extensions. Both actions could be undertaken manually or mechanically. The manoeuvring of the trolley for loading or stowing purposes was done pneumatically. The collapsible superstructure to accommodate the aircraft atop the trolley was provided generally by Chatham or Farnborough.

of a low beam about 50ft long, 3ft high and 3ft wide mounted on a central pivot about which it can be rotated to launch the aeroplane in any direction into the wind.'

Salmon continues to describe the control system used for operating the catapult, which detail – as for the Carey catapult – is passed over here. Further descriptions are given in the section 'The Jersey Brow Catapults'.

The launching carriage/trolley supporting the aircraft was attached direct to the free end of the front, smallest, ram. Air pressure was applied to the rear of the largest ram which, in being propelled forward, displaced hydraulic fluid ahead of the piston. This fluid in turn passed from the annular space around the tube via calibrated orifices into the space inside the ram behind the next smaller piston and cascaded in turn into the second and third rams. This ensured equal displacement of all rams at the same instant and gave the final ram a terminal speed of three times that of the first ram. Displacement of air ahead of the final ram was into an air reservoir, the pressure from which was used to retrieve and reset the system after each launch.

Relatively few of these machines went on shipboard. MacTaggart Scott built two under contract, the first of which was tested at

Hawker Nimrod II (K3654) at the point of release from the catapult at RAF Leuchars. (Royal Aeronautical Society Library)

Ships' catapults of other types – apart from the telescopic tube units (viz RAE (Salmon) and Vickers) – including those manufactured by Brown Brothers, and Ransomes & Rapier, adopted the same Carey principle of a self-contained gas-pressurised jigger driving a trolley running along the top of the beam through a rope-haulage system.

THE RAE (SALMON) TELESCOPIC TUBE CATAPULT

Salmon favoured the telescopic tube type of power unit for catapult operation and spent much of his early years at Farnborough in developing the principle. The RAE catapult differed from the Carey and any other early catapult system insofar that wire ropes, crossheads, pulleys and so on were abandoned in favour of a direct-acting, telescopic tube system. Perhaps it is best for the inventor, himself, (#87) to describe the machine. It was built to the same specification as the contemporary Carey catapult above.

'In this type of catapult the motive plant, which is operated by compressed air, consists of a number of tubes or rams arranged like a telescope, the outer one being fixed to the rear end of the catapult structure in such a manner as to allow the other tubes to be projected forward; the smallest tube pushes the launching carriage on which the aeroplane is mounted. The structure consists

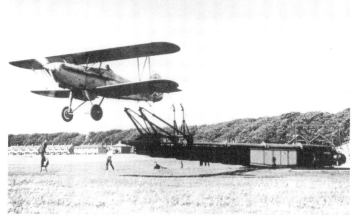

Hawker Nimrod II (K3654) immediately after its launch from the catapult showing the trolley in its collapsed attitude. (Royal Aeronautical Society Library)

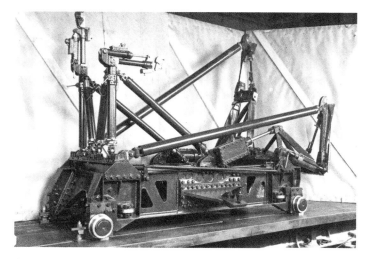

Catapult trolley, reported on the original print to be for a D.III.H catapult. This machine represented the last word in trolley sophistication. (Fleet Air Arm Museum)

Farnborough and then, in 1927, installed in HMS *Frobisher*. The other machine was sent to HM Dockyard, Chatham. In 1928 experiments were undertaken to see if the products of combustion from cordite could replace compressed air as the source of power. These trials being successful, breeches replaced air-pressure vessels. One of these catapults, with a folding beam, was tested at Farnborough and then installed in HMS *York*. Another was destined for HMS *Hood*.

The two permanent catapults of this type installed on Jersey Brow were locally identified as the Farnborough Mark I and Mark II. They are believed to have been:

Mark I – the original F.I.H machine, installed in December 1924. Its performance was 7,000lb at 45mph (39kts) {3,200kg, 20m/sec} at 2¼g within a stroke of 34ft {10m}. It was a telescopic ram machine, pivot-mounted on a circular track of 50ft {15m} diameter, total weight 8 tons. Originally it was powered by compressed air but was converted to cordite operation in 1928. This is likely subsequently to have been replaced by the F.IV.H machine.

Mark II – a 'one-off', non-standard machine, non-folding, fixed orientation, telescopic ram type, hydro-pneumatic powered. The trolley ran in a plated, steel joist track at ground level. It was a robust machine, well suited for experimental work. Its design never went to sea. During its life it catapulted aircraft at up to a maximum weight of 14,000lb {6,300kg} (Fairey Barracuda with torpedo). This catapult was powered by compressed air throughout its life. It continued to give good service up to the time when it was superseded by the C.I Accelerator. By this time aircraft no longer were fitted with catapult spools; all future launches were to be exclusively 'tail-down', employing towing strops (bridles).

A fuller description of these two installations appear later in this text. Crowfoot, in one of the issues of *'RAE NEWS'* (May 1953, p.17), refers to the first Navy catapults having been built at RAE as well as the RAE Mark I, F.IV.H and F.I.L machines.

The hydraulic fluid normally used was a 50/50 mixture of glycerine and water. A variation was introduced early in the design when cordite was used instead of compressed air as the primary power source. All shipborne catapults which were required to be folded when stowed were, of necessity, to this design. Whenever a new design or performance requirement of a telescopic catapult was to be introduced, the Stress Office in the Farnborough Designs Department was kept very busy in the laborious task of step-by-step calculation of the sizes, shapes and locations of the ports in the nested tubes. Only a slight miscalculation

Fairey IIIF, Mark III (S1336/F1076) loaded onto F.IV.H catapult. The catapult beam is hinged mid-way along its length which may infer the installation to have been either aboard, or destined for, HMS *Hood*. (Fleet Air Arm Museum)

could introduce a dangerously high hydrostatic pressure within the tubes which could have been disastrous.

Cronin *(#23)* continues to explain the minutiae of the workings of this hydro-pneumatic system and closes with a description of one of the experimental trial launches of this machine. He writes:

'Concern was expressed about the amount of acceleration that a pilot could endure, so before the tests on human guinea pigs were conducted a series of diverting catapult trials was initiated by the RAE Farnborough, which nowadays would draw some attention from animal rights "activists".

'So the RAE catapult was arranged to discharge its missile into a kind of slide. The missile was the fuselage of an aircraft, and strapped into the cockpit, in an intrepid and professional attitude, was a large sheep – because that was the nearest approach to a naval pilot that the RAE scientists could think of, or that the Admiralty would permit.'

SHIPS' CATAPULTS

The progress and development of ships' catapults was almost an ongoing feature of Navy flying through two decades from the mid-1920s. Almost exclusively the machines were developments of the Carey catapult, employing a cordite-powered jigger with a wire rope system, sometimes with as high as 6:1 reeving. Each type was identified by a class, its mark number (Roman numerals) within that class and a qualifying suffix, viz D.I L; F.IV.H; S.II.T and so on.

The Classes were:

C – thought to be the identification for either 'Carey' or 'Chatham', the Royal Dockyard, Chatham, generally being the place of manufacture. One of these 'improved' machines was installed (1927) in the ill-fated HM Submarine, M.2, for launching a Parnall Peto floatplane.

D – was a 'double-acting' catapult. The catapult beam was installed athwartships, the intention being to launch abeam, either to port or starboard, depending upon prevailing wind conditions. It was located normally between a pair of funnels with a hangar to house a folded aircraft each side of the for'ard funnel. The aircraft trolley was mounted upon a turntable to provide this alternative port / starboard deployment. This turntable was propelled by one of a pair of hydraulic jiggers – one for each direction of launch – and each with its own combustion chamber for the cordite charge. The rope system was reeved through both jiggers, fleeting through the idle unit as appropriate. The design was produced by the Engineer-in-Chief's Department of the Admiralty in collaboration with Brown Brothers, Edinburgh, who manufactured thirteen of these machines. (Another record places the manufacture of some of these machines with MacTaggart Scott, also of Edinburgh.) The 'D' Class of catapults, sometimes were known as 'Forbes Catapults'. It is known that with some of the machines in the 'D' series a catapult beam was not employed. In these cases the trolley rails were fixed direct to the deck with the machinery installed in a ram-room below. While this reduced the top-hamper at deck level, it isolated the operating team from the top-sides scene of action.

E – was an 'extension' catapult. In this instance the catapult beam was often mounted on a spigot at deck level (supported by rollers on a circular track) or atop a gun turret. In its retracted position the beam was kept within the confines of the ship but for launching, an extension – either at one or both ends – was deployed (firstly manually but later mechanically) to provide the full length required for launching. The rope reeving pattern for these machines will have been interesting to accommodate both stowed and deployed configurations without unshipping the ropes. This series was again an Admiralty design but the hardware was manufactured exclusively by MacTaggart Scott. Salmon had some whimsical ideas that the extension to the beam could be adapted to act as a crane for recovery of floatplanes from the sea, and for loading them onto the catapult trolley, also as a framework to support a canvas hangar. The illustrations remind one of a genetically modified praying mantis insect. These latter proposals were never put into practice.

F – was the designation allotted to Farnborough for machines designed and sometimes part-manufactured within the RAE. These were exclusively telescopic-tube catapults to Salmon's design. Some of these used compressed air as the powering fluid on the back end of the primary piston, others adopted cordite, but both used hydraulics for the extension of the inner tubes. The hydraulic fluid from the displacement of the smallest tube was collected into a pressure accumulator mounted upon the tube, with the residual pressure used for recovery of the system after a launch had been made. The hydraulic fluid was a basic mix, in equal proportions, of water and glycerine. It would be messy if a leak occurred in the system! The design of this system permitted the catapult beam to be folded when the tubes were in the retracted position. Folding was achieved both manually and pneumatically.

R – refers to a series of machines recorded in one list but with neither explanation of its nature nor its performance characteristics.

S – refers to a 'slider' system – an Admiralty design. This comprised a main catapult beam supporting another sliding beam upon which, in turn, was mounted the aircraft launching trolley. The cordite combustion unit propelled a rope jigger system with a 2:1 or 3:1 reeving to haul the slider along the track. The slider in turn carried a second rope system with a further 2:1 reeving for propelling the trolley. A system was incorporated to enable the trolley to be manoeuvred to any position. The whole arrangement was mounted upon a central spigot and supported by a circular rail track for deploying in any direction. It seems that Messrs Ransomes & Rapier of Ipswich held the monopoly in the manufacture of these machines.

V – is thought to refer to Salmon type machines designed by Farnborough but manufactured by Vickers. Only three of these are believed to have been built; one spent its life at RAF Leuchars, Fifeshire.

The Mark numbers of each type (inevitably at that period, Roman characters) are self-explanatory and further descriptions are not needed here.

The Suffix letters – only three are recorded – are:

H – refers to the catapult being a 'Heavy Duty' type. Its performance against the other machines of that Mark was relatively more robust than its compatriots.

L – conversely, this designation is for 'Light Duty' machines.

T – represents the machine as being mounted atop a gun turret.

The logic could not be simpler! Performance details, dimensions and other relevant information of a technical nature – where available – are reproduced as Appendix 5.

A final word on the subject of ships' catapults. Almost every type of the above shipborne catapults was installed and subjected to proof trials and flight testing in HMS *Ark Royal*[14] (*Pegasus*). This ship was commissioned in 1914 as a seaplane carrier. She was fitted with a clear foredeck to facilitate the handling of aircraft and was equipped with two steam cranes for this purpose. The ship was renamed HMS *Pegasus* in December 1934 to release the name 'Ark Royal' for the new aircraft carrier then in build at Cammell Laird. Over the period 1930–35 some 15 different marks of catapults underwent trials in this ship.

By 1939 some 114 machines, comprising 28 marks of seven types of catapult – mostly in the 'D' and 'E' series – had been manufactured. Around 90 of these were shipboard installations, in some instances early models being replaced by updated units. Although his list is incomplete, Sturtivant (#92) provides details of the equipment in 64 named warships, all of which were removed by the end of 1944.

As if this were not enough for the veteran HMS *Ark Royal*[14], over the period May/June 1933 she was further fitted with a temporary fore-and-aft launching track surrounded by a dummy deck to represent an accelerator in a carrier flight deck environment. This equipment was designed to launch 7,800lb at 56kts at 2¼g {3,500kg, 29m/sec}. It had a single track and was powered by compressed air with 4:1 under-deck reeving, providing an acceleration stroke of 62ft with retarding stroke of 18ft {19m, 6m}. At least 140 launchings were carried out on this rig using both dead-loads and manned aircraft, the latter involving five aircraft types including the Fairey Flycatcher & IIIF, Hawker Osprey and Blackburn Ripon. A subsequent development of the above-deck gear became the prototype for the accelerator installations in subsequent refits of the carriers, HMSs *Glorious* (1934), *Courageous* (1936) and *Argus* (1938), and in HMS *Ark Royal*[38] during her build. In retrospect these installations became known as the BH.I accelerators, predecessors to the ubiquitous BH.III systems of World War 2.

In 1939, as HMS *Pegasus*[34], this ship was used as an aircraft ferry to the Northern Isles and in 1940 was fitted with a P.II Catapult and equipped with three Fairey Fulmar aircraft to operate temporarily as a catapult fighter ship – a type of CAM-Ship.

LAUNCHING TROLLEYS

The basic method of launching by catapult was by means of a bridge-like, Warren-type girder structure containing the power mechanism for propelling the aircraft launching trolley (originally termed a cradle, sometimes a carriage) which, in turn, ran along rails fitted to the top of this structure. With a variety of aircraft requiring to be launched, such a trolley – while structurally rigid – needed to have inherent flexibility and the ability to be readily adaptable for each marque of aircraft to be launched. In the case of battleships and cruisers (ships-of-the-line) this latter facility was of little importance in that the same aircraft type was used repeatedly, but with seaplane and aircraft carriers coming into prominence such flexibility was becoming of increasing importance. The trolley was a carriage running between steel channels as rails, the upper and lower flanges of these channels acting as supports for the vertical forces imposed on the system during acceleration. At the four corners of this carriage were attachment points for the aircraft supports; at the top of each support unit – generally an adjustable tripod – was a hook forging with a spherical seating which accommodated its related aircraft spigot. This last eventually became known as a 'spool'.

Concerning the early design of catapult trolleys, Salmon (#87) has written:

'The type of launching carriage ... was first introduced and developed at the Royal Aircraft Establishment, Farnborough. It consists of a base frame mounted on four wheels running on the top of the catapult structure, and prevented from tilting up or lifting by small wheels or rollers running inside the channel forming the top member of the structure. Upon this truck is mounted a cradle which carries the aeroplane.

'The cradle is capable of being adjusted to suit all aeroplanes usually catapulted, and is arranged to hold the aeroplane at four points, two in the region of the rear spar of the lower plane and the other two about 9 feet aft of these. The forward points take the whole of the thrust given by the catapult and the rear ones the vertical reactions due to the centre of gravity and the centre of pressure of the aerodynamic loads being considerably above and in front of the forward points.

'Since the cradle is adjustable over a considerable range, both transversely and longitudinally and in the case of the rear legs vertically as well, the points on the aircraft are provided with spherical spools to accommodate the various possible inclinations of the legs, the jaws in which the spools are held being also made spherical. As the aeroplane leaves the cradle at the end of the launching run the front arms fall forward to permit the tailplane passing over them. At the same time the jaws holding the rear points open to release the rear spools.

'The forward arms of the cradle are hinged at their lower ends on the front end of the carriage and have jaws at their top ends. The inclined stays which transmit the catapulting thrust are hinged at the bottom end on to the rear of the truck and also at the upper ends at points close to the jaws. These members extend as the forward arms fall forward and are therefore telescopic, use being made of this arrangement to bring the forward arms to rest at the end of the fall, by making the telescopic members act as hydraulic buffers.

'The jaws at the rear points have weights attached to them and arranged in such a way that under the accelerating forces they keep the jaws closed, but swing forward to open the jaws when retardation commences.'

Salmon continues to describe the finer details of the trolley, including the method of retention of the aircraft in the trolley and its release, together with the means of holdback of the trolley up to the point of launch.

Having introduced the subject, the earliest trolley of which details are available was that provided in the 1917 Armstrong Whitworth catapult, introduced earlier in this chapter. This was a 'one-off' system in which the aircraft (floatplane) was supported with its floats sitting on small platforms, with arms rising from the base of the carriage to a pair of attachments on the aircraft to provide the forward motion. Subsequently, all aircraft flown from ships-of-the-line were supported in four-point trolleys.

With the earliest trolleys (1932) the front hooks (the units which held the front spools of the aircraft) were adjustable laterally within small limits, the main adjustments all took place in the supports for the rear hooks. The adjustments each side of the aircraft could be within the ranges of 23 inches {580mm} vertically, 12 inches {300mm} in the fore-and-aft direction and 14 inches {350mm} laterally. Within a few years, RAE issued a Standard Instruction Sheet making the fore-and-aft dimension between front and rear spools as 9 feet {2□7m} with the rear spools in a horizontal plane 4½ inches {114mm} above the front spools with the aircraft within a specified aerodynamic attitude. The spherical diameters of the spools were fixed as 3·34in {84·8mm} for the front units and 2·24in {56·9mm} for the rear spools.

Within a short period the basic form of trolley was developed, with varying depths of ingenuity, into more sophisticated forms offering greater flexibility. Mostly this development was applied to the adjustability of the front supports and improving the pattern of collapse immediately following the release of the aircraft from the trolley. Much of the experimental adaptations were carried out within the RAE workshops.

Within the four-spool system, the port and starboard hooks forming the forward pair (which transmitted the horizontal accelerating forces to the aircraft spools) were of suitably massive construction and each was mounted upon a complex tripod assembly with linear adjustment provided in all three legs. This triangulation allowed the appropriate settings for spool width and height to be obtained for a particular aircraft; apart from this adjustment, all early trolleys were rigid in construction. The hooks were provided with a detent system which retained the aircraft in the trolley and held it against forward thrust when the engine was running. The rear supports similarly were designed as tripods with adjustment in all three legs. These were required to absorb the vertical reaction of the aircraft when the engine was running as well as during acceleration. In this instance the accelerating forces were comparatively light and a latch-type detent – released by inertia forces during the launch stroke – was adequate. After a launch the trolley was retracted to its start position and reset to suit the next launch.

In spite of this attempt at standardisation, in 1935 a deviation was produced by Chatham Dockyard for mounting a three-point superstructure on a standard E.III.H trolley base. In this instance the equipment was intended to launch a 'centre-float' aircraft in which the fuselage formed the water-borne hull; in other words, a Supermarine Walrus. This configuration provided a pair of substantial hooks on tripod legs at the rear of the trolley. These hooks were spaced to a width of 4ft 8in {1·4m}. They had limited adjustment, and collapsed forwards. The front unit was a single support which mated with a fitment on the keel of the hull some 7ft-8in {2·3m} forward and 16in {400mm} below the plane of the rear hooks. It was mechanically linked to the rear hooks structure and collapsed forward with it. It is doubtful if this trolley was adopted beyond the experimental stage. The mainstream development continued with the four-spool configuration to the standard RAE (9ft and 4½in) dimensions being introduced at that time.

There were occasions when the aircraft did not leave the trolley as cleanly as it might and damage was caused to the rear fuselage. An early attempt was made to provide additional clearance between aircraft and hooks as it left the catapult trolley. This design allowed the upper part of the front hook assembly to collapse, leaving the rear hooks rigid. A still further development entailed the complete collapse of the front hook system at the end of the launch run. This had to be rapid to allow the tail of the aircraft to pass through the trolley without suffering impact damage. This later development also allowed a part of the rear hook assembly to collapse on release of the aircraft from the trolley.

An illustration shows a Fairey IIIF floatplane (S1336/F1076) loaded onto the trolley of an F.IV.H catapult. This is an early type of trolley in which the forward hook assembly collapsed forward, pivoting at the bottom of the vertical legs with the long diagonal legs extending as the hooks moved forward and downward. The tripods for the rear hooks remained upright but the hooks themselves were pivoted to fall forward just below the attachment points to the aircraft. The bob-weight for inertial release of the retaining detent may be seen pointing forward, in line with the bottom of the fuselage. Positional adjustment of the tripod legs was obtained by screw threads at the bottom ends of the tubes. The vernier height adjustment of the rear hooks was made by steel pins passing through matching holes in the inner and outer tubes.

Although this system was not ideal, further experimentation was undertaken with more sophistication introduced in later trolleys. Two of these are discussed below. This basic trolley was found to be adequate for the RAE Mark II catapult on Jersey Brow. It was retained throughout its life, apart from its temporary removal for initial trials on the above-deck gear for ATOG – the parent and prototype of the BH.III Accelerator.

Improvements in trolley design were made; one of these is seen in an illustration in which a Supermarine Walrus is shown mounted in a CF Mark II trolley on a D.I.H catapult, believed to be aboard HMS *Pegasus* (previously HMS *Ark Royal*[14]). This shows the typical arrangement for the tripods mounting the front hooks but the rear hooks were pivoted lower down. These had a linkage system to control the collapse with a telescopic damper to arrest the fall of the hooks.

Further sophistication is seen in the illustration of a later trolley, identified as for a D.III.H catapult, in which the front tripods were articulated. In this system the diagonal tubular struts did not extend and the collapse of the forward hooks was controlled by damper units. This workshop illustration is valuable in that it shows the port legs erected to receive an aircraft, while the starboard legs are in the collapsed, post-launch configuration.

The aircraft was always loaded on to the trolley by an adjacent crane. This activity was often undertaken while the ship was under way, sometimes on recovery from the sea after an operational flight. The trolley would be retracted to, and held in, its launch position with all legs erected ready to receive the aircraft. There would be a flight deck rating at each leg, ready to fit the hook to its relevant catapult spool on the aircraft. He would hold each detent away from the hook permitting the aircraft to settle into the trolley and then allow the detent to close as soon as all four spools were engaged within their hooks.

This was never an easy procedure, particularly if the ship was under way with some movement upon her. In 1940 the dockyard at Chatham, which seemed to specialise in the design of shipboard catapult trolleys, devised a rig that was given the most appropriate name of 'Shoe-Horn Loading Gear'. It comprised a readily demountable structure with vertical guide rails located immediately adjacent to the front legs of the catapult trolley. These rails had flared tops to receive spigots fitted to the aircraft at the catapult spool points. In fact, these spigots were made to retract into the structural tubework supporting the catapult spools.

For loading onto the catapult trolley, the aircraft – while suspended from its crane and with a sailor holding on to a rope slung around the tail, and perhaps another holding on to a bow rope – would have its spigots extended. It would then be lowered into the shoe-horn gear with gravity undertaking the guiding process until the aircraft eased into and bottomed within the rig. While supported in the shoe-horn, the aircraft was then in a reasonably stable situation in which the catapult trolley could be brought into position and the aircraft loaded simply by attaching the trolley legs. When this was accomplished, the shoe-horn spigots were retracted and the gear itself lowered into its stowed position. From this point the launch could proceed. This equipment was used with Supermarine Walrus amphibian aircraft; it is not presently known if the Supermarine Sea Otter and the Fairey Sea Fox were similarly adapted. An illustration shows a Supermarine Walrus (K5773) sitting in a CF Mark II trolley aboard HMS *Pegasus* with the riveted shoe-horn device in the loading position.

On Jersey Brow, although a gantry was provided for loading dead weights, a small mobile crane was used for loading aircraft onto the Mark I catapult. For the Mark II catapult, a different loading procedure was adopted. A staging was provided on both sides along the length of the catapult track to accommodate the main undercarriage wheels of the aircraft to be launched. Through the centre of the trolley a further timber runway, ramped upwards from loading level, allowed the passage of the tail wheel/skid to pass through the trolley. This incline was made adjustable. Removable decking ran the length of the catapult between the trolley rails. The aircraft was wheeled to the track, tail first. Its tail wheel/skid was located on the ramp and a towing wire, from a winch at the rear of the catapult, attached to it. The aircraft was hauled backwards until it reached the trolley, the trolley legs were then attached and in so doing the aircraft was raised above decking level. The loading exercise was completed when the hook detents were secured, the winch disconnected, the internal staging removed and the trolley hauled back to its ready-to-launch position.

The detent system – as detailed earlier – was provided on the front hooks to prevent the aircraft from moving forward when its engine was running. This had to be released while the trolley was moving along its track soon after acceleration was applied. A small ramp was fitted to the track. This mechanically released the retaining detent through a roller-operated linkage as the trolley passed over the ramp. In the case of the rear hooks, with the trolley nearing the end of its run and undergoing retardation, inertial forces acted on out-of-balance weights to release the rear detents.

It was this type of trolley that was used on the Farnborough Mark I and Mark II catapults and its shipboard derivatives. Further examples were developed for radio-controlled pilotless aircraft known as the 'Larynx' and the 'Queen Bee/Queen Wasp' series. In concept, these latter trolleys were less complex in design as only a single marque of aircraft was operated from each relevant catapult and universal flexibility in adjustment was not a necessary requirement.

By the time the fleet carriers, HMS *Ark Royal*[38] (1938–1941) and those typified by HMS *Illustrious*[40], were being commissioned a similar process for launching was still in use. Improvements to this system were felt to be necessary for operation from carrier flight decks. These and the development of tail-down launching are discussed in later chapters. The trolleys used for the Direction Controlled Take-Off system (DCTO) and the high Speed Track (HST) – not being FAA orientated – are also discussed in their related chapters.

THE PORTABLE CATAPULT

Having elsewhere described the distinction between a catapult and an accelerator, it is worth examining a piece of airfield equipment which, basically, was an accelerator but was known as the Portable Catapult. This had nothing whatever to do with the Fleet Air Arm but it was one of Salmon's schemes for mechanically assisting the take-off of heavy aircraft. Although Caunter (*#222 p.77*) states that it made its debut in launching a Vickers Virginia, Mark III prototype aircraft (J8236) at Farnborough on 4 June 1930, the RAE Flight Book records the date of the first catapult launch as 13 June 1930. A further nine launches were made at Farnborough during the ensuing 12 months.

In June 1931 the installation was transported to Hendon Aerodrome for the annual RAF flying display. On each of the four days of the display a catapult launch of the Virginia aircraft was made but at its conclusion the system was dismantled, returned to Farnborough and forgotten. An interesting sideline is that at the same display this aircraft was shown in an early public demonstration of in-flight refuelling – a double-first for Farnborough!

This catapult launched the aircraft by hauling it on its normal undercarriage along the airfield until it attained take-off speed. This was

Vickers Virginia, prototype Mark VII (J8236) was converted to Mark IX with a Mark X tail wheel. The aircraft in this picture of the Portable Catapult is shown ready for launching. June 1930. (© Crown Copyright/MOD)

The Vickers Virginia aircraft approaching the end of its launch from the Portable Catapult. The man with the large red flag is indicating to all and sundry, who may be on Farnborough Common, that flying is in progress. The vehicle to the right, with the driver watching the scene, is the Hucks' engine starter mounted on a caterpillar Crossley chassis. June 1930.
(© Crown Copyright/MOD)

Supermarine Walrus amphibian (K5773) loaded on the athwartships, 'D' type catapult – mounting a CF.II turntable trolley – aboard HMS *Pegasus* (ex-*Ark Royal*[30]). August 1939. (© Crown Copyright/MOD)

achieved by a power unit containing a winding drum (located behind the aircraft) with a steel cable run to, and around, a ground anchor well ahead of the aircraft and back to the power unit. With the aircraft positioned at its starting point, the cable was taken from the top, starboard side of the winding drum, around the ground-anchor pulley sheave ahead of the aircraft and its return hooked to the landing gear (or other suitable location) on the aircraft. (How this was achieved, with a divided undercarriage, has not been explained. The photographs are not sufficiently clear to show this detail.) A short, adjustable length of cable attached to the hook was connected to a truck supporting the tail wheel and raising the aircraft into its flying attitude. From this trolley, an arresting cable was continued on the return leg back to the lower, port side of the winding drum. A pneumatic shock absorber was fitted between the haulage and trailing cables. It will be seen that all the equipment was located on the centre line of the flight path, the

machinery and operators at the rear having to succumb to the slipstream and all the Aeolian detritus it generated.

For a launch to take place, the power plant rotated the winding drum, thereby hauling in the launching cable and paying out the trailing, arresting cable. When the aircraft had travelled a predetermined distance, a pneumatically-operated band brake was applied to retard the drum and power unit. This transferred the pull from the launching cable to the arresting cable thus decelerating the launcher moving parts and releasing the aircraft for take-off.

The performance of this equipment was to launch a weight of 18,000lb at 60mph {8,200kg, 27m/sec} with a maximum acceleration of 1·1g. The length of run was 120ft {37m} for the launching phase and 40ft {12m} for the braking distance. The overall length of the installation was 250ft {76m}. The system was driven by a pair of spheroidal, sliding-vane (swashplate type) engines, running on air at 300psi {20bar} developing a peak power of 1,500hp {1120kW} at 2,500rpm. (Salmon (#87) quotes the output as 4,000hp {3MW}, possibly the maximum output working at 38 per cent efficiency.) Maximum speed was reached in 2¾ seconds – this being the duration time for a launch. As the total weight of the equipment amounted to 5 tons, perhaps it would have been better named as a 'Transportable Catapult'. It can well be imagined that the ground-anchor unit must have been quite massive in construction and presented a somewhat disconcerting obstruction in the line of flight at take-off. The machinery, which the author saw rusting away on Jersey Brow a decade later, consisted of a carriage on wheels supporting compressed air receivers (ex-torpedo air vessels) and the spheroidal motors connected to the winding drum. The equipment was said to have been capable of operation by cordite. This was not in evidence at the time I saw it. I have seen no detail of the ground-anchor unit or of the towing hook itself; this latter piece of engineering could have been interesting.

Although the Fleet Air Arm Museum at Yeovilton holds no brief for displaying or researching flight deck machinery, or other launching devices, its archive contains a large notebook, bearing the date of 1926, in which are set out all the Farnborough performance calculations for this project.

A FLYING BOAT CATAPULT

With the employment of large flying boats for intercontinental travel becoming popular, attention was given to the adoption of an assisted take-off system. In 1937 MacTaggart Scott of Edinburgh were invited to prepare a design study for such an installation. These were the days of the Short 'C' Class 'Empire Flying Boats', with the flight of the first Short Sunderland only a few months away.

The designers submitted proposals for a track of 11ft {3·4m} gauge, having an overall length of 354ft {108m}. The acceleration stroke was stated as 290ft {88m} and the trolley retardation stroke as 28ft {8·5m}. The trolley to support the aircraft would be 20ft {6m} in length, twice that of a ship's catapult. With this relatively short accelerating stroke it is assumed that the trolley would be provided with an external means of propulsion. Such details – as well as performance figures – do not appear in the meagre records available. Needless to say, the project did not proceed further than its feasibility study.

There was one other, non-naval, launcher in which Farnborough was directly involved. This was the RAE Mark III Catapult, under construction in the late 1930s at RAF Harwell. It was the ultimate in the original Salmon telescopic tube design, being a one-off project of some magnitude. It is treated in a chapter devoted to this project.

AIRCRAFT 'NAVALISATION'

The term 'navalisation' is an all-embracing description of the equipment and facilities required to be provided in an aircraft for carrier-borne operation, particularly if it happens to involve the adaptation of a land-based machine, or marque, already in service. Primarily, it requires the provision of suitable attachments for catapult/accelerator launching and for deck arresting together with wing-folding capabilities. Complete navalisation may involve the further provision of deck handling gear, including defuelling facilities, and methods of lashing the aircraft to the flight deck or hangar floor when stowed. Special radio, radar and homing equipment may also be needed. In this book consideration is given only to the launching and landing aspects of carrier operation, the speciality of Farnborough's Naval Aircraft Department. The particular features are identified as catapult spools, towing hooks and holdbacks for powered launching from catapults and accelerators and arrester hooks for deck landing.

Port front catapult spool on Fairey Albacore (N4389). The complex forged mounting may be seen attached to the fuselage. The stub of the lower main plane is to the left, with the wing folded back alongside the fuselage to the right. (Author via FAAM)

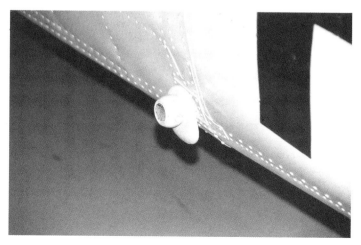

Fairey Albacore (N4389), detail of port rear catapult spool. (Author via FAAM)

Salmon (*#87*) recognised the catapult/aircraft interface as requiring special attention. He states:

'When considering the strength of an aeroplane in relation to the loads imposed on it, some idea is necessary of the operations involved in catapulting. Normally the aeroplane is mounted on the launching carriage in such an attitude that the wings are at a fairly large angle of incidence. Just before the catapult launch is made the aeroplane engine is set running at full speed. The thrust of the airscrew tends to drag the aeroplane forward so that provision must be made to stop the aeroplane from leaving the carriage and also to prevent any tendency to rotation on it. At the same time an adequate restraining force must be applied to the carriage if it is not to move forward until the launching force is actually applied. The launching force after overcoming the restraining force on the carriage is communicated to the aeroplane through the points of attachment to the carriage and these points must also be capable of preventing any tendency on the part of the aeroplane to rotate due to the application of the catapulting force below the centre of gravity.

'As the carriage and aeroplane together gain speed, the air loads on the latter increase until finally, at the end of the launching run, the aeroplane is airborne and leaves the carriage. Where possible it is customary to launch the aeroplane head into relative wind, the wind speed at the end of the run being slightly greater than that required to give a lift at the catapulting incidence equal to the weight of the aeroplane.'

The first aircraft types designed for launching by catapult were supported in a cradle which was propelled along a track when, at a suitable point in its travel, the cradle was triggered to collapse and release the aircraft for normal flight. Each aircraft was fitted originally with one pair and later with two pairs of attachment points. These were known as 'spools' but in reality were 'spigots'. Eventually this trolley launching procedure was superseded by the tail-down method using strops (bridles) for towing the aircraft along the deck – more of this later.

The mechanical forces imposed upon the aircraft during launch were between 2¼g and 3½g applied in the horizontal direction. The 'g' symbol refers to the gravitational force applied. In basic terms 1g, in the case of an aircraft catapult, is the force equivalent to the weight applied in a horizontal direction. The basic engineering formula (in relevant units), purloined from Newton's Second Law of Motion – Force = Mass x Acceleration – applies here. A value of 3g indicates that the force imposed on acceleration is equal to three times the dead-weight. An aircraft weighing seven tons will be subjected to a force of 21 tons when a rate of acceleration equal to 3g is applied to it. The accelerating force is applied horizontally at a low point in the airframe via the front spools. The components of both aircraft and catapult have to be designed with this factored figure in mind.

During launching it was desirable for all pitching moments affecting the aircraft to be maintained in equilibrium as much as possible. The acceleration forces diagram showed this to best advantage when the forward line of thrust applied by the catapult is below a horizontal centre line through the centre of gravity of the aircraft. The thrust of the engine was above the same reference line and the resultant tail-heavy load was absorbed by the rear legs of a trolley launcher. (In the case of

Starboard front catapult spool on Supermarine Walrus (L2301). The spool and its mounting, located at the hull step, are made of stainless steel. (Author via FAAM)

a tail-down launch, the accelerator thrust line must still pass below the centre of gravity, and the imposed vertical thrust at the rear must be absorbed through the tail wheel assembly of the aircraft itself.) Extensive trials were undertaken both ashore and at sea to determine the extent by which cross-winds could be tolerated during accelerated launches, the results generally indicated that catapult launching could be undertaken with the relative wind direction coming from a wide angle away from direct ahead. Later systems, after 1945, enabled even wider margins to be adopted, particularly when the launching speed was considerably greater than the wind speed down the flight deck.

CATAPULT SPOOLS

The aircraft strong-points interfacing with the launching trolley in the early days of catapulting were something of a hotch-potch design. As equipment proliferated so the question of standardisation needed to be addressed. This quickly progressed from a two-point attachment in the days of the Carey catapult to a four-point system. The spigot attachment points on the aircraft were given the name of catapult 'spools'. (The name 'spool' conveys the impression of a reel with a concave surface to accommodate a length of thread or wire. In this case each spool is derived from a sphere. Perhaps 'trunnions' would have been a better description for these devices.)

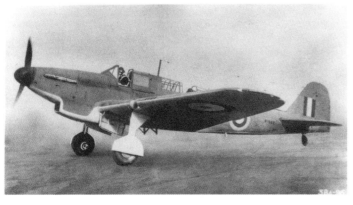

Fairey Fulmar I (N1957) at take-off showing catapult spools deployed. (Author)

The designs were adopted as standard to all aircraft launched from a trolley, whether catapult or accelerator, and continued well into World War 2. In the early days the width between each pair of spools was unimportant, provided they were contained within the space envelope dictated by the configuration of the mating 'hooks' atop the trolley supports. The spools protruded from the sides of the fuselage sufficiently far to provide clearance for the loading of the aircraft into the catapult hooks and to allow freedom from interference on release from the launching trolley.

Catapult spools were located in the lower fuselage of the aircraft, the larger pair, the front spools, being positioned one each side of, and standing proud of, the fuselage. The smaller units, the rear spools, were located some ten feet {3m} abaft the front spools. The connecting points in the launching trolley would be provided with matching, hollow spherical seatings.

Individual aircraft types each had their particular dimensions for the setting of the spools, and the catapult trolleys were designed to accommodate the resultant flexibility. This variation was of little consequence when a shipboard catapult installation would operate aircraft of only one type at a time. The trolley would be adjusted to the setting for its aircraft on embarkation and remain in this configuration for the duration of the ship's commission.

With the advent of accelerators on aircraft carriers, from which a variety of aircraft would be operating, rapid adjustment of the trolley to suit each aircraft type was essential in order that operational efficiency would not be impaired. This question of uniformity was solved by Farnborough in producing an SIS (Standard Instruction Sheet) specifying within dimensional limits the location of catapult spools on naval aircraft. In this Standard, both front and rear spools took the form of truncated spheres, the front spools being 3·34in and the rear spools 2·24in diameter {84·8mm & 56·9mm}. (The reason for the choice of these particular dimensions has been long forgotten.) The front spools were located at a low point on the fuselage, often at the junction of the rear spar of the lower mainplane with the fuselage. The front pair of spools was designed to take most of the weight of the aircraft in static mode, together with the horizontal accelerating forces applied in the catapulting mode. The rear spools were designed to accommodate the upward forces when the aircraft engine was running and the downward reaction from the airframe during catapulting mode. The spools were located low in the fuselage, 9ft {2·7m} abaft the front spools and on a parallel plane 4½in {115mm} above them. The attitude of the aircraft to which this geometry was applied was described as that providing a lift component equivalent to 33 percent of its weight at the launch speed of 66kts {34m/s}. (No wonder the aircraft often dropped uncomfortably near to the surface of the sea immediately on its release from the launching trolley!) The transverse dimensions between port and starboard spools were less critical than the fore-and-aft measurements. One early catapult trolley (F.I.L.) could accommodate front spool centres within the range 24–44in {0·6m, 1·1m} and rear spools at 12–40in {0·3m, 1·0m}. Similar dimensions for the D.I.H. trolley are recorded as front spools 30–44in {0·8m, 1·1m} and rear spools 15-40in {0·4m, 1·0m}.

This was fine for biplanes, such as the Blackburn Shark, Fairey Swordfish and Albacore. The Supermarine Walrus and Sea Otter amphibian aircraft carried their front spools on special brackets at the point of the step in the planing bottom of the hull. Their rear spools were cantilevered from the hull. An interesting modification was adopted on these aircraft in which a retractable spigot was fitted through the bore of each front spool. This was for use with a 'shoe-horn loading gear' to facilitate its deployment on the catapult trolley; not an easy task in a rolling ship and half a gale of wind blowing along the deck. (These spigots, unfortunately, are not fitted to the aircraft preserved in the FAA Museum at Yeovilton.). With the advent of monoplanes, for example the Blackburn Skua and the Fairey Fulmar, the situation became slightly

Blackburn Shark, K4352, Code 740 of 820 Squadron embarked in HMS *Courageous*. 1935. (Royal Aeronautical Society Library)

Hawker Sea Hurricane IB (Z7015). The 'V'-frame arrester hook is held in the stowed position by the snap gear (with additional lashings). The front catapult spools are seen each side of the radiator; the rear spools would have been located near the hinge point of the arrester hook. The arrester hook itself is not that normally fitted to this aircraft; it would have been a standard 10,500lb fitment. (Peter J. Cooper/Falcon Aviation)

marque, and AL247 was the first machine with spools to arrive at Farnborough in May 1941 for catapult trials. NAD was not slow in recognising the benefit of bridle launching and quickly adapted the C.I accelerator on Jersey Brow for tail-down experiments, making the appropriate 'Y' format strops for this purpose. Very soon, to accord with American practice, these strops became known as 'bridles'.

Following this introduction, in 1943 a Supermarine Seafire, NS487, at RAE was fitted with a relatively crude experimental installation employing a pair of towing hooks. In spite of the spindly nature of the undercarriage legs, trials were successful and the principle of tail-down launching became a new standard within the FAA.

The AH.II and AH.IV accelerators aboard the American Lend/Lease ships used a single towing hook (shuttle) in the deck. Throughout World War 2 British carriers employing the BH.III (and the single BC.II) accelerators were fitted almost exclusively with a twin-track system in the flight deck. (The first four installations in the 'Illustrious' Class ships and HMS *Unicorn* were single-track systems, the front slider readily accommodated an appropriate deck 'shuttle'.) To adapt these twin-tracks for bridle launching, a bridging plate, spanning the tracks, was fitted to the above-deck machinery. This required a pair of towing hooks on the accelerator trolley. An advantage here was that no special provision was needed for the clearance of the tail wheel as it passed

more complex. Front spools needed brackets to locate them beneath the monocoque fuselages or plated under-surfaces of wings. Early Fulmars supposedly had their front spools mounted within retractable goal-post structures. Contemporary photographs of these aircraft in flight inevitably show their spools deployed. (Disappointingly, front spools are non-existent on the prototype Fulmar displayed at Yeovilton.)

All this foregoing engineering disappeared with the introduction during World War 2 of tail-down launching with wire rope strops (bridles) forming the launcher/aircraft interface. This newer procedure is described in the chapter 'Tail-Down Launching'.

TOWING HOOKS

Towing hooks for tail-down launching were introduced to the UK when the first Grumman Martlet I aircraft arrived but at that time there was no launching equipment in British carriers to make use of this facility. The fitting of catapult spools was one of the early priorities to 'navalise' this

Fairey IIIF Mark I aircraft (S1182), the squadron code number 705 indicating its origin to be 822 Squadron embarked in HMS *Furious*. c.1933. (Author)

Fairey Seal (K3479): the first marque of aircraft to be fitted with a 'V'-frame arrester hook as standard. This aircraft, with Code 734, was from 802 Squadron embarked in HMS *Courageous*. *c.*1933. (Author)

through the flight deck machinery. The situation improved with the introduction of aircraft with tricycle undercarriages as the nose-wheel was forward of the towing hooks and deck shuttle.

The first post-war shipborne catapults (accelerators) were developments of the BH.V system. This used exclusively a single deck shuttle. When the BS.4 series of steam catapults replaced these earlier units, the single-track system was retained; the twin-cylinder configuration of the under-deck machinery was connected by a bridge-piece within the deck space to contain the single shuttle assembly.

From the earliest days, aircraft for tail-down (bridle) launching were fitted with a pair of towing hooks, located either beneath the fuselage or on the undercarriage legs. (The exceptions were the early Grumman Martlet/Wildcat and the post-war Hawker Sea Hawk. This latter aircraft

was fitted with a single towing hook.) In the detail design of these hooks NAD, apart from advisory procedures, had little involvement; such responsibility was left mostly to the aircraft manufacturers. Of course, any deficiencies would be discovered when the aircraft was subjected to catapult proofing trials; but seldom was there any problem here. The hooks, of necessity, were of robust construction and generally were proof tested at the suppliers' works. Problems mostly occurred with the proximity of the bridle to the airframe and its behaviour (tendency to whip, with the ends flaying) when the aircraft exited the towing shuttle.

HOLDBACK POINTS

With the advent of bridle launching, a means of constraint was necessary to prevent the aircraft from leaving the accelerator prematurely, particularly in the engine run-up prior to the start of the launch. This device was known in the US Navy as the 'holdback'. It consisted of a swinging hook restrained by a frangible element (a breaking ring which failed in tension), all mounted on a spring-controlled extensible shaft.

To accommodate this, a strong-point had to be provided at the rear of the aircraft. In some machines this appeared as a short bar held in a fork fitting in the rear fuselage. In others the attachment was formed by a shackle attached to the rear wheel assembly. The claw at the leading end of the holdback was hooked into this strong-point, with the trailing end attached to a rack in the flight deck.

The breaking ring was adequate to resist the thrust of the aircraft power plant but broke when the additional launching force was applied at the towing hook. The replacement of a new ring for each launch was a fiddly process on a cold, windy flight deck and this problem was addressed by NAD. Schemes were tried in which the aircraft was brought up from the hangar with its bridle and holdback attached, ready for connection to the launcher. But this did not solve all the problems.

NAD looked at this situation and one of its staff, Mollart-Rogerson,

Sea Hurricane IB (Z7015) at point of take-off. (Peter J. Cooper/Falcon Aviation)

Hawker Sea Hurricane IB (Z7015). The front starboard catapult spool is seen alongside the radiator cowling. The location of the 'V'-frame arrester hook is seen but the hook itself is not representative of that fitted to this marque of aircraft. The rear catapult spools are missing but would have been located at the arrester-hook pivot position. (© John M. Dibbs/The Plane Picture Company)

to slew during arrestment. Time and effort were spent investigating this occurrence, with instrumentation to record the behaviour of the aircraft. One advantage of this system was that at the end of arrestment, if the aircraft was allowed to roll back a fraction, the wire automatically became disengaged from the hook. No member of the flight deck party was required to do this little job.

The American Lend/Lease aircraft of World War 2 appeared with arrester hooks attached to the end of a shaft 'exuding' from the tail, immediately dubbed 'sting' type hooks. These installations were freely suspended in the tail, and allowed to articulate in pitch and yaw without any form of constraint. The tendency for these aircraft to yaw in an off-centre landing into an arrester wire was obvious but appeared to create no adverse effect on machines and aircrew. NAD experimented with a variety of similar configurations, with the sting type of arrester hook eventually becoming universal in its application in FAA aircraft. An added refinement was the introduction of powered retraction of the hook after arrestment. Many of the FAA applications were fitted with damper units to reduce the tendency of the hook to bounce on impact with the deck.

Two of the RAE experimental Supermarine Seafire aircraft (NS487 and NS490) throughout their time at Farnborough sported simultaneously both 'V'-frame and sting type hooks. Many 'special' aircraft passed through Farnborough's flight sheds for fitting and adaptation of arrester hooks. Apart from the usual run of naval aircraft, some of the machines fitted with arrester hooks by RAE included the Bell Airacobra (AH574),

produced a system in which a small spade-shaped lug became a permanent fixture at the rear of the aircraft. A more robust holdback was designed on the split-wedge principle and a larger breaking ring was handled more easily on a flight deck. This was adopted as a permanent feature in post-war aircraft. The design and working of the holdback unit is described in the chapter devoted to tail-down launching.

ARRESTER HOOK UNITS

During the war years NAD devoted a great deal of time and effort to addressing the problems of flight deck arrestment of aircraft. The development of the hooks themselves is a subject with its own chapter but the adaptations to the aircraft are treated here.

At the outset of World War 2 all FAA aircraft for carrier operation were fitted with the 'trapped' 'V'-frame system. The 'V' was pivoted to the fuselage at the top and was allowed to swing in the fore-and-aft direction with the hook itself at the apex. As the hook engaged the arrester wire it swung upwards to engage in a snatch block and retained in the fuselage for the arrestment of the aircraft. This system tended to reduce, but never completely avoided, any tendency of the aircraft

Hawker Hunter GA.11 (XE685) navalised with a 'sting' type arrester hook. This marque was not flown from carriers. (© John M. Dibbs/The Plane Picture Company)

Hawker Sea Fury FB.II (WH588). The single accelerator towing hook is located between the main undercarriage wheel fairings. (© John M. Dibbs/The Plane Picture Company)

Supermarine Sea Otter (JM987), North American Harvard (FT963) and Avro Manchester (L7246) aircraft. In the fitting of these arrester hooks all aspects of the adaptations needed to be addressed. Often the fuselage required to be strengthened in the area of hook attachments. One case is recalled in which, to accommodate the hook attachment, four layers of 22swg {0·71mm} of profiled duralumin sheets were added to the fuselage as patch plates to distribute the retarding forces and related side loads.

FARNBOROUGH AERODROME

To the children of the 1930s living in South Farnborough (of which the author was one) of a weekend or a summer's evening the aerodrome was a convenient playground and, during early September, a prolific harvest ground for blackberrying. It was also a convenient location for many of the field activities in which Boy Scouts of those days used to indulge. Occasionally there would be a polo game, or soccer match, on what used to be known as Sentry Hill (or would it have been the phonetically matching 'Centaury Hill'?), the promontory which later became the site of the Institute of Aviation Medicine. This name must have had a local origin; it does not feature on any Farnborough town plans and the Ordnance Survey does not identify this small escarpment.

At the foot of the northern face of this promontory was a brickwork structure, filled with a large bank of sand, used as a firing butt for the guns of the biplanes belonging to the local RAF Squadron. Close by was a wired-in compound containing three small, widely-spaced buildings; these were magazines holding the machine gun rounds for use in the range. Alongside the small gate was a red box with a notice asking those about to enter to deposit their matches and cigarette lighters within.

Hucks' aero engine starter on a Crossley chassis. March 1923.
(© Crown Copyright/MOD)

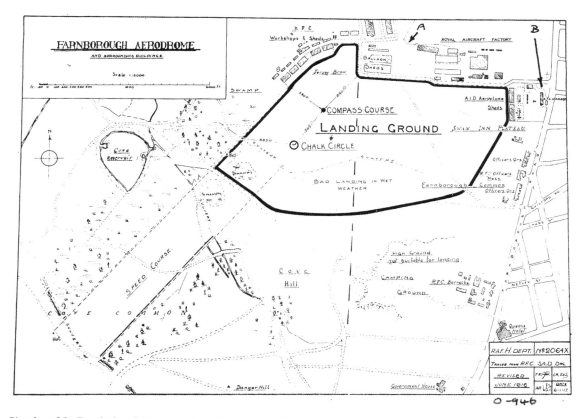

Site plan of the Royal Aircraft Factory and aerodrome in June 1916. The flying area is roughly trapezoidal in shape with the longest take-off distance of 3,700ft to the south-west of the Factory site. The north/south distance is considerably less at 1,700ft. South-west from Jersey Brow is the Speed Course of a mile or so in length (to local residents, the Mile Straight). This was the approximate orientation of the subsequent DCTO track and Runway 04/22. The two swampy patches became the sites of the experimental arrester pits. Cove Reservoir, in the early 1940s, was filled in to permit the construction of Runway 00/18 (18/36). Additions to the diagram by the author show the location of Jersey Cottage (A), which shared its name with Jersey Brow. The original headquarters building of the Royal Flying Corps is shown at (B). This latter, in RAE parlance, was known as Building G.1 and is now the headquarters of the FASTA organisation. The chain-dotted line running north-south is the approximate boundary between Farnborough and Cove Commons. 1916.
(© Crown Copyright/MOD)

These were prohibited in the enclosure. When the gate was unlocked a sentry was posted alongside; this may have given the name 'Sentry' to this locality. (The author's preference is for 'Centaury' in the supposition that the flora antedated the human fauna by a matter of centuries – no pun intended!)

The local Secondary School's 3½ mile cross-country run comprised a clockwise circuit of the aerodrome, entering near Queen's Corner, along the line of Lincoln Road, crossing to Jersey Brow, exiting via the Tank Corps playing field and through Pinehurst Cottages and Pinehurst Passage. There was no danger of aircraft interfering with the run. It could have been more than a coincidence that the Chief Superintendent of 'The Factory' – A. H. Hall – was also chairman of the board of school governors!

Returning to our subject, this was the environment at the time of the institution of the Catapult Section which, in 1945 – the seventh year of its existence – was to become the nucleus of the Naval Aircraft Department.

In the mid-1800s, during the development of Aldershot as the home base for the Army, the War Department identified a large area to the north of the Basingstoke canal as its property. This area was bounded by Ively Road to the north, Fleet Road to the west, the canal to the south and the Farnborough Road to the east. Within this purlieu were to be found Pyestock Wood, Ball Hill, Jersey Brow, Swan Plateau, Sentry/Centaury Hill, Cove Hill, Eelmoor, Jack Goddard's Moor, Bridge Hill, Berkshire Copse, Laffan's Plain and Watts' Common. These surrounded Farnborough Common, Cove Common and Cove Reservoir. This last, known locally as 'The Rezzer' was an army watering place and exercise ground for its cavalry. It was not a true reservoir, being the source of Cove Brook. Its site was to the north-west of the original aerodrome. It was drained and filled in at the outbreak of World War 2 and subsequently disappeared beneath Runway 18/36.

The ground enclosed by this perimeter was designated as War Department Land and defined by a large number of granite boundary stones, each bearing its WD-BS identification number and located on the OS maps of the period. These stones were numbered in sequence. BS.360 (adjacent to the Swan Hotel on the Farnborough Road) to BS.369 (on Jersey Brow) formed the eastern and northern boundaries to the original RAE 'Factory' site. Some of these stones were identified by local names. Those nearest to Norris, Eelmoor and Claycart bridges carried their respective names but other examples were known as Street's, Wheeler's, Porter's, Evelyn's, Burte's, Tichborne's, Watts' and Green Stone. Parts of this area were taken up by the Army golf course, west of Golf Lane (Government House Road) and Shoe Lane, and another site forming Elles Barracks spanning Ively Road at its junction

RAE and Farnborough aerodrome in 1937. In the ensuing 20 years, many buildings were erected in RAE including 'A', 'B' and 'C' Sheds and F1E aircraft workshop. Jersey Brow is to the west of 'A' and 'B' Sheds with the hangars of the WW1 RFC aircraft repair depot (by this time, hangars for the Royal Tank Corps) to the west and north. The long line running from the south-west towards Jersey Brow is the original Mile Straight, a 'measured-mile' for determining aircraft speeds. The rectangular area at the bottom centre of the picture is the army polo pitch atop Centaury Hill, also known as Danger Hill, a site later to accommodate the RAE Physiological Department which in turn developed into the RAF Institute of Aviation Medicine. The RAF Station forms the eastern boundary to the aerodrome. 1937. (© Crown Copyright/MOD)

Farnborough airfield, showing the fourth runway (18/36) and the third extension to Runway 07/25. The development of the Ball Hill site is almost complete; the new Structures Hangar for Concorde trials is at the left edge of the picture. April 1966.
(© Crown Copyright/MOD)

with Farnborough Road. At first, only a small portion of the remaining area was given over to the Royal Aircraft Factory. By the outbreak of World War 2 some 1,400 acres {570ha} were enclosed by security fencing, all of which eventually became absorbed into Farnborough Aerodrome. Of this, 55 acres {22ha} formed the 'Factory' site of the Establishment, expanding to some 90 acres {36ha} by the 1960s.

During the working day a fair amount of flying activity was to be seen on Farnborough aerodrome as it was called in those days. In residence, besides the RAE research aircraft, there was a Royal Air Force contingent forming an Army Co-operation Squadron equipped with Hawker Audax and Hector aircraft (later to be replaced by the Westland Lysander and Hawker Henley).

A pre-war aerodrome operational procedure at Farnborough was that every aircraft due for flight on a particular day would be wheeled out of its hangar to the hard-standing in front of 'A' Shed [Building P72] and its tail lashed to a length of railway line partly buried at the front of the apron, some 43 yards {40m} south of the hangar doors. There it would remain until its flight was due; on return it would be lashed down again to its picketing rail. At the end of the day it, with its fellows, would be wheeled back into the hangar. No aircraft was left in the open during the night. This daily activity was quite an involved procedure as almost every aircraft was man-handled into position, with 'tail-arm-Charlie' manoeuvring the tail wheel, or tail skid, by lifting the latter into a swivelling box mounted on the steering arm and guiding it at will. His was the master-mind in the procedure, and the man in the cockpit – holding the control column back to keep the elevators

raised – applied the brakes (when fitted) to his command. It was said of this 'chain gang' that 'If everyone pushed as hard as he should push, where he did push, then something on the aircraft would certainly break'.

When aeroplanes, as they were called in those days, were in the air, a large red flag would be displayed alongside the 'A' Shed hangar [Building P72], or even in mid-field, some unfortunate person having to hold it in position for as long as was necessary. This red flag also was a warning to those pedestrians or cyclists who were hoping to make a short-cut diagonally across the aerodrome between Rafborough and the Queen's Hotel corner (Cock-a-Dobbie hill). If 'Hookie' Spriggs (he lost both his hands in a World War 1 incident) was on duty as aerodrome warden in his cubby hutch at the corner of 'A' Shed [P.72], a further wrathful warning would be issued to any prospective wayfarer, with a strict and unmistakable injunction to remain on the roadway between the 'Factory' and the aerodrome.

The daily wheeling-out and in procedure was also adopted by the Royal Air Force (RAF) Squadron which occupied the eastern hangars. These, in later air-show days, became known as the 'Black Sheds' (they were green in my young days). A singular difference with the Service was that on a Wednesday the wheeling-in process occurred at midday. Like the Army nearby in Aldershot, Wednesday afternoons were reserved for sports activities. However, until the post-war period, both the RAF and RAE worked on Saturday mornings.

Servicing both RAE and RAF aircraft (mainly biplanes) were two 'open cockpit' Ford 'Model T' motor vehicles, to the chassis of one of

RAE Transport Flight of three Navajo Chieftains (ZF520, ZF521 and ZF522) and Douglas Dakota C.3 (ZA947) flying low over the Southern area of Farnborough airfield. This view covers many of the active NAD sites, the main being that of the arrester area on Runway 04/22. December 1986. (© Crown Copyright/MOD)

which was fitted a rectangular sided tank of some 400 gallons {1,800 litres} capacity of aviation fuel with a hand-operated semi-rotary pump. (This was an improvement on the man-handling of two-gallon cans of fuel up to a comrade at the aeroplane's fuel tank filler position.) Also loaded on the platform were a fire extinguisher and a two-gallon {9-litre} gardener's watering can. This was the official aviation fuel tanker, distributing the normal 87 octane petrol (known as DTD224) around the aircraft hangars and dispersal areas. When the more powerful aero engines started to appear, a precursor of the ultimate, formal type of cylindrical bowser was used to carry the enriched DTD230 (100 octane) fuel; this was sufficiently modern a vehicle to possess an electric fuel transfer pump.

The second 'Model T' chassis carried a structural contraption which supported a horizontal rotary shaft projecting forward from a power unit mounted on the rear platform, passing over the head of the driver, and terminating in a bayonet type fitting protruding ahead of the car. This vehicle was driven head-on to the front of an aircraft, the bayonet unit was engaged into a matching socket fitted to the hub of the propeller and turned by the power unit until the aircraft engine began to fire. The driver then carefully reversed his vehicle away from the rapidly rotating propeller immediately in front of him; any mistake in gear selection could have proved disastrous. (I wonder what the current Health and Safety Executive would have thought about this activity, and what recommendations and impositions would have been made by them.) Those in the know will recognise in this vehicle the basic elements of a Hucks' Starter, named after the inventor, Capt B. C. (Benny) Hucks, an aviation pioneer and a racing motorist (c.1910).

This second chassis also carried a fire extinguisher and similar watering can. The purpose of these watering cans had nothing whatever to do with operation of aircraft. On a dry, hot summer's day, when the

wooden spokes of the 'Model T' road wheels began to rattle, the driver, or his mate, simply dismounted and liberally applied water to the spokes in the hope that they would swell and thereby reduce the castanet-like clatter they were making. At one time, c.1922, a second Hucks' starter was provided on the airfield; this was mounted on a Crossley motor chassis in which, eventually, caterpillar tracks replaced the rear wheels. All this aerodrome equipment was still to be seen in full operation in 1936. It remained in use up to the outbreak of World War 2 when the intensity of aircraft movements rapidly increased, demanding the introduction of greater sophistication in methods of servicing aircraft – viz a small tractor.

Another airfield-related, time-honoured, feature during this period was that every RAE aircraft, whether parked outside on the hard-standing in rows fronting 'A' Shed, or in the hangars, carried a hard-back folder, tied to its ASI pitot head. This folder, somewhat smaller than the current A4 paper size, when folded in one way displayed a red colour with the initials 'U/S' inscribed on its face. When folded the opposite way it presented a green surface carrying the letter 'S'. This folder contained the flight service certificate carrying the inspector's signature releasing the aircraft for flight. The designations referred to its being 'unserviceable' or 'serviceable'. As virtually every aircraft at RAE was an experimental machine, every time it landed it was immediately declared to be 'unserviceable' (U/S) until further released for flight under a fresh signature from the inspector, when it became 'serviceable' only for that particular flight. A result of this practice was that no aircraft was declared 'serviceable' until immediately prior to its being flown. Airfield discipline dictated that only an aircraft inspector could remove the U/S card from the aircraft. During World War 2 the American air forces resented our use of the letters 'US' to represent something unserviceable. They retaliated by declaring their own aircraft as 'GB', suggesting the letters indicated the aircraft had 'gone bad'!

The major development of Farnborough Aerodrome from a small grass airfield began with its wartime expansion in 1940. In the autumn of that year, a number of strange-looking, caterpillar-tracked, earth-moving machines arrived at the south-western extremity of the airfield and began working on Laffan's Plain. There were bucket-excavators, back-acters, ditchers, scrapers, graders, levellers, dumpers, rollers and concrete batch-mixers in proliferation, busily engaged about their respective duties. The first area to be worked upon was the site of the junction of Runways 04/22 and 07/25 – with the western end of Runway 07/25 terminating at the position where it now forms the junction with Runway 18/36.

Progression was made northwards and eastwards along these alignments, with the construction of Runway 12/30 (subsequently 11/29) following upon the completion of the other two, crossing and being faired into the DCTO rail-track in the bargain. (The gaps in this track were filled with removable railway sleepers.) Completion by the summer of 1941 saw Farnborough with a standard, three-runway system having Runway 07/25 with 3,800ft {1,160m} of paving, and Runways 12/30 and 04/22, both of 3,200 feet {975m}. All runways were made to the standard width of 150 feet {46m}. It was about this time that, for some forgotten reason, the word 'aerodrome' was dropped in favour of 'airfield' and continued to be known as such. At the time of writing, an occasional reference to aerodrome reappears; perhaps the Greek derivative is regaining favour.

Within a few months, Runway 07/25 had been extended westwards to 4,900ft{1,500m}, making its intersection with the taxi-ways, now known as Spitfire Way and Vampire Way. At the same time Runway 04/22 had been extended northwards from its junction with Mosquito Way to 3,700ft {1,130m}, conveniently forming a continuous link with the Jersey Brow launcher site which also was given a concrete pavement.

With so much of the airfield being subjected to hard surfacing, a number of new access roads, particularly on the south side of the airfield, were being laid to form aircraft dispersal areas and to locate new hangars. Inevitably, for ease of identification, these were given names by those who needed to use them. Recorded examples are:

Puttock's Parade	now	Camel Way
Froud's Folly	now	Hurricane Way
Chandler's Chase	now	Halifax Way
Busby's Beat	now	Hunter Way
Denter's Dip	now	Barracuda Road
Towers' Turntable	}	a couple of aircraft
Tonks' Terrace	}	panhandle dispersal areas.

These all were named after various foremen and charge-hands of the hangars and flight sheds staff of that era, and appeared as such on site plans of that time. However, these alliterative titles were not to continue into perpetuity. Although more prosaic than their successors, they had to give way to later replacements commemorating many of the aircraft marques which had passed through RAE.

By the end of 1944, further extensions had been completed, giving Runway 07/25 a length of 6,600ft {2,000m}, taking it to the intersection with Stirling Way and Wellington Way. At this time the new Runway 00/18 (subsequently 18/36) was built to a length of 4,200ft {1,280m} and Runway 12/30 (subsequently 11/29) was extended also to 4,200ft (1,280m) to link up with, and cross, the new runway. This brought the airfield up to the 1942 Air Ministry 'Class A' standard for Bomber Command airfields. With the advent of Runway 18/36, Runway 04/22 was given over to Naval Aircraft Department. This permitted NAD to construct an arrester-gear proving base on that runway, its location occupying the intersections with Runways 07/25 and 11/29, and using its northern extension for catapult trials.

A third and final extension to Runway 07/25 was built, making it 7,814ft {2,380m} in length, with a further extension westwards for overshoot. Wellington Way was extended to provide a connection with the taxi-way at this end of the airfield. It is not known exactly when this work was undertaken but it was completed well before the spring of 1948, the first year in which the SBAC exhibition and air display was held at Farnborough. Although much work has been done to provide taxi-ways, peri-tracks and hard-standings to subsequent research and service areas, the runways themselves, at the time of writing (1999), have remained essentially as left in 1948.

The access roads within the perimeter of the airfield were divided into four categories: runways, ways, roads and tracks. (The author is indebted to Brian C. Kervell, one time Curator of the RAE Museum, for the following descriptive details.) 'Runways' were designated by their compass orientation while the other three grades of thoroughfare carried the names of aircraft or – in the Laffan's Plain area – commercial road vehicles. Each runway was used exclusively by aircraft in taking off or landing. All other traffic – apart from emergency service and maintenance vehicles – was barred from these areas. Vehicular crossings were provided with telephones and were controlled by traffic signals operated from the control tower. Movements of all vehicles and personnel within the vicinity of runways could be undertaken only by permission of the air traffic controller.

A 'Way' denoted a thoroughfare suitable for towed, or taxying, aircraft which had priority of access onto, along and exit from the route. Movements, again, were under the supervision of the control tower. A 'Road' is almost self-explanatory by definition: it described the main vehicular routes around the periphery of the airfield. A 'Track' was a route passable to ordinary vehicles with care, but four-wheel-drive was the more usual form of transport. Mostly, 'Tracks' were access to research areas or service routes across relatively undisturbed scrub and common land.

Eventually Naval Aircraft Department operated a number of sites scattered about the western half of the airfield, all carefully identified with building numbers allocated, though many locations subsequently have reverted to 'brownfield' areas. The sites of the catapults and accelerators on Jersey Brow carried the building identification of [O.9] with various suffixes. The office blocks were known as [O.1] with post-war additions of the Design Office [O.2] and the Model Testing Laboratory [O.3]. The adjacent compass platform was identified as [O.10]. The sites of the Flexible Deck project are identified as [O.19] for the preliminary trials (Stages 1 & 2); [N.6] for the site of the first full-scale landings (Stage 3); and [W.5] for the location of the final activities (Stages 5 & 6) of this project. The northernmost pulley sheaves to the prototype Brake Drum Arrester Gear (BDAG) installation occupy sites just west of [N.13] with the southern sheaves forming mirror images on the opposite side of Runway 07/25. The High Speed Track (HST) was laid on Laffan's Plain and identified as [T.16] group.

Within the post-war security restrictions of 1948, Pudney (#80 P.54), provided an overview of the happenings on Farnborough Aerodrome:

'If the visitor to Farnborough is lucky, he may see trials of naval aircraft taking place on a runway given over to this work. At one end is the catapulting base from which aircraft are launched, and near the middle is the deck arresting equipment where aircraft can be brought to rest from flying speed in a distance of 50 yards. There are several of these rigs which simulate conditions at sea, and all manner of experiments are carried out on them, for the Farnborough Establishment has collaborated with naval airmen since the earliest days.'

Also, it is not possible to mention Farnborough Aerodrome without making reference to the pioneer of British aviation, Colonel Samuel Franklin Cody. He made the first sustained, powered flight in Britain on

16 October 1908 during which he covered a distance of 1,390 feet {424m} from the Swan Plateau at the south-east corner of Farnborough Common, now just beyond the eastern termination of Runway 11/29. At this time many trees and shrubs were still existing on the Common, mostly along the boundary between Farnborough and Cove Commons, but a few – generally Scots Pines – were to be found in front of the 'Black Sheds'. To one of these, Cody would lash a large spring balance and attach it to one of his aeroplanes for measuring the engine thrust. After his death in an aeroplane accident on 7 August 1913, this tree was retained over the years as a memorial to Cody's exploits and thereby became recognised as 'Cody's Tree'. When the tree died, the trunk was treated with a preservative but this was not long-lasting. The trunk was removed and a replica was cast in aluminium alloy by the RAE Foundry using the orginal as a pattern. This replica was erected on a plinth in the original location. Ultimately the 'Tree' was removed from its long-established site and re-erected (in 1996) as a permanent and prominent feature of the new DERA/QinetiQ establishment beyond the western end of Farnborough airfield. The timber from the tree was retained in RAE and, from time to time, pieces were used for making special artefacts for presentation to senior staff members on their retirement, or to visiting VIPs.

(Much of the foregoing information has been abstracted from site drawings, photographs and occasional textual references. The information, relating both to distances and dates, unavoidably may not be completely reliable.)

JERSEY BROW

Jersey Brow is a small plateau situated at the extreme north-east corner of Cove Common, just to the west of 'A' and 'B' Flight Sheds, the erstwhile aircraft hangars, more recently identified as Buildings [P.72] and [P.71] respectively. The boundary between Cove and Farnborough Commons runs in an almost straight line southwards across the airfield, from the south-west corner of 'A' Shed [P.72] to the corner where Golf Lane meets Shoe Lane, latterly the site of Stonehenge Gate. Maps of the early 1900s show this boundary quite clearly delineated as a row of trees and shrubs, some of which feature in early photographs of the RAE and its antecedents. The Brow has a fairly level crest at OD 226·40ft {69·01m}, is roughly circular in plan and about 400ft {122m} across. The ground slopes slightly to the south-west, falling from the OD 225·00ft {68·58m} contour to OD 220·00ft {67·06m} within a

A view westwards across the apron in front of 'A' Shed towards Jersey Brow. Projecting beyond 'A' Shed is the forward end of the 'P' Catapult mounted on its gantry. The Jersey Brow catapults were located at the head of the runway, hidden by 'A' Shed. June 1947. (© Crown Copyright/MOD)

Vertical part of the RAE [AS2] and eastern end of airfield, showing Jersey Brow. The major sites used by Naval Aircraft Department for trials purposes are seen. In the centre of the picture, Jersey Brow with the launching area, contains the catapult/accelerator units pointing south-west along the alignment of Runway 04/22. NAD site offices and lab are on the north-west side of the area. Adjacent to the site, eastwards, may be seen the compass platform, and the two main hangars – 'A' and 'B' Sheds. The arrester gear area with the locations of five units is seen along Runway 04/22. The location of the machinery pit for the Brake Drum Arrester Gear (BDAG) is seen as the sites of the four fairlead pulleys spanning Runway 07/25. The remaining short section of the Direction Controlled Take-Off (DCTO) trackway which ran from Jersey Brow to an intersection with Runway 07/25 is seen local to Runway 12/30. The trials area for the early stages of the flexible-deck landing system is seen at the upper left of the picture. The lightly shaded patch shows the area of the launcher, the experimental carpet, the arrester gear, and the handling cranes. The location of the flexi-deck unit for the series of live landings is at the lower left. Across Runway 04/22 is the location for the early trials of VSTOL systems, the Rolls-Royce 'Flying Bedstead' etc. The sites of the High Speed Track (Eelmoor Gate), the slotted tube catapult (Ball Hill), and the final stage of the flexi-deck carpet (Laffan's Plain) are outside the range of this picture. Their locations were generally westwards of this view. c.1955. (© Crown Copyright/MOD)

distance of 300ft {91m}. From thence the ground is almost level for a further mile {1·6km} or so.

Jersey Brow as such is not identified on maps of the early twentieth century; its name was probably associated with Jersey Cottage which does feature on a map of 1873. This cottage stood within an area of some four acres {1·6ha} and would have occupied the site of the later RAE Building [P.160], with Buildings [P.4, P.69, P.70 and P.161] on the rest of the estate. A straight track from Farnborough Road gave access to Jersey Cottage on the alignment of the subsequent Fowler Avenue to West Gate (at one period known as North Railway Gate). The southern limit of the Jersey Cottage plot was marked by WD Boundary Stones 364–368.

To the south-west of Jersey Brow, rather more than a mile-and-a-half {2·4km} distant, lies Eelmoor Flash, a pound on the nearby Basingstoke Canal. A drawing, dated June 1916 (#321), shows that along this alignment (orientation 23) for a distance of a mile from the Brow a swathe some 200 yards {183m} wide had been cut through the woodland and scrub to form a clearing, given the name of Speed Course. About a mile {1·6km} below Jersey Brow a small hut was placed, with another hut a urther 880 yards {800m} towards the Brow. These are assumed to have been locations for personnel and equipment to record the speed of aircraft along a measured distance. Within this length, two areas of swampy ground are identified just about where, some 30 years later alongside Runway 04/22, the NAD arrester pits would be constructed. To the locals this part of Cove Common, when freely accessible to the public, was known as the 'Mile Straight'. Some twenty years after the date of this drawing a special track was to be laid along this stretch. This became the Direction Controlled Take-Off (DCTO) project. Both these areas feature again in this record.

By 1910 massive balloon sheds had appeared to the east of Jersey Brow, the most prominent being the Large Dirigible Shed built for the ill-fated Lebaudy airship. (Was this the first attempt at Anglo-French co-operation in the aviation world?) On its arrival, as the airship was being man-handled into its shed, its fabric was torn by the building structure, the gas envelope being larger than specified. This was of little consequence as, returning from a flight on 10 May 1911, it became a total loss by wrapping itself around some trees bordering the Farnborough Road. Subsequent to this the shed was increased in height by 15 feet {4·6m} becoming the Large Balloon Shed. By 1931 it was dismantled for 'A' Shed aircraft hangar [Building P.72] to take its place.

Early photographs (c.1915) portray the balloon sheds to the west of the 'Factory' in a light shade of colour. Such an enigma is solved in reminiscences of 'Bill' Davis (one of the NAD staff) writing in the May 1950 issue of 'RAE News'. He records:

'There were, in fact, two aerodromes, the ordinary one which ran from the Farnborough Road as far as the big balloon shed and another one on Laffan's Plain. By modern standards the "ordinary" aerodrome must have been very tiny; in retrospect it seems to have been little larger than a couple of football pitches. It wasn't even flat. There was a hill in the middle, so that a machine taxiing out from the sheds at the eastern end would sink almost out of sight before it took off. The northern boundary was the Factory, with the great yellow, domed cylinder of the balloon shed dwarfing everything.

'The other aerodrome was used by the Wireless School. It lay between two clumps of pines in the neck between Farnborough Common and Laffan's Plain.'

Until the late 1930s Jersey Brow, as part of the Cove and adjacent Farnborough Commons, was wide open to the public. In fact it was access and exit for RAE personnel who lived in the housing estate known as Rafborough – the Royal-Aircraft-Factory-Borough. Most of the work being conducted on Jersey Brow was overtly evident to all passers-by. Early in World War 2 the entire Establishment, including the aerodrome, was brought under close security with chain-link fencing erected around the area and regular policing patrols. Before long the installation on Jersey Brow began to expand. A small workshop was erected on the site and at the time when the airfield was being given its first concrete runways a forward-looking Naval Aircraft Department was planning an installation of six launcher units on an apron some 200ft {61m} wide at the northern end of Runway 04/22. Later, a couple of 'Seco' huts were added – Buildings [O.1] & [O.2] – into which the still expanding technical staff complement of the Catapult Section was decanted from the Main Drawing Office [Q.1]. Eventually a further building was provided as a Drawing Office, and a 'Robin' Hangar became a laboratory for experimental work on model scale.

THE LAUNCHER BASE

From earliest days, Jersey Brow became the site for full-scale experimental work, firstly with lighter-than-air craft then with winged aircraft. Caunter (#222 p.23) quotes the experimental work of Capt Broke-Smith and his falling-weight pylon in 1909 as the first mechanical appliance on this site. With the arrival in the 1920s of the first attempts at introducing more sophisticated mechanical means for assisting aircraft in take-off, the experiments were concentrated in the Jersey Brow area, with the colouring of the adjacent Balloon Shed reduced to a more sombre shade.

As some of the earliest photographs of the proofing of navy catapults were taken in this vicinity, perhaps it was logical that the specialist Catapult Section, set up in 1938, should claim right of possession of the site for its own development projects. The first installation was already in being: it was a small level patch of concrete forming what became known as the Catapult Proving Base. As its name implies, this was the test area for a variety of shipborne catapults, trials being conducted prior to the machines being installed in their various warship destinations. By the late 1930s an F.II.L machine had assumed a permanent installation on this pad, becoming known locally as the RAE Mark I Catapult. This was shortly followed by a second, purpose-built machine, making two hydro-pneumatic catapults seemingly installed as permanent features of the local landscape. This second machine, which was never adopted for shipboard use, logically became known as the RAE Mark II Catapult.

Other machines, including that for HMS *Hood*, had appeared before them for trials purposes but these two became virtual fixtures and subsequently were used in the development of the aircraft side of mechanical launching procedures. Launching was carried out towards the south-west, in which direction the contours of the ground fell slightly, by about seven feet {2m}. This topographical feature was exploited in that it permitted aircraft to drop a little when leaving the catapult trolley without fear of mishap in making premature wheel-contact with the ground.

In 1940, when the author was posted to the Catapult Section, the equipment on Jersey Brow comprised the Mark II catapult, the vestiges of the Mark I catapult having been removed with parts of the machinery stacked up behind its erstwhile location. In addition, the loading turntable at the start of the Direction Controlled Take-Off (DCTO) system was under construction nearby with the track running along an imaginary runway orientation of 05/23. This trackway (made from rolled-steel channel sections) was installed immediately prior to the construction of the concrete runways.

Over the years Jersey Brow had become a repository for redundant equipment – a veritable bone-yard – in company with the area immediately north of the Seaplane Tank [Building Q.120]. It seemed as if every bit of unwanted junk ironmongery, including old catapult

TABLE 1

LOCATION	EQUIPMENT	WEIGHT x 1000lb	LAUNCH SPEED	MAX ACCELERATION	STROKE
Plot 1 – Unit 1	Elevated P Catapult	to 10K	65kts	3¼g	70ft
Plot 2 – Unit 2	K.I Inertia Catapult	6K–14K	60kts	3¼g	60+5ft
– Unit 3	K.II Inertia Accelerator	6K–30K	75kts	3¼g	97+12ft
Plot 3 – Unit 4	C.I Cordite Accelerator	to 14K	60kts	3¼g	65ft
Plot 4 – Unit 5	30,000lb Accelerator	6K–30K	80kts	3¼g	90+16ft
Plot 5/6 – Unit 6	Ark Royal Accelerator	to 30K	75kts	3¼g	96ft

machinery from the earliest days of experimenting, was dumped in this supposedly out-of-the-way location. (In December 1940 this became a useful source of material for an 'emergency' project culminating in the 'P' Catapult.) Photographs of work conducted on the Brow invariably show a background of clutter which persisted until the mid-1940s when a compound was built locally to tidy up the 'junk store'.

An NAD Paper (#310) records that in the early days of aircraft carrier operation, when free take-off of aircraft was possible, naval aircraft were cleared for launching by catapult, accelerator or rocket motors at a relatively late stage in the series of trials designed to give clearance for operational use. The advent of aircraft whose low speed performance did not permit a free take-off within the length of a carrier deck demanded that clearance for catapulting should be established at an early stage, and before shipboard trials could take place. This led to the development of launcher machinery at Farnborough.

The first of these, subsequent to the provision of the Mark I and Mark II Catapults on Jersey Brow, was the adaptation of the Mark II machine to provide a proving base for the new above-deck equipment for the 'Illustrious' Class accelerator. The prototype, designed and made at Farnborough, was quickly found to be satisfactory and the further development of the 'production' models was undertaken at Chatham Dockyard, becoming known as the Assisted Take-Off Gear (ATOG). When rocketry began to usurp this title and the American escort carriers began to arrive, this system became known as the BH.III Accelerator.

Before long, development work started on the Jersey Brow site with the installation of the Cordite Assisted Take-Off Gear, designated CATOG and later renamed the C.I Accelerator; in reality, the first 'accelerator' to be installed on Jersey Brow. The Flywheel (or Inertia) Catapult, later known as the K.I (or Kinetic) Accelerator, followed. Both of these were constructed with the machinery located in concrete pits, all below ground level with only the flight deck equipment – trolley and/or shuttle – appearing above the surface. In connection with this development a concrete perimeter track was constructed to enable aircraft to taxi around the western end of the hangars giving access to the rear of the accelerator loading area. Even this concrete area was put to various uses. The rocket-powered launcher – the P.I Catapult – followed in two stages. Firstly the prototype, a proof-testing lash-up at ground level, was built within a few days. This was followed by the first production beam (a plated structure), accommodated for a short while at ground level, then almost immediately erected on a raised structure from which most development and training

activities for 'Catafighter' operation from CAM-ships ensued. All of these were wartime developments.

By 1941, the construction of the concrete runways on the airfield produced one of them on the alignment of Jersey Brow. This was designated Runway 04/22 and, although being of shorter length than the main Runway 07/25, was a useful adjunct to aircraft catapulting activities emanating from the northern extremity of the airfield. Coincident with the third and final extension of the main Runway 07/25 in 1944, the construction of the fourth runway on 18/36 orientation released Runway 04/22 for catapult and arrester-gear operations, a monopolistic privilege greatly valued and extensively used by the Department. NAD had already provided a scheme for the installation of six accelerator systems on Jersey Brow. RAE Drawing 2934, dated April 1941, depicts these parcels spaced at 30ft {9m} centres with the first, on an approximate alignment of the early Mark I Catapult, allotted to the elevated P.I Catapult. Moving westwards, the next became the site for the K.I/K.II Catapult; plot 3 became CATOG (the first name for the C.I Accelerator); next was a proposed 20,000lb {9,100kg} Accelerator. Unit 5 was declared a 'Spare' while the final space was reserved for a future 'Proofing Catapult'. This last plot happened to occur on the alignment of the earlier RAE Mark II Catapult.

RAE Drawing CE.1372/1A of 1945 tabulates these plots on the launching apron as accommodating the following launcher 'Units', summarised in Table 1.

By the late 1940s all five plots were occupied. The first, after the dismantling of the P.I Catapult (c.1952), became the site for the short-lived, cordite-powered Slotted Tube Catapult for the flexi-deck project. Plots 2 and 3 continued to accommodate respectively the K.II and C.I Accelerators. The fourth plot became the site for the short-lived 30,000lb {19,100kg} Launcher. Plot 5 became the location of the deck gear forming the BH.V track with its subterranean power-house alongside at an angle occupying Plot 6.

Alongside Unit 1, under the crane serving the P.1 Catapult, were a couple of concrete slabs. One of these – 12ft square {3·7m x 3·7m} – was used for experimental work on the rebound of arrester hook systems on impact with a deck. The other, 30ft square {9m x 9m}, supported an experimental rig to determine the impact effects of a weight when dropped on to a flexible suspended deck: the embryo of the entire flexi-deck programme. Plots 2–6 each comprised a pit/cavern containing the under-deck machinery for the accelerators which they served. The pit

TABLE 2

PLOT	MACHINE	LENGTH feet {m}	WIDTH feet {m}	DEPTH feet {m}	AREA sq ft {m²}
2	K.I/K.II	62 {19·0}	27 {8·0}	7½ {2·3}	700 {65}
3	C.I	55 {16·2}	12 {3·6}	6¾ {2·1}	523 {49}
4	30k	134 {40·8}	14 {4·3}	8 {2·4}	1532 {142}
6	BH.V	105 {32·0}	17 {5·2}	9 {2·7}	1575 {146}

sizes for the under-deck machinery occupying these plots are given overleaf. The pits were not necessarily completely rectangular in form; the sizes given are the outside dimensions within which the relevant floor areas were to be found. These are shown in Table 2.

All that remains of this area of high activity is the underground cavern of Plot 6. This is now used as a smoke chamber for the airfield rescue unit whose vehicles and fire tenders are located in an adjacent building on the erstwhile launcher apron.

During the war years, development of flight deck machinery was continuous and by 1945 NAD was addressing the situation of launching and landing shipborne aircraft of up to 30,000lb AUW {13,600kg}. The 30,000lb arrester gear was under development by one of the naval engineering contractors in Edinburgh. Messrs Brown Bros were engaged in the design and manufacture of the new BH.V accelerator for Jersey Brow. As this was likely to be a lengthy development project, NAD – in what appeared to be the usual hand-to-mouth way – began looking around for resources from which to construct a temporary facility. The search was not prolonged as redundant material from the abandoned Mark III catapult at Harwell was available. The structural steel and decking from the large turntable was recovered to form the roof of the new machinery pit. Some of the larger telescopic tubes were retrieved and modified to suit the new installation and the main hydraulic control components were brought back to Farnborough. The system was installed but getting it to work was a rather greater problem than anticipated; the control system was unproven and was the cause of many a frustrating headache. Perhaps it was for the best when the erection of the BH.V machinery overtook the attempted solving of the teething troubles of the 'Thirty-Thousand' and development of the temporary rig was abandoned. The Mark III catapult in either of its forms at Harwell or Farnborough never did launch an aircraft.

THE ARRESTER DECK

The central portion of Runway 04/22 – between its junctions with Runways 07/25 and 12/30 (11/29) – soon became the location for the installation of deck-arrester gear. This was a stretch of 839ft {250m} of runway 150ft {45m} wide within which any span of carrier flight deck could be replicated. By early summer of 1944, NAD had initiated an extensive programme for the provision of five pits to accommodate arrester-gear systems. All the under-deck machinery of various shapes and sizes were located in these pits alongside the western edge of this runway.

The same RAE drawing in the above section provides a summary of the arrester units as Table 3.

It appeared that the Unit (last) location was reserved for development of the Brake Drum Arrester Gear (BDAG) should it have continued as a viable project.

The arrester pits were given the designations as below:

Pit A.1 – Unit 11. Standard 'Illustrious' Mark 4 arrester unit (this was readily adaptable to accommodate Unit 12, the 'Implacable' Mark 6 type of unit by exchange of a couple of small hydraulic components and their controls). Job DES.2011.

Pit A.2 – Unit 13. Location not used.

Pit A.3 – Unit 14. American Mk 4 (5A) arrester unit. Job DES.2311.

Pit A.4 – Unit 15. The vertical 30,000lb prototype Mark 10 arrester unit for use in 'Ark Royal' class. Job DES.2621.

Pit A.5 – Unit 16. The original Mark 8 Light Fleet Carrier arrester unit ('Activity' Class). Job DES.2577. Replaced in 1952 by Mark 12 gear.

Pit A.6 – Unit 17. 30,000lb Experimental Mark 12 arrester unit. Job DES.2606.

No explanation exists for the absence of pit A.2. A later drawing indicated a space between pits A.1 and A.3, adequate for the installation of a further unit. Pit A.3 for the American unit was little more than a covered, concrete-lined trench of about 4ft square section, 43ft in length {1·2m x 13m}. Pit A.4 is of special interest in that it was designed to accommodate the Mark 10 arrester system, in which the hydraulic jigger unit was installed vertically instead of the normal horizontal orientation. It quickly became known as the 'mine shaft'. In 1952 the 'Activity' gear was removed from pit A.5. After strengthening work to the foundations, including the provision of a 15ft long pre-stressed concrete thrust beam in the machinery pit to accommodate a design load of 176 tons, a Mark 12 unit was installed as a replacement.

At that time each system provided two, non-fleeting, across-deck wires with its attendant pulleys, sheaves, fairleads and so on, with return cables running through under-deck ducting. This followed standard shipboard practice. At pit A.4 the pitch of these across-deck wires could be adjusted to 25ft, 50ft or 75ft {7m, 15m, 23m} and on at least one occasion four wires were rigged, using a dummy trolley running in an underground duct to obtain the appropriate reeving, into a single arrester unit.

Problems were anticipated in the construction of pit A.4 and NAD was not surprised when the first of these loomed large. This pit was required to have finished internal dimensions of 10ft diameter by 43ft depth {3m x 13m}. The location happened to be the spot where swamps existed before the runways were built. Because of this a trial bore was made in September 1944 to a depth of -50ft {-15m} (fifty feet below ground datum). This indicated that the first 18 inches {450mm} of the core sample was top-soil, then clay was found down to -6ft {-1·8m} below the surface. Below this was a variety of sand strata with the water level at -12ft {-3·7m}. Such was the case on a specific date! The excavated depth was required to go down to -43ft {-13m} signifying that some 30ft {9m} of the workings would be in wet conditions.

TABLE 3

LOCATION	EQUIPMENT	WEIGHT x 1000lb	ENTRY SPEED	MAX DECELERATION	PULL-OUT
Pit A.1 – Unit 11	Illustrious Gear	5·5K-11K	60kts	1½g	155ft
– Unit 12	Implacable Gear	to 20K	60kts	1½g	155ft
Pit A.2 – Unit 13	(Spare)	Site reserved but not used			
Pit A.3 – Unit 14	American Gear	to 16K	75kts	2g	150ft
Pit A.4 – Unit 15	Ark Royal Gear	to 30K	75kts	2g	165ft
Pit A.5 – Unit 16	Activity Gear	to 13K	60kts	1½g	155ft
Pit A.6 – Unit 17	Experimental Gear	10K-30K	75kts	2g	165ft
Unit (last)	Brake Drum Gear	Site reserved but not used			

In the civil engineering contract, provision was made for six bores to be sunk around the site to a depth of -55ft for de-watering as the excavation proceeded. Without entering into a detailed description of the civil engineering procedures, excavation of the main pit proceeded reasonably well down to -35ft {-10m} using sheet piling and close-boarded timber waling. At -30ft {-9m} an unpredicted water course was encountered, this gave continuous trouble throughout the period of construction. The contractor, finding that a second stage of de-watering was necessary, proceeded to sink a further set of bores. Additional de-watering submersible pumps were brought to the site and, in spite of these pumps running continuously at the bottom of bore-holes 70ft {21m} deep, an embarrassing inflow of water persisted. (During the period of construction the local water table varied between levels -6ft and -15ft {-1·8m, -4·6m} below ground surface.) Eventually it was decided to attempt to freeze the ground – virtually to create a local permafrost – to enable the lower levels to be excavated. Fourteen months after the test bore-hole, and ten months after striking the water course, the civil engineering aspect of this work was brought to a successful close with the sowing of 7,800 yards super {6,500m2} with grass seed!

Although this freezing of subsoil was thought to have been only the second attempt in the UK to carry out such a procedure – the first job being of lesser magnitude – no record of this unique exercise seems to exist. Being war-time, secrecy was paramount in almost every sphere of life and it may not have entered the minds of those involved to record this activity for posterity. (It has been suggested that this permafrost experiment at Farnborough was instrumental in generating Prime Minister Churchill's brain-child for a large flat top iceberg to be constructed and used as a giant aircraft carrier. This was not proceeded with!) Success at Farnborough ultimately was achieved, the arrester machinery installed, many trials conducted and the result was that HMS Eagle, in her 1951 commission, was equipped, and operated for many years, with this system; later to be replaced by Mark 13 equipment. (Although this Mark 10 equipment was known as the 'Ark Royal Gear' the construction of this latter ship was lagging behind her sister ship, the 'Ark' ultimately was equipped with an entirely different arrester system, developed at RAE Bedford.

All this arrester equipment, apart from the 'Activity' gear in pit A.5, was put to good use throughout the ensuing decade. In 1952, the 'Activity'/Light Fleet installation in this pit was replaced by a Mark 12 system. By 1958, when the Department had completed its transfer to Bedford, all this machinery is thought to have been dismantled and sent for scrap. (The new site was furnished mostly with 'state-of-the-art' experimental plant and equipment but the existence of both Marks 12 and 13 units at Bedford would indicate that this equipment may have been transferred from Farnborough.)

The site at Farnborough sourced and hosted the early trials on the carpet deck (flexi-deck) project as well as a number of other research activities, including the development of nylon rope safety barriers and investigation into the behaviour of highly stressed wire ropes moving at speed. At the time of writing (1999) the western side of this runway still displays a number of rusting steel plates of various sizes, embedded in areas of concrete paving, covering who-knows-what!

One of the early post-war features on Jersey Brow was the provision of a laboratory building dedicated for the use of NAD. This, a rebuilt 'Robin' hangar together with the new drawing office, was located behind the office block and was used intensively for impact trials, on a model scale, of aircraft with arrester gear and safety barriers. These models normally were at one-eighth scale. Martin (*#271*) describes this particular facility as a simple catapult set-up for launching the aircraft model onto a model deck rigged with relevant arrester systems. (This catapult is thought to have been adapted from the unit originally used for projecting model aircraft into a water tank for 'ditching' trials the latter being the cooling pond for the High Speed Wind Tunnel [R.133].) The model aircraft was propelled into the arrester system by a gravity-operated catapult in which a falling mass extended a 6:1 reeving system which hauled a trolley along the underside of a launching beam. The aircraft model was suspended from the trolley during the acceleration stroke and released when the trolley struck a transverse retarding strop made of 'bungee' rubber cords. Appropriate adjustments for angle of approach, speed and attitude were incorporated. Typically the arrester wire stanchions in the model were located nine feet apart, representing 72ft full-scale {22m, 2·7m}. Similarly the model of the de Havilland Sea Hornet weighed 28·3lb, representing 14,500lb full-scale {13kg, 6,600kg}. (It is obvious that while linear dimensions would be direct in proportion, other parameters would be dependent upon relevant factors of dimensional theory in scaling.)

From the date of the installation of the first catapult on Jersey Brow in 1921 until the last mechanically assisted launch in September 1957 from the BH.5 catapult, some fifteen launcher systems had been developed and operated on the Farnborough airfield. All except three of these were installations on the Jersey Brow launcher site. The exceptions were the rocket trolley systems for the high speed track on Laffan's Plain, the trolley system for launching the Hotspur gliders in the early flexi-deck trials, and later for the contra-prop Seafire safety barrier trials, at the Meadow Gate site. The slotted tube catapult, latterly occupying Plot 1, had been transferred from the Ball Hill site where it had undergone its development programme. Neither the Mark.III catapult at Harwell nor any RATOG systems have been included in the above data.

Similarly, from the date of the first arrester-gear installation on Runway 04/22 in 1944 until the completion of the transfer to Bedford in 1957, some 17 arrester systems had been operated at Farnborough. Of these, ten installations were on or alongside Runway 04/22. One (BDAG) was on Runway 07/25, two were on the High Speed Track and another two on the main DCTO system, both on Laffan's Plain. Three were on the flexi-deck installations and the last was on Ball Hill for the slotted tube catapult trials.

In many of the illustrations depicting the activities on Jersey Brow, groups of pitched-roof buildings of corrugated steel sheets appear on the northern boundary. These 'tin sheds' originated in World War 1 when they were erected for the construction and repair of aeroplanes by the set-up known at one period as the Hampshire Aircraft Parks. Soon after World War 1 these buildings became the hangars for the Army Tank Corps, accommodating a variety of tracked, armoured vehicles. Later, as the Royal Tank Regiment, a home was found for them in Dorset and on Salisbury Plain, with the research facilities provided at WVRE, Chobham.

THE JERSEY BROW CATAPULTS

The RAE Mark I and Mark II Catapults were hydro-pneumatic devices for launching aircraft by means of a trolley. These designations were applied at Farnborough to indicate the two sizes of equipment for many years in regular use on Jersey Brow, the Mark I being virtually an ancestor of the larger Mark II machine.

THE RAE MARK I CATAPULT

This machine, being an early, 'F' telescopic tube type, was developed in stages by RAE through various performance requirements from the mid-1920s through the 1930s. A technical description is given in the chapter 'Catapults – Pre-1939'. It and its fellows were designed for launching aircraft mostly from warships, generally while under way. The aircraft, inevitably, were observation-type floatplanes mostly employed for gunnery spotting, although a few were armed. Each adaptation for shipboard installation carried its own designation, either related to the ship itself or to its class. One typical designation has been recorded by Caunter (*#222 p.71, fig.42.2*) as 'Farnborough Heavy Type IV' (F.IV.H). (Friedman (*#34 p.106*) states that, for launching in still air, the requirement was for accelerating a 7,000lb aircraft to 55kts {3,170kg, 28m/sec}. The chief candidate for this, according to Friedman, was a Vickers Model, 73ft-9in {22·5m} long, intended for HMS *Hood*.) The specialist feature in this instance was that the catapult runway structure, for stowage purposes, could be folded, a vertical hinge having been provided halfway along its length. The ranging and stowing activities were achieved pneumatically with manual override.

Although not the first equipment to be located on the site, the 'formal' Mark I catapult was preceded by at least two others of RAE design. That installed on Jersey Brow in December 1924 was situated at ground level, mounted on roller outriggers allowing directional orientation in plan through rotation around a vertical pivot, later modified to run on a circular track formed by a pair of concentric rails. Its location was adjacent to the erstwhile public access road to the west of 'B' Shed

[Building P.71]. This catapult was of the telescopic tube type, hydraulically operated with compressed air as the primary driving element in the first cylinder. Caunter (*#222 p.56*) states that in September 1925 the first launch from this catapult was with a Supermarine Seagull amphibian with Fl/Lt R de Haga Haig as pilot. (The record of RAE aircraft movements gives the date as 21 July 1925 for this launch of Seagull (N146).) He lists the names of those involved as P. Salmon, J. D. H. Pritchard, S. Child, C. Crowfoot and A. R. Crossfield, most of whom were still involved in these activities a couple of decades later, and who feature in this book. (Andrews (*#2 p.80*) dates this occurrence as 19 May 1926.)

This catapult is recorded by Caunter (*#ibid. pp.56/57*) as being rated at 7,000lb AUW, producing a terminal velocity of 39kts within a launching distance of 34·5ft {3,170kg, 20m/sec, 10·5m} with an acceleration of 2g within 1·04 seconds. This weight was at maximum rating

Fairey IIIF (S1317) leaving the RAE Mark I catapult on the Jersey Brow launcher site. 1930. (© Crown Copyright/MOD)

An early Salmon type (telescopic tube) catapult – possibly an F.I.L – in an excavation on Jersey Brow. This is thought to have been the first catapult to have been installed at Farnborough. August 1926.
(© Crown Copyright/MOD)

Supermarine Seal II (N146) appears to have come to grief on the RAE Catapult on Jersey Brow. This aircraft, bereft of its ailerons, was possibly used as a ballast vehicle for proof-testing the catapult. (Fleet Air Arm Museum)

but the design end speed for a reduced AUW was a maximum of 45kts {23m/sec}.

Let Salmon, the designer of the machine, speak for himself (*#87 p.720*):

'The RAE type catapult was built at the Royal Aircraft Establishment and by the end of December 1924 had been installed in a pit on the Common adjacent to the Aerodrome at Farnborough, where it remains to-day *[1932]*. It is now used for catapulting type trials on aeroplanes and for training pilots.

'This catapult was designed to launch aeroplanes not exceeding 7,000lb AUW at a speed of 45mph, in a distance of 34ft, the mean acceleration being 2g. In this type of catapult the motive plant, which is operated by compressed air, consists of a number of tubes or rams arranged like a telescope, the outer one being fixed to the rear end of the catapult structure in such a manner as to allow the other tubes to be projected forward; the smallest tube pushes the launching carriage on which the aeroplane is mounted. The structure consists of a low beam about 50ft long, 3ft high and 3ft wide, mounted on a central pivot about which it can be rotated to launch the aeroplane in any direction into wind.

'In this catapult four tubes are used, one fixed to the structure and three moveable ones, so arranged that they all move relatively to each other simultaneously. This is accomplished by making each tube such an amount smaller than the one in which it telescopes, that the area of the annular space between them is equal to the area of the bore of the smaller tube of the pair, and enclosing this space by a piston fitted at the rear end of the smaller tube and a bush at the front end of the larger, the resulting space being filled with fluid. Ports are cut through all the moveable tubes near their inner ends.

'Now, if the largest moveable tube is pushed out of the fixed tube the fluid trapped between these two tubes will be displaced; it will be forced through the port cut in the wall of the moving tube and in turn will force the piston attached to the next smaller tube to move forward. This will cause a similar motion between the next pair, and so on, until the smallest and last tube of the series is reached, in which the fluid from between it and the next larger tube is accommodated. The fluid which passes into the smallest tube is used to push a piston compressing air at its forward end, use being made of this compressed air to close the rams when the pressure on the large piston is released.

'As it is necessary to bring the truck and rams to rest at the end of the launch, the telescopic tubes are arranged to act as hydraulic buffers for the last foot of each of their strokes, the port cut in the walls of the moving tubes being covered by passing into the bushes through which the tubes slide, the fluid thus trapped being forced through a number of small holes drilled through the tubes between the main ports and the pistons.'

It is regretted that very few photographs of this early equipment can be found. The original glass plate negatives, like those of many another early project, have disappeared from the DERA archive – possibly removed long ago as being of little interest. The author has traced a couple of images of this machine as it existed in August 1926, sited to the west of the Large Dirigible Shed. The first, dated early in the month, shows the catapult from the west side with the rams in the extended position. The second picture shows the rather depleted Supermarine Seal II (N146) amphibian aircraft having come to grief midway along the track. A companion picture from the east side shows the catapult beam to be in a sorry state with part of the steel framework in a tangled mess. Obviously some mishap in the launching procedure had occurred. (Was this at the occasion of changeover from compressed air to cordite

operation?) The aircraft itself probably was being used as a dummy as its ailerons are missing and no sign of an engine and its tractor propeller can be seen. The photographic record for 25 August 1926 refers to a 'catapult accident'. It is thought that in the subsequent rebuild of the catapult beam the circular track was incorporated.

Another aircraft to have been launched from this installation was the Parnall Peto (N181) destined for the ill-fated HM Submarine M.2 (*Treadwell #113*). Six successful launches with this aircraft in landplane configuration were made at Farnborough on 6 July 1926. (These launches do not appear in the RAE record of aircraft movements.) Another early aircraft was the Fairey Ferret (N190) which was launched at 4,424lb at 45·3mph (39kts) {2,200kg, 20m/sec} and 2½g, this taking place on 9 July

The F.I.H. catapult on Jersey Brow with a test weight loaded in the front hooks of the trolley. This is an early, Salmon type, telescopic tube catapult to have been installed at Farnborough. It is fitted with one of the first of the RAE designs of collapsible launching trolleys. May 1929.
(© Crown Copyright/MOD)

The prototype Supermarine Sea Otter (K8854) has been loaded onto the trolley of the RAE Mark II catapult on Jersey Brow for initial proofing trials. Jack Marsh (wearing a trilby hat), the assistant foreman to the Fitting Shop, stands beneath the aircraft number. To the right is Bill Davies (back to camera), catapult fitter, while Ted Spooner, catapult rigger, with arms akimbo and wearing a cloth cap – stands behind the tripod. The chap on the extreme right is thought to be Sid Vincer, the chargehand of the launcher site. May 1939. (© Crown Copyright/MOD)

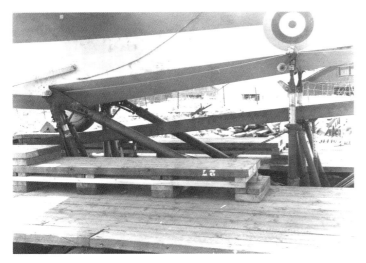

Supermarine Walrus (X1045) on the RAE Mark II Catapult. The spigot passing through the front catapult spool for use with the 'shoe-horn' loading system may be seen. This was a device incorporated in shipborne catapults to facilitate loading; in this picture the spigot is retracted. July 1940.
(© Crown Copyright/MOD)

1926. Months later, in December 1927, ten launches were made with a Fairey IIIF (S1148) aircraft. (Again, these do not appear in the RAE record of aircraft movements.) These last experiments were terminated through the aircraft breaking its back, having been subjected (for an unexplained reason) to an acceleration force exceeding 5½g.

It is not clear if this was the catapult which later was installed in HMS *York*. Andrews (*#ibid. p.80*) suggests this to have been the case. Caunter (*#ibid. p.67*) records the catapult having been first used at sea in 1928 on board HMS *York*. (Other records indicate this to have been pre-dated by sea trials in late 1925. Caunter's reckoning is to be preferred as he would have had official records readily available for his research.)

The subsequent, permanent Mark I installation on Jersey Brow is thought to have been located about 30 yards further north of the initial catapult pit shown in the illustrations. (There must have been horrendous problems with wind in close proximity to the large balloon sheds.) The Mark I had a busy life for a decade or so but was rarely used during the author's time. It was retained mostly for the launch of both de Havilland Queen Bee and Airspeed Queen Wasp radio-controlled target aircraft. A few ballistic 'dead-shots' of obsolete Blackburn Shark aircraft were also undertaken. It is possible that the final launch with this equipment was that of the Airspeed AS.30 Queen Wasp (K8888) on 21 April 1938. This was the second of two prototype machines scheduled to replace the de Havilland Queen Bee, radio-controlled target aircraft. This machine, as a floatplane, previously had been successfully catapulted from HMS *Pegasus* but the launch on this particular day – in land-plane configuration – resulted in its becoming a wreck. (See 'Pilotless Aircraft – The Three Queens'.) This catapult was dismantled shortly before the outbreak of World War 2 to permit aircraft access from the airfield into the western end of the newly constructed 'B' Shed hangar [Building P.71].

THE RAE MARK II CATAPULT

The Mark II catapult on Jersey Brow was essentially a robust, land-based installation, unique to Farnborough, adopting a long-stroke, telescopic tube system of operation. Although Salmon has not provided

a specification for this machine, it was a logical development of, and larger in capacity than, the Mark I, becoming the maid-of-all-work over many years. Air pressure was applied to the rear of a piston fitted to a tube sliding in an outer cylinder. This initiated forward motion to the piston. Fluid (water/glycerine mix) contained within the annular space to the front of the piston was displaced to pass through ported orifices to move forward a second piston, within the first tube. This system was repeated throughout the nested assembly of four tubes. To the front end of the smallest tube was attached the trolley mechanism similar to that referred to above. As each tube was displaced, so its speed of travel became relatively faster until the launch speed of around 66kts {34m/s} was attained by the smallest tube. In this case the prime mover was high-pressure air. It has not been possible to trace a record of performance details of this installation but it is thought its original maximum capacity was for an aircraft of 10,500lb AUW accelerated to 66kts over a distance of 45ft {4,800kg, 34m/sec, 13·7m} at a rate of 2·75g. (However, a Fairey Barracuda aircraft (P1770) with torpedo, weighing some 14,000lb {6,300kg}, was launched from this catapult in 1941.) In the 1930s one of the local aircraft dedicated to catapult work was the Fairey IIIF (S1781) which also was used for early shipboard trials in HMS *Courageous*.

The Mark II catapult was orientated in a fixed direction towards the south-south-west. It comprised a trolley running along a track made of deep-webbed rolled steel joists, these being above ground level. A wooden loading ramp was built each side of the structure; these were adjustable in height to suit the type of aircraft to be launched. This allowed the aircraft to be winched backwards along the ramp towards the trolley with its tail wheel running in another adjustable ramp

Supermarine Walrus (X1045) loaded onto the RAE Mark II Catapult. July 1940.
(© Crown Copyright/MOD)

contained within, and passing over, the trolley structure. This procedure avoided the need of a crane for loading. The aircraft remained fully supported within the trolley cradle throughout its launch even though it may have conveyed the impression to the pilot that he was being trundled along the ground rather than being thrown into space from an already elevated position.

Mishaps, although rare, occasionally did happen with this catapult. One instance of this occurred when an early Grumman Martlet I aircraft (to which catapult spools had been fitted) was released from the trolley hooks prematurely during its launching stroke. It was thought that an obstruction along the track caused the trolley to hesitate after the retaining detents had been retracted, thereby permitting the aircraft to separate from the trolley and continue its forward motion without attaining flying speed. It was often the fear of a pilot that this should happen and that the machinery would be careering along behind him without his plane having been launched properly! Cooper (#22 illn.134) carries a photograph of this incident.

A more devastating incident occurred at the close of a busy day after a number of successful aircraft launches. The rear cylinder head of the outside tube cracked with a bang as the extended tubes were being retracted to the closed, pre-launch, position. It was thought that the high pressure air within the outside tube ignited spontaneously ('dieseling'), probably through the presence of oil mist originating from lubricant. (This machine was powered by compressed air throughout its life.) The detonation fractured the cylinder head beyond repair. Fortunately, by this time (1944) other accelerator facilities were being provided on Jersey Brow and the loss of the Mark II was not as serious as it might have been. Up to this incident all trolley launching was carried out on the Mark II: at its demise tail-down launching was coming into prominence and the only subsequent full-trolley launches at RAE were carried out on the 'P.I' and 'K.I' Catapult installations and limited to Fairey Fulmar, Barracuda, Hawker Sea Hurricane and Supermarine Seafire aircraft.

The Mark II catapult had been used as the power unit in the early development of the 'Assisted Take-Off Gear' (the embryonic BH.III Accelerator), with the collapsible trolley systems devised for the flight decks of the six World War 2 Fleet Carriers, HMS *Illustrious*[40] and her ilk. In this instance, timber decking was provided to simulate the flight deck environment; the accelerator above-deck gear was attached to, and propelled by, the main telescopic tube nest.

For a short while in the late 1920s Jersey Brow was host to a third catapult. This was used for the development of the 'Larynx' project involved with early, unmanned, autopilot systems. This equipment was a Carey type of machine containing a short stroke jigger unit as prime mover, with the cylinder lying horizontally within the catapult beam and a rope reeved along its length as a multiple sheave system for propelling the aircraft trolley. This machine was cordite-powered. By the nature of the project, it is thought that this installation on Jersey Brow was used only for launching dead-weight dummies. The Larynx aircraft was neither fitted with an undercarriage nor did it possess a pilot's cockpit for test flying.

For a while, the rocket-powered, P.I Catapult appeared on Jersey Brow as a spectacular, pyrotechnic interloper, often used as a display feature for visitors to RAE. Details of these later installations appear in this text under their respective titles.

Supermarine Seafire IIC (MA970) was developed from the Spitfire VB by the addition of catapult spools and a 'V'-frame arrester hook. June 1942. (© Crown Copyright/MOD)

Supermarine Seafire IIC (AD371) on the RAE Mark II catapult trolley. February 1942. (© Crown Copyright/MOD)

Fairey Barracuda (P1770), the second prototype fitted with dummy torpedo, is seen loaded into the Mark II catapult trolley. September 1941. (© Crown Copyright/MOD)

Date	Cat & Accelerator	ATO & RATOG	DCTO	Total Launches	Arrest	Flexi-deck
1925	4			4		
1926	0[1]			0		
1927	0[2]			0		
1928	0			0		
1929	3			3		
1930	15[3]			15		
1931	9[4]			9		
1932	18			18		
1933	58			58		
1934	24			24		
1935	52			52		
1936	15			15		
1937	7			7		
1938	2			2		
1939	19			19		
1940	19		2[5]	21		
1941	22	98		120		
1942	49	21	1[6]	71		
1943	101	34		135	6	
1944	94	55		149	18	
1945	42	30	2[7]	74	25	
1946	102	57		159	9	
1947	90	33		123	27	20
1948	66	38		104	71	172
1949	305	10		315	72	142
1950	138	83		221	51	26
1951	194	21		215	99	93
Totals	1448	480	5	1933	378	453

Notes 1 – Other records show seven launches were made
2 – Other records show ten launches were made
3 – Includes five launches of the Airfield Portable Catapult
4 – Five launches of the Airfield Portable Catapult at Farnborough and four at Hendon
5 – Handley Page Heyford aircraft
6 – Avro Manchester aircraft
7 – Taylorcraft Auster aircraft on 'mini' trackway

CATAPULT LAUNCHES AT FARNBOROUGH

The above table comprises abstracts from the RAE Flight Books, the daily log of aircraft movements from the Farnborough aerodrome. It is thought that all launches by the Jersey Brow catapults have not been recorded, particularly those in the early catapult development days of the late 1920s. Prior to 1952 the records included one-line summaries of the sorties/operations under which the flights were made. In the case of operations from Jersey Brow, only the actual airborne flights were recorded; ballistic launches, taxying runs and deck arrestments were not listed. At no time were launches of dummy aircraft, or 'iron birds', recorded as airborne – they just did not fly! Differentiation between launches from catapults or accelerators was not made. The column marked 'Cat & Accel' is a combination of both procedures. Similarly, the designation 'Assisted Take-Off' (ATO) included launches from an accelerator and from the 'P.I' rocket catapult as well as 'Rocket Assisted Take-Off' (RATOG) – the system in which rockets were attached to the aircraft and jettisoned after take-off. DCTO refers to the 'Direction Controlled Take-Off' systems in which the aircraft was guided by a trackway, take-off from a trolley being effected solely by the aircraft's engines without additional external power sources. The 'Arrest' column is a record of live landings into an arrester system; again, taxying and towed trials are not included. The 'Flexi-deck' is a record of the many landings into the flexible (carpet) deck systems on the Farnborough aerodrome. Many of these landings were into a dummy set-up not involving the arrester system – what the Navy termed 'ADDLs' (Aerodrome Dummy Deck Landings).

PILOTLESS AIRCRAFT

Although not essentially for the Fleet Air Arm, Farnborough from its early days was involved to a great extent in the development of radio-controlled pilotless aircraft. The first incursion into this field, according to Child (*#223 p.25*), was undertaken in 1916, only five years after the first successful air-to-ground radio transmission had been achieved from a military airship.

Of the following cases, the first three concern light duty catapults while the fourth, and last, refers to an elementary arrester-gear installation.

system was brought to rest simply by allowing the weight (water bag) to drop into the sea alongside the ship! (See 'Launching by Gravity' chapter.)

The first installation aboard HMS *Stronghold* was with the catapult beam inclined upwards at a shallow angle of about 5°. Before long the beam was re-positioned at an angle of elevation of some 30° in which attitude the gravity operation was retained. Caunter says that subsequently these aircraft were launched from a cordite powered catapult. It is thought that cordite operation of a catapult was not introduced at

RAE PILOTLESS AIRCRAFT (1917)

This small machine was known by this lengthy title well before the advent of the multitudinous acronyms and mnemonics currently in vogue for aviation matters. Officially, the aircraft was an aerial gunnery target but undeclared inferences implied that it was to be considered also as an experimental guided missile. Contemporary records, not surprisingly, are reticent on this subject. The subject was treated by Sir George Gardner, Director of RAE 1955–59, in a paper read to the Royal Aeronautical Society in May 1958. (Thirty years earlier Gardner, as a young scientist, was engaged on this project and others of a similar nature.)

Caunter (*#222 p.43*) picks up the story, stating that in 1922 work continued on this project with an improved aircraft, known as the RAE 1922 Target, fitted with a small two-cylinder Armstrong-Siddeley engine developing 45hp. The control system was similar to that employed in the earlier experiments with the aircraft flying off the deck of HMS *Argus*. The trials were disappointing. Cronin (*#23 p.76*) says that two attempts both ended in failure, with the aircraft crashing into the sea on take-off. Attention was then diverted to launching the aircraft by catapult from the bows of a destroyer. The catapult was gravity operated, its power generated by a water-filled bag which, in falling, launched the aircraft from an inclined runway beam. The launching

HMS *Stronghold* with the high-angle, counter-weight, gravity-operated catapult in her bows. The aircraft mounted on the trolley is the RAE 'Pilotless Aircraft (1917)', an early guided missile of 22 feet wing-span. *c.*1924. (National Archives)

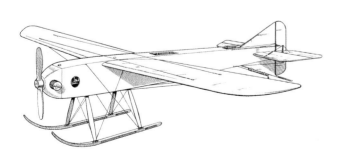

R.A.E. PILOTLESS AIRCRAFT — 1917

Line drawing of the first RAE 'Pilotless Aircraft (1917)' with which basic catapult trials were conducted. (© Crown Copyright/MOD)

RAE 'Pilotless Aircraft (1917)' being made ready for launch from the high-angle, gravity-operated catapult in HMS *Stronghold*. Percy Salmon (designer), John Grosert (head of Flight Sheds) and Jack Marsh (foreman of the Fitting Shop) of RAE staff feature among the civilian personnel. September 1924. (© Crown Copyright/MOD)

Farnborough until a few years later (1926 is one recorded date). Such an installation in turn was fitted aboard HMS *Stronghold* for the 'Larynx' experiments – see below.

Through 1923–4, some twelve launches were accomplished with this marque of pilotless aircraft attaining varying degrees of success. In Caunter's words (#*ibid.*):

'At the conclusion of these trials it was considered that remote automatic control of an aeroplane had reached a safe practical stage.'

It is thought that this basic principle of a gravity launching technique, following a patent specification No.3333 of 1911, was never used again.

As a postscript, Taylor (#*95 p.157*) records, with an illustration, that an RAE Target plane was flown pick-a-back upon a Fairey IIIF from Gosport in 1933; this experiment was not repeated. Circumstantial evidence, based upon the Farnborough Flight Records, could produce a

Larynx pilotless aircraft on the low-angle catapult in HMS *Stronghold*. The men forming the research team are recognised as (left to right) A. Stratton, P. A. Cooke, G. W. H. Gardner (later to become Director of RAE), G. J. R. Joyce, P. Salmon (Head of Main Drawing Office), W. Andrews, J. Grosert (Manager of Flight Workshops), R. King, C. Crowfoot (Manager of Fitting Shop), E. Cox. October 1927. (© Crown Copyright/MOD)

Larynx pilotless aircraft on the low angle catapult, HMS *Stronghold*. October 1927. (© Crown Copyright/MOD)

scenario as follows. On 4 December 1931 Flight Sergeant Simpson ferried a Fairey IIIF, Mark IIIB (S1789) from Heath Row (the manufacturer's aerodrome alongside the Great West Road) to Farnborough, arriving at 14.45. On arrival, the aircraft was wheeled into F1E hangar/workshop [Building P67] where it was fitted with a cradle, designed, manufactured and installed by RAE, to support an RAE 22 Target plane as a pick-a-back, in flying attitude. This cradle assembly was secured atop the centre section of the upper mainplane.

At 15.20 on 7 January 1932 the aircraft was taken for a 15-minute test flight by test pilot F/Lt Purdin. The following day, 8 January 1932, the aeroplane was flown for 20 minutes, from 15.10, on a Glider test. This time it was flown by test pilot Fl/Lt Maitland with Mr C. Howarth as observer and possibly operator. It is assumed that this was the experimental flight in which the glider should have been launched for free flight, probably over Laffan's Plain. It is known that the experimental project was abandoned after this flight, whether through the glider not having been released, or its release being 'untidy', or even the subsequent flight of the glider being unsatisfactory.

It is thought the IIIF aeroplane was returned to F1E for removal of the top-hamper and restored to its original format, for the next record is that it was collected by Flight Sergeant Simpson who was flown up from Gosport on 2 February 1932 as aircrew in Fairey IIIF S1523. At 11.20, both aeroplanes left Farnborough together for return to Gosport. From this time neither aircraft re-appears in Farnborough's flight records.

THE LARYNX PROJECT

The operating of pilotless aircraft was taken a stage further in the development of the Larynx machine with a new project which started in July 1925. LARYNX was a rather low-key attempt at an acronym for Long Range Gun with Lynx Engine. This machine was larger than the RAE Pilotless Aircraft (1917 and 1922 vintages) but the techniques learned from earlier experiments were adopted, extended and embodied in the new project. The control and instrumentation systems were more elaborate than those of its predecessors. The Larynx aircraft, designed by RAE, was a pleasant-looking monoplane with a closely cowled engine of 200hp {149kW} without an undercarriage. It was designed to carry a 'pay-load' of 250lb with a range of 300 miles at 193mph {110kg, 480km, 86m/sec}. From the first it was intended to be launched by catapult from a ship; in this case HMS *Stronghold* once more hosted the trials programme.

In this installation, the catapult beam again was angled upwards with 5° of elevation. The aircraft was mounted on a trolley connected to a telescopic tube prime mover, power being provided by a cordite gas generator unit. This catapult was designed by Salmon and built at Farnborough. It was an 'F' type installation; the illustrations show two types of trolley in use. The first launch was achieved in the Bristol Channel in July 1927, subsequent trials being in Cardigan Bay and the English Channel. A few launches were made from a shore installation on Portland Bill with further land-based trials continuing through 1929 in a high temperature environment in the Mesopotamian Desert, now Iraq. Turnill (#*101*) records the number of aircraft involved to have been six, none of which reached the design performance. Caunter (#*ibid. p.62*) states that these tests were moderately satisfactory but by the end of 1929 the project was abandoned. However, the development of pilotless aircraft continued at a low level of priority.

THE THREE QUEENS

During the 1930s three aircraft marques were 'ennobled' in being dubbed 'Queen'. These were the Fairey IIIF, the de Havilland DH.82B (a variant of the company's Tiger Moth) and the Airspeed AS.30. These were known respectively as the 'Faerie' Queen, the Queen Bee and the Queen Wasp. All were pilotless, radio-controlled, gunnery target aircraft associated with the four Anti-Aircraft Co-operation Units (AACUs), two of which were in the UK, a third was in Malta and a fourth in Singapore.

The first group of these aircraft was based upon the Fairey IIIF of which three machines were adapted as experimental, auto-stabilised, radio-controlled gunnery target aircraft. In October 1930, RAE was given

The de Havilland Tiger Moth aircraft was selected in 1935 to replace the Fairey IIIF for experimentation on pilotless, radio-controlled target aircraft. Some 380 of these aircraft were produced; this machine (K5114) is a Mark II version fitted with floats and spools for catapult launching from land or ships. August 1935. (© Crown Copyright/MOD)

Fairey IIIF, Mark IIIB (S1536/F1379) was the first of three aircraft of this type, specially modified for radio-controlled pilotless aircraft experimental work. It was fitted with a four-bladed propeller on the standard Napier Lion engine. The project carried the name of 'Queen' and the IIIF machines were affectionately known as Faerie Queens. This aircraft is being prepared for a launch from the catapult in HMS *Valiant*. March 1935.
(© Crown Copyright/MOD)

Experimental de Havilland 82B Queen Bee (K5114) radio-controlled aircraft being loaded onto its catapult at Jersey Brow. August 1935. (Fleet Air Arm Museum)

the task of developing such an aircraft which, after catapulting, would carry out certain attack manoeuvres and, if surviving the defensive anti-aircraft fire, would alight on the sea for recovery by the crew of the host ship. The RAE programme was named 'Target' and four aircraft, S1317, S1490, S1497 and S1536, were selected to support these experiments. The flying trials at Farnborough, including catapult launches – according to the aerodrome Flight Book – began in September 1931 and continued until May 1933. Early in the programme S1317 was diverted to other projects. S1536 was the first machine to be completed and on 21 January 1932 was subjected to its final shore-based catapult launch at Gosport; thence it disappeared from Farnborough's records. Work on the next machine, S1497, began in February 1932 and it, in turn, was sent to Gosport, with its final shore-based catapult launch on 19 April 1932. Machine S1490 was retained at Farnborough for the entire programme, often being sent to Lee-on-Solent for catapult trials. It was subjected altogether to some 50 flights, a few being catapult launches. Finally, in May 1933, S1490 was despatched to Malta for its final series of trials. All flying trials at Farnborough were conducted with the aircraft in landplane configuration. At Gosport each was fitted with floats for shipboard operation.

A catapult was installed in HMS *Valiant* and the first trial launch – this time as a floatplane – was made in January 1932. The aircraft (S1536?) crashed into the sea after some 18 seconds of flight. The second aircraft (S1497?) was prepared for a repeat test in April 1932. This was marginally more successful as the aircraft flew for 25 seconds before falling into the sea! The third, and final, Faerie Queen (S1490?), was ready for trials in September 1932 with its first launch and flight being entirely successful. As a result, HMS *Valiant* sailed for the Mediterranean where between January and May 1933 this aircraft tested the skill of naval gunners in target practice until, ultimately, it was destroyed by accurate shooting.

The experience gained in the operation of the Faerie Queen led to the development of the de Havilland Queen Bee which, in due course, became the production version of gunnery target aircraft. This was a hybrid machine, developed from the firm's Tiger Moth marque with undercarriage interchangeable between wheels and floats. RAE became deeply involved with this work from providing and proofing the catapults to the trials installation and operation of the aircraft in its entire

De Havilland Queen Bee floatplane mounted on a cordite-powered, type F catapult trolley. *c.*1935 (© Crown Copyright/MOD)

flight envelope. The first Moth under the 'Target' project was K1876 which was flying at Farnborough in November 1932, while the 'real' prototype, K3584, was in the flight programme at Farnborough by June 1933.

The airborne control system was fairly complex with a host of Post Office relays and pneumatic operating devices proliferating. An early control console was seen by the author, gathering dust in 'A' Shed hangar, during his apprenticeship. This, before the days of chip technology and printed circuit boards, was a pedestal filled with yet more PO relays and Strowger distributor switches, having a telephone dial among the desktop push-button controls. Later models resorted to push buttons but the electrical system remained the same.

Flight trials from Farnborough aerodrome, both in rolling take-off and catapulted launches (1936), were conducted with the landplane version. As anti-aircraft gunnery ranges were located on the coast so that when an aircraft was hit it would drop into the sea without hazarding life and property inland, live launches of Queen Bees were conducted from coastal catapult installations. The first of these was at Watchet on the Somerset shore of the Bristol Channel, the launch taking place with K8661 in July 1937. Another catapult was installed at Gosport with further installations at Hal Far, Malta and Seletar, Singapore. Recovery of aircraft from Watchet was effected by a trawler – SS *Radstock* – which hoisted the aircraft on to its foredeck and conveyed it back as deck cargo to a quayside in the Minehead locality. The propeller, wings and floats would all be removed for road transport back to Watchet by a 'Queen Mary' (a road transporter, not another 'Queen' aircraft) or its equivalent.

At least six F.I.L/F.II.L catapults for Queen Bee aircraft were built; their locations are not known to the author but one record states that five sites were in the UK. One unit was temporarily installed on Jersey Brow for initial launching trials. As the holding depot for these aircraft was at Manorbier it is almost certain that one catapult was installed there. Another is said to have been built on Portland Bill in 1932 (probably on the site of the earlier Larynx unit) for use in launching gliders –

no record of this unit has been found. Gardner, in his RAeS paper, records that exercises were conducted overseas at Malta, Alexandria, Singapore and Hong Kong.

Layman (*#58 p.73*) records that, on her re-commissioning in May 1938 as a deck-landing training ship, HMS *Argus* was fitted with a catapult to operate wireless-controlled de Havilland Queen Bee anti-aircraft target planes. Although Brown (*#14 p.33*) supports this, the author has come across no other evidence to confirm this installation. Hobbs (*#45 p.41*) records that arrester wires and an accelerator (a BH.I) were installed during the 1938 refit when HMS *Argus* was engaged as a Queen Bee tender. This makes sense as a bulky F.I.L/F.II.L catapult on the flight deck would have inhibited all other flying activities in the ship. HMS *Argus* was used either for rolling take-off of Queen Bee aircraft with wheeled undercarriages, or a special adaptation was employed using a platform for launching floatplanes from the accelerator. Layman (*ibid.*) records that only a few target planes were flown from this ship. Jackson (*#ibid. p.318ff*) states that some 380 Queen Bee aircraft were built, production ceasing in 1944; Gardner (op.cit.) quotes 420; Caunter (*#222 p.85*) puts the figure at 500. At least 48 Queen Bee aircraft were routed through Farnborough for calibration trials . This project was terminated in November 1946.

While, from the outset, the performance limitations of the Queen Bee were recognised, the later development of a faster aerial target for both naval and army anti-aircraft training was initiated. This aircraft, built by Airspeed as its marque AS.30, became known as the Queen Wasp, having a top speed almost double that of its predecessor. Two prototypes were built, one each as a landplane and a floatplane. Piloted catapult trials with the float plane (K8888) were conducted aboard HMS *Ark Royal*[14] while the landplane (K8887) was received at Farnborough for catapult and radio-controlled flight trials. K8888 was fitted later with a wheeled undercarriage and arrived at Farnborough for similar procedures.

The RAE Mark I catapult on Jersey Brow was used for these trials. Following a series of experimental flights with K8887, all of them

A picture of a de Havilland Queen Bee radio-controlled aircraft immediately after its launch from a shipborne catapult; possibly aboard HMS *Pegasus*. The catapult was a deck-mounted, 'D' type machine with the trolley incorporating a turntable sub-structure. *c.*1935. (© Crown Copyright/MOD)

The trawler, SS *Radstock*, has recovered a de Havilland Queen Bee from a successful alighting and is drawing alongside a wharf local to Watchet for off-loading ashore. *c.*1935. (© Crown Copyright/MOD)

piloted, K8888 was to be launched on 21 April 1938 (the day after its arrival at Farnborough in landplane format). The pilot, Fl/Lt MacDougall, entered the cockpit through the door on the starboard side, clambered across the lowered backrest to his seat and settled down for the flight. Unfortunately he did not flip up the backrest sufficiently far for the seat-retaining catches to engage properly. The catapult was fired, the aircraft rose from the trolley and immediately turned over onto its back, impacted the ground and became a wreck in the process. The standby rescue team was at the scene within a moment and quickly located the pilot (through an outburst of colourful language) who was curled up in the tail. The back of the seat had collapsed at the application of the forward 'g' forces and the pilot – of ample proportions – was propelled ignominiously down to the rear of the fuselage. Thetford (#96) states that five more aircraft were built but the Queen Wasp project was abandoned after this mishap. Future gunnery target practice was enabled by the use of towed drogues. It was shortly after this incident that the Mark I catapult was dismantled. For the next four years, all the local catapulting procedures on Jersey Brow were left to the remaining Mark II unit.

LLANBEDR

Naval Aircraft Department did not become completely detached from radio-control of aircraft by the termination of the 'Three Queens' projects. By 1958 a number of redundant Fairey Firefly aircraft were fitted with radio-control systems to provide targets for early guided weapons research, particularly Firestreak and Seaslug missiles.

Most of these were flown from the airfield at Llanbedr, Merionethshire (Gwynedd), the airfield affiliated to the missile range along the coast at Aberporth. As the aircraft were already fitted with arrester hooks, advantage was taken by providing an arresting facility on the main runway at the home station. This gear was a modified replica of the drag chain system which, some 15 years earlier, had formed an experimental set-up at Farnborough. The installation was basic in design with little or no sophisticated engineering content. Six wires were installed, five as a cluster in the main landing area with the sixth as an overshoot-stopper. If an aircraft entered this last wire it was generally off-centre and a damaged undercarriage would be the result, requiring the attention of what Llanbedr called the 'plane doctors'. A couple of tractors were used to recover and reset the chains after each landing. The time interval for such activity was of no significance. The system was in constant use for the duration of Firefly U.9 operations. Further details of the equipment used at Llanbedr will be found in the chapter describing the Drag Chain Arrester Gear and the section under the same title in the chapter on Airfield Overshoot Arresters.

Perhaps through financial constraints, development of the concept of pilotless aircraft continued since these early experiments has been low key. A short article in *Professional Engineer* – the journal of the Institution of Mechanical Engineers, dated 19 April 2000 – shows that the field of unmanned aerial vehicles (UAVs) remains active with a Systems Association (UAVSA) holding annual international conferences. Currently, the market devolves upon military requirements but a wider field of application is envisaged. Questions to be addressed relate to the crucial, practical issues emanating from the operation of UAVs in conjunction with not only scheduled military and civil aircraft movements but with the myriad unscheduled flights of private planes, microlites and even balloons.

These UAVs generally are launched by catapults of varying complexity, often mounted atop road vehicle chassis. Such basic equipment remain the sole survivors of the complex machinery and systems developed by NAD in its heyday.

BH.III ACCELERATOR – (ASSISTED TAKE-OFF GEAR)

Biggs (#7 p.24) states the first installation of an accelerator to have been made in HMS *Glorious* in 1928 as a trial set-up. Apparently the reaction from the operators, including Commanders-in-Chief, was not very favourable and the device was removed. This comatose state could not be tolerated for too long as the weights and performance of new aircraft were increasing. Ultimately, in the 1934 refit, an accelerator adopting the hydro-pneumatic principle was installed in HMS *Glorious*. Trials were conducted using an adaptation of the existing type of catapult launching cradle for accommodating the aircraft. A number of proving launches were made using pilotless Fairey IIIF aircraft. A contemporary illustration in Biggs (#*ibid.*) shows the track to be atop the deck with the collapsed trolley adding to the above-deck obstructions. This time the launching device, in fact a pair of them, would remain aboard.

An ingenious trolley system had been developed to cater for the requirements of launching aircraft from a carrier flight deck. With the earlier type of catapult trolley the aircraft was fully supported in the trolley and, in essence, was 'projected' into the air. Development of the trolley method, via the partial collapse of the launching hook assemblies, eventually resulted in a fully collapsible trolley for use on carrier decks. In essence the system was based upon the linkage geometry for the current trolleys used in ships' catapults. In these instances, the aircraft was still positioned in a trolley system with its wheels lifted marginally above the deck in the catapulting attitude, with the ensuing complete collapse of the trolley at the end of launch. A disadvantage was that the entire trolley structure, although collapsed, remained above deck level and formed an obstruction to following aircraft when flying off unaided. The ideal was to reduce further the above-deck silhouette of the collapsed trolley in order to create an absolute minimum of obstruction. Initial installations of what later became known as the BH.I accelerator were fitted between 1936 and 1938 to the earlier carriers HMSs *Courageous* (two units), *Glorious* (two units), *Argus* (one unit) and *Ark Royal*[38] (two units). The first two ships possessed the capacity for launching aircraft of 10,000lb AUW at 52kts, {4,600kg, 27m/sec}; the last two were upgraded to 12,000lb AUW at 56kts {5,500kg, 29m/sec}.

Refinements to this original design were introduced by RAE in 1939 with a view towards its installation aboard ships of the 'Illustrious' Class. The Jersey Brow Mark II catapult (with a specially constructed false deck), was prepared for testing the new system. A set of contemporary photographs shows the basic installation and depicts the staged loading of a Blackburn Shark ballistic dummy as a sequence. The first picture shows the aircraft positioned at the head of the Mark II Catapult, ready for the loading procedure to begin. The wheels are seen against a pair of chocks and the trolley, with legs collapsed, beneath the aircraft. (At Farnborough the aircraft could not approach the trolley from the rear, hence the provision of a tail-wheel ramp over the front bracket.)

A second picture shows the telescopic front legs, fully extended, with the hooks located on the aircraft front spools. The rear legs are in a partly raised condition (by what means is not apparent). It appears that the aircraft has been hauled rearwards a few inches to achieve this. Another picture shows the aircraft loaded in the trolley, ready for launching. The towing cables to the front hooks are tensioned, the front legs are fully compressed and the aircraft wheels are just clear of the decking. These pictures show there to have been no detents in the front and rear hooks to restrain the aircraft when its engine was running. (It did not matter in this case as the aircraft was fitted with a 'concrete engine' – a ballast-box in lieu of an engine.) The rear hooks were also bereft of damping devices to control their collapse at the end of the launch.

The initial, experimental stage of the Assisted Take-Off Gear – later to be developed as the BH.III Accelerator – with the prototype RAE trolley rigged on the Mark II Catapult. A number of date-expired Blackburn Shark aircraft were held at RAE, most of them for use on one-way trips as catapult dummies. A few were used for engine-on launches, suitably restrained by wire ropes, but most were fitted with a wooden box ballasted with concrete to replicate mass and cg position. March 1939. (© Crown Copyright/MOD)

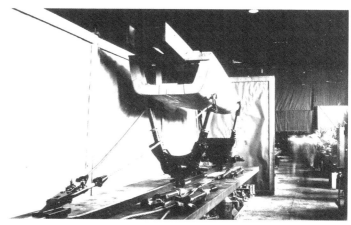

A workshop assembly of the above-deck trolley (original design) for the BH.III Accelerator as fitted to wartime fleet carriers. November 1941. (© Crown Copyright/MOD)

Front bracket and legs of an early shipboard BH.III single-track accelerator. The aircraft is a Fairey Barracuda set up for trials carrying a torpedo which had been fitted with a monoplane air tail. (Fleet Air Arm Museum)

The series of ballast trials of this prototype installation on Jersey Brow having proved its concept to be feasible, manufacturing drawings of a modified set of machinery were prepared for the installation in HMS *Illustrious*[40] itself in May 1940. The new installation was developed to be universally adaptable to all existing and projected FAA aircraft at that time and was given the title of Assisted Take-Off Gear (ATOG) – a title later to be adopted for rocket launching. The sea trials were comprehensive, starting with the launch of a dummy Blackburn Shark, followed by a couple of dummy Fairey Swordfish aircraft which, on launching, were lost overboard. A number of squadron aircraft were used for 'live' trials to cover a range of marques, weights, and a variety of 'stores' replicating service equipment and weaponry. The trials aircraft, all from the Fairey stable, included Swordfish L2791, P4085, P4154, P4220, P4224; Albacore L7119; and Fulmar N1873, N1881, N1882, N1884 and N1886. Flying trials at sea were completed in August 1940 with the equipment immediately being taken into service. The manufacturing drawings for this later equipment – subsequently known

as the BH.III Accelerator – were prepared at the Royal Dockyard, Chatham, under their project number 8000.

Although at first the aircraft was fully supported in the trolley, with its main undercarriage wheels just free of the deck, it was felt that no advantage was to be gained from this carry-over from earlier practice. Trials aboard HMS *Illustrious*[40] in 1942 showed that the aircraft could be loaded more speedily by adjusting the trolley legs to engage the catapult spools with the wheels remaining on the deck throughout. This quickly became standard practice. The significant difference in the new development was that the aircraft remained on its main undercarriage throughout the launch, the tail being supported in a take-off attitude and, in effect, the aircraft was 'trundled' along the deck until take-off speed was attained. The finer point in this principle was that the term 'catapult' was replaced by 'accelerator', the latter term being a better description of the operation of the system. (Occasionally the term 'booster' was used by the flight deck party.)

This installation was designed to enable the aircraft launching device, still termed a trolley, to collapse completely to deck level at the end of its accelerating stroke. This was partially attained by reducing the amount of above-deck 'ironmongery' in the new system. The top surface of the

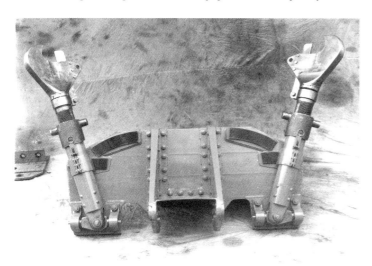

The special design of front bracket for the BH.III Accelerator, built at Farnborough. This component was the substitute for the normal bracket adapted to accommodate, first, the Hawker Sea Hurricane. The same assembly was later used for launching the Grumman Martlet and the Supermarine Seafire. The standard front legs for other aircraft were fitted inboard of these. July 1941. (© Crown Copyright/MOD)

A detail of one of the rear hook units (starboard) for the BH.III Accelerator. On launching, with the hook travelling to the left, the balance weight at the bottom of the swinging arm would be carried rearwards by inertia forces. This allowed the rear portion of the hook to fall backwards, and to be held there by the swinging arm. As the trolley was retarded, the hook itself would fall forward (under damper control) and release the rear spool of the aircraft from the trolley. November 1940. (© Crown Copyright/MOD)

track was faired into the flight deck level. It was also designed so that the trolley superstructure could be completely and rapidly unshipped from the under-deck gear (it was light enough for the flight deck handling party to man-handle this when required) to present a near flush deck for normal flying-off operations. A third feature was that the trolley installation was flexible in that immediate adjustment of the leg supports for the aircraft spools was available to suit any type of aircraft requiring to be launched. (For a short while there was an exception to this. This was for launching the Hawker Sea Hurricane and Grumman Martlet aircraft but an effective, universal unit was quickly designed and produced for ships that operated these machines and the method of operation was not impaired.) Caunter (#222 Fig.43.1) portrays a Swordfish aircraft loaded for launching on an early BH.III (with 'Hurricane' modification) installation.

Fairey Albacore on BH.III Accelerator for trolley launching. c.1942. (© Crown Copyright/MOD)

Grumman Martlet I aircraft (AX825) loaded onto a British BH.III Accelerator trolley (single track) in either HMS *Formidable* or HMS *Illustrious*[40]. This Martlet would have been fitted with catapult spools – one of many modifications – on arrival in the UK. August 1942. (© Crown Copyright/MOD)

This trolley system was mounted upon a series of three, in-line sliders which ran in a pair of steel channels, faired at two levels into the deck surface, the above-deck attachments being fitted to small, low-profile, tee-section projections from the under-deck gear passing through a slot in the deck. The operational flight decks of fleet carriers were supported by and located some six inches {150mm} above the armoured roof to the hangar deck. All immediate in-deck running gear had to be contained within this depth.

The main hydro-pneumatic, jigger type prime mover for the accelerator was installed below the armoured deck within the machinery flats adjacent to the hangar deck. This jigger was basically similar to those fitted to ships' catapults, comprising a multi-reeved unit, normally installed with the cylinder in the horizontal position with a fixed crosshead at the back of the cylinder and a moving crosshead attached to the ram assembly. The unit would be in the closed position prior to a launch and, when triggered, would extend lengthwise and, in taking up the rope, would haul the deck trolley (or shuttle) forward along the deck. In fact, the system was rather more complex than this. All moving parts had to be brought to a stop after the aircraft had been launched. This necessitated the use of a braking system.

This was achieved within the cylinder itself with machined fluid transfer ports in the (hollow) ram being progressively covered towards the end of the stroke. With the jigger itself undertaking the braking process the trolley could be brought to rest by connecting it to the jigger by tail ropes. This required a separate multi-reeved rope system to operate alongside the launching cylinder. In fact it was carried by the cylinder. A second stationary crosshead with rope sheaves was fixed to the structure downstream of the cylinder and the moving crosshead attached to the ram carried a second duplicate set of sheaves. The reeving was such that, as the launching wire ropes were taken up, the trailing ropes were paid out, these latter being fully extended at the end of the launch. It was a simple matter for these trailing/braking ropes to be used to retract the system back into its loading and launching position. A further complication was introduced in that both the hauling and trailing ropes were duplicated, a pair of ropes being used at both ends with their independent reeving all accommodated on the jigger assembly. Below deck there were ropes everywhere. All accelerators in the AH.II, BH.III, AH.IV and BH.V series adopted this principle. It was only relinquished on the introduction of the BS.4 series of Steam Catapults but, even here, a jigger was adopted as a holdback pre-tensioner and shuttle retrieving system.

The three slider units forming the in-deck running gear were linked together to form a 'train' having an overall length of some sixteen feet {4·9m} and weighing about one ton. Each slider unit was provided with a tee-form projection passing through the deck slot. An oil reservoir was provided within each slider and at every launch the track was lubricated through the accelerating forces driving a free piston within each reservoir, the inertia of which forced oil into a series of Michell bearings forming the rubbing surfaces between sliders and track.

The leading slider had a smooth surface on the top face of the 'tee' to allow a slipper crosshead to slide its full length of four feet {1·2m} or so. The second and third sliders respectively carried the front and rear hook support structures and mechanism; these were designed to permit rapid adjustment for all types of spooled aircraft. The front hook support comprised a pair of extendible nested tubes, with a breech type interrupted screw thread for rapid adjustment, pivoted at the lower end and having the front hook in a sliding fitment at the upper end. The restraining detent was fixed to the inner tube, sliding within the hook unit. A wire rope was shackled to each front hook with the leading end connected to the crosshead on the top surface of the front slider. This conveyed the accelerating forces from the front slider to the front hooks, thence to the aircraft.

The rear hook support structure was similar, being a further pair of telescopic tubular legs in which adjustment was obtained by pegs fitted into holes spaced as a vernier for fine setting. Suitable locking devices were embodied to prevent inadvertent adjustment when once set. Both pairs of legs were pivoted on brackets attached to the in-deck slider system. Serrated pads attached to both the brackets and legs provided a measure of lateral constraint to the system. The brackets themselves were pivoted to the slider and retained by catches to hold the aircraft when its engine was running. Movement of the brackets was controlled by bell-crank levers to which dampers were attached for constraint

Grumman Martlet I (AX825) loaded onto a British BH.III Accelerator trolley, showing the very short accelerator legs supporting the aircraft. August 1942. (© Crown Copyright/MOD)

The prototype, twin-track BH.III Accelerator in HMS *Indefatigable*. The identity of the two officers – blanked out by the wartime censor – is not known. March 1944. (© Crown Copyright/MOD)

when, towards the end of the launch stroke, the legs were permitted to collapse forwards.

The accelerating forces were transmitted to the aircraft from the under-deck gear by wire ropes from the front slider reaching up to positions on the front hooks. Prior to a launch these ropes would not be fully tensioned to take the accelerating loads. The engine thrust from the aircraft would be absorbed by the front legs, retained in position by the catches holding the front bracket. Vertical reactions would be accommodated by the rear legs, also restrained by catches to the rear bracket. During launch, the accelerating forces would be taken up by the automatic tensioning of the ropes, applying the towing load to the front hooks. Part way along the track the catches would be released by their riding upon ramps. In this situation the aircraft was held in the trolley solely by the applied accelerating forces.

As soon as the accelerator began to decelerate towards the end of the launch the aircraft would be free to leave the trolley, the towing ropes would slip forward along the front slider and the brackets supporting both pairs of front and rear legs would collapse forwards under control of their damper units, the whole coming to rest at the end of the track. The motion of the front legs falling forward would allow the hooks to lift in their housings, thereby freeing themselves from the detents. The forward motion during launch would allow the detents in the rear hooks to be released through the inertia effect upon a swinging pendulum. This removed a dead-beat mechanical lock which would then remain open throughout the launch. The entire trolley would come to rest in a fully collapsed attitude, ready to be retrieved to the starting position for a subsequent launch.

With the operational experience gained in the first four 'Illustrious' Class ships and HMS *Unicorn*, it was felt that a redesign of the trolley system using two trolleys running in a pair of tracks would introduce a more stable set-up. The method of loading the aircraft on to the trolleys would be the same. This principle was adopted, with minor adjustments, for HMSs *Pretoria Castle*, *Implacable* and *Indefatigable*, and all the ships in the 'Colossus' Class. All the foregoing were adaptable for tail-down launching. Trolleys were not fitted to subsequent ships; these employed deck shuttles running in single tracks with relevant strops/bridles for tail-down launching only.

In the case of the single-track installation the entire above-deck trolley system was mounted on the single trolley 'train'. When the twin-track system was introduced, the support legs were separated into the port and starboard components. The development of the above-deck gear was undertaken at RAE using the C.I Accelerator (already a twin-track installation) for proof-testing the system. When the time arrived for its adaptation for tail-down, strop/bridle launching the early twin-track units were connected above deck by a bridge to which the towing hook was fixed. Later installations sported a pair of towing hooks, one on each of the in-deck slider units.

With the changeover from a single-track to a twin-track system, the slider unit for the trolley was superseded by a pair of wheeled trolleys, still comprising three components in each train. This also meant that the brackets supporting the front and rear legs would have to be abandoned and a separate damper unit be substituted to control the collapse of each leg. Apart from these refinements the latter system differed little from the original. An interesting feature was adopted uniquely in HMS *Indefatigable* in which the entire twin trolley assembly was stowed in a cavity at the forward end of the accelerator track. This was covered by hydraulically operated doors.

The method of loading an aircraft into the accelerator trolley was different from that of the earlier ships' catapult – neither crane nor shoe-horn loading unit being involved – and from that of the Jersey Brow prototype. The aircraft taxied from the rear up to the launch position, coming to rest against a pair of wheel chocks, sometimes held in position by a couple of naval ratings. (Before long, this was automated with a set of hydraulic chocks built into the flight deck. These were

Fairey Firefly about to be launched from the BH.III (twin-track) trolley in HMS *Indefatigable*. May 1944. (© Crown Copyright/MOD)

controlled from the accelerator conning position at the port deckside adjacent to the accelerator.) The trolley was retracted to the launch position – it was normally stowed forward in the collapsed position – and then inched forward with a rating at each hook/spool position. The fully extended rear legs, pre-set in height to suit the aircraft to be launched, would be located first on to the rear spools and their detents allowed to close. The trolley would continue its progress, inching slowly forward, and as the rear legs began to support the rear spools the fuselage would gradually lift, pivoting around the main undercarriage. At a suitable stage the front hooks (again, pre-set but slightly extended by internal springs) would be located on to their relevant spools, the front detents were allowed to close and the support brackets firmly located in their catches. At this stage the aircraft was supported on its main undercarriage and the trolley, the wheel chocks would be removed (or retracted into the deck) and the launching procedure could begin. With a skilled flight deck handling crew all this could be achieved within a very short time, intervals between launches occasionally being down to two minutes.

A subsequent adaptation to the BH.III accelerator permitted the fixing of a direct towing hook to the front slider of the trolley train, replacing the crosshead, thereby permitting complete tail-down launching of later types of aircraft, both of British and American origin. This method required a holdback unit to be incorporated into the system whereby the aircraft engine could be run up and the aircraft retained on the accelerator until the moment of launch. This tail-down – or deck shuttle – method of launching offered many advantages when operating

under service conditions and it quickly superseded the trolley system. Ultimately it became the standard practice for deck launching from subsequent aircraft carriers with all new Fleet Air Arm aircraft being fitted with towing hooks instead of catapult spools.

The initial conversion of a BH.III accelerator for tail-down launching was undertaken by RAE. A towing hook (deck shuttle) was designed and made in the RAE workshops. All it required was for the hook to be clamped to the top flange on the leading slider of the in-deck trolley-slider system, the remainder of the trolley top-hamper having been removed. The problem with a single-track accelerator – as the BH.III was at that time – was that the tail wheel of the aircraft would probably impact with the deck shuttle in passing, unless it was partly airborne by the time it reached the end of the launch. The rear face of the deck shuttle was chamfered to reduce any effect of this nature. This problem disappeared in later installations where the twin-track system was adopted. Here a single hook would be positioned on a low bridge spanning both tracks and offering minimum obstruction to the aircraft or, alternatively, there would be a towing hook on each of the two leading sliders, offering no projection in the way of the aircraft tail/nose wheel as it passed between them. With either system, individual towing bridles would require to be made to suit; interchangeability between aircraft marques was not possible.

The first BH.III installations were designed with the capacity to launch aircraft of 11,000lb at 66kts {5,000kg, 34m/sec} when using the launching trolley. The total length of the trolley was about 16ft {5m} and ran in a track length of 142ft {42m}. The depth of the track was 6in

{150mm} and stood proud of the flight deck by some 3¼ in {83mm}. The deck in the vicinity of the track was ramped and faired-in to accommodate this difference.

Before long, the installation was upgraded to 12,500lb {5,700kg} using the launching trolley; for launches in the tail-down mode the AUW could be further increased to 14,000lb {6,300kg}. During later refits this upper limit was increased to 20,000lb {9,100kg} and the holdback capacity raised to 10,000lb {4,500kg}. (Details are given in Appendix 6.) Brown (*#14 p.49*) states that this was achieved by making the installation a twin-track system. It is not known if the doubling-up of the flight deck track required duplication of the under-deck accelerator machinery. Marriott (*#63 p.15*) states that the speed of launch at this weight was also increased to 75kts {39m/sec}. This last specification was solely for tail-down launching. By this time the length of the trackway in the deck had reached 151ft {46m}. The rate of acceleration was designed as a mean of 2·6g with a short duration peak not exceeding 3·25g. One record states that the abandoning of the trolley, which weighed about one ton, enabled the launching speed to be increased by 4 knots {2m/sec}.

It is significant that the BH.III was the only accelerator used in British Fleet Carriers throughout World War 2, a total of 23 units having been built. This design progressed through a number of development phases to match the increased capacity required of it but the principle and the general mechanics of the system remained constant. It was inevitable that the machinery was being called upon to extend its operational range upward as the weights and speeds of new aircraft increased. Much of the wartime research bore fruit in the immediate post-war era.

The succeeding BH.V accelerator surpassed the BH.III specifications with the ability to launch an aircraft of 30,000lb AUW at 75kts {13,600kg, 39m/sec}, again at the rate of 2·6/3·25g (Marriott – *#ibid. p.56*). Ultimately the steam catapult (accelerator) of greatly improved performance was installed on all operational carriers. (*vide* 'Developments at NAE/RAE – Bedford – BS.4. Steam Catapult'.)

Hobbs (*#45*) provides comprehensive details of flight deck and other machinery on a ship-by-ship basis, giving dates of refits and operational service careers of all British and Commonwealth aircraft carriers. Friedman (*#34*) is a valuable source, resulting from a wide-ranging research, presenting an in-depth appraisal of historical and technical features of all British carriers, both actual and proposed. A summary of the performance specifications for all shipboard accelerators is given in Appendix 6.

TAIL-DOWN LAUNCHING

THE PRINCIPLE

With the arrival of aircraft from the USA during World War 2, under the 'Lend/Lease' arrangement, a number of adaptations were necessary before these aircraft could be operated from Royal Navy carriers. The first aircraft to arrive in the UK were the Grumman (F4F-3), to which the FAA gave the name Martlet, and Grumman (TBM-1), similarly named Tarpon; their respective US designations Wildcat and Avenger were restored in January 1944. Both types were designed for tail-down launching. (Also both were fitted with 'sting' type arrester hooks instead of the 'V'-frame type universally found on Fleet Air Arm aircraft.)

The development of the principle of tail-down launching of aircraft involved an in-depth reappraisal of the means of applying the external accelerating forces to the aircraft. Hitherto, Royal Navy practice for mechanically assisted launching of aircraft entailed the provision of two pairs of 'spools' fitted to the fuselage. These were seated into complementary supports ('hooks') on a catapult trolley which, on being mechanically propelled, enabled the aircraft to be projected – virtually thrown – into the air.

The elements of tail-down launching comprised the aircraft, on its undercarriage, being towed along the deck by a strop ('bridle' in American parlance) connecting a hook – or hooks – on the aircraft to a shuttle operating in a slot in the flight deck. The aircraft was restrained during the pre-flight engine run-up by a holdback unit (containing a frangible link) attached to the deck. This link was broken when the launching force was applied through the shuttle. This system dispensed

with a collapsible trolley – the basis of UK accelerators for the previous two decades.

The problem for catapult launching was that at that time the Royal Navy had no accelerator immediately capable of tail-down launching. This made it necessary for both types of American aircraft to be fitted with the catapult spool configuration. Such an expedient was adopted as an interim measure but it was felt that the tail-down method of launching should be examined as an alternative, particularly when the Royal Navy began to receive (March 1942) from the USA escort carriers equipped with AH.II tail-down accelerators. Fleet Air Arm pilots rather liked these robust aircraft from America and many

Experimental cuff attachment fitted to the throat of the deck shuttle of AH.II & AH.IV Accelerators in American Lend/Lease carriers. This was to ameliorate excessive wear on the launching bridles by increasing the bight radius in the throat. December 1943. (© Crown Copyright/MOD)

Grumman Martlet II (AM991) rigged on the AH.II Accelerator aboard one of the early American CVA Lend/Lease carriers (either HMS *Avenger* or HMS *Biter*). This shows the single bridle connecting the aircraft launching hook to the deck shuttle. July 1942. (© Crown Copyright/MOD)

Supermarine Seafire (MB141?) has been fitted with a jury rig for tail-down launching. This crude installation was to assess the suitability of the aircraft for such experimentation. March 1943. (© Crown Copyright/MOD)

Bell P.39D, Airacobra I (AH574), adapted by RAE workshops for carrier trials, is seen rigged on the C.I Accelerator on Jersey Brow. This is thought to be the first aircraft with tricycle undercarriage (nose-wheel) to be equipped for bridle launching. July 1944. (© Crown Copyright/MOD)

The Grumman Hellcat, one of the Lend/Lease marques from the USA, is seen on the newly installed BH.III Accelerator in HMS *Indefatigable*. The RNVR Commander to the extreme right, wearing a soft cap, is Cdr Colin C. Mitchell, acknowledged expert in matters of flight deck machinery, and patentee of the steam catapult. March 1944. (© Crown Copyright/MOD)

squadrons embarked in subsequent 'Woolworth' Escort Carriers (in the US Navy, 'Carrier Vessel, Escort' – CVEs) and flew them to good effect. A number of advantages were to be seen in adopting the American system and a research programme was quickly initiated to study the practice. This technique of launching was not new to the RAE. The Department had been investigating methods to simplify the launching trolley in an effort to reduce obstructions at flight deck level. This was developed, in association with Chatham Dockyard, as the Assisted Take-Off Gear, subsequently the BH.III Accelerator. (It was deemed to be more expedient to modify the accelerator installations in a relatively small number of aircraft carriers than to adapt a much larger number of American aircraft to 'trolley' launching.)

The term 'tail-down' was convenient and appropriate for all naval aircraft operating through World War 2 as, without exception, they were equipped with tail wheels. With the post-war advent of nose-wheel, tricycle, landing gear, this description became less apposite and was replaced by the term 'bridle launching' whether applied to nose-wheel aircraft or the earlier 'tail-draggers'. Yet, having said that, with the

Scimitar, Buccaneer and Phantom aircraft the 'nose-high' launching attitude became virtually 'tail-down'!

The first British aircraft to be adapted for tail-down launching was a Supermarine Seafire NS487. (The same aircraft was used for early trials with 'sting' type arrester hooks.) The installation was basic in the extreme, perhaps best described in Navy terms as a 'jury rig'. It was, however, suited to its purpose and enabled development of a bridle system to proceed on the Farnborough C.I Accelerator.

An advantage of the tail-down launching procedure was that the aircraft, from the outset, was in a take-off attitude – with a better angle

Front view of the previous illustration. The aircraft nose-wheel can be seen on the bridging plate spanning the twin-tracks of the accelerator. The cylinders ahead of the towing hooks are the barrels for the trolley spike and barrel-braking gear. The board between the port wheel and trolley is calibrated for ciné film analysis of the start of the launch. The open hole behind the port wing is the pit accommodating the machinery and control position for the K.I Accelerator. July 1944. (© Crown Copyright/MOD)

By 1944 the Grumman Martlet had become the Wildcat, one of which is shown on the twin-track BH.III Accelerator aboard HMS *Indefatigable*. The first Martlets on arrival in the UK were designated Mark I and fitted with catapult spools for launching by trolley. With the ensuing development of tail-down launching, aircraft spools were no longer required and launching was by means of a strop (or 'bridle' as it later became known). There is little activity on the deck but, as the ship is not steaming into wind, the bo'sun may have piped a temporary 'stand-easy' for the flight deck party. March 1944. (© Crown Copyright/MOD)

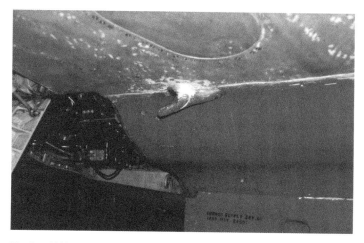

Hawker Siddeley Sea Vixen FAW.2 (XP924). Detail of starboard accelerator towing hook. (Author)

Towing hook attached to port undercarriage leg on Grumman TBM-3E Avenger AS.6B (XB446). With the proximity of the hook to the fairing, the latter could be damaged should the towing bridle not leave the hook cleanly. (Author via FAAM)

the deck, the skid remaining in contact with the deck throughout the launch. The Royal Navy Phantom had a specially extended nose leg which remained on the deck throughout the launch stroke. The tensioning against the holdback was achieved by the hydraulic jigger manoeuvring the piston assemblies forward.'

THE TOWING STROP / BRIDLE

The American style of tail-down launching used a single-slot track in the flight deck, similar to that on some British carriers. While early development at RAE concentrated on the continuing use of single-slot deck gear on existing carriers, as soon as a refit was required by the carrier the accelerator was modified to a twin-track shuttle with a bridge spanning the two trolleys to permit single or twin-deck hook configurations to be adopted. All BH.III accelerators were subsequently modified

Grumman Martlet II (AM991) rigged on the AH.II Accelerator aboard one of the early American CVA Lend/Lease carriers (either HMS *Avenger* or HMS *Biter*). This shows the spring-tensioned holdback located in the rack, set in the deck. The American name 'Wildcat' was adopted for the Martlet by the Royal Navy in January 1944. July 1942. (© Crown Copyright/MOD)

of attack – and there was no necessity for it to be configured to clear any obstruction from the trolley structure. This was fine for the wartime tail-draggers and the technique continued into the nose-wheel undercarriage, beginning with the experimental Bell Airacobra (AH574), and the first operational jets from the de Havilland Sea Vampire onwards. Although from this period the term 'tail-down' became a misnomer, it continued in use for a number of years until gradually it was superseded by 'bridle-launching'.

By the time of arrival of the larger, high performance jet aircraft, particularly the Supermarine Scimitar, Hawker-Siddeley Buccaneer and McDonnell Douglas Phantom, a further development was introduced into the bridle-launching procedure, particularly related to the steam catapult. George Ray (another RAE ex-apprentice) writes:

'The nose-high launching of Scimitar, Buccaneer and Phantom was indeed to ensure that the aircraft had the correct angle of attack when it left the catapult. The end speed required to permit the aircraft to fly in a level attitude was way beyond that which the catapult could achieve. The aim was for the aircraft to fly from the end of the launch stroke without height loss or pilot intervention. This was achieved by setting the trim before launch and using the already high angle of attack to provide the correct lift component. For Scimitar and Buccaneer, the aircraft was rotated by tensioning the bridle forward against the holdback until the tail skid touched

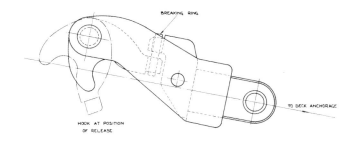

Drawing of American type of holdback for bridle launching, originally used on AH.II and AH.IV accelerators in Lend/Lease CVA carriers. April 1947. (© Crown Copyright/MOD)

Basic design of holdback for use on AH.IV Accelerators in Lend/Lease carriers. This is shown mounted in the adjustable holdback rack in the deck of one of these ships. A sample breaking ring is seen to the right with its containment muff to the left. December 1943. (© Crown Copyright/MOD)

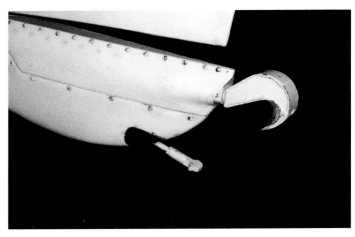

Supermarine Seafire, showing both the RAE experimental, W48 welded-plate, 'sting' type arrester hook in its stowed position and, below it, the spade unit for the RAE holdback attachment. c.1945. (Author)

Supermarine Seafire with experimental, fabricated 'sting'-type arrester hook. This machine is rigged on the C.I Accelerator using the American standard, spring-tensioned holdback unit but mounted on a bottle-screw adjustment. The strong-point on the aircraft for the arrester hook is used for attachment of the holdback; in this case the aircraft attachment is a shackle. June 1946. (© Crown Copyright/MOD)

to this operational format. The Aeronautical Research Committee in 1942 (*ARC 6,510 'Agreed Recommendations – Minutes of Meeting of 30th December 1942, February 1943'*) directed that the catapulting requirement (ie trolley launching) for carrier-borne aircraft should be abandoned. The three-point, tail-down launching was to be developed for all future designs of carrier-borne aircraft for use with the bridle system of launching.

As a result of this, a new field of research was initiated. The necessity was felt for an investigation to be made into the design of the aircraft/deck shuttle interface, its towing bridle. To British eyes the scantlings of the towing hooks on US aircraft and the details of the bridles appeared rather flimsy for the duty expected of them. The first of these aircraft to be seen in the UK, the Grumman F4F-3 Martlet/Wildcat, arrived with a single hook fixed to the fuselage between the main undercarriage units.

The bridle itself was of interest: it was made from a six-strand rope

on a fibre core but only two of the strands formed the splice. A section of the parent rope, rather more than three times the required length to form a grommet, was cut and two adjacent strands were carefully unwound. These were next formed into a coil with three turns completed, four strands of which were complete circles and two with free ends; these latter were spliced together. The grommet was then formed into a long, narrow loop, with both sides served together along its length, leaving a thimble-less eye at each end. One end formed the attachment to the launch shuttle and the other was hooked to the aircraft towing point. The Americans provided a bridle catcher, a device designed to retrieve the towing strop at the end of the launch, but it was rarely successful in doing its job.

All other World War 2 tail-down aircraft used linear bridles with a pair of towing hooks fitted to the aircraft adjacent to the main undercarriage strong-points. Each bridle was a single length of wire rope with a bight around an open thimble spliced into each end. One feature introduced early into the development programme was the replacement of the splicing activity with a cold-swaged termination. (This facility was already being introduced to the across-deck wire elements for the arrester gear. It avoided the use of the normal capels [white-metal sockets] and expedited their manufacture and rigging times on board ship by being smaller in size and therefore lighter in weight.) In these instances, the eyes at each end of the rope fitted the aircraft towing hooks with the mid-point of the bridle being matched to the deck shuttle. This type of bridle was more easily retrieved at the end of the launch and could be re-used time and again. In fact, in some experimental shipboard installations the bridle was physically retained in the deck shuttle.

At this point there was a serious development connected with man-handling the bridles. The operatives were brought very much closer to the front of the aircraft and its whirling propeller than in the case of trolley launching. Various aspects of mechanically loading the aircraft on to the accelerator and engaging the bridle and holdback units were tried. The advent of jet propulsion considerably reduced the danger although engine intakes provided a new hazard.

LAUNCHING HOOKS

The US Navy tail-down method employed either a single towing hook located on the underside of the fuselage between the undercarriage units

RAE design of improved pattern of holdback. The eye to the right is the spade attachment fitted to the aircraft. January 1945. (© Crown Copyright/MOD)

Blackburn Firebrand TF.4 aircraft (EK670) showing the RAE holdback system. The holdback tensioning-spring unit lies on the steel decking. December 1946. (© Crown Copyright/MOD)

or a pair of towing hooks located on strong-points on the sides of the fuselage, beneath the inboard section of the wings or on the undercarriage legs. The bridle connected the aircraft hook installation to the deck shuttle. Once loaded on to the accelerator, the aircraft was attached to a holdback unit anchored to the deck. The geometry of the forces generated by the launching process involved the provision of very sturdy undercarriage units. This was not necessarily a disadvantage in the case of operating aircraft from and to a carrier flight deck.

The earliest accelerator hooks seen in the UK were on Lend/Lease American aircraft, the Grumman Martlet and Avenger. The first of these was the Martlet I (Grumman F4F-3). On the earlier deliveries to the UK a number of modifications were carried out on each machine, including the fitting of the standard catapult spools necessary for their use on Royal Navy accelerators.

However, later marks of Martlet (renamed Wildcat in January 1944) retained their towing hooks – of more substantial dimensions – and were not equipped with spools. Similar modifications were carried out on the Avenger aircraft (Grumman TBM-1), originally named Tarpon by the Royal Navy but also in January 1944 reverting to the American name. A few early models were fitted in the UK with catapult spools but these were abandoned as soon as tail-down launching became possible with Royal Navy carriers. Similar comments on the towing hooks as for the Martlet are made here but in the case of the Avenger a pair of hooks were provided, fitted to the inboard sides of the main, outward retracting, undercarriage legs.

By the time of the arrival of the first of these American aircraft in the UK, Naval Aircraft Department had already been researching all aspects of principles and procedures for arresting aircraft on landing, both for shipboard and airfield installations. (Particulars of these are discussed later.) Much of this investigatory work was found to be relevant in application to hooks used for accelerating aircraft.

Whereas for accelerated launching some aspects of hook design were less sophisticated, those applicable to both conditions included:

a) investigations into the shape of the throat of the hook,
b) its diameter relative to the size of the wire forming the bridle,
c) the interaction between the bight of the rope and the hook profile,
d) the reaction between bridle and throat surface relative to the application of soft/hard coatings to the throat,

Supermarine Seafire XVII (SX311) is loaded onto the C.I Accelerator using twin towing bridles. The early type of holdback is attached to the arrester hook strong-point. June 1946. (© Crown Copyright/MOD)

De Havilland Sea Hornet on the Jersey Brow launching area. This is the second prototype F.20 aircraft (PX214) without folding wings but fitted with accelerator and arrester hooks. It is rigged on the C.I Accelerator using a special bridle to accommodate the twin towing trolley at deck level. January 1946. (© Crown Copyright/MOD)

e) the angle of wrap-round the hook at the application of maximum loading.

All the experience gained in these investigations was incorporated into the design configurations of the towing hooks both for the deck shuttle and the aircraft launching hooks covering all future naval aircraft specified for tail-down launching. In the case of towing hooks, there was no requirement to cater for relative movement of the bridle through the throat of the hook as the geometry was relatively stable throughout the launch. The main essentials were ease of attaching the bridle when loading the aircraft onto the accelerator, and similarly for the attachment of the holdback unit, and to ensure a clean fall away of the bridle from the aircraft itself and its flight path immediately after launch.

With the accelerating forces applied during launch expected to be in excess of 4g and the minimum design parameters required to contain a factor of four on the ultimate tensile stress (or a factor of two on the 0·2

Blackburn NA.39, prototype of the Buccaneer. Detail of the starboard towing hook with sample bridle attached. (Author)

A Blackburn Buccaneer S.1 (Code 105 of 800 Squadron, XN968) being prepared for launch from the forward BS.4 catapult aboard HMS *Eagle*[51]. The figure in the central foreground wearing a yellow surcoat, inscribed FDO, is the Flight Deck Officer, holding his launching flags behind his back. (Royal Aeronautical Society Library)

Blackburn Buccaneer S.1 (XK529) exiting the BXS.4 raised catapult at Thurleigh with a duplex bridle falling away from the aircraft. July 1961. (© Crown Copyright/MOD)

De Havilland Sea Vixen FAW.I (XN657, Code 491 of 899 Squadron) exiting the waist BS.4 steam catapult in HMS *Eagle*[51]. September 1964. (Royal Aeronautical Society Library)

The bridle catcher aboard one of the American Lend/Lease CVEs of World War 2. The spider is rigged across the track of the AH.II/IV Accelerator with its mousing-latch ready to accept the lower eye of the bridle as it arrives from the right. The cable connects the device to its own arrester – a pneumatic tube lying along the starboard side of the track. (Author)

The highly successful van Zelm bridle retrieval system being installed for trials on the BXS.4 catapult at Thurleigh. Of the four sliders in the tracks, the starboard rear unit is that connected to the under-deck braking system, located at the far end. February 1964. (© Crown Copyright/MOD)

Farnborough's accelerators prior to their adoption on board ship. It was one of these experiments which led to a more sophisticated bridle retention/recovery system which ultimately was put to good effect in HMSs *Eagle*[51] and *Ark Royal*[55].

THE HOLDBACK UNIT

One of the features of tail-down launching was the provision of some form of restraint to prevent the aircraft prematurely moving forward from the launching position under its own engine power. This was a device anchored to the deck abaft the aircraft with a quick-release facility allowing the aircraft to proceed on the point of initiation of the launch. The first of these came over with the initial CVE from the USA (HMS *Archer*) and was known as a 'holdback', a term which remained throughout the life of flight deck machinery.

Basically, the holdback was a tensioning unit containing a heavy duty coiled spring with spigots at the deck anchorage end and a swinging, spilling hook secured by a frangible link to a strong-point in the tail of the aircraft. The deck anchorage comprised a rack, set flush in the deck, into which the holdback spigots were located. The frangible link itself was a specially calibrated breaking ring, designed to be of sufficient strength to accommodate the pre-tensioning load of the accelerator and the engine thrust but which would break as soon as the launching force of the accelerator was applied. The strength of this breaking ring, which was manufactured from high-tensile steel drawn tube, was calculated for each type of aircraft to be launched. Extreme care was taken to ensure the correct ring was matched to its aircraft.

The procedure was for the aircraft to taxi forwards to the loading position on the accelerator. The bridle was then attached while the aircraft was against the wheel chocks; the holdback was next inserted into the deck rack. Then the whole system was 'tensioned-up' by the operator slowly inching the shuttle forward to a 'loaded' position indicated by a red band. When this was done, the system would be ready for the launch to proceed, after which the holdback would be removed and the exercise repeated with another holdback fitted with a new breaking ring (probably with a new bridle as well) for the next-in-line aircraft.

A development of the holdback adopted in the AH.IV accelerator abandoned the use of a tensioning spring. In this instance the holdback

percent proof stress) of the material to be used, a hook designed for a working load of some 45 tons would be required to launch an aircraft of 10 tons all-up-weight at this rate of acceleration. A commercially available hook for this load (for example, for a crane) would be of massive proportions and would be completely out of place in an aviation application.

The detail design of aircraft launching hooks almost inevitably was based upon a steel forging. It goes without saying that high tensile steels were universally specified for these components as weight becomes important in such applications. In later years, aircraft of greater weight than that quoted above have been launched by the tail-down method, and their launching hooks have been designed accordingly. Samples of these may be seen on various carrier-borne aircraft on display at the Fleet Air Arm Museum at Yeovilton.

The procedures involving attachment of the bridle between aircraft and accelerator hooks (and that of the holdback between aircraft and deck), particularly under operating conditions, were fraught with both difficulties and danger. These matters were continually addressed until some obvious improvements upon the basic US Navy equipment had been implemented, making the tasks of the flight deck party less onerous and less hazardous. On their ships, during World War 2, the US Navy naïvely provided a recovery device ('bridle catcher') for retrieving the wire rope bridle as the aircraft was launched. In practice the recovery of the bridle was seldom successful and ultimately the attempt was abandoned. These developments were proof-tested on

Hawker Siddeley Buccaneer S.2C (XV868, Code 020 of 809 Squadron) in the final stage of its being loaded onto the waist BS.4 catapult in HMS *Ark Royal*[55]. The van Zelm bridle retrieval system is clearly seen. (© Crown Copyright/MOD)

The final word in bridle launching. An unidentified Fairey Gannet AS with fuselage-mounted towing hooks rigged on the BXS.4 steam catapult at Thurleigh using the van Zelm bridle retrieval system. September 1964. (© Crown Copyright/MOD)

rack in the deck was adjustable and the operator tensioned up the system by control of the rack position against hydraulic pressure.

The fitting of new breaking rings to the holdback on a cold, windy flight deck was not the easiest of activities. The rings were almost as small as a gentleman's signet ring. Before long, the problem was addressed by Mollart-Rogerson (*#276*) and by 1946 a spring-loaded, split wedge, latching device was developed using a larger-sized breaking ring – about the size of a lady's bracelet – as its weak link. This involved the replacing of the stirrup by fitting a spade-like terminal to the aircraft to match the holdback. A number of variations on the original principle were tested over a period of years. One scheme was tried in which the aircraft was ranged from the hangar with its bridle and holdback attached, ready for connection to the launcher. The advantage here was that, through the fitting of these attachments in the hangar 'free' time, the time taken on the flight deck to load the aircraft onto the launcher was reduced. But this did not solve all the problems.

Eventually, in the case of the steam catapults, the holdback became a relatively fixed location on the flight deck and an auxiliary under-deck jigger unit was used to draw the aircraft forward, once the bridle had been attached, to tension up the bridle and holdback system into the ready-to-launch configuration.

BRIDLE RETRIEVAL

Another problem which confronted tail-down launching was the recovery of the towing bridle at the end of the launch. In the early days, bridles were fairly simple in concept and the frequent loss of one overboard, although inconvenient, was of little consequence. In the post-war years, as the system developed and became the norm for launching high

performance aircraft, bridles became somewhat more sophisticated in both design and manufacture. The requirement for regular retrieval of the bridle after each launch became important.

The first bridle retrieval system was to be seen aboard the early American CVEs which arrived in Britain (1942) under Lend/Lease. A small spider-like device was placed at the end of the launching track; this was restrained by a lanyard passing through a tube (pneumatic damper unit) laid alongside the track. The bridle was supposed to be caught by this spider, retained within a mousing-latch and arrested by the paying out of the lanyard, to be recovered manually by hauling back the lanyard and resetting the system.

Many other devices were tried but the general outcome was disappointingly unsuccessful. For many years the practice was to expect to lose the bridle overboard in each launch. A deck-top feature of a launching programme was the array of bridles laid alongside the accelerator track, or stacked in the deck-side walkway, for instant availability as replacements through loss overboard or deterioration in use.

When, in post-war years, the principle of catapult launching all aircraft for a sortie was adopted, and the bridles became much more cumbersome, the problem needed to be addressed in depth. An urgent need was declared in an NAD Paper of 1956 (*#317*) for a device for catching launching bridles to reduce expense and logistic problems by using each bridle for more than a single operation. Bridles weighing about 200lb {90kg} needed to be arrested from speeds of about 130 knots {67m/sec}, without any risk of their rising (flailing) to strike the aircraft on release at the end of a launch. An important qualification was the need for such equipment to be cycled operationally with the successive rapid launching of aircraft.

The risk of the flailing ends of the bridle striking the aircraft was a feature requiring investigation. Although this seldom occurred in practice, it remained a threat. Beard, in 1950 (*#206*) had already recognised this and referred to a proposal using a solid link (in lieu of a rope bridle)

McDonnell Douglas Phantom FG.1 (Code R012, either XT863 or XT870) of 892 Squadron, exiting the waist BS.4 catapult in HMS *Ark Royal*[55]. The bridle is being retained by its van Zelm retrieval system. (© Crown Copyright/MOD)

between deck shuttle and aircraft hook. At the moment of release, this link would be restrained by a lanyard passing through an eye in the shuttle to retain the link with the above-deck gear during its retardation. Beard was fearful also of the steam catapult requiring heavy duty bridles which would counter their flexibility, thereby incurring the need for a solid-link system. As it proved, this would not happen for a couple of decades.

Various schemes of bridle capture were tried out at Farnborough, some more successful than others, but none being completely satisfactory. Whichever scheme was in use the problem of rapid, automatic retrieval and resetting for the succeeding launch was always to the fore. Johnston (#263) in his review identified the research projects investigating this feature and reduced them into four channels, viz:

a) A hook mounted on a small 'roller skate' held by wires.
b) A hook attached to a spring-loaded arm placed forward of the catapult. The arm could rotate about a vertical spindle which was able to move in a track offset from the catapult centre line. This track was curved so that it guided the arm, complete with hook and bridle, to one side of the catapult where they were arrested against a hydraulic buffer at the end of the track.
c) A modification of the above system using a straight track and with friction pads to absorb most of the kinetic energy.
d) A bridle retaining strop which was wound round a spring-loaded drum; this drum was to be mounted on a spindle and housed in the catapult shuttle.

Johnston then asserts that, at the time of his report, the form that the catcher would take should be left until more was known about the flight of a bridle. This was the objective in subsequent research at Farnborough and in trials at sea. One of the areas to receive attention was to increase the chances of the aircraft becoming clear of the bridle

before the latter was arrested, hence eliminating the danger of the bridle striking the aircraft. The ideal was to ensure that the tendency of the towing eyes of the bridle would be made to fall away from the aircraft immediately following its release from the aircraft, causing the eyes to drop and prevent their rebounding into the path of the aircraft. Whatever engineering solution was found to be effective, five parameters needed to be met. These were:

a) The amount of structural modification which could be tolerated in the ship forward of the catapult installation.
b) The extent of permanent structure that could be built into the bows of the ship.
c) The extent of retractable structure that would be permitted forward of the bows.
d) The delay in retrieval which could be tolerated during aircraft loading onto the catapult.
e) The efficiency (statistical success) of the installation.

These features were addressed in detail by NAD.

The nearest to an ideal was that in which the bridle continued its travel – after the aircraft had been released, and with the deck shuttle itself under retardation – following a curved track in the deck to divert the bridle ends away from the aircraft. An experimental track, running alongside the shuttle track at the release end, was some 25ft {8m} in length, following a quadrant to port of about 10ft {3m} radius. Beard (#206) refers to this system which, although reasonably satisfactory on the 30,000 Accelerator on Jersey Brow, was not adopted on shipboard for a number of reasons, the obvious one being that deck space was not available for such an installation. The linear development of this system was the next logical step.

Attempts were made to restrain the forward motion of the bridle at release by attaching it to the shuttle by a pair of elastic (bungee rubber) lanyards. The problem was that, at the end of the launch, release of the kinetic energy absorbed by the bungees in the forward motion of the bridle threw the bridle back onto the end of the catapult track with some violence. A simple solution for the control of this feature was not found.

Even with the installation of the steam catapult in HMS *Ark Royal*[55] at the time of her commission, after five years' service the bridle was still expendable at each launch. Greaves (*#246*) describes an experimental installation at Bedford in 1960 in which the bridle – on release from the shuttle – was retrieved by a pair of undrawn nylon straps retained by the shuttle and permitting the bridle to run forward against the tensile resistance of the nylon elements. A protrusion forward of the catapult shuttle track was designed to accommodate this system; it appeared to be a forerunner of the extensions to appear later on British and American carriers. The bridle was retained but, for every succeeding launch, the nylon straps had to be renewed. This was still a labour-intensive and time-consuming feature of the launching procedure which could only benefit from further development.

Eventually a system of some sophistication was installed in both HMSs *Eagle*[51] and *Ark Royal*[55] associated with their steam catapults.

A McCormick/International Farmall H 'Rowcrop' (WRF361) agricultural tractor. This piece of Lend/Lease farm equipment was used as a stand-in 'iron bird' for an aircraft of equivalent weight and tricycle undercarriage for experiments on a nose-wheel-tow catapult launching system. August 1966. (© Crown Copyright/MOD)

The Farmall 'iron bird' with the experimental nose-wheel towing link at the entry point to the catapult loading rig, located behind the flush catapult at Thurleigh. September 1965. (© Crown Copyright/MOD)

The evidence is seen where sloping proboscis-like protrusions appear on the flight deck ahead of the catapult tracks. These contained an additional pair of tracks running alongside and extending forward of the catapult, each of which carried a pair of sliders. The bridle was attached to these sliders by lanyards, allowing them to be drawn along with the launching shuttle. These sliders were connected by a steel tape to an independent water-cooled, brake-drum system. At a predetermined point in the launch stroke a secondary pressure was applied to tauten the lanyard system. This made ready the force to pull the bridle away from the underside of the aircraft at the end of the launch stroke. Prior to the end of the power stroke (at approximately 15ft {4·6m}), the primary brake pressure, depending upon the speed of launch and the weight of the bridle, was applied to the steel tape connected to the bridle sliders. (The term 'shuttle' used in the official documentation is confusing as it applies to both the main launching shuttle and the subsidiary shuttle [slider] controlling the bridle arrester unit.) This was the point of release of the aircraft from the catapult. At this stage the bridle would continue forward, away from the slowing launching shuttle, a third brake pressure was then applied and the bridle would smoothly come to rest some 35ft {11m} beyond the end of the power stroke. On completion of the launch, the bridle along with the launching shuttle was mechanically retracted to the standby point at the head of the catapult for re-use.

For the foregoing, the author is indebted to pages from an instruction manual (the source of which he has not been able to identify) and to Lewis (*#60 p.48*) who gives a detailed description of the installation. This facility allowed bridles to be re-used for a number of launches, instead of losing them overboard on every occasion of use. Frequent renewal of the bridles themselves was necessary by virtue of the repeatedly heavy stresses generated during the launching procedure. The diagram which shows the installation for use with McDonnell Douglas Phantom and similarly with Hawker Siddeley Buccaneer aircraft, together with photographs of the system in action, will be self-explanatory.

Although, according to Lewis (*#ibid.*), the name of van Doorn is connected with the above system, another not dissimilar title of van Zelm has appeared and, through a similarity in operation, could cause some confusion in understanding. This second system, developed in the USA, adopted the principle of coiled steel tape as its *modus operandi*. Official RAE reports and other records define this as the van Zelm system.

Dyer-Smith (*#235*) refers to an interesting arrester system adopted for airfield over-run situations which employed the paying-out of steel tape through a 'torture chamber' of loaded rollers to effect the braking system in the arrestment of a landing aircraft. The method of pay-out and configuration of the rollers was such that the steel tape was severely deformed in the process and was not retrievable for re-use after pull-out. The residual, twisted, unmanageable mass had to be disconnected from the system and a new coil of tape installed. This was no easy procedure as the container was some 6ft {1·8m} diameter, holding 1,000ft {300m} of tape and weighed about 1,500lb {680kg}, requiring mechanical handling gear. The duration between one engagement and the resetting for further use was far too long for carrier operation and in no way could have been used for bridle retrieval.

NOSE-WHEEL TOW

Notwithstanding the provision of systems for bridle retrieval, the flailing of the ends of the bridle coming into contact with, and damaging, the aircraft on release from the deck shuttle was always a possibility. This problem was addressed from time to time at

McDonnell Douglas F/A-18F ready to be launched from the catapult at the Naval Air Engineering Station, Lakehurst, New Jersey, USA. This is the two-seat version of the F/A-18E Super Hornet. The solid link tow-bar interface between deck shuttle and the nose-wheel unit is seen, together with the holdback hooked to the same aircraft strong-point. (United States Navy)

Farnborough with no ready solution having been found. Thought was given to the replacement of the bridle by a solid link but the problem relating to rapid disengagement of aircraft from shuttle always seemed to be much in evidence.

It was with the Mark III catapult at Harwell that this particular feature was first addressed (c.1940) and designs prepared. It was felt here that the towing device should be a rigid link between the under-deck gear and the aircraft attachment point. In this instance the problem, with aircraft of the size of the Short Stirling and its four-engined siblings (all 'tail-draggers') was to ensure that the collapse of the link was absolute and completely out of the way before the tail of the aircraft crossed the release point. In spite of many draft schemes this rapid release problem remained unsolved, becoming a contributory factor to the eventual cancelling of the Mark III project.

A decade later, the proposal came to the surface again when the general run of naval aircraft began to appear with nose-wheel undercarriages. (Hardly a tail-down situation, nevertheless involving bridle launch techniques.) With a strong-point of the airframe being at the forward undercarriage attachment, it was argued, why should this location not be used for launching and holdback purposes? Three possibilities appeared to suit the bill:

a) a push from behind with a direct face-to-face contact between shuttle and undercarriage;
b) a pull from in front with a collapsible link as an attachment of the launcher;
c) a front pull but with the link as a retractable component of the aircraft – a stowable arrester hook in reverse.

The early considerations were for the contact point to be at the top of the nose-wheel undercarriage leg, but the advantages of lowering this point to wheel-axle level soon became apparent and obvious when the nose-wheel assembly became a twin-wheel unit.

A lead had already been given by the US Navy and there was a common desire for cross-deck operation between the USN and the Royal Navy. NAD Bedford picked up the feature and began developing a system for incorporation in the proposed new British carrier CVA-01. Such an installation would have been retro-fitted to all British carriers in commission at the time of its adoption. (One proposal employed a thrust pad fixed to the deck shuttle applying the accelerating forces to the lower end of the aircraft nose-wheel unit. It is thought that this principle was also being investigated in the UK for the CVA-01 carrier project.) The system favoured for development was that using a towing link attached to the deck shuttle but trials proceeded with the link attached to the aircraft – it was simpler to simulate at the experimental stage. Without an incursion into the details of these experiments, it is sufficient to record that full-scale experiments – using a dummy, 'iron bird' – were encouraging, but all work was brought abruptly to a close with the cancellation of CVA-01. The author is indebted to George Ray (who was engineering this project) for details on its development and its untimely end.

RAE, with its penchant for employing whatever redundant material was conveniently available for initial experimentation, used an agricultural tractor as an 'iron bird'. With a minimum of adaptation, this became a dummy representation of a nose-wheel aircraft. The wheel arrangement, although a little out of scale, was thought to have been a fair geometric representation of an aircraft undercarriage footprint. The

Detail of McDonnell Douglas F/A-18F – Super Hornet – nose-wheel in the ready-to-load configuration. The tow-bar is engaged in the deck shuttle and the spring-tensioned holdback is located in the deck rack. (United States Navy)

linkage was a short tow-bar attached to the front wheel assembly. This, perhaps, was the strangest of all 'iron birds' to be found at Thurleigh. The tractor has been identified as an International Farmall H 'Rowcrop' machine built in the USA between 1939 and 1953, registered as WRF 361. With the tractor already possessing a pair of narrow track front wheels, it was ideal for developing hardware attachments and investigating operational details of the system, ironing out any snags as and when they occurred during test runs. These were all dead-load trials, completely unmanned by tractor crew!

The US Navy has developed this principle still further in providing a solid, retractable link, permanently attached to the nose-wheel unit of its high performance aircraft. It is now widely used in carrier operations. The illustrations show a McDonnell Douglas F/A-18F aircraft, the two-seat version of the F/A-18E Super Hornet, being prepared for launch during its development testing at the Naval Air Engineering Station, Lakehurst, New Jersey.

DECK HANDLING

The manoeuvring of aircraft about the flight deck of a carrier – ranging and striking – presented no problem in the quieter periods of operations. When it was brought up from the hangar, each machine would be positioned on the flight deck by a handling crew, latterly assisted by a tractor. Problems arose when the activity became more intense, particularly with propellers turning and rapid launching or recovery in progress. There would also be a gale blowing down the deck with the ship steaming at speed into wind.

An early example of mechanical assistance was the provision of hydraulic wheel chocks at the accelerator loading point. The aircraft

De Havilland Mosquito FB.VI (RS657) appropriately at the corner of Mosquito Way on Farnborough airfield demonstrating a sideways transfer system for deck handling. October 1946. (© Crown Copyright/MOD)

Supermarine Seafire (possibly SX314) on the automatic positioner, BH.5 Catapult. The aircraft – well offset to port – is approaching the powered rollers. The rollers will centralise the machine as it taxies forward up to the tubular bars which guide the undercarriage wheels and will be held against the hydraulic wheel chocks for attachment of holdback and towing bridle. April 1947. (© Crown Copyright/MOD)

would taxi up to the loading point, bringing its wheels hard against the chocks, the trolley system would be attached to the aircraft, the loaded trolley would then be retracted the short distance sternwards to the launching position, the chocks retracted and the launch procedure would follow.

It was difficult to position the aircraft accurately on the centre-line of the accelerator. A lateral tolerance of only a few inches was permissible; if this limit was exceeded the mechanical constraints of the trolley prevented its being attached to the aircraft. Attempts to ease this situation were made, the more successful being the provision of balks of timber slotted into the deck (parallel to and straddling the accelerator track) against which the aircraft wheels would slide and be guided into the correct position. These balks were made smooth and carried a liberal application of grease on the interfaces with the wheel tyres. Care had to be taken by the flight deck officer to direct the pilot into an alignment from which the aircraft could make the correct approach into the wheel guides. A further problem was that repositioning of the guides on the deck was necessary for each aircraft type to be launched.

One attempt towards a solution was the provision of a series of rollers onto which the aircraft would be taxied and centralising would be effected manually. The first effort comprised a large number of needle rollers, about ¾in {18mm} diameter, held within a cage framework with a steel sheet placed over the top, the whole system being free-standing. The aircraft would taxi on to this sheet and then be slewed by manpower into the correct position. The first two or three experiments showed some reasonable chance of success but before long the top plate became deformed through the weight of the aircraft, the rollers started bending and problems in manoeuvring over the welds in the flight deck plating were not easily overcome. Furthermore, this device often fouled the hydraulic chocks to the extent of preventing their functioning.

Later systems used a series of rollers of some 3in {75mm} diameter set into the deck. The outer rollers were inward-rotating, power-operated while those in the centre were free running, the selection of those to be powered being made to suit the wheel spacing of the aircraft to be loaded. This was more successful but made access to the trolley rather hazardous for the handling crew when loading the aircraft. The advent of tail-down launching made this part of the exercise a little easier to handle.

A wholly satisfactory solution had not been achieved by the time the author left the Department (1946) but, eventually, a good working system was developed and adopted for carrier use after extensive shore-based experiments had been carried out at both Farnborough and Bedford. Garstin (#35) describes the system:

Supermarine Seafire F47 (PS948) on the BH.5 Catapult, Jersey Brow, rigged ready for launch. The RAE holdback unit is attached and the 'V' type towing bridle is linked to the deck shuttle. October 1947.
(© Crown Copyright/MOD)

'It is necessary, in order to launch an aircraft by means of a catapult, to position it laterally in relation to the catapult track with precision, greater than the taxying accuracy of the average pilot. ... A device known as a roller mat positioner was therefore developed to do this job mechanically. The positioner consists of two frameworks, set either side of the catapult track, in the loading area, in which are mounted rollers, with their axes parallel to each other and to the track. The tops of the rollers are flush with the deck and their surfaces are fluted, to improve grip on the tyres. These rollers can be rotated so as to produce translation (of the aircraft) towards the catapult track. In addition, they can be clutched into or out of, or released from, the driving mechanism. If the pitch between driven rollers on either side of the track centre line is selected to be the same as the track of the aircraft main wheels, then an aircraft coming up off centre will be fleeted sideways by the driven rollers on the side to which it is off centre, the other main wheel moving freely on (intermediate) free rollers, until it is accurately positioned. At the same time, the wheels come to rest against chocks, to make sure the aircraft is aligned to the catapult.'

Eventually a measure of automation was introduced in which the launching bridle and holdback were fitted in the hangar in 'slow time' and carried to the accelerator by the aircraft. They were dropped into position by the release of lanyards by the deck handling party.

HMS *Eagle*[51], in 1952, was the first carrier to be equipped with all these post-war mod cons – a full set of automatic loading gear. Garstin (*#244*) lists them as:

a) Automatic loading positioner
b) Roller-faced loading chocks
c) Plain-faced safety chocks
d) Resilient automatic holdback loader
e) Bridle engaging gear
f) Forward tensioning
g) Fittings on the aircraft to support the bridle and holdback.

With the arrival of jet engines the question of heating of the deck plating by the exhaust efflux needed to be resolved. After a limited series of tests on Jersey Brow, using a 'Whittle' W2B/23 engine fixed in a frame, Kell (*#264*) reported a significant temperature rise over a long period which could be countered by spraying the deck with water. In service this heating effect was found not to be of great importance. One feature that did assume prominence in later years was the erection of jet

screens immediately abaft the accelerator launching positions. These screens would deflect upwards the full blast of jet engines, together with reheat at take-off conditions. The examination of this problem at Bedford produced a design of mechanically retractable, water-cooled, blast deflectors.

For the deck recovery of aircraft there were no problems of this nature; all movements about the flight deck during landing were made by the aircraft taxying under the direction of the flight deck officer. In the event of a bad landing in which an aircraft was damaged, the flight deck rescue party would be brought into action and often the use of 'Jumbo' – the Ransomes & Rapier flight deck mobile crane – was the most effective way of dealing with the situation.

Reference is made elsewhere to the deck handling of aircraft not fitted with folding wings, for example Hawker Sea Hurricane and early marks of the Supermarine Seafire. These aircraft were loaded on to small trolleys with castoring wheels which were guided by steel battens welded to the deck on the lifts and in the hangars. The disadvantage of this was the intensity of man-effort required to load each aircraft on to its handling trolley and traverse it sideways; fortunately this was a short-term expedient.

Grumman Avenger III (KE436) in the loaded position on the BH.5 Catapult with holdback engaged, bridle tensioned and mechanical wheel chocks stowed, all ready for launching. September 1949.
(© Crown Copyright/MOD)

De Havilland Sea Hornet PR.22 (VW930) on the BH.5 Catapult on Jersey Brow. It is centred on the mechanical loader system with its bridle lying across the accelerator track. March 1949. (© Crown Copyright/MOD)

Grumman Avenger III (KE436), showing the RAE holdback unit connected to the tail wheel assembly for automatic loading onto the BH.5 Catapult. September 1949. (© Crown Copyright/MOD)

LAUNCHING PROCEDURE

It may be opportune at this point to introduce the methodology of the process of launching from an accelerator. The aircraft will have been ranged on the after portion of the flight deck and, on the pilot receiving a signal from a nearby flight deck officer, would taxi forward. The aircraft would be held on the centre line of the deck some ten yards {9m} to the rear of the accelerator loading point. It would be signalled forward and directed by the catapult (sic) officer on to the accelerator-loading chocks. At this point the catapult officer closely watches the loading of the aircraft on to the trolley or towing unit. The catapult officer then takes up position forward of the aircraft and on the starboard side in full view of the pilot. He will be holding a small green flag in his right hand and a red flag in his left. By this time the ship will have turned into wind and all flight deck personnel will be 'leaning into the wind' as half a gale blows along the length of the deck.

When the catapult officer is satisfied that the aircraft is loaded correctly – all with the engine running – he will await a signal from the catapult operator that he in turn is satisfied with the operational state of readiness with the launching gear. At this point the catapult officer will hide his left hand, holding the red flag, behind his back; he will then raise his right hand and rotate the green flag above his head. With this signal the pilot will increase the engine speed of the aircraft until take-off conditions are achieved. He will hold the aircraft throttle in his left hand with his elbow hard fast against a stop (to prevent inertia forces from closing the throttle during the launch), raise his right hand above his head to indicate that he is ready to be launched, and then smartly drop it to take hold of the control column. Three seconds after the pilot drops his hand, the catapult officer will sharply lower his right hand (with green flag) to the deck level. On seeing this signal, the accelerator operator will depress a pedal to unlock the firing lever and move the lever to its launch position, at which point the launching of the aircraft is initiated.

If, during this sequence, a hang-up of any sort occurs immediately prior to a launch, the catapult officer – if his right hand is already raised – will hold the green flag still, then slowly raise his left hand (with the red flag) to the same level and cross the flags. The pilot accepts this as a sign to throttle back the engine to idling conditions; the accelerator operator removes the firing control lever from its socket and lays it on the deck in view of the catapult officer. This is a sign that all equipment has been de-energised and brought to a safe condition. The catapult officer then slowly lowers both flags together, following which the cause of the hang-up may be investigated and appropriate action taken.

This procedure was meticulously followed on board ship and at all launches undertaken at Farnborough. During World War 2 there was no voice communication between the flight deck party and aircrew – all signalling was visual. This arrangement quickly improved in the late 1940s.

Comments from a couple of pilots on catapulting are apposite here. Quoting from Cronin (#23 p.76), this is a comment from Cdr Peter Bethell RN, an FAA pilot of the early 1930s:

'Personal experience of being catapulted seems too dull to relate. It seemed prudent to hold the control column in a normal position and to hope for the best, not forgetting a brief prayer to St Lawrence, Deacon of Rome, who was broiled on a gridiron in AD 258.'

(So that is how 'Winkle' Brown and his colleagues managed it!)

The author's preference is the account by Wellham (#106 p.29), one of the 'Taranto' pilots and a contemporary of 'Winkle' Brown, who writes more convincingly of his late-1930s experience in being launched from the training catapult at HMS St Vincent, RNAS Gosport:

'Another frolic to which we were introduced was being catapulted into the air. This lengthy and complicated process involved the Swordfish being lifted by crane and lowered, with a great deal of shouting and pushing, on to a trolley which ran on rails along the top of a structure resembling the jib of a dockyard crane. This device was fitted on a turntable to allow it to be swung into wind. On the inboard end of the jib was a breech like that of a large gun, into which was inserted a sausage-shaped paper bag containing explosive, which we were told was used to fire the shell of a battleship's 15 inch gun. When the aircraft was adjusted on the trolley with its catapult points engaging in the hooks provided, the engine was started and the pilot ran it up to take-off revolutions and locked the throttle, laid his head against the head-rest, put his left hand on the stick and raised his right; when ready in all respects he dropped his hand, whereupon the cartridge was fired.

'There was then a violent explosion and the trolley, complete with aircraft, was hurled along the rail until, on reaching the end, it stopped dead, leaving the Swordfish to shoot off into the air with the pilot shaking the stars out of his eyes and attempting to gain control.'

Caspar John, in the biography written by his daughter, Rebecca (#52 p.111), speaks of his experience in HMS Exeter in 1932:

'She carried two seaplanes, mounted on catapults. The aircraft, held in place by the cradle, was wound back to one end of the cata-pult; the pilot then gave full power to the engine, and signalled the torpedo officer, who dropped a flag. The cordite charge which followed threw the machine into the air with tremendous force just above the stalling speed. If the charge failed, the aircraft tumbled into the sea. At the end of a flight, the seaplane was hoisted aboard by a crane.'

Safety and rescue procedures were implemented at all times when shipboard flying was in progress, the rescue party being stationed alongside the island. At Farnborough, the presence of such teams was always in evidence when the airfield was in operation. One team moved up to the vicinity of the launching base or arrester deck when either of these facilities were in use. Adjacent to the flying operations a crash and rescue tender – basically a fire engine – was positioned, and prominent among its crew was what the navy called a 'firesuit-man'. He was a fireman, fully clad in an asbestos suit, whose duty it was to enter an area where a crash had generated into a fire and to attempt the rescue of the aircrew. Often he had a mate similarly attired who carried a foam nozzle into the fire area to douse the flames, hopefully to the point of extinguishment. Other colleagues would direct a spray hose on to the firesuit-men to keep their suits cool. (Needless to say, this was an era when the health hazards of asbestos were not realised!) These rescue parties, operationally, were always at key pitch and undertook their hazardous duties well.

THE 'C' (CORDITE) ACCELERATOR – (BC.I/II)

For the first deck-level, accelerator launching system to be installed at RAE the hydro-pneumatic power unit, standard on board ship, was not adopted. Naval Aircraft Department, in conjunction with the ballistics people at the Royal Arsenal, Woolwich, had developed a cordite operated 'gun' as a power unit. At the outset, this equipment was to be given the title of CATOG – Cordite Assisted Take-Off Gear – but this was soon changed to the C.I Accelerator. This equipment was designed to be simple in concept, manufacture and operation and formed the prototype for an accelerator (BC.II Accelerator) in the new trials carrier, HMS *Pretoria Castle*. The installation at Farnborough was a twin-track system quickly adaptable for use with a BH.III type twin trolley or a deck shuttle for tail-down launching. The twin trolley was a development of the BH.III system in that the port and starboard spool attachments were separate units. The total length of the BC.I trolley was greater than the BH.III system, with an overall dimension of 28 feet {9m}. For tail-down launching, a pair of deck shuttles could be fitted independently to the deck gear or a single deck shuttle mounted on a bridge spanning both tracks could be fitted. The alternatives would require different lay-ups of bridle for the same aircraft.

The under-deck machinery installation at RAE was housed in a covered pit on the Jersey Brow site. The 'gun' was installed horizontally in this pit (its shipboard fellow was mounted vertically) and comprised a single ram within a cylinder carrying a multiple-sheaved system of pulleys at each extremity with a pair of wire ropes reeved through the system; again a jigger-type unit.

A rapid-action, sideways-operating breech unit was provided at the closed end of the cylinder. Into this a shellcase of about 12 inches {300mm} diameter, containing the appropriate cordite charge and percussion primer, would be loaded. The products of combustion from the explosive – as a gas generator – would extend the piston and the rope would pull a system of sliders in a pair of tracks on the surface of the ground above. The sizes and details of the various cordite charges were designed by Woolwich Arsenal and verified by trial. The maximum weight of propellant charge for the heaviest aircraft to be launched (14,000lb {6,400kg}) was around 18lb {8kg}. Prior to a day's launch programme, a dummy run of the deck machinery was made using a light charge of less than three pounds {1·4kg} in weight.

The cordite charges were made up immediately prior to a day's programme of launches, a small cordite-preparation magazine building being provided on Jersey Brow. The making up of the charges for a day's launch was an involved process using a variety of forms of cordite arranged around a central primer in specific formats to obtain the required control of the burning rate of the charge. The ready-to-use charges, silk-laced in their cartridge cases, were held in a steel cabinet

Fairey Albacore I (BF658) ranged on the C.I Accelerator on Jersey Brow. This accelerator was the first to adopt the twin-track system with modified BH.III trolley units. October 1942. (© Crown Copyright/MOD)

The C.I trolley seen towards the end of its launching stroke. This was the first twin-track trolley used for this system and it shows the adapted BH.III front and rear legs for supporting the aircraft. Normally the legs would have been collapsed by this stage, with the towing shackle and ropes well forward. The retarding spike on the centre line of the trolley is about to penetrate the frangible water-retaining disc in the above-deck hydraulic barrel, to bring the trolley to a halt. October 1942. (© Crown Copyright/MOD)

Hawker Sea Hurricane I (Z7082) loaded on the twin-track trolley of the C.I Accelerator, ready for launch. If this picture is compared with that of the Albacore, the adaptability of the support legs may be realised. February 1943. (© Crown Copyright/MOD)

alongside the loading breech. A formal drill was adopted in which the charge was loaded into the breech chamber, the storage cabinet closed, the primer fitted and the breech locked before the 'ready' signal was given. On completion of a launch, the breech was opened, the cartridge case partly ejected and removed by gloved hands, then dunked – with a plop and a sizzle – into an adjacent cold water tank.

The above-deck machinery was designed to accommodate any type of trolley or towing hook likely to be required. As there was no tail-rope braking system, the moving parts were retarded at the end of each launch stroke by means of a group of simple hydraulic devices. These devices in reality are 'spike-and-barrel' systems. (In hydraulic parlance, the terms 'ram and cylinder' generally refer to a composite pair of precision machined components as an integral unit. The cylinder forms the static component and the ram the moving element, typically as in a jigger system.) The C.I Accelerator was the first to adopt this new braking system, one unit each for the deck trolley and under-deck jigger.

Grumman Tarpon/Avenger I (FN792) ready for tail-down launching from the C.I Accelerator. The early stages in the proof testing of the K.I catapult are seen in the foreground. June 1943. (© Crown Copyright/MOD)

Blackburn Firebrand TF.4 (EK670) on the C.I Accelerator. This shows the twin bridle set-up for launching and its relationship to the dummy 18in torpedo with its air-tail attachment. The experiment was to determine the extent of interference – if any – between the bridles and the torpedo air-tail unit at the moment of release from the accelerator shuttle. December 1946. (© Crown Copyright/MOD)

This 'spike-and-barrel' system comprised a pair of normally separate components. The barrel was a cylindrical unit in which the entry was machined to take a removable nozzle; the bore could remain in rough condition. This removable nozzle was of bronze (or brass) with an orifice, itself accurately machined to a simple profile, screwed into the cylinder against a shoulder with the mating faces designed to accommodate a blank, frangible disc – often made of 'Hallite', an engine gasket material. A couple of plugs were fitted to the top surface of the barrel to permit its filling with water and venting.

The spike was a tapered steel plunger, machined accurately to a calculated profile, designed to pass through the orifice of the barrel and, after puncturing the retaining disc, expelling the water in the barrel through the annulus between spike and nozzle. The gradual taper on the spike caused the annular space to become progressively smaller as it passed through the orifice in the nozzle, the resistance of the water in its expulsion through the annulus creating the retarding effort. The principle of operation was straightforward, the machining of replacement components presented no complications, and the servicing and operational techniques were simple. The initial setting-up needed careful attention to ensure that the spike was presented cleanly to the barrel but, with adequate clearance at the tip of the spike, problems were virtually non-existent.

At the end of each run, with the retraction of the trolley, the spike and barrel would be parted, the nozzle unscrewed for a new frangible disc to be inserted, the nozzle replaced and the barrel refilled with water. On each occasion the nozzle would be inspected for signs of mechanical damage through impact by the spike and for evidence of 'wire-drawing' – the phenomenon associated with the passage of fluids at high speed through narrow gaps. There is no need to state that, after a number of launches, the environment around these units became rather damp.

In its original format the C.I Accelerator, although a twin-track system, used a single 'spike-and-barrel' braking system, located above deck. The two trolley elements were connected by a bridging piece which carried a single spike. This was fine for aircraft supported on a trolley. With tail-down launching the centre line barrel became an obstruction for the aircraft tail wheel. In the remodelling of both systems – at Farnborough and aboard HMS *Pretoria Castle* – a pair of spike-and-barrel units were installed in line with the tracks, thereby creating a clear path for both tail and nose-wheel undercarriages.

This spike-and-barrel system was employed on a number of projects at Farnborough. It was universal in that either the spike or the barrel

The second prototype de Havilland Sea Hornet F.20 (PX214) is set up for launching from the C.I Accelerator on Jersey Brow. The twin towing bridles, and the holdback unit at the tail are seen. The chequered bar alongside the port wheel is for photographic purposes. June 1946.
(© Crown Copyright/MOD)

could form the moving component. Obviously the length of the composite unit dictated the braking pressures and forces applied. Generally the braking distance was between two and six feet {0·6m, 1·8m}, producing retardation values across the range of 30g-120g, depending, of course, upon the speed of entry. The spike diameter varied between three and eight inches {75mm, 200mm}. Control of the retardation forces could not be achieved within fine limits, but this was of little consequence within the applications for which the systems were adopted. With the C.I Accelerator such retardation units were required at both the under-deck machinery position and on the deck surface for the trolley gear.

The control position was in a small pit abeam to the port side of the accelerator, simulating the shipboard location. As for all catapult/accelerator launches, a standard procedural drill, strictly following the naval pattern, was adopted and meticulously followed. This accelerator quickly became the standard workhorse for the majority of launches at Farnborough, its main advantage being the rapidity with which it could be brought into operation and the ability to adjust its scantlings to accommodate all types of FAA aircraft.

On one occasion, October 1942, a number of Fairey Albacore aircraft from RNAS 827, disembarked from HMS *Indomitable*, arrived at Farnborough for a series of rapid-launching tests using this accelerator. This proved to be a successful trial; Caunter (*#222 p.133*) records seventeen launches having been made within twenty minutes. (This figure is more likely to have been seven launches within twenty minutes!) A photograph shows one of these aircraft (BF658) in the four-point, tail-up launching position on the trolley. This particular machine, on 9 October, undertook eight flights from the accelerator, W/Cdr H. J. Wilson and Lt K. J. Robertson RN sharing the piloting. A couple of months later, on 15 December, a Fairey Barracuda (P9644) completed 11 accelerated launches within one day's flying, Lt Robertson again being the pilot.

When the conversion of the Union Castle liner RMS *Pretoria Castle* into a trials carrier was being considered, the question of providing a launcher was raised. With the success of the elementary trolley for the 'P' (rocket) catapult having been established, some of the Admiralty engineers suggested the adoption of similar rocket-propelled trolleys for use in the new carrier, having one or more trolleys provided for each marque of aircraft. The aircraft would be loaded onto the trolley in the hangar and brought up to the flight deck ready for launching. The only flight deck equipment would be a pair of rails as track for the trolley, with a buffer-stop at the end. Apart from the multiplicity of trolley sizes involved, the question of loading and handling, together with the fitting of rockets below deck, posed too many problems for the scheme to proceed.

RAE demonstrated that the success of the BC.I cordite accelerator on Jersey Brow would indicate a similar installation for HMS *Pretoria Castle* to be adopted. This proceeded with the shipboard BC.II under-

Hawker Sea Fury (SR666 second prototype) positioned on the C.I Accelerator. The 'Y' shaped bridle for launching from a twin-track trolley/shuttle is readily seen. July 1946. (© Crown Copyright/MOD)

deck gear being installed vertically within a trunk of some eight feet {2·4m} diameter, dropping through the decks in the bows of the ship to an extent of 36ft {11m}, rather like an enlarged anchor chain locker. (An interesting feature in this installation was that the retarding barrel was the stationary element located at the bottom end of the equipment with the spike element of the jigger descending into it. In this situation, the barrel did not require a water-retaining disc and reminded one forcibly of an over-full WC toilet pan. This impression obviously had appealed to the dockyard engineering staff who provided a header tank, an enlarged version of the standard two-gallon flushing cistern, alongside to refill the barrel after each firing.)

The operational capacity of the BC.II accelerator in HMS *Pretoria Castle* matched the BH.III of that period with the ability to launch aircraft of 14,000lb AUW at 66kts {6,400kg, 34m/sec}. Friedman (*#34 p.186*) states that other auxiliary carriers, HMSs *Nairana* and *Vindex*, although smaller than HMS *Pretoria Castle*, were also fitted with the BC.II accelerator. The author challenges this. He cannot recall any trials having been held by NAD on either of these two ships; neither has he found any reference to this elsewhere. Hobbs (*#45 pp.139&201*) states that neither ship was fitted with an accelerator. Poolman (*#75*), in his detailed record of the wartime activities of HMS *Vindex*, has no reference whatever to an accelerator in this ship. Photographs of the flight deck offer no evidence of such an installation. In the author's opinion, the shipboard installation of a BC.II unit in HMS *Pretoria Castle* was unique.

THE 'K' (INERTIA) ACCELERATOR – (BK.I/II)

During the early months of World War 2 much RAE effort was devoted to upgrading the performance of existing aircraft in squadron service, both for the FAA and the RAF. One of the fields for investigation was improvement in take-off, particularly as few, if any, airfields had hard surfaced runways. This was addressed by Tye and Fagg (#305) in their paper of 1940 which reviewed various methods for getting aircraft off the ground. One of the important factors was to increase the ratio of 'aircraft thrust/aircraft weight' available during the take-off run.

The authors proposed four fields of investigation, indicating that some months of development might have been necessary for this. The suggestions were:

1. a second wingless aircraft as a tug
2. tugs of the tractor or motor-car type

The K.I Catapult track (first stage) as it appeared in the summer of 1943. This was the above-deck phase with a basic, non-adjustable trolley similar to those built for the 'P' Catapults. The view is of the braking end of the catapult track, showing the trolley approaching the 'spike-and-barrel' retarding unit. July 1943. (© Crown Copyright/MOD)

3. winding gear, similar to that used for lifting the cages in coal mines
4. rocket propulsion.

For system (1) the authors had calculated that a Handley Page Hampden aircraft (two Bristol Pegasus, nine-cylinder radial engines), devoid of its outer wings and military load, weighing 11,000lb {5,000kg} would be adequate to tow an Avro Manchester (two Rolls-Royce Vulture, 24-cylinder 'X' type in-line engines) at 45,000lb AUW at 100mph to a take-off point 650 yards {20,400kg, 44m/sec, 590m} from the start of the run – a free take-off run being some 800 yards {730m}. The authors continued by saying that system (2) was unlikely to offer any advantage over (1). Both methods would require a tow rope of some 150 yards {140m} in length. After the towed aircraft had taken off, a further run of some 150 yards {140m} would be required to bring the tug to rest. To the bystander this would have been hilarious. To see a tug, be it emasculated aircraft or tractor, careering along at the end of a grass airfield at 100mph {44m/sec} (seemingly being pursued by a heavy bomber) would have been spectacular, to say the least. The authors continued to write, perhaps rather wistfully, that:

'... scheme (1) seemed preferable to (2) but without doubt an airscrew-driven tug designed specifically for the purpose would be even better than a converted aeroplane.'

These two schemes were finally dismissed by the thought that:

'... although the ground run of the towed aircraft would be reduced, the total aerodrome space required may be actually increased!'

In discussing scheme (3) the use of winding gear was proposed but it is obvious that the context was from the mining industry. Against the proposal was the time required to manufacture the equipment, with a further disadvantage being its 'vulnerability'. Rocket propulsion for scheme (4) was considered but such an application was very much in its infancy. Some doubts were expressed although development seemed feasible.

The authors appended a further system using what was termed as the RAE Trolley. This was a look at the development on the western edge of Cove Common known as the Direction Controlled Take-Off system (qv) undergoing installation at the time and two years ahead of the actual launch from this system. Again the disadvantage of vulnerability was expressed. Their final selection was for either the winding gear or rocketry.

An addendum accompanies Tye's paper (#ibid.) bearing the title *Assisted Take-Off by Means of a Flywheel*. The principle of this method was to:

'... use the stored energy of a large freely-rotating flywheel, the energy being transferred to the aeroplane by means of a tow rope which wraps round the flywheel, and this can be arranged by making the flywheel in the shape of a cone.'

The authors suggested such an installation be designed to assist the take-off of an aircraft of 60,000lb AUW at a terminal speed of 150ft/sec within 350 yards {27,200kg, 46m/sec, 320m}. The machinery to attain this performance would be a flywheel of some 10 tons mass running at 600rpm. This proposal surely must have been triggered by awareness of the scheme for the 'Inertia Catapult' (the 'K' Catapult) the installation of which began on Jersey Brow a few months later.

The Naval Aircraft Department was already investigating types of prime mover for powering catapults and accelerators other than the prevailing hydro-pneumatic machinery. One of these was to adopt a system withdrawing kinetic energy ('K') from a suitable source to propel a deck trolley, the most favoured being an installation in which the energy stored in a flywheel could be usefully extracted. Brown (#14 p.52) makes an interesting reference to a single inertia-type (electrically driven flywheel) catapult having been installed in the American carriers, USSs *Lexington* and *Saratoga*. These were located to starboard of the centre line of the flight deck but could not have given good service as they were removed in about 1931. It was thought that NAD was unaware of this project when the 'K' Catapult was under consideration.

Calculations indicated the flywheel system to be a feasible proposition and its development was thought not to offer any difficult points for solution. The scantlings would not be as demanding as for launching a heavy bomber from an airfield. The design parameters for an installation on Jersey Brow would be for equipment suitable for placing on shipboard to launch an aircraft of about 11,000lb AUW at 66 knots {5,000kg, 34m/sec}.

The design was based upon a flywheel of seven feet {2.1m} diameter, having a mass of 2.8 tons and spinning at 2,100rpm as the power source. This flywheel was brought up to operational speed by a petrol engine, similar to those used extensively at that time to power buses. The drive was through a fluid coupling which allowed slip (free-wheeling) between the two – drive and driven – components. The flywheel itself was enclosed within a sheet steel casing to reduce windage drag, thereby providing an effective safety guard. The drive from the flywheel was taken through a Sinclair 'synchro-self-shifting', scoop-controlled fluid coupling to a pair of rope drums. These grooved drums reeled in a pair of ropes taken forward via pulleys and attached to the leading end of the launching trolley (latterly the deck shuttle). The core of the control system was the fluid coupling. The depth of immersion of the adjustable scoop into the annular layer of the fluid whirling within the rim of the coupling through centrifugal force controlled the quantity of fluid being

K.I Catapult machinery pit. The casing to the large flywheel is seen in the centre, the hydraulic coupling is next, with the twin rope-winding drums in the foreground. At this time the flywheel was brought up to speed by a 'Meadows' bus engine, driving through a fluid coupling, both on the far side of the flywheel. July 1943. (© Crown Copyright/MOD)

Fairey Fulmar I (N1925) loaded onto the initial version of the K.I 'Inertia' Catapult. 'Pincher' Martin is holding the flag (for what purpose is not known) with Stan Chisman on his right. The chap in the control pit is probably Norman Skelton, the usual 'button-pushing operator' for the Jersey Brow launchers. August 1943. (© Crown Copyright/MOD)

transferred across to the driven element of the coupling, and thereby the amount of power transmitted through the coupling. The adoption of a fluid coupling enabled a freewheeling and over-run facility to be incorporated in the launching system.

The project initially was known as the Inertia Catapult. While metal was being cut in the RAE, its name 'Inertia' became adapted phonetically by those in the work-force to 'Errsher' Catapult. The matter was simplified when it was decided to adopt the symbol 'K', in engineering parlance as representing kinetic energy.

The system originally comprised an above-deck track of rolled steel channels, accommodating a basic, non-adjustable trolley similar to those used on the P.I catapult. The trolley was brought to rest by a spike-and-barrel system with the water barrel forming a constituent part of the trolley. Operation of this equipment began in the summer of 1943. After successful launches of a Fairey Fulmar aircraft (N1925), the catapult was modified the following summer for tail-down bridle launching. All the top-hamper was removed and the track converted to an in-deck rail accelerator system. This was a twin-track system in which the shuttle at the end of the launching stroke was brought to rest again by the basic 'spike-and-barrel' system. In this instance the barrels were affixed to the shuttle, impacting static spikes which penetrated frangible discs fronting the barrels. This provided the braking effort also to the under-deck rope drums.

Later (1945) the installation was upgraded to become the K.II Accelerator when an electric motor replaced the petrol engine. The unit continued to give useful service as a facility on Jersey Brow alongside its neighbour, the C.I Accelerator. This type of inertia launcher was interesting but, in spite of its inherent simplicity in design and operation, it was not adopted for shipborne use. Problems may have been anticipated from the gyroscopic interactions (precession) between the flywheel and its bearings resulting from the wave-induced motion upon

a ship at sea. Also, the system was not capable of being quickly re-charged with energy this would 'put up a black' against the service requirement for rapid launching of a series of aircraft. However, it became a useful adjunct to the equipment on Jersey Brow for many years.

Boddington (#209) in 1945 was investigating some of the problems associated with increasing the end-of-launch speed from accelerators to 130 knots. The flywheel system at Farnborough, which had already shown signs of promise, would provide a basis for further investigation. Construction of an experimental flywheel accelerator had been approved for a performance envelope of 30,000lb at 75 knots {13,600kg, 38m/sec} and 3¼g. In July this was revised to 12,500lb at 130 knots {5,700kg, 66m/sec} and 5g. Boddington (#ibid.) suggested that a system adopting an endless chain drive for the tail-down shuttle should be examined. It is not clear how this was expected to operate; needless to say, the scheme was abandoned without further experimentation. One reference states there to have been both a K.III and K.IV Inertia Catapult but no evidence exists for either of these having been built. This project may have been the intended K.III machine.

Early in 1946 correspondence was exchanged between E-in-C's Department at the Admiralty and Power Jets Limited on the subject of employing gas turbines as power units for a flywheel catapult. The people involved were Cdr Colin C. Mitchell (RN) and A/Cdre Frank Whittle (RAF) respectively, both of whose names featured in Percy Salmon's proposals of May 1939 (#283), detailed in an earlier chapter. It was thought that with 'a requirement for driving an aircraft carrier acceler-ator, it seems that a gas turbine drive should be considered'. This proposal may well have been the basis for the proposed K.IV Accelerator.

Mitchell's specification called for a prime mover to be coupled permanently to a flywheel, and energy to be taken from the system during the launch over a period of two seconds by clutching the acceleration mechanism to the flywheel, and declutching when the launch was complete. (This was the basis of the Jersey Brow installation.) Thereafter, the prime mover was to restore, within fifteen seconds, the kinetic energy taken from the flywheel, and on completion to maintain the flywheel at constant speed for 13seconds, after which the cycle would be repeated. A load/time diagram was included with the specification based upon the following assumptions:

Governed idling speed of flywheel	2,500rpm
Drop in speed during launch	500rpm
Power required when idling	500hp {370kW}
Power output when restoring speed	6,000hp {4,500kW}

The specification was sent to Whittle with the statement that it had occurred to its compiler that some existing design of jet unit may be available to provide the required power source and that the design and development of the turbine and gearing would not present great difficulty. Whittle's response was that these conditions could be met by the adoption of four gas turbines; the Admiralty thought that only two would be needed. This was at a time when the feasibility of a power take-off from a jet engine was at only an embryonic stage. Bowden and Jefferson, in their paper published by the Institution of Mechanical Engineers in June 1948 (#109), gives the date of the initial run of the first base-load gas turbine to be manufactured in the UK, by C. A. Parsons & Company, to have been in December 1945. This was only a month before the issue of the Admiralty specification. The Parsons machine was designed for a shaft output of 500hp {370kW} at 8,000rpm its recorded performance was the delivery of 340hp {250kW} at 6,100rpm.

Thomlinson (#292) records post-World War 2 research having revealed that Germany had successfully launched a 30,000lb aircraft at 110mph within a distance of 330ft {13,600kg, 49m/sec, 100m} using a 'portable' flywheel catapult mounted on a large road vehicle. This too had proceeded no further than a research exercise.

The 'K' Accelerator (in its K.II format, powered by an electric motor) on Jersey Brow continued in use for many years without undergoing further major modification.

THE 'P' (PROJECTILE / PYROTECHNIC) CATAPULT

During the Battle of the Atlantic an urgent requirement arose for air cover to be provided over the central stretch of the ocean, beyond the range of land-based fighter aircraft. One proposal, submitted from an unidentified source, was for a long-range Consolidated B24 Liberator aircraft to be adapted to carry either a Hurricane or Spitfire aircraft suspended in a redesigned bomb bay (a major exercise in itself) to be released for engagement when an enemy aircraft approached the convoy. From the first (apart from the logistics involved in the provision of such a facility) this was fundamentally impractical as the Liberator could not be guaranteed to be in the vicinity of the convoy when the enemy aircraft was around. It also would be a valid target itself for the raiding Focke-Wulf Condor.

The RAE proposal was for a selection of merchant ships, which normally sailed in convoy, to be each fitted with a catapult carrying a fighter aircraft to be launched when needed. At the end of its sortie the aircraft would be abandoned near to its host ship, or an escort destroyer, with the pilot either ditching the aircraft or parachuting into the sea for recovery from his life-saving dinghy. It was thought that the sacrifice of a single aircraft at the end of its one-way trip was preferable to the loss of one or more merchant ships.

Thomlinson (#293) quotes the requirements for such a catapult as the ability to launch an aircraft of 10,000lb AUW at 65kts within a stroke of 70ft {4,500kg, 33m/sec, 21m} to represent the smallest weight of equipment having the simplest of installation features. Briefly, the aircraft, in the wheels-up configuration, was to be carried on a trolley running in a simple track. Mounted on the base of the trolley would be a battery of rocket motors to propel the trolley and the aircraft. At the end of the runway track the trolley would be arrested, or buffered, releasing the aircraft for its onward motion at take-off speed plus. The aircraft was to be mounted on the catapult while the ship was in harbour and remain in the launching position until the need arose for its sortie. All maintenance and servicing would be undertaken with the aircraft in this position. The trolley was to be as light as possible because rocket effort was required to accelerate the trolley as well as the aircraft.

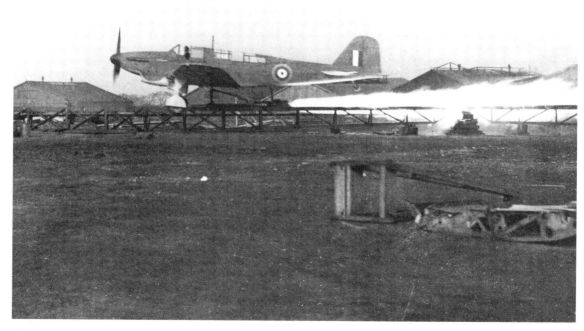

The initial launch of the Stage 1 rocket-fired P.I Catapult on Jersey Brow. The catapult beam was rapidly cobbled together using two redundant girders from a previous catapult and erected on old railway sleepers for this demonstration. The trolley was one of three made in the RAE workshops for launching respectively Fairey Fulmar, Hawker Hurricane and Supermarine Seafire aircraft. January 1941. (© Crown Copyright/MOD)

Lightness would be a bonus in that the retarding effort applied to the trolley at the end of its stroke would be kept low. From its inception, this equipment was known as the RATOG (Rocket Assisted Take-Off Gear) but its title was quickly usurped by the system of launching aircraft by rocket cradles attached to the aircraft itself and subsequently became its permanent name. The less prosaic title of 'P Catapult' was substituted and either 'Projectile' or 'Pyrotechnic' were used as names. The reader may make his choice.

A design was quickly prepared for a rocket-propelled trolley to carry a catapult-spooled aircraft. This was built from standard structural steel sections (angle iron and the like) to be driven by a package of 3in UP (Unrotating Projectile) {76mm} rocket motors. (At this time the existence of such devices as rocket motors and projectiles was a closely guarded secret.) Thetford (*#97 p.145*) records the first firing of a UP missile (as a weapon) having been made from a Hawker Hurricane on 25 October 1941, some nine months after the first use of rocket motors on this catapult.

First of all, within a fortnight, an experimental ballast trolley was built from scratch in RAE workshops for running in the starboard rail of the redundant DCTO track. This was loaded to 10,675lb {4,840kg} weight and charged with 18 3in rocket motors. Test runs were successful, the concept was established and incorporated in the production trolleys. These initial trials formed the basic parameters for the High Speed Track (HST) which is discussed elsewhere in the text.

Initially, three trolleys were made, one each for specific use with Fairey Fulmar, Hawker Hurricane and Supermarine Seafire aircraft. The first two trolleys were used for operational demonstration and crew training purposes and were rigged as appropriate for the particular aircraft to be used. That for the Seafire was built solely for RAE use as it was thought that this aircraft marque would not be embarked on any catapult aircraft ship, primarily because the ditching characteristics (emergency alighting on water) of the aircraft were somewhat 'dicey'. Fortunately, on Jersey Brow there were two surplus lattice framework catapult runway structures remaining from the 'Larynx' project of the 1930s. These were retrieved from storage, cleaned up and joined nose-to-nose to form a runway structure of some 80ft {24m} in length. This was mounted at ground level on old railway sleepers.

The first trolley was designed to run in the tracks of the old structure and was used for trials with a Fairey Fulmar aircraft (N4016). The trolley, designed to carry a maximum of 18 rocket 'motors' in two banks

of nine, was made in the RAE workshops within a very short time and trials began on Jersey Brow. (It was rumoured that the raw material for manufacturing the first trolley was obtained from structural steelwork recovered from the bomb damage to Building [Q.17] incurred during the daylight raid on RAE on 16 August 1940.) The trolley weighed less than half a ton {<500kg}, comfortably achieving the requirements of the design specification. The method of firing the rockets was based on a bank of 12 to be fired to start the launch, three more about half a second later, and the final three a half second later still. The actual number of rockets fired in the first bank would depend upon the weight of the aircraft to be launched and the environmental temperature, fewer rockets being required if the ambient temperature was relatively high. A simplified detent system was used to restrain the aircraft against its engine thrust, and the trolley itself was held by a weak link similar in

A rear view of the first 'production' P.I Catapult trolley – that for the Fairey Fulmar aircraft – in the RAE workshops. March 1941.
(© Crown Copyright/MOD)

Supermarine Seafire IIC (MA970) on the P.I trolley ready for launching. The port retardation spike and the detent system to the front hooks for restraining the aircraft during engine run-up are readily distinguishable. June 1942. (© Crown Copyright/MOD)

An experimental ballast trolley for testing the rocket propulsion system for the prototype P.I Catapult. This was the first installation in which the rocket motors of a new 'secret weapon' had been adopted. January 1941.
(© Crown Copyright/MOD)

Fairey Fulmar I (N4016) loaded on the P.I Catapult beam in its temporary, low-level location. October 1941. (© Crown Copyright/MOD)

shape to a metallurgical tensile test specimen. The hooks on the trolley, forming the aircraft interface, were rigid and did not collapse at the end of the launch. The trolley was retarded at the end of its track by the 'spike-and-barrel' system, similar to that used on the 'C.I' type accelerator. This comprised a pair of tapered spikes attached to the trolley penetrating water-filled barrels fixed to the runway structure. The normal braking effect on the trolley was equivalent to some 40g. Work began on this project within RAE in December 1940 and the first launch of a Fairey Fulmar aircraft took place 25 days later, on 17 January 1941, with a second flight the following day, Sqn/Ldr Wilson being the pilot. Almost immediately after, a heavy snowfall prevented flying from Farnborough for almost a week.

Fairey Fulmar I (N4016) on the P.I Catapult beam at ground level. This shows the disposition of fifteen, 3in rocket motors (three spaces are left blank) for propelling the trolley. October 1941. (© Crown Copyright/MOD)

From the first the experiments were highly successful, accelerometer records indicated a satisfactory build-up to a near-constant rate of acceleration of around 2½g with a peak value approaching 3½g. The aircraft exited the trolley cleanly and the braking of the trolley at the end of the launch presented no problems. The situation was favourable for contracts to be awarded for a number of production units to be manufactured. Shipyards built and installed the catapult structures while the LMS Railway took over the responsibility for making the trolleys in its workshops at Derby and Crewe. One of the first production runway beam structures was constructed with solid plated side panels (unlike the shipboard beams which were of the open 'Warren' type). This was delivered to Jersey Brow, replacing the temporary rig.

This new beam at first was mounted at ground level, again on old railway sleepers. It was felt from the start that the RAE installation, as it was to be used for training pilots and ground crew, should replicate as far as possible the set-up to be expected on board ship. To this end, the new catapult runway beam, which was 80ft {24m} long, was mounted on an elevated structure giving a launch level some 20ft {6m} above ground, equivalent to the height above sea of a weather deck on board ship. The aircraft was loaded on to its trolley by a seaplane crane located adjacent to the structure. For this type of operation, its elevated position would show that the pilot had nothing to fear if, in training, his aircraft was to sink slightly from the catapult as it picked up flying speed. All pilots, both RAF and FAA, engaged upon this exercise, together with the aircraft deck handling and pyrotechnic maintenance crews, were trained on this installation at Jersey Brow. Whilst with shipborne catapults the aircraft would be launched in the wheels-up configuration, most launches at Farnborough were conducted with undercarriages down. This allowed the aircraft to recover quickly should it sink on take-off and momentarily touch the ground. Fortunately this never happened.

Only one mishap occurred during the period of experimentation and training. This was with a Supermarine Seafire (MA970), flown by the senior naval pilot at RAE, Lt/Cdr E. M. 'Winkle' Brown. When the aircraft was loaded on to its trolley the pre-launch check failed to disclose that the trolley retardation barrels at the far end of the track had not been

Supermarine Seafire IIC (MA970) on the P.I trolley at ground level. The Mark II Catapult is seen to the extreme left, and an experimental test rig – relevant to the K.I Catapult – between. June 1942. (© Crown Copyright/MOD)

An early, 'spooled' Supermarine Seafire IIC (MB299) exiting the high-level 'P' Catapult with the trolley in hot pursuit and rockets still burning! The barrels of the spike-and-barrel trolley brake system had not been charged with water. The only braking force available was that ensuing from the failure, in shear, of the spigot attachment bolts to the trolley, allowing it to career away from the track. The aircraft was not in danger, being clear of the trolley (the track-side detent release mechanism had operated satisfactorily). July 1944. (© Crown Copyright/MOD)

Supermarine Seafire IIC (MA970) having already been through a series of trials on the RAE Mark II Catapult on Jersey Brow is shown here on the stage two, plated, beam of the P.I Catapult. June 1942. (© Crown Copyright/MOD)

filled with water. The procedure was to fit the frangible ('Hallite') disc and fill with water as soon as the trolley was withdrawn from its previous launch. This had not been done. The Seafire was started up, the launch procedure initiated and the firing switch closed. The aircraft sped along the track and as it left the trolley the latter continued to travel with the aircraft – having 'crashed through the buffers', slightly below and behind the aircraft to the extent that for a short while bystanders thought the trolley was still attached to the aircraft. (This was a mechanical impossibility; the design of the equipment embodied adequate 'fail-safe' procedures, which functioned correctly.) Fortunately, the trolley dropped away very quickly and became a tangled mass of metal on the ground. There being no water in the retardation units, the spikes punched into the empty barrels, sheared the former from the trolley and allowed the trolley to pass right through the end of the structure. Turnill's (#101 pp.48–49)

photograph of this incident carries an incorrect caption. Two photographs compare this incident with a good launch. For future runs, the filling plugs to the braking units carried indicators to show the controller at ground level that all was well aloft.

The P.I Catapult installation at Farnborough remained in use as an experimental rig for some time. As a demonstration feature it became a 'show-stopper' with its noisy pyrotechnic display. After a decade of service and having fulfilled its purpose, this installation on Jersey Brow was removed by 1952. It was dismantled and all traces of its existence have long since disappeared.

CAM-SHIPS

The first three ships commissioned to operate 'Catafighter' aircraft and equipped with a 'P' type catapult (December 1940) were known as 'Auxiliary Fighter Catapult Ships' and these located the catapult beam athwartships, abaft the forward funnel. These Royal Navy ships, HMSs *Ariguani*, *Maplin* and *Patia*, were equipped by April 1941. (With the almost immediate loss of HMS *Patia* through bombing while at anchor, HMS *Springbank* and the old faithful HMS *Pegasus* were added to the complement.) Although these ships were known as Fighter Catapult Ships and carried the appellation 'HMS' they are not recorded as such in Thomas (#98). It is thought that these ships operated Fairey Fulmar aircraft, HMS *Pegasus* having been equipped with five machines.

Subsequently some 36 merchant ships were fitted with the catapult beam accommodated in the fore and aft orientation at the bow, with the firing position offset to port. These continued sailing as Merchant Vessels under the Red Ensign and were known as Catapult Aircraft Merchantmen (CAM-ships). The first CAM-ship was ready by the end of May 1941 and for two years, until July 1943, these ships were used effectively, only being superseded for a short while by the newer Merchant Aircraft Carriers (MAC-ships). Although experiments at Farnborough were conducted on three aircraft marques, it is thought that only Hawker Sea Hurricanes were embarked on CAM-ships. RAF

The pilot is climbing aboard his Hawker Sea Hurricane I (V6756 - Code NJL showing its origin in the Merchant Ship Fighter Unit), mounted on a rocket-fired 'P' Catapult aboard a CAM-ship, thought to be MV *Empire Tide*. (Fleet Air Arm Museum)

personnel took over the responsibility for the catapult and its crew. Records indicate that of the CAM-ships, four were lost by enemy action: HMS *Patia* was bombed before entry into service in April 1941; *Michael E* was lost in June 1941, and *Empire Lawrence* in May 1942. HMS *Springbank* was lost in September 1941. A list of these ships appears in Appendix 4.

MAC-SHIPS

Although not equipped with a catapult, each of these Merchant Aircraft Carrier ships was a bulk-carrier (six grain ships and 13 tankers) fitted with a basic flight deck and arrester gear. Messrs MacTaggart Scott, who provided the equipment, record the installation of a single safety barrier to at least three of these tanker ships. The CAM-ships and MAC-ships were both operated by the merchant service, sailing under the Red Ensign, with service personnel taking responsibility for the equipment and operation of the air department and airborne weaponry. The MAC-ships' complement of Fairey Swordfish, and later Sea Hurricanes, were designated as flights within a Royal Naval Air Squadron. The Merchant Navy crew were jealous of their privileges in operating these ships and in some cases liked to identify the aircraft as belonging to their branch of the sea-going services. Their possessiveness extended to painting the Merchant Navy insignia on the tails of the aircraft embarked with them. The identities of these ships are recorded in Appendix 4. Of the 19 MAC-ships, *Empire MacAlpine* was the first to sail in May 1943. A total of 4,447 days at sea are recorded, with MAC-ships escorting 217 convoys. In only two of these were any of the escorted ships lost, without the loss of a single MAC-ship. Ultimately, the auxiliary Escort Carriers (CVEs – 38 in number, and again Royal Navy ships), produced in quantity in the USA, took over the FAA support role at sea.

A Hawker Sea Hurricane at the moment of release on launching from a shipborne 'P' Catapult. The flame from the rockets is mingling with the water spray from the spike-and-barrel trolley braking system. Neither the location, nor the aircraft can be identified but it is thought to be a record of one of the first shipborne trial launches aboard either HMS *Ariguani* or HMS *Maplin* during April 1941. (© Crown Copyright/MOD)

ROCKET ASSISTED TAKE-OFF GEAR (RATOG)

Military use of rockets – 'Unrotating Projectiles' (UPs) – was becoming an established technology and the wider employment of such simple propulsive elements was being encouraged. In 1939 there were two sizes of rocket motors, 2in and 3in {52 & 76mm}, these dimensions being the diameters of the metal casings containing the solid propellant. The 3in UP was about four feet {1·2m} in length, with the smaller unit somewhat shorter. The casing was a simple steel tube with a plug (obturating head) at the forward end and a venturi, forming the discharge nozzle, at the tail. The charge was a hollow cordite plug, filling the casing and held in position by the venturi. The electrically operated firing primer was located in the throat of the nozzle. The fins, normally used for free flight as missiles, were not fitted. The 3in unit would provide a maximum thrust of about half a ton {500kg} during a burn time of about 2½ seconds.

The performance of these devices, considered to be of use as solid propellant motors, was quickly assessed and proposals for and calculations in relation to their non-projectile use were already formulated by the end of 1939. Practical application with initial flight trials for boosting the take-off of bomber aircraft were conducted in 1941 but interest was diverted to the FAA requirements where the greater need was thought to exist.

Experiments were conducted using groups of two, three or four rockets, generally the 3in {76mm} units, housed in a jettisonable framework attached to the aircraft. The catapult spools on naval aircraft were useful attachment points for early trials, permitting an installation on each side of the fuselage. Taxying trials were first conducted to examine interaction between the hot gases from the rockets and the aircraft itself. The reaction against the failure of one rocket per side was also investigated. The normal process was not to fire all rockets at once but to control the time of ignition so that the second rocket would start its burn about 0·9sec after the first and thus seriatim. This was achieved through the use of a timer mechanism. It was during such experimentation that Lt K. J. ('Robbie') Robertson, the RAE Naval Test

Pilot, lost his life in January 1944. The accident occurred on recovery of the RATOG Seafire (MB141) to HMS *Chaser*.

Early flight trials were conducted ashore to investigate the fall-away of the rocket carriers as they were jettisoned from the aircraft once airborne. It was inevitable that the rocket carriers would be expendable through damage on impact with the ground or being lost at sea.

RATOG came into its own with the advent of the American CVEs. Some of these auxiliary carriers embarked Swordfish squadrons which, not being fitted with towing hooks, could not be launched from the AH.II and AH.IV accelerators. 'Boosted' take-off was achieved by the use of RATOG but a drill was introduced to reduce the fire hazard on the timber flight decks. The Swordfish started its take-off run and by the time it had reached some speed (perhaps 20 knots) the pilot triggered his rocketry and all was well. One wag was heard to refer to this as being the first jet-propelled biplane to be used in the FAA. On one occasion HMS *Biter*, returning to the Clyde for refit during the late summer of 1944, needed to fly its Swordfish contingent ashore. It was felt that, through using RATOG, this could be done without putting to sea. Wellham (#106 p.168) records that this was achieved successfully. This antedates by nine years the claim by Hobbs (#45 p.143) for the launching of Sea Furies by RATOG from HMS *Ocean*[45] at anchor, at Sasebo on 1 June 1953, as being the first occurrence of such a feat. Within a few years such rare incidents of launching while the ship was not under way became a familiar procedure.

A further field of exploration was the case of assisting gliders at take-off under tow. Records exist of experiments having been undertaken during 1943 using an Airspeed Horsa (DP749) and a GA Hamilcar respectively; it is thought that the system was not adopted to any extent for airborne activities.

While this launching procedure, at the time, was not adopted in any great depth, subsequent developments along a different channel, employing 3in {76mm} rocket motors, proved highly successful. (Early

Blackburn Shark II (K5656) with RATOG at the point of lift-off. April 1941. (© Crown Copyright/MOD)

Fairey Swordfish II (either DK747 or HS671, more likely the latter) carrying a torpedo with an air tail in RATOG trials at Farnborough. The audience between this aircraft and the Grumman Avenger are NAD research staff; the chap wearing the light raincoat and holding the bicycle has all the mien of R. J. Nixon, the Department's engineer for RATOG projects. December 1943. (© Crown Copyright/MOD)

A lightly loaded Fairey Barracuda II (P9791) with wheels well clear of the runway under a RATOG launch. Each carrier was fitted with two 5in rockets, fixed to strong-points on the fuselage. The high-level 'P' Catapult gantry on Jersey Brow is seen to the left of the picture. December 1943. (© Crown Copyright/MOD)

experimentation in this field was the development of the 'P' Catapult, described elsewhere in this book.) Two further instances are recalled in which extensions of the principle were employed: one was thought to have been successful, the other was decidedly disastrous. The 'successful' occasion was a case of an American Boeing B17 Flying Fortress having been emergency landed in a field from which it was impossible for it to be recovered by flying out again; dismantling seemed to be the only alternative. A scheme was proposed by NAD in which a pair of rocket carriers were to be fitted, one under each wing, and an outline design was prepared. The American forces undertook to conduct the entire exercise from their own resources and to recover the aircraft intact. Nothing more was heard and the exercise is thought to have been successfully completed.

The second case was part of a project designed to assist four-engined heavy bombers to become airborne without the inevitably long take-off runs. Experiments with a Whitley twin-engined aircraft had been successful and the project was extended to include the 'heavies'. A Short Stirling IV (N3635, the first production model) was made available and adapted for these trials. The proposal here was for a pair of jettisonable rocket carriers to be fitted, one under each wing between inboard and outboard engines. Each carrier contained twelve 3in UPs, wired up to fire serially in groups of one or two pairs at a time. Those who witnessed the early runs of this set-up recorded that the sight of a Stirling during and after a RATOG climb was an extremely impressive event.

The rocket installation was rigged so that all rockets could be subjected to a pre-flight electrical check together, rather than test each one individually, a master changeover-switch being provided for the purpose. On 18 August 1941 a special demonstration in front of VIPs was arranged at A&AEE, Boscombe Down. The pre-flight check was completed but the mechanic failed to return the test switch to its normal operating position. The aircraft was taxied out to the head of the runway, the engines were run up and the aircraft began its take-off run. With engines at quarter-throttle, the test pilot, Sqn/Ldr B. O. Huxtable,

Port side cordite rocket pack for RATOG fitted to a Supermarine Spitfire. At this early date the rockets are the 3in size. The installation is basic, using bungee rubbers to jettison the casing after firing. April 1943. (© Crown Copyright/MOD)

Blackburn Firebrand III (probably DK387) was used for experimental work on Rocket Assisted Take-Off Gear (RATOG). The project involved the fitting of jettisonable rocket carriers to strong points on the aircraft fuselage. This illustration shows a cluster of three 5in cordite rocket motors, fitted to the starboard side of the fuselage; an installation on the port side would match this. June 1946. (© Crown Copyright/MOD)

selected ignition for the rocket units. He saw both rocket carriers cartwheeling ahead of him, then back over the wing, all rockets burning, and carrying parts of four aircraft propellers with them. The aircraft was swinging wildly and an emergency shut-down and personnel evacuation was carried out.

Another pilot, 'Sandy' Powell (#77 p.155) witnessed the event:

'When the Stirling was ready an impressive gathering of important visitors arrived on the aerodrome, including a general, a cabinet minister, and many senior RAF officers. Huxtable taxied the great aeroplane out. The brass hats watched in silence. The Stirling advanced slowly, it was approaching the rocket firing stage and then there was the loudest, longest and most satisfactory explosion yet heard on Salisbury Plain. The scene immediately round the Stirling was confused due to smoke, flames, spent pieces of rocket and so on. When it all cleared away we saw the aeroplane. It had come to rest, one undercarriage partially collapsed, engines pointing in all directions, three propellers missing, and bits of blades here and there, but no one was hurt. At quarter throttle all the rockets had gone off at once, applying an acceleration which the Stirling's designer had not even dreamed of.'

The Accident Branch at RAE is understood to have been puzzled over this occurrence for a long time. Jones (#54 pp.64–65) portrays the wreckage of the rocket cradles from this incident. The plain fact was that the rocket cradles and their attachments were not designed to operate under an applied force of some 12 tons – an overload of 600 per-cent, or more! Mason (#64) records that the mishap ended further interest in the project.

Subsequent developments in rocket technology enabled a single 5in {125mm} long-burn rocket motor to be fitted each side of the aircraft, thus making the installation and servicing of the units a simpler process from both aspects of man-handling and pyrotechnology. Thomlinson (#292) records that these larger rocket motors produced a similar thrust of about half a ton {500kg} over a burn time of four seconds using a cordite charge weight of 26lb {12kg}. Very roughly, this resulted in the use of a pair of rocket motors for each 6,000lb {2,700kg} weight of aircraft. This would enable an aircraft to take off within a distance of 500ft {150m}. The run would begin in the normal way then after a travel of some 50ft {15m}, when a speed of 20kts {10m/sec} had been attained, the rockets would be fired and would continue to burn for a brief period after 'unsticking', giving an acceleration of slightly less than 1g. The spent rocket cases would then be jettisoned over a convenient dropping area.

French Navy Aquilon 201 no 05 (F-WGVT) (de Havilland Sea Venom) fitted with two pairs of 5in rocket motors being prepared for catapult proofing trials on the BH.5 catapult at Farnborough. This was a 'CAT-RAT' system employing RATOG to assist in the catapult launch. October 1954. (© Crown Copyright/MOD)

Supermarine Attacker (TS416) with a ventral drop-tank and fitted with RATOG carriers, seen on Jersey Brow. It is apparent that eight 5in rockets were required for this aircraft when operating at maximum weight. September 1950. (© Crown Copyright/MOD)

A rare picture of Short Stirling I (N3635) undergoing RATOG trials with twelve 3in rockets beneath each wing. This launch was successful; a later demonstration at Boscombe Down was catastrophic with the aircraft written off! August 1941. (© Crown Copyright/MOD)

In these cases the rocket motors were located alongside the fuselage in above-wing positions; appropriate strong-points were provided adjacent to the engine mountings. This type of installation became the norm whenever the facility was required on shipboard. For a short period some carrier aircraft were launched from a combined powering of accelerator with RATOG. This was quickly dubbed as the CAT-RAT system. It could not have been popular with pilots as there was no cut-back feature should a launch need to be balked at the last instant. The demise of shipborne RATOG began with the introduction of jet aircraft, and catapult launching became the standard policy for FAA shipboard launching. The use of re-heat introduced new features and new problems to be solved.

NAD continued to use rocket motors for special applications, mostly for propulsion of trolleys in experimental rigs. The 3in rocket motor adopted for the P.I Catapult was used in only a few subsequent applications. Rocket technology had advanced during World War 2 and the 5in rocket motor latterly became the maid-of-all-work. The physical difference, apart from size, was that the 3in rocket carried a screwed 'obturating head' at the front, to which the normal war-load was attached, and its venturi at the stern was contained within the rocket casing. The 5in rocket used a smaller size spigot at the head and its venturi unit was exposed. The technicalities of the cordite formation (hollow and/or cruciform) and its ignition were kept 'under wraps'. No attempt has been made to unravel these features.

LANDING-ON

The Fairey Seal was the first aircraft in squadron service to be fitted with the purpose-made arrester hook. This was suspended below the fuselage on a swinging 'V'-frame for use with the new transverse system of arrester wires. Taylor (*#95 p.221*) reports this aircraft to have appeared in 1933. It was a development of the Fairey IIID aircraft via the IIIF that (at the Mark VII stage) was renamed 'Seal'. This aircraft was also the first to be provided with a tail wheel instead of a skid, and to be fitted with wheel brakes as standard. All these new features had been developed on various IIIF aircraft as prototypes. (In the process, a few of a late mark of Fairey Flycatcher aircraft were fitted experimentally with wheel brakes.)

The main essential of an arrester gear for aircraft carriers is that it shall be capable of bringing to rest, within a minimum distance, an aircraft returning to the carrier from its flight. The term 'aircraft recovery' has been adopted for this process. From the earliest trials relating to mechanically assisted recovery (the experiments with a Fairey Seal aircraft in HMS *Courageous* in 1933 with the Mark 3 arrester gear), the principle adopted has been that in which a number of athwartships wire ropes are spaced at intervals across the after section of the flight deck. These are connected to under-deck braking systems almost exclusively of the hydraulic 'jigger' type. (At least, this was the case until 1970 when HMS *Ark Royal*[58] was equipped experimentally with an alternative system.) Each aircraft is provided with a hook

Hawker Sea Hurricane landing on an unidentified Lend/Lease carrier. The aircraft is fitted with a standard 'V'-frame arrester hook which is about to engage an early arrester wire. September 1941. (© Crown Copyright/MOD)

suspended below the fuselage which, on landing, engages one of these wires, which brings the aircraft rapidly to a halt.

Willis, in 1945 (*#313*), examined the whole process of landing-on in his lengthy treatise on this subject. Much of his dissertation relates to detailed technical points within specific areas which would be tiresome to repeat here. His introduction perhaps indicates the scale of the problem:

'Landing an aircraft on the flight deck of an aircraft carrier involves many factors that are not present, or present to a lesser degree, in the case of land-based aircraft. These factors are all due, to some extent, to the fact that the area in which the aircraft must alight is only about 400ft long by not more than 100ft wide {120m, 30m}, and may have considerable movement due to the motion of the ship. The small dimensions of the flight deck necessitate the use of

Supermarine Seafire, possibly Mark XVII, landing on an 'Illustrious' type carrier. The aircraft carries the 'sting' type arrester hook located behind the tail wheel and replacing the lower part of the rudder in its configuration. The guard in front of the tail wheel is to prevent its snagging the arrester wires as the aircraft passes across them. October 1947. (© Crown Copyright/MOD)

Fairey GR17/45 – prototype Gannet (VR546) – with 'sting' arrester hook deployed, about to pick up No.3 arrester wire in landing on HMS *Eagle*[51]. February 1955. (© Crown Copyright/MOD)

De Havilland Sea Venom NF Mk20, (WK376) with 'V'-frame arrester hook deployed, landing on HMS *Eagle*[51]. February 1955. (© Crown Copyright/MOD)

arrester gear, consisting of six or more ropes, spanning the deck at a height of about 6ins {150mm}, reeved to hydraulic ram units. On landing, one of these ropes is engaged by an arrester hook on the aircraft, reducing the landing run of the aircraft, after engagement, to about 160ft {49m}, or less.

'A successful landing depends partly on the skill of the pilot and, to a greater extent, on the suitability of the design of the aircraft for deck landings. The operation of landing may, for convenience, be divided into three stages, the approach, touchdown on the deck and arresting. Each is governed by a set of requirements that are inter-dependent in a successful deck-landing aircraft. The approach covers the period up to the instant that first contact of the wheels on the deck, or of the hook with an arrester rope, is made. Requirements for this period deal with the aerodynamic and handling qualities of the aircraft, together with the pilot's view of the deck. The touchdown involves the capability of the undercarriage to absorb the vertical energy of descent that may be aggravated by the method of approach and the moving deck. Arresting requirements cover the strength and method of operation of the arrester hook installation and also the positioning of the hook to provide good stability of the aircraft under arrester gear loads.'

Thomlinson (#303), rather prosaically, offers a clinical appraisal of the art of recovering aircraft on to the deck of an aircraft carrier when (in 1964) he writes:

'In aircraft arresting gear practice, the problem can first be divided into two parts:

a) that of catching hold of the aircraft, and
b) that of taking from the aircraft its kinetic energy.

In general, the two parts can be considered separately and any preferred catching system can usually be linked with the chosen system of energy dissipation.'

While the mechanical arresting of an aircraft features both (a) and (b) above, the following paragraphs address the art and practice of deck arresting, dealing primarily with the extraction and dissipation of kinetic energy from the aircraft. Catching hold of the aircraft is treated later.

It is a logical requirement that naval arrester gear should be capable of bringing to a stop all types of aircraft (with regard to weight and landing speed) that are capable of being handled by the ship, without subjecting the aircraft (and its crew) to an unreasonable braking force, within a prescribed distance, generally some 165ft {50m}. This must be achieved automatically without the need for complicated adjustments to be made to the system by virtue of variations in the weight of, and speed of engagement by, the aircraft. Under operating conditions, 'speed is of the essence'. Not only is it desirable but it is essential that an arrester gear in service should be capable of efficiently recovering all aircraft in use at that time and still have something in hand to cope with immediate future increase in performance of naval aircraft. Like most features in naval aviation, the state of development of flight deck machinery needs to be ahead of the operational needs of the aircraft.

A hook suitably located on the aircraft is arranged to engage a steel wire rope stretched across its landing path. The rope is arranged to pass over two sheaves located at some distance on either side of the point of engagement and tension in the rope is developed in this particular case by an under-deck hydraulic jigger unit. It is desirable for this engagement to occur at the mid-point of the base wire spanning the flight deck. (This is an ideal which is not always achieved in practice.) In this way, equal amounts of wire are paid-out on each side of the system and the resulting appearance of the wire, in plan, is an isosceles triangle with the aircraft hook at its apex. Any deviation from this induces unequal forces on both the gear and the aircraft, often with resultant sliding of the wire through the throat of the arrester hook causing abrasion problems with both wire and hook. Investigation into the reactions at this interface was also necessary.

The under-deck machinery for producing this braking effort generally is of one type, the hydraulic jigger unit, with a rope reeving ratio of 12:1, the stroke of the hydraulic ram being ten feet {3m}. Alternative forms of arresting machinery have been considered over the years, including a rotary brake in the form of a fluid dynamometer, an electrical unit using eddy currents as a retarding force, both of which were discarded without undertaking research into their probable usefulness. Other mechanical means were examined, including the use of a large brake drum controlling the pay-out from an integral rope drum. A second field of investigation related to a linear friction track. An adaptation of this principle was tried out in two formats with the DCTO system (qv). One had a direct spring-loaded friction shoe running along

a smooth slide. The second had a trolley carrying a hydraulic system which ran along a secondary track. A series of sinusoidal plates were carried which achieved braking by pumping the hydraulic fluid against a back pressure. Yet another 'friction track' device was tried. This entailed the dragging of heavy steel chains along the ground, known as the Drag Chain Arrester Gear (qv). None of these last projects progressed beyond the experimental stage for shipboard use. The hydraulic jigger seemed to win 'hands-down' over any alternative and it progressed through many stages of development, only to be succeeded by the water spray unit (Direct Acting Arrester Gear) in the final refit of HMS *Ark Royal*[55].

Garstin, (#35) lecturing on *Hydraulics in Flight Deck Machinery* in 1966, commented:

'This type of gear has great potential and in principle it is possible to use it to stop the largest aircraft yet envisaged, with an installation that is itself relatively insignificant in cost. What emerges is the capability of simple hydraulic machinery to perform functions of a very specialised nature, in which the supply and precise control of large amounts of power are involved, and to do this without excessive development cost. The practical difficulties have nearly always loomed much larger than the theoretical. The extent machinery components are large and the scale effect thus introduced has presented many manufacturing problems. Over the years, development has been progressive so that improved techniques have been used as they became available, to improve both performance and reliability simultaneously. Only a handful of engineers work on flight deck machinery and it could be quite impossible for development to have gone at the speed it has, if the general advance in the technology of hydraulic machinery had not also been rapid during the last twenty years.'

By the early 1940s, with the increasing speed and operational weight of aircraft designated for naval flying, Thomlinson (#294), in his appraisal of the situation, showed that some fundamental features began to assume importance in the design of arrester gear. These were related to the impact of the aircraft with the arrester system and the physical inertia of the system itself. On first engaging the arrester gear the moving parts of the gear are snatched, or accelerated into motion and so, without any external forces being applied to the rope ends, a rope tension is developed which in turn applies a retarding force to the aircraft. At some stage in the process, acceleration of the gear ceases and its own motion has to be arrested by resistance generated within the gear itself; in other words, the braking effort has to arrest both the aircraft and the gear. It has to be borne in mind that the weight of the wire rope itself in one arrester installation is of the order of one half of a ton {500kg}. To produce a mathematical model of the constituent elements in the arrestation process would have been laborious. The variables to have been considered would comprise:

a) the tension forces through stretch of the arrester rope on initial impact of the aircraft,
b) the inertia forces in accelerating the rope,
c) the inertia forces in accelerating the machinery,
d) the inertia forces in accelerating the hydraulic braking fluid,
e) the resistance imposed by the hydrostatic pressure of the braking fluid, then, when everything is in motion,
f) the hydro-pneumatic pressure applied in killing the momentum of all the foregoing, to bring the aircraft and arrester gear machinery to rest.

A half-century ago this problem might have been given to a computer. At that period, a 'computer' was not a set of programmed chips in a desktop processor but more likely to have been a young girl (having just left school with a good grade in Higher Schools Certificate

Maths) driving a large slide rule, or probably an Otis-King or Fowler calculating cylinder, even perhaps one of the rare Monroe 10x10 electro-mechanical machines chuntering away noisily in an office corner. The man-power (woman-power) component was labour-intensive even in the 1940s!

To investigate these phenomena at a practical level, it was decided (c.1941) to equip Farnborough's Runway 04/22 with a standard, jigger-type arrester gear. While the situation was being addressed by the Catapult Section, other equipment of advanced, superior capability quickly followed this first installation. From this point Farnborough became progressively involved in the mechanics and dynamics of ship-borne arrester gear.

By 1955, the date when NAD began the transfer of personnel and equipment to its new location at Bedford, there were five arrester installations along the western edge of Farnborough's Runway 04/22. These included the early Mark 4 system (progressively updated), the Mark 10 vertical unit and the Mark 12 system. Details appear in the chapter 'Jersey Brow – Arrester Deck'. NAD did use a duplexed American Mark 4 unit of its own concoction in at least three locations on the airfield: at Meadow Gate the flexible deck installation alongside Runway 07/25; and at the High Speed Track on Laffan's Plain. It is not known whether these were three separate installations or one unit rotated into each location in turn. It is thought that a single American Mark 4 unit was at Farnborough for a period and a passing reference to a British Mark 6 arrester occurs in one report. Details of these various – and other arrester gear – installations appear in the section 'Arrester Gear, Shipborne'.

Thomlinson, in 1952 (*#298*) and later in 1964 (*#303*), records some of these developments to which reference is made elsewhere in these pages. In the former paper he sums up the current situation in assuming:

'. . . that the principle of arresting follows the lines now in use, ie a hook attached to the after structure of the aeroplane engages a rope stretched transversely across its path. This rope is arranged to pass over two sheaves located one on either side of the path of the aircraft. The hook thereafter draws out rope over these two fixed sheaves from a store of rope, and the desired rope tensions are developed and transmitted to the aircraft – thus arresting it. No alternative is foreseen to this principle when retardations of upwards of 1g are required or more particularly for retardations of the order of 3g to 4g. There are, however, various ways of storing the rope, and developing the rope tension.

'In working up the design of any new gear, in addition to knowing the weight range of aircraft and the maximum engaging speed, it is necessary to know how the gear functions and to have knowledge of the mechanics of these functions. Whilst the general principles of the functioning of the gear are simple, the details are far from simple and present many difficulties in mathematical treatment and in engineering design. These . . . are most important at speeds of 90 knots upwards. As present knowledge stands, a limit is foreseen to maximum engaging speeds, the value of which will depend to a large degree on design details and physical properties of the materials of construction – particularly of the rope.'

(This last paragraph was prophetic in that it formed one core of investigatory experimentation for the Naval Aircraft Department over the next decade or so.)

Fairey Firefly AS.6 has made a good landing, having picked up No.5 arrester wire (aboard HMS *Indomitable*?). The flight deck hand has disengaged the wire which is about to be retracted and all wire bow-spring supports have been lowered to deck level. DLCO ('Bats') at the top of the picture, behind his windshield, is ready to bring on the next aircraft. c.1951. (© Crown Copyright/MOD)

ARRESTER GEAR - SHIPBORNE

The application of longitudinal arrester ropes with Bryer hooks on the aircraft as landing-on aids has been described above. The first attempt in which mechanical assistance was given to the arresting of aircraft in shipboard deck landing was aboard HMS *Furious* in 1918. Biggs (*#6 p.14*) states that this was on the occasion of the ship having been equipped with a separate landing-on deck abaft the funnel and bridge structures. Arrester ropes again were stretched in a fore-and-aft direction some nine inches {230mm} above the deck. These, in turn, supported further transverse ropes with sandbags attached to each end. 'V'-shaped hooks fitted to the aircraft undercarriage would pick up one or more transverse ropes while the longitudinal ropes guided the aircraft in a straight line as it was brought to rest. Palisades were erected along both port and starboard deck edges to restrain the aircraft from going overboard should the arrester system fail in its purpose.

This procedure proved to be awkward and was not liked by anybody who had to operate it. As soon as carriers were built with continuous flight decks, from stem to stern – a flush-deck carrier – this form of arrester gear was abandoned. All arrester-gear equipment was removed from these ships by 1927. In fact for a period of six years, arrester gear was thought to be unnecessary! HMS *Argus* was the first full-deck carrier to be re-equipped, followed by HMS *Furious* in 1931.

It was in the early 1930s that the all-up weights of aircraft were approaching 7,000lb {3,200kg}; landing speeds began to increase and, as a result, their flight performance envelope began to nudge the constraints obtainable from 'natural' resources. By virtue of these increases, the need was felt for alternative means of arresting aircraft on deck. Rawlings (*#81 p.36ff*), in his well-illustrated and lively history of the Fleet Air Arm, gives an insight into the early attempts at mechanical arresting of aircraft. He relates that in 1928 HMS *Courageous* was fitted with athwartships wires on the after end of the flight deck. An arrester hook was suspended under the rear fuselage of each aircraft which would engage one or another of these wires and bring the aircraft to an abrupt halt.

Biggs (*#ibid. p.25*) states that in 1931 HMS *Courageous*, in the first instance, was fitted with a single across-deck arrester wire, the ends of which were connected to friction drum devices, similar to a winch. This was not too successful as the inherent difficulty in matching the combined performance of the two drums as each tended to pull the aircraft to one side or the other. This system passed through a few modifications and eventually emerged as a group of wires, still attached to drum units but fitted with hydraulic braking devices. Hobbs (*#45 p.18*) refers to work being undertaken at RAE at this time in the design of a suitable arrester gear, investigating electric and hydraulic retardation systems with winch and piston operation, including effects of off-centre landing. Unfortunately, the author has been unable to discover any record of these trials. Subsequently, after a series of experimental landings, the real answer was found in 1933, still aboard HMS *Courageous*, when the same system of transverse wires was linked to hydraulic jigger units that produced a smooth, balanced pull-out of the arrester wire. The rating of this equipment was to accept an aircraft of 8,000lb AUW at 61kts {3,600kg, 32m/sec}. This became known as the Mark 3 Arrester Gear. This principle was universally adopted and developed over the next four decades, throughout the life of the fleet carrier, ultimately attaining Mark 13.

Thetford (*#97 p.132*) states that a Fairey IIIF aircraft (S1781) was the first to have been fitted with a swinging, triangular steel frame with a spring-loaded hook located at the apex, beneath the rear fuselage.

Research Blackburn Buccaneer S.1 landing on HMS *Hermes*[59]. The targets painted on the side of the fuselage and engine nacelle were for photo-analysis purposes. October 1966. (© Crown Copyright/MOD)

Flying trials took place aboard HMS *Courageous* in 1931. Thetford (*#ibid. p.136*) records the Fairey Seal – successor to the IIIF – to have been the first aircraft to be fitted, as standard during manufacture, with a similar arrester hook. (The Seal aircraft also was the first to be fitted with wheel brakes, an additional benefit for deck-handling.)

A comprehensive list of the development of the arrester gear is provided by Friedman (*#34 p.110*). The first two systems aboard HMS *Courageous* were designated Marks 1 and 2. The Mark 3 arrester gear was the first to use the hydro-pneumatic principle in a multi-reeved jigger unit. The Mark 4 system was installed in the 'Illustrious' Class and it is here that the Naval Aircraft Department at Farnborough first became involved as an ongoing exercise. It was one of these units which formed the primary installation on the dummy deck runway south-west of Jersey Brow.

The system comprised a series of wire ropes laid athwartships, spaced at intervals along the landing area on the after end of the flight deck and connected to an under-deck braking system. Aircraft were fitted with an arrester hook, located at the free end of an arm swinging below the fuselage forward of the tail wheel, or skid, which would pick up one of the across-deck wires. This wire would pay out along the deck and its mechanism would bring the aircraft to a halt. Of the seven carriers in commission at the outbreak of World War 2, Hobbs (*#45*) records each – with the exception of HMS *Ark Royal*[38] – having been equipped between 1932 and 1938 with a four-wire installation, without safety barriers. These systems allowed entry of an aircraft of 10,000/11,000lb AUW at 53kts {4,500/5,000kg, 27m/sec}. HMS *Ark Royal*[38] had a comprehensive system of eight wires, graded in braking intensity, and a safety barrier installation.

The six armoured deck fleet carriers specified and laid down before the outbreak of World War 2 – HMS *Illustrious*[40] and those of her ilk – generally were fitted with Mark 4 arrester units providing up to nine wires, with two or three safety barriers in addition. The barriers were rigged amidships, forward of the landing-on area and, generally, abreast the island. Adjacent wires were reeved in pairs to under-deck brake units. This equipment was designed to receive aircraft weighing

11,000lb, landing at speeds up to 60kts {5,000kg, 31m/sec} with a maximum induced retarding force not exceeding 2g. (Higher rates of deceleration would be expected if contact with the safety barrier was made.) This speed of 60kts {31m/sec} was the entry speed of the aircraft relative to the deck. When landing aboard ship an additional benefit would accrue in that the wind speed down the deck with the ship steaming into wind would be equivalent to the ship's speed plus any natural wind speed.

Throughout World War 2 all fleet carriers and the new light fleet carriers were equipped with this type of arrester gear. In 1942 the Catapult Section (Thomlinson, *#289*) computed that the scantlings of the system would permit it to be upgraded, without modification, to receive aircraft of weight of 16,500lb at 65kts entry speed {7,400kg, 33m/sec} and a similar rate of deceleration. An extensive series of experiments on the dummy deck installation at RNAS Arbroath (HMS *Condor*) was conducted to substantiate these speculations. Other Navy airfields equipped with a dummy deck were used on an occasional basis, two examples being RNAS Crail (HMS *Jackdaw*) and RNAS Henstridge (Yeovilton – HMS *Heron*). From these, the case for future upgrading was confirmed enabling the equipment to be used by all naval aircraft in shipboard operation during World War 2. At subsequent refits all ships, through adaptation or renewal, had their arrester-gear installations progressively upgraded in parity with the accelerator gear to match the weights and performances of the types of aircraft to be operated.

Almost invariably each carrier was equipped with its complement of arrester wires and safety barriers, totalling an even number, the reason being obvious. Two wires/barriers were rigged to one under-deck arrester unit with the above-deck wires (centre spans) in a non-fleeting arrangement. The system of reeving these units is discussed below. In essence, each brake unit was a hydraulic jigger. This was a hydraulic ram complex with a set of 12 wire rope pulleys (sheaves) at each end ram unit. The stroke of the ram was 10ft {3m} which, within the constraints of the installation, allowed a pay-out of the deck cable of some 156ft {47m} linearly along the deck. Each unit was self-contained, comprising a steel 'Warren'-type beam in which the whole of the machinery for a pair of arrester wires was assembled. This beam was about 6ft wide by 3½ft deep and 30ft long {1·8m, 1·1m, 9·1m}. Generally these beams were located in the roof of the aircraft hangar, immediately below the armoured flight deck, or occasionally on the deckhead below the hangar floor.

According to Griffith (*#248*):

'Basically the gear consists of a ram which moves into a cylinder and forces hydraulic fluid through a small orifice, thereby creating a pressure in the cylinder and hence providing resistance to the aircraft via a flexible steel wire rope which passes round a number of sheaves on the outboard ends of the cylinder and ram unit. The number of sheaves is chosen to produce a suitable gearing down between the movement of the aircraft and the movement of the ram and its associated sheaves. The fluid displaced is stored in an air-loaded accumulator; the displaced fluid compresses the air in this accumulator thus storing a small amount of the energy absorbed from the aircraft arrested. This energy is used to reset the gear to its original condition so that it is then ready to receive another aircraft.

'The unit framework consists of two side beams constructed mainly of channels and plating, connected at top and bottom by cross bracing panels. The framework is not intended to withstand or to transfer to any external structure the main rope loads which occur during an arrestment. As normally rigged in a ship, the system of ram, cylinder and their associated crossheads form a balanced system with two ropes leaving each end of the unit (in a double rove system) ie all arresting forces are internal. However, in practice, due to an off-centre engagement of the centre span by an aircraft, the tensions in the two sides of the main framework have to withstand the difference between the two. In ships, the units are normally slung from the deckhead. There is a three point suspension system which allows for expansion and contraction.'

The ropes reeved around these pulleys were taken up to the flight deck through fairleads and deck sheaves ranged each side of the landing-on area. The ends of the ropes were terminated in shackles. Each pair of shackles was connected by a further wire rope spanning the flight deck (the across-deck element, later known as the 'centre span'). This was the wire which the aircraft arrester hook would pick up on landing. In the ready-to-land configuration, the jigger would be extended fully, thereby tensioning the across-deck wire. On picking up the wire, the aircraft would pull it out along the deck against the jigger which would be closing under hydraulic pressure. A suitable control system, including adjustable orifices within the fluid flow path, was incorporated to enable the whole activity to be undertaken smoothly and for the whole unit to be rapidly reset to permit a following aircraft to land-on with minimum delay. Further particulars of this machinery are given in a chapter below (vide 'Land Trials').

The system of reeving was designed so that the across-deck element, which would be subjected to all the wear and tear of aircraft landing into it, could be readily replaced; such replacement could be effected within minutes. The type of reeving adopted in this installation was known as a 'non-fleeting' system; that is, the reeving was such that if one of a pair of across-deck wires was being extended by the aircraft arrester hook, the complementary wire remained stationary at the flight deck level. This was achieved by cross-connection of the arrester wires within the jigger unit.

When the gear was in use, each centre span rope was supported about six inches clear of the deck by a pair of bow-spring supports. At other times, when non-operational, they were held flush to the deck by a system of pneumatic or, in later marks, hydraulic, pull-down devices. The bow-spring was a nine feet {2·7m} length of ¾in {19mm} diameter spring steel bar. It was anchored to the deck at the after end with the forward end free to slide in a grooved deck fixing. (In some of the later ships these bow-springs were duplexed, with two bows to each fitment.) Two of these units were used for each centre span, spaced about 40ft {12m} apart, the alignment of each pair being staggered between adjacent wires along the deck. The mid-point of each bow spring – the area which supported the centre span – was attached to an under-deck fitting which retracted the spring to deck level. This comprised a pneumatic/hydraulic plunger employing a 2:1 reeving which depressed a wire rope, or a rigging chain, against the spring of the bow to which it was attached. Each unit was neatly housed within a casing which projected two feet {600mm} below the deckhead.

A typical format for World War 2 fleet (armoured) carriers was for six under-deck units to be installed. (It is thought that at one time the first three of the 'Illustrious' Class carried 11 wires plus three barriers, requiring seven under-deck units.) Numbering from the aftermost position (No.1 wire), each pair of wires up to No.8 would be reeved to one arrester unit and the retardation setting of each unit would be common to each other. No.9 wire, often known as the 'trickle' wire, would be more fiercely trimmed to give a shorter pull-out and would be reeved to No.1 barrier. Some ships graded the arrester wires progressively more severely as the identification of their location moved up the deck. For instance, wires Nos. 1–4 would be rigged for a normal approach, Nos. 5 & 6 would have the hydraulic back pressure increased slightly, whilst Nos. 7 & 8 would be set fiercely. No. 9 wire and No. 1 barrier would be 'vicious', and the final two barriers, reeved together and provided as a last resource, were 'impossible', giving a pull-out as low as 40 feet {12m}!

The intention was to arrest the aircraft by the wires in a minimum distance to prevent its entering the barrier. Should the aircraft miss all wires, then the barriers, hopefully, would halt its forward movement into the forward deck park. If the aircraft reached these final barriers it was in serious trouble; the barriers were intended to hold it against further forward movement, thereby preventing devastating, possibly fatal, consequences should there be aircraft ranged in a deck park forward of the barriers. (With the angled deck adopted in post-war carriers the number of wires was reduced to six, or even to four, and the need for barriers as protective devices disappeared.)

Returning to the wartime situation: the deck coverage, or spread, of the arrester wires along the flight deck was kept to a fairly consistent standard. The fleet carriers carried eleven wires across a spread of 211ft {64m}, the later installations being nine wires within an average of 220ft {67m}. The Lend/Lease CVEs possessed nine wires over 195ft {59m} while the light fleet carriers had ten wires in 237ft {73m}. The trials carrier HMS *Pretoria Castle* was one exception: her spread of wires were six, within a short range of 101ft {34m}. The earlier carriers of the 1920s all had four wires with no barrier, but HMS *Audacity* – the first British escort carrier, and among the smallest of all carriers – was fitted with three wires and one barrier. For later ships with angled decks and visual landing aids, fewer wires were required in the landing area; the spread comprised four wires within 78ft {23·8m} – equivalent to a half-second of landing time! Details of these systems are given in Appendix 7.

The reduced spread of HMS *Pretoria Castle* accounted for many occasions when the approaching aircraft was forced to do a 'bolter' (fly round and try again) and was responsible for many a landing into the barriers. The author recalls the occasion in September 1945 when during three days of trials, three Blackburn Firebrand aircraft (DK395, EK601 & EK657) were written off through barrier entries. In deference, a total of 96 successful landings had been achieved in that same period (Cooper, #226 & #227). On each occasion, touching down with its hook striking the deck abaft No.1 wire, the rebound of the undercarriage

caused the aircraft to float over the entire spread of the arrester area and make contact with the barrier without its wheels touching the deck again. These impressive incidents probably would have been avoidable if the landings had been undertaken on a fleet carrier with the greater, fore-and-aft, deck spread of its arrester installation.

A number of records state that the World War 2 Fleet Carriers all had arrester-gear installations at the forward end of the flight deck. The initial designs may have indicated this, but in very few circumstances were bow installations ever used. Wallace (#103 p.52) records that early in January 1942 HMS *Indomitable* effected a single landing into the bow arrester wires. The author has not come across any record of this having taken place on any other British ship. It is difficult to imagine the occasion when an aircraft would need to land over the bows of the ship, which – of necessity – would be steaming astern at high speed to provide adequate wind down (or up) the deck. Operationally, such an installation would interfere with the functioning of the accelerator and the forward lift.

Some early American carriers were fitted with bow arrester installations but they were seldom used. The policy of the US Navy in carrying some of the aircraft complement in a deck park would mean that at times the aircraft would be ranged aft. It was not expedient for them all to be moved forward to allow one or two isolated aircraft to land on so bow wires were installed for such occasions. They were not put to regular use. In none of the author's visits (six in total) on board the armoured carriers did he see any provision for bow wires. As the matter stood, often the wire in the way of the after lift – perhaps No.3 or No.4 – would remain unrigged during flying operations, especially if aircraft were to be ranged via the after lift. Rigging and replacement of the across-deck (centre span) wires was a simple process and could be carried out within a matter of minutes.

The conning (control) position for the arrester system was located on the port side in the catwalk abreast the deck landing area. As soon as the

A classic picture of a Supermarine Seafire with 'sting' type arrester hook about to engage No.2 barrier, having overshot all nine arrester wires in HMS *Illustrious*[40]. No.1 barrier has not been rigged; its port and starboard tripod support structures are seen in the collapsed position. This pattern of the barrier net was developed early in World War 2 and remained standard until the mid-1950s. August 1950. (© Crown Copyright/MOD)

aircraft had picked up a wire the operator lowered all across-deck (centre span) bow-spring supports and safety barriers to provide a clear access for flight deck personnel and to allow movement of the aircraft forward or for striking-down into the hangar below.

Frequent landing into the cross-deck wires caused them ultimately, through stretching, to distort into a coiled configuration similar to a spring; this disappeared as soon as tension was applied for landing-on. This feature was inherent from the interaction between hook and rope, abrasion was inevitable and both hooks and across-deck wires were subjected to inspections every few landings; frequently the latter needed to be replaced. A rule-of-thumb check was to count the number of visible broken strands of wire within a distance of a yard run or so. If the tolerance was exceeded, the centre, cross-deck span of the rope would be renewed. Technical details of the wires used and their reeving are given in the chapter 'Wire Ropes'.

Early problems with the landing of American aircraft were expected through the US Navy using smaller arrester ropes, mostly with a diameter of ⅝in or ¹¹⁄₁₆in {16mm, 17·5mm}, with their aircraft arrester hooks correspondingly designed to match these dimensions. The behaviour of these hooks was closely watched and no serious trouble was experienced. With their propensity for flowery language, our American friends gave the name of 'pendants' to the across-deck wires. (It, perhaps, was picturesque to imagine a 15-ton aircraft hanging – horizontally as it happens – on its arrester wire as a jewel suspended from a necklace!)

With increases in aircraft performance, creating in turn a demand for increases in the performance of flight deck machinery, NAD later conducted in-depth investigations into the behaviour of wire ropes moving at speed within the entire arrester system. These covered the movement of wire ropes around pulleys and sheaves, the tendencies to whipping of long unsupported runs and the adoption of swaged terminations to shackles and dead-eyes to supersede the current capels – white-metalled sockets – and splicing applications.

Towards the end of World War 2, a new project was initiated on the development of an arrester-gear system for landing aircraft up to 30,000lb {13,500kg} all-up weight on to aircraft carriers with preparations for the first installation to be made at RAE. This is thought to have been the prototype for the shipborne Mark 11 arrester gear that was further developed into the Mark 12 hydraulic installation, the final unit to have been installed in the Farnborough complex. By 1955, its successor, the Mark 13 equipment, was installed at Bedford and remained there as the basic arrester unit until the Naval Air Department was disbanded.

Throughout the life of the hydraulic jigger type of arrester gear, RAE was seldom involved in the actual design and manufacture of under-deck machinery. This seemed to have been the prerogative of the Admiralty DNC department and their contractors, generally Messrs MacTaggart Scott of Loanhead, Edinburgh. The speciality of Naval Aircraft Department was research into the performance of such gear, hence the installation of typical units at Farnborough. Particular attention was given to the design and performance of arrester hooks, their installation on aircraft, together with the investigation of the hook/rope interfaces, and the rope behaviour when running at high speeds within the system.

Having said that, NAD was not averse to adapting the systems installed on the Farnborough Runway 04/22 and elsewhere, particularly the experimental installations at the Shoeburyness ranges! Cases in point are the merging of two Mark 4 American arrester units to form a single composite for special trials on both sites at Meadow Gate (flexible deck) and Laffan's Plain (High Speed Track and the latter stage flexible deck), to which references are made elsewhere in this text. RAE was also involved in the ultimate development of shipboard deck-arrester gear originally designated as the 'Water Spray Arrester Gear' but later known as the 'Direct Acting Arrester Gear' (DAG). This was a complete departure from the long-established hydraulic jigger principle.

The performance of an arrester system depended almost entirely upon the rating of the under-deck equipment. Throughout World War 2 there appears to have been only one major upgrading from the universal installation of the Mark 4 gear to a selective installation of the Mark 6 equipment. It appears that a Mark 5 system was in the design stage but proceeded no further. During this period the performance of the basic Mark 4 equipment was gradually extended to match the demands of each new aircraft until the physical limits of the hardware were reached.

Records of post-World War 2 arrester-gear installations are given in some detail in Whipp (*ARC 15,233 'A Note on Recent Developments in Naval Arresting Gears, October 1952'*), Chapman (*#20*), Loynes (*#61*) and Friedman (*#34*). It appears that the series of hydro-pneumatic, jigger type, arrester units passed from the prototype stage at Mark 3 through to Mark 14. Performance and engineering details of these installations and the ships in which they were fitted are to be found as Appendix 7 'Shipboard Arrester Gears'. It will be noted that the performance of the hydro-pneumatic machinery increased 11-fold over a period of 22 years from Mark 3 through to Mark 13. Figures quoted by Loynes (*#ibid.*) raises the performance factor of the Mark 13 even higher, to almost 15-fold through the following decade – this was expected to have formed the Mark 14 gear.

As the speed of engagement and the all-up weight of aircraft increased, particularly with the advent of jet aircraft, a new phenomenon appeared. It manifested itself in the first flexi-deck landings of the de Havilland Sea Vampire aboard HMS *Warrior* in November 1948 when a rapidly travelling kink appeared in the arrester wire.

After close investigation, Lewis (*#60 p.73*) reports:

'The engagement of the aircraft hook set off a tension wave along the centre span and caused the wire to stretch locally. Even at moderate entry speeds an effect of this could be seen, in that a small vee formed in the wire with its apex at the hook engagement point. Instead of allowing the wire to form a progressive vee, rendering from the deck edge sheaves in two straight lines, this strain-induced vee simply widened its base, until it reached the deck edge sheaves, the wire appearing to pay out around the inner angles. Arriving there, it partly reflected back along the centre span and partly travelled onwards into the reeving. A more or less violent oscillation was set up with peak to peak variations in rope tension reaching up to 10 tons at the natural frequency of the system. The reflected waves added to the tensions in the centre span and could triple them momentarily. Travelling onwards, the disturbance arrived at the unit crosshead just when it started to move, under the influence of the wide orifice area provided for that part of the arrest. The relative slackness in the wire under these conditions induced yet more impact loading, which added to the oscillations. These tension waves, travelling at about 9,000ft/sec, damped out as the arrest proceeded and the designed fluid pressure control was only then able to come into play. But as speeds got higher and higher these tensile fluctuations became more persistent and showed signs of overwhelming the pressure control altogether. Sheave bearings, wire slippage, wire guides, piston and gland friction all added their quota and the net effect was to add to the instabilities and to cause high, unpredictable 'g' loadings on the aircraft.'

A new system of under-deck machinery – the Direct Acting Arrester Gear – was developed at RAE Bedford. It was brought into use in 1967 with HMS *Eagle*[51] being the first ship to be fitted with this equipment (a single experimental unit) with HMS *Ark Royal*[55] following in 1970 with a full installation. This system is discussed in a later chapter.

SAFETY BARRIERS

The first safety – emergency – barrier to be found on board ship was that erected on the after landing-on deck in HMS *Furious*. This was during the early years (*c.*1918) when this ship retained her bridge and funnels complex between her forward and after flight decks. The intention was to prevent aircraft from landing into all this midships hardware in the case of overshoot. The structure was a large goal-post arrangement, spanning the width of the deck, with a number of ropes (hemp?) suspended from the crossbar. An illustration of this system is given in Rawlings (*#81 p.95*). It was effective but any aircraft sliding into this arrangement would suffer serious damage to the fabric and structure of its airframe and wings.

References to safety barriers in the previous chapter relate to carriers with full-length flight decks. Their introduction was to protect aircraft

Model of a fleet carrier of the 'Illustrious' Class in the FAA Museum. This view clearly shows the early 1940s type and rigging of No.1 safety barrier with No.2 barrier collapsed while the Seafire taxies forward. (Author via FAAM)

De Havilland Vampire F.1 (TG285 – a non-flier by this time), adapted for use in barrier trials on Farnborough Runway 04/22. The aircraft is facing Jersey Brow and is abreast arrester pit A.4. The bow-spring supports for the arrester wires serving pits A.5 and A.6 can be seen collapsed behind the aircraft. The tail rope is hitched to the arrester gear in pit A.6. The twin tow ropes are connected to the shuttle of the BH.5 Catapult some 400 yards distant. April 1950. (© Crown Copyright/MOD)

which had already landed on and had been taxied forward into a deck park – a holding position prior to being struck to the hangar below. The problem was caused through the rapid recovery of aircraft from a sortie. With aircraft arriving at 30-second intervals, and the operating cycle of the lift as one minute, it does not need a mathematical mind to recognise the magnitude of the traffic jam accruing at the forward end of the flight deck. The advantage of a barrier to protect the forward deck park becomes obvious.

First, a description of the situation. On returning to its carrier, each aircraft followed a routine circuit of the ship and would be guided into the approach by the DLCO (Bats to his friends). The pilot would follow Bats' instructions meticulously and, hopefully, would be brought in to touchdown, picking up an arrester wire at the after end of the flight deck. As the aircraft picked up the wire, it was brought to rest within a distance of some 150ft {45m} from the point of contact. This was the norm for the first half dozen or so wires. If for one reason or another a pilot missed the earlier wires he could snatch one of the later wires and his landing would be rougher, the retarding force would be greater and the pull-out of the arrester wire would be something less. If he missed all the wires, there could be a catastrophe.

The method of operation was for a barrier to be used only when work was proceeding on the flight deck forward of the landing area when flying was in progress. If the deck was clear of machines and personnel, only the arrester wires would be deployed, the barriers would be left in their stowed position, lying flat on the deck. This would permit a 'bolter' to go round again to make a good landing in a second attempt. However, if more than one aircraft were in the landing circuit, as soon as the first had been recovered and taxied forward of the landing area and should another be following immediately, at least two of the barriers would be deployed. This would provide a measure of protection for those working on the first aircraft, either 'spotting' in a for'ard deck park or striking down into the hangar. Regrettably it has to be admitted that on occasion even these precautions were inadequate and much thought was given to improving the procedure.

When barriers were in operation, the sequence of working was for all arrester wires to be deployed with at least two of the three barriers also in service. No.1 barrier would be situated aftermost, a little forward of the last arrester wire, with the remaining two in sequence. If there was adequate space for'ard of the barriers, Nos 2 & 3 would be deployed, giving maximum space between wires and barriers. A deft operator at the barrier controls would collapse the barrier as soon as he saw that an aircraft being recovered had hooked a wire, allowing the aircraft to taxi forwards over the barriers. As the aircraft passed, so each barrier would be re-deployed for the next landing.

If an aircraft impacted the barrier – perhaps even after engaging a late wire – emergency procedures would be implemented immediately and the recovery of aircraft remaining airborne would be delayed. The condition of the aircraft impacted with the barrier would be rapidly assessed with the wreckage either hauled forward for repair or, if badly damaged, immediately pushed 'over the side'.

The safety barrier in 1940 was a three-wire net supported above deck level and connected to a standard under-deck, hydraulic jigger type arrester gear. It next went into the phase where different designs of wire net were produced, often to meet needs related to a particular marque of aircraft. Experiments were not completely satisfactory and attention was drawn to the properties of nylon webbing for the net, but still connected to a jigger under-deck gear. From this proceeded the final,

Hawker Sea Fury of the Netherlands Navy (RNNAS) landing onto HMNLS *Karel Doorman* (ex-HMS *Venerable*) in her first commission, is about to become a nasty accident. With arrester hook deployed, the Sea Fury has failed to pick up an arrester wire and appears to be about to snag No.2 barrier with its hook. No.1 barrier has not been deployed. *c.*1949. (© Crown Copyright/MOD)

Fairey Firefly AS.5 (WB378) coming to grief in HMS *Vengeance*. The arrester hook has not been deployed and the aircraft, having already engaged No.1 safety barrier, is about to impact No.2 barrier. This close-up shows the construction of the barrier net. March 1950. (© Crown Copyright/MOD)

satisfactory system in which the nylon webbing net was attached directly to undrawn nylon energy absorber packs at deck level.

In the field of flight deck machinery almost the entire development of barriers was the responsibility of NAD. In general, the normal run of arrester gear and accelerators/catapults – especially the hydro-pneumatic equipment – were the responsibilities of the Admiralty and their contractors, particularly Brown Brothers and MacTaggart Scott, both of Edinburgh. Exceptions were dealt with at Farnborough. Catapult trolleys and other above-deck machinery were handled both by RAE for prototypes and research, and Chatham Dockyard for production units.

PRE-WAR RIGS

The fitting of safety barriers, or crash barriers, in aircraft carriers was first discussed at a meeting of the Joint Technical Committee of Aviation Arrangements in HM Ships in June 1934. An experimental barrier was ordered for trials to be carried out at a dockyard prior to the introduction of such a barrier in HMS *Ark Royal*[38]. The specification was for the arrestment of an aircraft of 6,000lb-10,000lb {2,700kg-4,500kg} with an entry speed of 40kts {21m/sec} and within a pull-out of 25ft {8m}, giving a mean retardation of 2¼g. This barrier consisted of a pair of wires stretched across the flight deck. These were cross-connected by a wide mesh net and the whole supported by stanchions near the deck edges. The ends of these wires were coupled together and led to an under-deck hydraulic braking system. The stanchions served to hold the net in the raised position until it was engaged by the aircraft. Thereafter, the net was pulled clear of the stanchions and the arresting wires rendered from sheaves at deck level. (This was the system subsequently adopted by the US Navy as its standard.)

By June 1937 such a barrier was installed for trials purposes on Farnborough airfield (hardly a dockyard!) on a site which remains to be identified. The first experiments were not wholly satisfactory and a number of changes became necessary, the principal ones being:

a) re-rigging of the system so that the barrier net was kept taught throughout its travel, and

b) rendering of the net cabling continuously from the top of the support stanchions, rather than at deck level.

On completion of these modifications further trials produced the desired results within the upper loading limit, the maximum retardation being recorded at 4½g. This barrier, with a slight adjustment to accept aircraft of AUW of 11,000lb {5,000kg}, was installed as a single unit in HMS *Ark Royal*[38] in readiness for her commissioning in 1938. For HMS *Illustrious*[40] and all subsequent ships, similar installations with stanchion rendering were adopted but with the pull-out increased to 40ft {12m}.

This first mechanically operated safety barrier consisted of a simple transverse net, rendering from collapsible tripod stanchions of the type which continued to be seen in RN ships throughout World War 2. It was

Blackburn Firebrand TF.5 (EK731, Code A 103 of 813 Squadron in HMS *Indomitable*, having damaged its undercarriage and floated over the arrester wires has engaged No.2 barrier and is brought to rest against No.3 barrier which is still deployed. September 1951. (© Crown Copyright/MOD)

designed originally to engage the nose of a 'bolter' (runaway aircraft) and to bring it to rest within a short distance, more often than not causing a fair amount of damage to the aircraft – particularly the engine and propeller – hopefully less to the aircrew. The impact area was a net formed of a pair of parallel, horizontal ⅞in {22mm} diameter, steel wire ropes, spaced 3ft {0·9m} apart by a rigid spacer bar, with some wire interlacing, spanning the flight deck. It was connected at each end through load-equalising swivel fittings to the wire reeving of a hydraulic jigger unit. This was a repeat of the under-deck arrester gear machinery, but was set at a heavier rating with reduced reeving, thereby producing a shorter pull-out. The height of the topmost wire above deck was set at 6ft-4in {1·93m} which became the standard for subsequent installations. The arrangement was simple and effective and had the advantage of extreme economy of deck space since the maximum pull-out of 40ft {12m} represented the total space required. The ends of the barrier were supported by collapsible (retractable) steel stanchions, forming tripods of substantial dimensions. Occasionally the stanchion on the starboard side, if adjacent to the island structure, was replaced by a quadrant fixed to the wall of the island. This particularly was the case in the light fleet

carriers of the 'Colossus' Class. There would still be a fairlead and a channel for the arrester cable to disappear below deck level, as for the port side. These stanchions would remain deployed throughout the arresting process; they were not designed to collapse automatically on impact – although the Engineering Department in HMS *Fencer* is reported to have introduced a measure of automation into this activity. Being pneumatically actuated, they were rapid and very noisy in operation with their characteristic 'hiss – clunk'.

POST-WAR RIGS

Such was the situation up to and during World War 2. It was not until 1946 that a serious need was felt for further attention to be given to barriers. Boddington (*#213*) in January 1947 had been looking at developments over the war years. By this time the conditions for arrestment had increased to:

a) the maximum weight of aircraft to be arrested 30,000lb {13,600kg}
b) the maximum speed of entry 75kts {38m/sec}
c) a new requirement for accepting twin-engined aircraft with either normal tail-wheel or tricycle undercarriages.

Consideration was given to the adoption of the American system – a heritage of cross-deck working with the US Navy during the war years in the Pacific – but, in weighing up all the factors and experimental evidence, it was decided to retain and adapt the existing system. The upgrading would need to include:

a) the barrier to engage the aircraft at the top of the undercarriage
b) the barrier to have maximum pay-out of 100ft {30m}
c) no objection raised to using the American system provided there would be no slack in the initial pull-out (as was the case in current USN ships)
d) arrangements to be devised to accommodate twin-engined and nose-wheel aircraft.

Fairey Firefly AS.6 has lost its tail, possibly by striking the round-down on approach to HMS *Illustrious*. With propeller broken, its nose is gathering four arrester wires. *c.*1948. (© Crown Copyright/MOD)

Fairey Firefly 5 (WB267 probably of 814 Squadron embarked in HMS *Vengeance*) with arrester hook deployed has missed all arrester wires and, with port undercarriage collapsed, engaged No.2 barrier which has brought it to rest against No.3 barrier. September 1949. (© Crown Copyright/MOD)

The equipment for flexible deck trials at the Meadow Gate site [O.15]. The installation was readily adapted for experimental work into the behaviour of contra-prop aircraft when engaging shipboard safety barriers. A hybrid Supermarine Seafire F.45 dummy has been adapted and is ready for launching into the experimental safety barrier. October 1950. (© Crown Copyright/MOD)

Provision also was to be made for the future operation of jet aircraft. The policy was to match the barrier requirements to those for the companion arrester-gear installation. These were envisaged as being 100kts entry speed for 15,000lb aircraft {51m/sec, 6,800kg} to 85kts at 30,000lb {44m/sec, 13,600kg}, the pull-out to be subjected to experimental investigation.

A research programme was drawn up in which studies, firstly on a model scale, were to be conducted and attempts made to correlate them with full scale experiments. The next phase was to include the immediate problems facing barrier engagement of the Sea Hornet twin-engined aircraft. Finally, barrier engagements with all types of aircraft, including full-scale trials, at Farnborough were to be carried out. These, and many more, investigations were made throughout the following decade and continued at Bedford after the NAD transfer in the mid-1950s; this developed into a virtually continuous research programme. The NAD project engineer overseeing this exploratory work at Farnborough was P. R. ('Pincher') Martin whose comprehensive reports (#274 – #275) form the basis of the following account.

Subsequent to the installation in HMS *Ark Royal*[38] all Fleet Carriers were equipped with three (rarely two) safety barriers located forward of the landing area, generally positioned abreast the island. Two barriers were the norm in the light fleet carriers and in the Lend/Lease wartime CVEs from the USA. If there were aircraft or personnel ahead of the island during the recovery of aircraft, the barriers would be deployed; the operator would collapse them when the aircraft had made a good arrested touchdown. If this were not the case, the barriers would be left up to bring the aircraft to rest.

Some ships mounted an odd number of arrester wires with three barriers; in these cases the last wire – generally No.9 and known as the 'trickle' wire (or by other blasphemous, pilot-induced epithets) – would be reeved to the same under-deck unit as No.1 safety barrier. Nos.2 & 3 barriers also would be rigged to a common unit with a reduced pullout. Throughout the war years the only material attention given to the system was the inclusion of a third horizontal wire in the construction of the net itself, and increasing the sizes of wires used in the net. Whenever the capacity of the main arrester gear jiggers were upgraded during a refit, so the matching units serving the barriers were likewise upgraded.

With the post-war introduction of high performance aircraft, a need for the development of this equipment was foreseen. The first remit was to accommodate increases in both weight and speed of entry, then to accept twin-engined aircraft. Large single-engined machines with counter-rotating propellers (contra-props) were introduced and, finally, came the jet era. Each of these cases required and was given attention in its own particular field, sometimes involving a new development or procedure in its progress. For the first case, that of a single-engined machine of advanced performance, there appeared to be no need to depart from current practice, apart from the ability to accommodate a dimensionally bigger aircraft. With the current physical size of the barrier net, the principle of arresting the aircraft was not to pick up the nose as before but to look at the arrestment taking place at the top of the undercarriage legs. With these conditions, the likelihood of greater damage being caused to the aircraft became evident.

One of the most common forms of damage to an aircraft was the collapse or removal of its undercarriage and, in order to be sure of arrestment, two barriers, 40ft {12m} apart, were always deployed. If the failure of an undercarriage occurred in impacting the first barrier, the

A successful barrier engagement with a contra-prop Seafire F.45 (LA439). The barrier net is entangled around the undercarriage and the propeller is a write-off. The machine would probably be used again; there were four aircraft (LA439, LA442, LA448 & LA450) and five sets of contra-props, so one or another airframe was employed more than once in these trials. October 1950. (© Crown Copyright/MOD)

A contra-prop Seafire F.45 has engaged the barrier, overturned and landed inverted on the concrete hard-standing along the starboard side of the trials dummy deck. The contorted tangle of the barrier net around the undercarriage is apparent. October 1950. (© Crown Copyright/MOD)

A contra-prop Supermarine Seafire F.45 (LA448) having failed to pick up the experimental safety barrier – or lost its undercarriage in the process – does some gardening on the verge of Mosquito Way. December 1950.
(© Crown Copyright/MOD)

aircraft dropped on to the deck and was halted by the second which engaged the nose. In the immediate post-war years the first barrier to be encountered was rigged with a longer pull-out which produced lower retardation forces and thereby reduced the incidence of damage.

With the significantly increased performance of high-powered aircraft, it was necessary to examine the behaviour of aircraft and barrier under high speed impact. The pilot of a twin engine or a jet engine aircraft, facing a barrier with no substantial protection ahead of him, felt very exposed and vulnerable. With the introduction of a powerful single-engined machine with a large five-bladed metal propeller – such as the Hawker Sea Fury – the problem was different. Previously, aircraft with wooden propeller blades, on entering a barrier, inevitably smashed the blades; this feature helped in bringing the aircraft to rest. This condition no longer would prevail, the propeller blades would be damaged, probably badly bent; they could even become fragmented with bits of metal scattering all over the flight deck. This called for the increase in the size of the wires forming the net. The experimental equipment installed at Farnborough for this purpose was

further developed for similar work with jet-engined aircraft. This was due to the absence of a large mass of engine in front of the pilot, as in a single, piston-engined aircraft.

TWINS & PROP-JETS

With twin-engined machines the danger of two propellers entwining the wires around themselves, particularly if one or more wires were severed in the process, had to be addressed. The now highly vulnerable aircrew had become of paramount importance. The solution adopted was to replace the net with a single heavy duty centre span which would impact the aircraft at the top of its undercarriage. This would reduce any tendency for the propellers to pick up the wire and throw it through the cockpit. During the late 1940s a considerable amount of practical research went into addressing this situation concomitantly with the introduction of the Sea Hornet aircraft with its wooden construction. This twin-engined aircraft featured 'handed', metal-bladed propellers, no protection of the aircrew, and a single-leg, spindly undercarriage with a tendency to pitching and overturning if the barrier was trapped by the undercarriage. All these conditions were foreseen with the experimental use of its Mosquito predecessor. There followed, at Farnborough, a series of trials at one-eighth model scale but, through inadequacies, due mostly to the scale effect, these tests were inconclusive and the need for full-scale trials became apparent. Throughout these tests attention would need to be given to the performance or stability of the aircraft during engagement and the development of the barrier net itself. The basic investigation, for a start, was to major upon the use of a 'long pull-out' system, ie a pull-out of 80ft {24m} or more as against the standard 40ft {12m} of existing systems.

The equipment at Farnborough comprising a special barrier support and various cabling was erected on Runway 04/22, using the Mark 6 arrester-gear installation, suitably reeved for barrier use. The aircraft was positioned south of the arrester area and hauled into the barrier by a ⅞in {22mm} diameter wire rope taken up to Jersey Brow and connected through a 2:1 reeving to the towing shuttle of the BH.5 cata-

The model scale catapult and deck-landing set-up in the NAD laboratory. A Sea Hornet model is loaded for engaging a standard World War 2 type of barrier net. June 1950. (© Crown Copyright/MOD)

A de Havilland Sea Hornet has been towed into a nylon cord safety barrier net system on Runway 04/22. In this experiment the engines would not have been running. The cable across the rear of the cockpit canopy would not have been welcomed by the pilot! August 1947.
(© Crown Copyright/MOD)

A non-airworthy de Havilland Hornet F.1 (PX211) has been towed into an experimental development of the safety barrier. The original barrier nets were made of unyielding steel wire ropes but a system adopting flexible nets as replacements was developed. This shows one of the interim experiments – on a propeller-less aircraft – using vertically hung bunches of hemp or nylon ropes, and its results. August 1947. (© Crown Copyright/MOD)

pult, some 500 yards {460m} distant. To avoid overshoot, although the trials aircraft was linked to one of the arrester gear units on Runway 04/22, a further catchment net was provided as a 'long-stop'. This last-ditch facility, of some 200 yards {180m} length, spanned the runway at the end of the pull-out of the arrester system. It was not connected to any arrester system.

Dummy runs were made using redundant Barracuda and Firebrand airframes suitably ballasted. With the technique having been proved, a Hornet airframe was next on the list. At this time the RAF was phasing out its Hornet aircraft and one machine (PX211), with unusable engines, was diverted to RAE. For a while, this was used as a grounded dummy, being repaired each time it incurred damage during an engagement.

The view being that the standard barrier net was unsuitable, early experiments with the Hornet concentrated on the entrapment of the aircraft by its impacting a single wire barrier at the top of the undercarriage. In addition, a further wire at high level was allowed to deploy above the cockpit to engage the tail of the aircraft as it pitched forwards; a tendency displayed early in the trials series. This worked reasonably well but it was feared to be not too successful if the aircraft had lost, or had failed to deploy, its undercarriage at the time of engagement with the barrier.

The trials reached the stage (May 1948) where engagement with engines running was required. Having witnessed so many 'dry' runs with the aircraft finishing perched up on its nose, this problem needed to be addressed. There was no lack of ingenuity here. The engine ignition

Experimental safety barrier rig on Farnborough airfield. This is titled as the Mark 5 net with two undrawn nylon packs per side as retarder units, assembled ready for test. March 1957. (© Crown Copyright/MOD)

De Havilland Sea Venom FAW.20 (WM503) aircraft arrested by a barrier using undrawn nylon packs. March 1957. (© Crown Copyright/MOD)

system was wired to a telephone jack socket, fixed to the nacelle, with its contacts held open by a plastic plug. This plug was tied to the ground by a long lanyard which, as the aircraft taxied over the anchor point, was dragged out and the ignition short-circuited by the closing of the contacts in the socket and, hey presto, the engines stopped! Three more ex-RAF Hornet aircraft (VA964, VA965 & VA966) were used for these experiments with a variety of nets – curtains – being deployed. Lewis (#60 p.83) records that a dozen redundant RAF Hornets were allocated to RAE for trials into a variety of barrier nets. These 'dummies' were bereft of their outer wing sections and were either taxied or towed into the barrier systems with their engines running. Eventually it was found that a curtain comprising nylon cords suspended from a high-level wire was the ideal substitute for the original steel wire net.

A barrier unit, matching that at RAE, was installed in HMS *Implacable*, which was about to operate Sea Hornet squadrons, with sea-going trials arranged for January 1949. Dummy runs again were made using the ship's BH.III accelerator to tow the aircraft into the barrier. In September 1949 a Sea Hornet 20 (TT202) was flown on to the ship and, after its engine ignition systems had been adapted to give automatic cut-out, was used for a 'live' trial. The aircraft was wrecked but, from all relevant counts, the system was considered to be suitable for its purpose and was taken into use. During the following nine months' commission the barrier was engaged six times, performing correctly on each occasion. This process was the forerunner of the all-nylon barrier subsequently adopted for operation of jet aircraft.

Concurrently, the advent of the prop-jet aircraft was investigated, particularly in the case of the Fairey Gannet that had its cockpit well forward and the behaviour of an entry into a barrier could not be predicted. A hybrid was produced at Farnborough using a Westland Wyvern TF.1 (VR133) as a base aircraft. A plywood mock-up to represent the forward cockpit position was fitted and an assembly using two Spitfire tail wheels attached to the engine casing simulated the Gannet nose-wheel. The towing system was identical with that adopted for the Sea Hornet trials.

To cope with the contra-prop situation, involving the Gannet and Wyvern aircraft, the steel wire net was retained but the vertical spacing of the horizontal wires was reduced. A great deal of research was devoted to this. Model scale trials exposed the problem but the results were not capable of extrapolation; full-scale research was found to be necessary. There was concern about the possibility of a scissors action between the two propellers and their combined destruction of the net. For this purpose, three redundant Supermarine Seafire 45 aircraft (one of them is recorded to have been LA448) were made available, together with five sets of Rotol contra-props. These machines were reasonably similar in profile to the Wyvern, but at ⅔ scale, allowance was made for this scale effect during the trials. The aircraft were adapted for catapulting and were launched, with engines running, into a stanchion-supported barrier net. The location for these trials was the redundant Stage 2 installation (glider launching) in the development of the flexible deck project on the Meadow/Ively Gate site. The barrier element was connected to the existing duplexed Mark 4 American arrester gear on this site.

Each aircraft was provided with 'add-on' fitments to the fuselage, embodying catapult spools to suit the 'Hotspur' launching trolley on the Stage 2 rocket catapult. The trolley was modified to be capable of launching the Seafire in a 'datum horizontal' attitude on to a temporary steel deck. Martin (#273) records that sixteen 3in {76mm} rocket motors were required to give the aircraft an entry speed of about 60kts {31m/sec}. The aircraft was set up on the trolley with its engine running to simulate a real recovery situation. The actual flight path between catapult and barrier was less than 20ft {<6m} but this was adequate to provide impressive and decisive results, most of which were unpredictable from the preliminary experimental work.

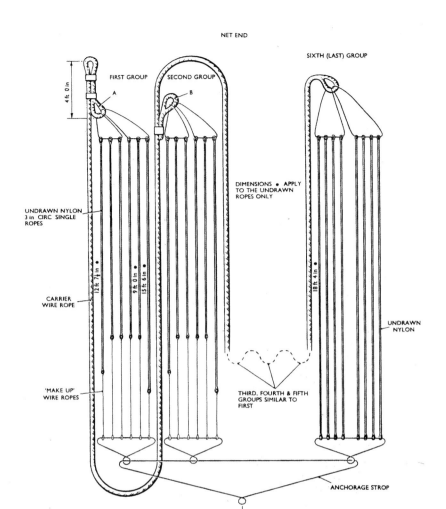

4 ft 0 in

A

B

DIMENSIONS ● APPLY
TO THE UNDRAWN
ROPES ONLY

UNDRAWN NYLON
3 in CIRC SINGLE
ROPES

CARRIER
WIRE ROPE

12 ft 7¾ in ●

9 ft 0 in ●

15 ft 6 in ●

18 ft 4 in ●

UNDRAWN
NYLON

'MAKE UP'
WIRE ROPES

THIRD, FOURTH & FIFTH
GROUPS SIMILAR TO
FIRST

ANCHORAGE STROP

TO ANCHORAGE

DECK ANCHORAGE END

Diagrammatic arrangement of the undrawn nylon packs used for barrier arrestment in carriers with angled decks. In operation, the four shortest lengths of nylon in the first group would be the first to fail in tension, followed by the next two, then similarly in series for the next four groups. If the aircraft had not been brought to rest by then, the eight full-length cords in the sixth group would complete the job. (© Crown Copyright/MOD)

JETS & NYLON

The later stages of safety barrier development were interesting in their employment of new materials to reduce to a minimum the damage to aircraft (and aircrew) through a barrier engagement. An early stage in this project was the adoption of a series of ten vertical strips suspended from a wire stretched high across the deck. These tapes were of thin steel about 6 inches {150mm} in width and disposed to permit five strips to impact with each wing of the aircraft. The results, although not ideal, were encouraging, with further development proceeding from this point. The first of these changes was the substitution of the suspended steel tapes by nylon webbing tapes of about 3in width and ¼in thickness {75mm x 6mm}.

At this time, attention was drawn to the advantages offered by the inherent physical properties of undrawn nylon. Not only was nylon used in the barrier net itself but the elastic properties of the material were employed as the arresting element in its own right. In adopting the use of packs of undrawn nylon the under-deck hydro-pneumatic jigger of history was superseded.

The first change to be introduced was for the across-deck catchment 'net' to be replaced by a multiple-strand nylon (or equivalent) barrier, but still connected to the existing under-deck decelerating machinery. Its pull-out distance and rate of retardation were the same as before. The new net was much softer and flexible, amenable in yielding to the contours of an aircraft fuselage.

Further development was necessary with regard to the barrier entry of jet aircraft. There was neither a large engine nor a propeller to act as a protective shield to the aircrew. Something drastically new was required and Martin (#275) in 1953, reporting on an intensive series of wide-ranging experiments, pointed the way to the adoption of undrawn nylon as an energy absorbent. These trials were more complex than the 'contra-prop' Seafire trials. In this case a mock-up of an aircraft fuselage was towed into a barrier system built across Runway 04/22 at Farnborough, adapting the installation previously used for the Sea Hornet experiments. Here the first nylon units were demonstrated using this material only for the net itself. The new net was some 14ft {4·3m} in height with vertical tapes loosely suspended between, but firmly attached to top and bottom horizontal wires which, in turn, connected to the under-deck jigger. The system was designed to permit a pull-out of 200ft {61m}. The principle was to snatch the aircraft at the leading edges of the wings, with the tapes parting to each side of the cockpit *en passant*. This was found to inflict minimum damage to the aircraft and the system became the norm.

Coupled with the development of arrester gear in that the pull-out of the wires was being extended from the old standard of 150ft to somewhere nearer 200ft {45m, 61m}, and with a similar increase in the launching stroke of the accelerators, flight deck space aboard ship was becoming rather congested. The desirability of extending the pull-out of the safety barriers could only aggravate the situation. If the barriers were moved further aft, then the pull-out of the later arrester wires would encroach upon the barrier area insofar that an aircraft picking up a late wire inevitably could not avoid a barrier contact. This was a situation incapable of being resolved with a straight deck carrier without restricting the forward deck park.

The project was short-lived; the introduction of the angled deck in 1955 obviated the primary need for safety barriers and the installations were removed in a subsequent refit of each carrier in turn (vide 'Post-War Developments – Angled Deck'). However, the last three fleet carriers, each when fitted with angled deck, carried a single safety barrier of a multiple-nylon-strand net enabling the qualities of undrawn nylon to be exploited as an energy absorption device. The system was no longer adopted to protect aircraft in a deck park forward of the landing area but for recovery of an aircraft which was damaged to the extent that deployment and use of its arrester hook was inhibited.

At this stage the barrier systems project was transferred to the new site at RAE Bedford with its purpose-designed test facilities. Further development of the net took place until it appeared as a spread of some 24 vertical tapes of about 30 feet {9m} in length, strung along a 'rope' suspended across the deck on a pair of stanchions 24ft {7·3m} in height. The upper end of each tape was suspended from the top 'rope', the lower end being secured to another 'rope' which lay on the deck, each tape having about six feet {1·8m} of its length lying loosely along the deck.

In the normal process of making synthetic yarn for conventional man-made fibre ropes, the nylon filaments are stretched, or drawn, until the required physical properties are attained. This has certain similarities to the drawing of wire for steel wire ropes. By eliminating this drawing process the plastic extension of the filaments can be used to provide an energy absorber. The undrawn nylon filaments are spun into yarns of 90-denier; these are then twisted into rope yarns from which rope strands and finally ropes themselves are formed. For example, a 3in {75mm} circumference undrawn rope will consist of three strands each comprising 8,500, 90-denier yarns of 10 filaments. The energy absorption capacity of a 3in {75mm} rope is approximately 50×10^3 ft.lb per pound weight {16×10^3kg.m/kg}, and the extension at break generally is between 23 percent and 400 percent (This compares favourably with conventional nylon rope which has only one-third of this energy absorption and a greatly reduced, down to 50 percent, extension.)

For undrawn nylon, 6 percent of the energy during drawing is not released as heat but is absorbed as elastic strain energy. This is returned when the load is removed; therefore it is not possible to combine all the nylon into one large rope as this would give an unacceptable rebound energy, tending to propel the aircraft backwards at some 20–30kts {10–15m/sec}. This is overcome by arranging several groups of rope to be drawn in sequence. When the aircraft has come to rest, most of the groups of ropes would then either have been broken, or not yet loaded, but the aircraft would still tend to move rearwards at the end of the engagement. However, the residual rebound was minimised and partially nullified by idling engine thrust and rolling resistance.

A de Havilland Sea Vixen 'iron bird' dummy has been towed into a nylon safety barrier and successfully brought to rest. The rectangular plate fitted to the nose of the dummy formed an interrupter to a light ray which triggered the recording equipment as it passed through a datum point along the track. March 1957. (© Crown Copyright/MOD)

Supermarine Attacker dummy aircraft, 'iron bird', being hauled by the accelerator into the experimental nylon safety barrier aboard HMS *Eagle*[51]. April 1952. (© Crown Copyright/MOD)

McDonnell Douglas Phantom 'iron bird' at the moment of engagement with the nylon net safety barrier. (© Crown Copyright/MOD)

Hawker P1127 (XP980) experimental vehicle, under radio control, at the moment of impact with the nylon net safety barrier at Bedford. *c.*1976. (© Crown Copyright/MOD)

different aircraft types, more than one pack could be used each side of the net. The pay-out of packs depended upon actual energy to be absorbed and the number and size of particular packs used. The number of packs was determined to allow ample margin of pull-out to cover the performance variations between packs and possible excess engaging speeds. The packs were located at the edges of the flight deck and were attached at the static end to deck anchorages. No part of this system – apart from the ready-for-use store – was deployed below flight deck level.

The illustration shows a Mark 2 pack – a typical, middle-of-the-range unit – which used 3in {76mm} circumference undrawn nylon ropes arranged in six , sometimes eight, groups. Each intermediate length in turn was looped to a 2¾in {70mm} circumference steel carrier rope which connected to the barrier net and gave a maximum pull-out of 200ft {61m}. The nylon ropes, the carrier rope and all accessories were assembled together in pockets and sewn up in a rectangular bag, with the carrier rope and anchor strop just protruding at opposite ends. This valise also had two sisal tie ropes for fastening the cover to a deck anchorage to prevent its being drawn along the deck in an engagement.

The complete unit, provided as a flat pack, was 29½ft long by 4ft wide {9m, 1·2m} and between 4in and 9in {100mm, 230mm} thick. It was deployed in a folded state as a package approximately 8ft x 4ft x 1ft {2·4m, 1·2m, 0·3m}. Its weight was 750lb {340kg} and it was fitted with eight carrying handles along each edge, requiring sixteen men ('with a yo-ho-ho and a bottle of rum') to handle it. The barrier nets themselves were made to a 75ft span and were about 30ft in height {23m, 10m}. The lighter of the two more popular sizes (the Mark 5) weighed 600lb {270kg} while the heavier, multiple-packaged units (Marks 7/9) for arresting Phantom aircraft, totalled 1,700lb {770kg}. In spite of its bulk and weight, the whole barrier system could be rigged within five minutes – the best time achieved was under three minutes – well within the time for a 'bolter' aircraft to fly around again before making an emergency landing. Obviously, a large flight deck party was

In essence, a pack consisted of a number of groups (generally six) of undrawn nylon ropes. Each successive group was picked up in turn by a 'carrier' rope when the preceding group had been pulled out to its full extension. This carrier rope was shackled to the safety barrier net during rigging and transferred the arresting forces to the nylon packs. Thus not more than the equivalent of one group at each end of the barrier net was under load at any one time, although the second group began to come under load before all ropes in the preceding group had broken. The individual ropes in each group were of varying lengths to give as smooth as possible load/extension characteristics. To cater for the range of

A pilot's-eye view of the angled deck in HMS *Ark Royal*[55] with the nylon safety barrier net rigged and connected to a pair of undrawn nylon arrester packs positioned on the flight deck. The port and starboard projector landing sights are seen. (Author)

129

De Havilland Sea Vixen FAW.1 (XN699?) on impact with the nylon safety barrier aboard HMS *Ark Royal*[55]. This emergency landing was expected – probably through a fault in the deployment of the arrester hook – as the arrester wires remain in the stowed position. (© Crown Copyright/MOD)

necessary, and available, to rig the system, some 30 men of the flight deck party working on the net alone. Needless to say, due to its lengthy pull-out, this nylon barrier system was employed only in carriers with angled decks. More recently, a similar system has been widely adopted for use against overshoot on airfield runways.

Inevitably, by its very nature, a barrier engagement produced an interesting, perhaps exciting, spectacle, particularly during the research and development of the relevant equipment. A procedure was established for a ship's photographer to be present during deck landing activities at sea to record any recovery of an aircraft which appeared to be barrier-bound. Such records, ciné or still, were of value during the experimental stages of barrier design; copies were sent to Farnborough for evaluation. At the demise of NAD, many of these photographs were passed across to the FAA Museum archive.

ARRESTER HOOKS

STANDARD

The first deck landings into shipboard mechanical braking systems were undertaken with Fairey IIIF aircraft. Taylor (*#95 p.153*) writes of a Mark IIIM aircraft (S1317/F1057) having been fitted with a 45lb {20kg} arrester hook then being used for deck-landing aircraft. (It would be interesting to have details of such a heavy hook. Those in use by the early 1930s weighed less than 8lb {3·7kg}). Taylor (*#ibid.*) continues in saying that Fairey IIIFs were not fitted operationally with arrester hooks but that a Mark III machine (S1781/F1518) was used for early experiments with a rear fuselage installation. This was a swinging, triangular steel frame carrying a spring-loaded hook at its apex. It normally lay flush with the fuselage and was released by the pilot. This became the prototype of the 'V'-frame system.

From the early 1930s, arrester hooks and their installation in naval aircraft followed this standard pattern. The hook was affixed to the apex of a 'V'-frame (simulating the letter 'V') pivoted downwards from anchor points in the fuselage, forward of the tail wheel. This method of suspension allowed the hook to swing freely in a fore-and-aft direction below the fuselage, permitting it to engage an across-deck wire (centre-span) when landing into an arrester system. Of the nine aircraft types listed in the 'Introduction' the first seven – all except the two Supermarine amphibians – were fitted with units of this pattern. Some Sea Otter aircraft subsequently were fitted with hooks of the 'sting' type. These will be discussed later.

A latching device was fitted to the underside of the fuselage that would allow the arrester hook to be housed firmly within the airframe when a cross-deck wire had been picked up. An illustration shows the latch unit (snap gear) as installed in a Fairey Swordfish aircraft. Another

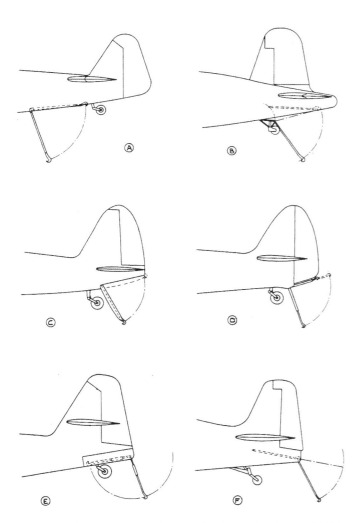

Arrester hook locations on typical aircraft. Samples are based upon: (A) Fairey Firefly; (B) Chance Vought Corsair; © Westland Wyvern; (D) Hawker Sea Fury; (E) Fairey Spearfish and (F) Grumman Hellcat. Application (A) was appropriate for both 'V'-frame and single arm installations; all the remainder were single-arm units. (© Crown Copyright/MOD)

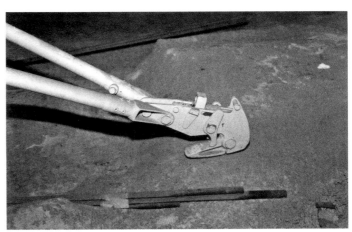

The 7,000lb 'V'-frame arrester hook on Fairey Swordfish I (P4139) shown deployed. From the 1930s, this hook was the standard fitment for all FAA aircraft designed to be arrested within that weight at 2g deceleration. (Author via FAAM)

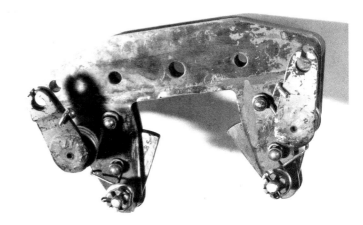

Snatch gear for the 'V'-frame arrester hook. This was the standard fuselage fitment for accommodating the 'V' frame arrester hook in all FAA aircraft at the outbreak of World War 2. February 1944. (© Crown Copyright/MOD)

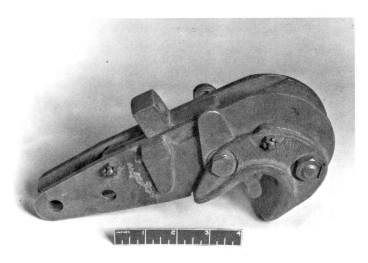

The performance of the standard 7,000lb hook was increased by 50 per cent through the replacing of the throat insert. This unit was increased in size, manufactured as a high-strength steel casting and identified as the 10,500lb hook. October 1945. (© Crown Copyright/MOD)

Hawker Sea Hurricane IB (Z7015). While the installation of the 'V'-frame suspension and the hook snatch gear appear authentic, NAD would never acknowledge an actual hook of this profile! (Peter J. Cooper/Falcon Aviation)

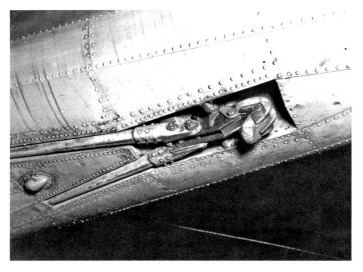

Standard 10,500lb arrester hook 'V'-frame installation seen in the stowed position. The aircraft is thought to be a Fairey Fulmar. October 1944.
(© Crown Copyright/MOD)

shows this unit – displaying its retention pawls and the Bowden release system for deploying the hook – removed for inspection after a heavy impact during a deck landing. Failure to pick up a wire would prevent the pawls from retaining the hook, permitting it to continue swinging free until a subsequent wire was engaged. The idea of latching the 'V'-frame into the fuselage was to provide a rigid system between the aircraft and the external retardation forces to which it was being subjected. This would prevent the hook unit from being exposed to side-load forces should it pick up the wire in an off-centre attitude, also it ensured the line of retardation force passing below the centre of gravity of the aircraft. In general the 'V'-frame was allowed to swing freely in the down position. Some aircraft ultimately were fitted with a damper unit to prevent the swing from being excessive and to avoid bounce if the hook struck the deck before picking up a wire.

The standard hooks came in two sizes, basically similar but with differences in the size and format of the 'throat' and 'beak', and the size of the bolts attaching it to the 'V'-frame. (There is an element of anthropomorphic indecision here in that the point of the hook, where it

picks up the wire, has been known variously as 'toe', 'nose' and 'beak'. Probably through its proximity to the 'throat' of the hook, the term 'beak' appears latterly to have been the preference, and is therefore used in this chapter.) The hooks were identified by the weight of the aircraft to which they were to be fitted, the smaller was known as the 7,000lb {3,200kg} hook and the larger the 10,500lb {4,700kg}, the latter being a logical development of the former. One illustration shows the smaller hook on a Fairey Swordfish aircraft while another is a close-up of the heavier pattern. The main difference is the throat insert,.Basically the standard hook comprised a pair of parallel steel plates, of about ¼in {6mm} thickness spaced about 1in {25mm} apart, with the hook element sandwiched at the free end. The hook itself was a high-tensile steel casting or forging with a throat marginally larger than the diameter of the across-deck wire, the throat (or 'bight') around which the rope would be stretched being about 1in {25mm} in diameter. The beak of the hook, which would pass under the wire when picking up, was about 3in {75mm} in length. On each side of this hook a pair of spring-loaded, pivoted triggers were located with the lower ends in the way of the rope access into the throat. When a wire was picked up, these triggers would be drawn back into the throat by the wire and would permit the hook to be engaged into the fuselage latch mechanism (snap gear), thereby trapping it into the fuselage. If the triggers were not constrained by the wire, the block on the top of the hook would move into position for it to strike a fixed anvil, thereby preventing the hook from being housed within its snap gear and allowing it to drop again to pick up another wire.

The basic hook was designed for arresting aircraft of all-up-weight around 7,000lb {3,200kg} at a maximum rate of retardation of 2g. For heavier aircraft (AUW of 10,500lb {4,700kg}) the diameter of the bight within the throat of the hook was increased from one inch to three inches {25mm, 76mm}. This was achieved by the fitting of a hardened steel insert between the standard side plates of the 7,000lb {3,200kg} hook. With this latter design the hook was provided with slots to accommodate the trigger latches, thus introducing breaks into the hook surface against which the across-deck rope (centre span) would have to contend. This severely aggravated an abrading wear and tear problem to both hook and wire at their interface.

With the war-time adaptations of the Hawker Hurricane and Supermarine Spitfire aircraft for deck-landing, the principle of using the 'V'-frame installations ahead of the tail wheel continued. In the case of the Hurricane the normal, open 'V'-frame was adopted; that for the Spitfire (the 'Hooked Spit') used a solid 'V', in that the gap between the

New design of cast-steel arrester hook, fitted with latching device, shown deployed. May 1945. (© Crown Copyright/MOD)

Supermarine Sea Otter I (JM987). Its predecessor, the Walrus, was able to land on carrier decks without the need for an arrester hook. The Sea Otter, with its enhanced performance, was thought to present a border-line situation. The installation shown is a 'jury rig' mounting an RAE forged type arrester hook at the end of a 'V'-frame. June 1945.
(© Crown Copyright/MOD)

two limbs was plated across to preserve the streamlining of the fuselage under-surface when the hook was in its stowed position.

On the arrival of the Lend-Lease Grumman Martlet/Wildcat and Tarpon/Avenger aircraft from America came the new concept of the 'sting' arrester hooks. The word 'sting' was used because the hook, attached to the end of a length of round bar, was extended from the rear of the aircraft in a similar way to the sting of an insect. It dangled in space until the aircraft landed and picked up a wire. The hook was a cast or forged unit rigidly fixed to the outer end of a freely swinging arm. The inner end was attached to the fuselage by a twin axis, pivot assembly that, when deployed, allowed the arm freely to swivel laterally as well as vertically. In operation there often was appreciable yawing of the aircraft as it picked up a wire, but that did not seem to adversely affect either the machine or the pilot. The hook remained extended on landing; its stowage after the aircraft came to rest was originally a manual process carried out by the flight-deck handling party. This feature was continued with the subsequent Grumman Hellcat and Chance-Vought Corsair aircraft, except that with the Corsair the arrester hook was pivoted from the tail-wheel assembly.

One of the interesting features of these different systems was that in cases of failure, the British system mostly was faulted by the hook at the extremity of the 'V'-frame jamming up into the fuselage latch gear without picking up a wire; the trigger latch system having been damaged in the process. Conversely, failure in the American sting units was mostly through the whole arrester hook installation being pulled completely out of its housing by the retardation forces. The incidence of this occurrence was disappointingly high, particularly with the Martlet/Wildcat and Tarpon/Avenger aircraft. This problem seemed to have been resolved in some of the later models in which the machines had powered hook retraction with cockpit control.

The principle of allowing the hook this freedom of movement was queried by NAD who conducted a series of experiments the results of which proved convincing enough for future FAA aircraft to be specified without the requirement for latching. The aircraft designers had to ensure that their installations were adequately stressed to contain the transverse retarding forces imposed by off-centre landings. At the same time NAD was conducting an in-depth series of trials on a wide variety of arrester hooks. The following section provides an insight into this field of wartime research.

Of the new British aircraft brought into service during World War 2 the Fairey Fulmar, Barracuda and Firefly all adopted the 'V'-frame system, pivoted ahead of the tail wheel. The Blackburn Firebrand was the first British aircraft to deviate from the norm. The hook hinge point for this aircraft was on the underside of the fuselage but, abaft the tail wheel, it possessed a damper device to stiffen up the system. An illustration shows this system as in 1946; its development in 1951 with experimental hydraulic hook retraction is seen deployed in another picture.

Without the historic requirement for the constraint of the hook within the fuselage during arrestment, the field for a wider development in hook design and suspension was opened up.

NEW DESIGNS

At the beginning of World War 2 it was felt that the existing 'standard' patterns of arrester hooks, although adequate for the aircraft in use at that time, were far from satisfactory. The basic hook was limited to aircraft of all up weights of 7,000lb {3,200kg} and, with a modification, at 10,500lb {4,700kg} AUW. With the advent of aircraft possessing much higher performance characteristics than hitherto, the mechanics of the entire procedure of aircraft arresting needed to be addressed. It was the undesirable (but unavoidable) abrasion occurring at the interface between hook and wire, together with the severe bending of the across-deck (centre span) wire, which prompted an intensive programme of research into hook forms to be started within NAD. The problem may be understood when it is realised that, with a deck span between sheaves of some 80ft {24m} and a rope pull-out of 150ft {46m}, the included angle of the rope centre-span at its apex was 30°. This meant that a rope of _" {22mm} diameter experienced – while under load – a wrap-around of 330° on a hook with a throat diameter of 1" {25mm}. This could hardly have been considered an elegant engineering situation.

An entirely new approach to the subject, irrespective of previous practices, was encouraged and the full co-operation of the Fleet Air Arm was assured to the extent of permitting use of the dummy-deck arrester gear installation at RNAS Arbroath (HMS *Condor*) and the involvement of the pilots of No.778 (Trials) Squadron in the experiments. Trials began in April 1942 and continued throughout the summer. (This work eventually was concentrated at Farnborough once the first of the arrester deck installations on Runway 04/22 was completed.)

Investigations were made into all aspects of the shape and profile of the throat, the material used in its manufacture; the procedure for manufacture; the attachment to the aircraft; experimentation with hard or soft coatings to the throat of the hook; the advantage of making the throat of low-friction material (for example, aluminium-bronze), even to the employment of a roller insert. All of these were carried out to determine the best means of reducing abrasion to both hook and rope at their interface. Tests were made also with various damping systems for eliminating undesirable hook motion during arrestment, particularly in the vertical (bounce) mode. Additional investigation was made into the relationship of the hook to the arrester wire both at the instant of pick-up and at the end of its pull-out, particularly its change in geometry; the change in stress throughout its pull-out; its behaviour under side loading; abrasion consequent upon the rope fleeting through the hook; the reaction of the deck wires and their ability to withstand repeated use.

One line of development involved the fitting of a roller within the throat of the hook. An example of a hook, for a 'V'-frame installation in 1942, is seen in an illustration. This was fabricated from steel plate and was fitted with lugs for housing in the snap gear. The roller was lubricated via a 'Tecalemit' grease nipple. Experiments in this field were conducted to investigate the swing of the tail of the aircraft in picking up an arrester wire when landing well off-centre, particularly to discover whether any adverse affects could be eliminated and to compare the rates of abrasion between the wire and the surface in the throat of the hook. In one series of trials, this roller was free to rotate allowing the rope to pass across the hook in case of off-centre landings. An alternative allowed friction pads to be incorporated in the roller assembly to restrict to some extent its ability to rotate. The question of fitting damper mechanisms into the hook suspension arms and an investigation into the need for the 'V'-frame type of arm to be held rigid in the airframe were both addressed. The first investigations into the shape of the hook were conducted on Fairey Swordfish and Albacore aircraft.

At this stage, a basis of rationalisation was adopted in that the hooks would form a series of four sizes, each covering a range of aircraft weights. These were decided as ranges with upper limits of 8,000lb, 12,000lb, 16,000lb and 20,000lb {3,600, 5,400, 7,200, 9,000kg}; their designations being respectively '32', '48', '64' and '80' (representing a factor of four times their working load, divided by 1,000!). The scope was doubled in that both 'V'-frame and sting types of attachment would be catered for; this enabled existing aircraft to be readily adapted to accommodate experimental hooks developed for one or another of the practical aspects of the investigation. The range of arrester hooks produced in the departmental research programme is listed in Appendix 8.

To obtain hooks for experimentation as quickly as possible the first designs were fabricated from steel plate, formed into the appropriate profiles as required. These were produced in RAE workshops.

Supermarine Sea Otter ASR.II (RD876) was fitted with a sting type arrester hook for deck landing trials. The picture shows the prototype installation with part of the rudder removed and a rope guard fitted ahead of the tail wheel. February 1946. (© Crown Copyright/MOD)

The second prototype de Havilland Vampire I (LZ551/G) was the first jet aircraft to be fitted with an arrester hook and is seen here with it deployed. It was later converted to Sea Vampire and is now at the Fleet Air Arm Museum. October 1945. (© Crown Copyright/MOD)

Prototype Supermarine Seafire XV (NS487) showing both the experimental, plate welded, sting hook deployed complete with its fairing, and its normal 'V'-frame hook in the stowed position. September 1944. (© Crown Copyright/MOD)

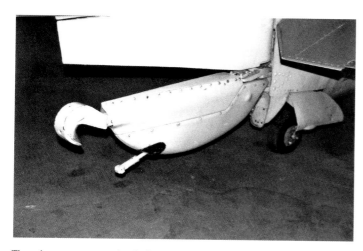

The sting type arrester hook fitted in the stowed position to a Supermarine Seafire. The spigot below the hook is the RAE pattern of holdback spade attachment for use when launching by bridle. (Author)

Prototypes were statically tested to destruction within the RAE Materials and Structures Laboratory and modified as desired. Some of these were retained for trial purposes 'as manufactured', others were prepared with a lining of the throat with a soft material, eg bronze, still others were hard-faced with a chrome-based alloy, for example 'Cromaloy', and those of another experimental batch were given case-hardening treatment.

The RAE foundry had no facilities for producing cast steel hooks so these were made to RAE patterns by a contractor. One of these hooks is illustrated. Orders were given for specimen hooks to be made in alloy steels, including cast chromium/molybdenum (CrMo) steels, and drop-forged in high-tensile and case-hardening steels respectively. These were to be adapted further with throat linings if necessary. Manufacturers were required to produce six hooks of each type for trials assessment, particularly in the case of drop-forged hooks. Two random samples would be tested progressively to destruction in a Denison tensile test machine; two would be sectioned – one longitudinally and the other laterally – for micro-etching to examine the metallurgical properties as well as the grain flow structure; while the remaining two would be fitted to appropriate aircraft for full-scale deck-landing trials.

Following the preliminary experiments, another aircraft used on these trials was the prototype Supermarine Seafire XV (NS487) – a later version of the 'Hooked Spit'. Originally this aircraft was fitted with the standard 10,500lb {4,700kg} arrester hook set in a 'V'-frame with latch mechanism in the fuselage, forward of the tail wheel. An additional experimental sting type installation was fitted carrying a modified, plate-welded W.48 hook protruding from the tail. This required the lower part of the rudder to be cut away to accommodate the assembly. The unit was extensible and freedom of movement in yaw and pitch was provided, the latter motion being controlled by a damper unit. Trials with either hook could proceed at will and direct comparisons made between the two patterns. With the sting hook, manual stowage by the flight deck party was required after each landing. A second Seafire XV (NS490) was fitted, first with a welded fabricated hook in the sting position; this was replaced with a forged hook for comparative trials.

In developing these hooks and their attachments the earlier principle of latching the hook into the aircraft fuselage on picking up a wire was abandoned. It was felt that no virtue accrued in making the hook and aircraft a relatively rigid feature by retaining the hook in an automatic latch. In fact, as a result of these trials, a certain amount of flexibility in the wire/hook relationship during arrestment was thought to be advantageous. The findings of these trials were recorded in various reports and appropriate recommendations were made and generally adopted.

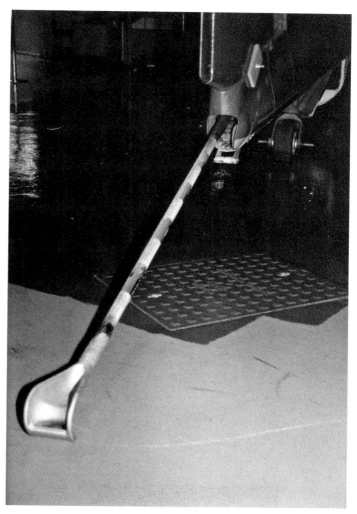

The sting type arrester hook fitted to Grumman Avenger AS.6B (XB446) shown deployed. (Author via FAAM)

The intention was the standardisation of a range of hook sizes adequate for covering all foreseeable requirements. The variety would include patterns for attachment to 'V'-frames by bolts in shear, or by end-screwed fittings, both patterns being subjected to rigorous testing. While the intention was commendable, the reality was the reverse. There seemed to be reluctance by aircraft manufacturers to adopt a standard pattern of hook attachment. Perhaps they were not keen on interchangeability, or the requirements may have been too demanding. Each seemed to prefer to design his own arrester system incorporating the recommended design of hook into its own suspension unit, sometimes making the entire unit as an integral forging. There was no problem here, the only requirement being that the profile and contours of the throat and beak were mandatory and that the whole unit was adequately tested to ensure its fitness-for-purpose. The penalty for going-it-alone was in the replacement of a damaged hook where the whole fitment would need to be removed from the aircraft. The sting type of articulated installation became virtually universal in its adoption, although many instances occur in which the hook was hinged abaft the tail wheel rather than its being extruded from the rear fuselage.

Subsequent experimentation was undertaken on the problems of hook bounce and yaw. One feature needing to be addressed was the behaviour of the arrester system through the increase in the speed of impact of the aircraft into the arrester wire. Another disturbing feature was a tendency – fortunately rare but a feature to be resolved – for an aircraft to pick up two ropes, effectively doubling the retardation force imposed upon the aircraft. The drill for retrieving an aircraft was for the deckside operator of the arrester gear, as soon as he had observed the aircraft to have picked up a wire, to lower all the wire supports, thereby dropping the remaining wires on to the deck out of the way of the machine. Accepting the mean distance between adjacent wires on all service carriers to have been 25ft {8m} and an aircraft approaching at a relative speed of 45kts, {23m/sec} the time interval between crossing a pair of adjacent wires would be one-third of a second. This was a short enough time but adequate for an accomplished deck operator to manipulate his controls. With only the doubling of this speed the time interval reduces to one-sixth of a second, which was too fast for immediate, spontaneous reaction on the part of the operator. With even higher closing speeds the situation worsens accordingly. The result was that other wires would still be deployed, standing proud of the deck on their bow-spring supports, as the aircraft passed across them.

With NAD having already determined the ideal profile for the throat of the arrester hook, attention next had to be given to some adjustment to prevent this inadvertent snatching of two wires. Various shapes of the throat and beak profiles were examined and the outcome was a reduction in the length of the beak. The subsequent risk from this was that the hook might be too small and miss all the wires. The entire mechanical aspect of the hook/wire interface was examined with investigation into the point of contact of the wire being somewhere along the suspension arm instead of the hook itself. The aim was to guide the wire into the throat of the hook without its missing the throat entirely. These short-beaked hooks can readily be seen on the various marques of FAA aircraft preserved in aviation museums.

As an aside, the author's personal contribution to the early part of this project was mostly in the seemingly never-ending exercise of meticulously calculating the 'modulus of section' of the throat elements in their various designs to determine the actual stress loads being imposed on the hook. The author's tools for this were a planimeter and a 20in slide-rule, long before the introduction of number-crunching by computer programs became the norm. This activity applied to all forms of design to which the various hooks were manufactured. The author also had responsibilities in the servicing and calibrating of accelerometers used in these and sea-going trials, together with the post-trials analysis of the graphical records. The layout of new aircraft for launching and deck landing was also his responsibility to ensure that deck machinery arrangements fell into line for sea trials; this required the provision of suitable launching strops/bridles and satisfactory hold-back frangible links. A further occupation involved the analysis of high-speed ciné films to evaluate relative movements of aircraft and hook assemblies and the rates of retardation – 'g' loads – induced by the arrester gear.

Whilst the majority of aircraft adapted for these trials were standard production machines (or prototypes), a few aircraft, not normally considered for deck arresting, were also modified for experimental purposes. One particular machine was the Bell P-39D Airacobra (AH574) which had been used at RAE for some other specific trials and had become redundant. At this time all naval aircraft were fitted with tail wheel assemblies (known as 'tail-draggers'). Tricycle undercarriages – as fitted to the Airacobra – with nose wheels were still in the future for deck-landing aircraft. The experimental use of this aircraft is discussed in 'Aircraft Adaptations'.

One installation of interest was the 'Heath Robinson-ish' jury rig affixed to the Supermarine Sea Otter (JM987) in 1945. While the predecessor of this amphibian – the Walrus – did not require an arrester hook when deck landing, this new marque was considered to be a borderline case. The grotesque, add-on attachment, was quickly prepared in RAE workshops for taxying trials, by fixing a forged hook to a scaffold-like, undamped, non-swivelling, 'V'-frame. Trials must have been satisfactory as the production version on Sea Otter (RD876) – some six months later – is seen in another illustration. A further illustration shows Sea Otter (RD872) landing into the wires of an 'Illustrious' Class carrier. However, only a selection of this marque were fitted with arrester hooks.

Reference is made earlier to a Taylorcraft Auster aircraft possessing an arrester hook. This was never seriously used as this small aircraft was capable of operating from, and onto, a carrier deck without mechanical assistance.

Experimental hooks of a variety of forms were fitted to the Avro Manchester aircraft (L7246) for the Brake Drum Arrester Gear experiments in 1941 (vide 'Brake Drum Arrester Gear'). This was an extension of the development of arrester hooks for naval aircraft, the only difference being that of magnitude. One of these was a heavy-duty unit, of welded steel plate construction matching the smaller, experimental versions used on shipborne aircraft. A second model was of similar construction but incorporating a large roller in its throat and a long beak, a feature supposedly to assist the picking-up of the arrester wire by the aircraft. It would not be out of place to record that these designs, originally for the Brake Drum Arrester Gear project, were basic to the extreme. Its constraints laterally and against bounce were achieved by liberal adoption of 'bungee' rubbers. The hook was fitted to a single, sting-type arm, attached to the aircraft just forward of the tail wheel. It was manually stowed and not held captive during the arrested

The sting type arrester hook on Chance Vought Corsair IV (KD431) shown attached to the retractable tail-wheel unit. The hook is seen deployed and the shackle hanging between hook and tail-wheel forms the catapult holdback attachment. (Author via FAAM)

Forged steel, sting type arrester hook fitted to Hawker Sea Hawk F.1 (WM961). (Author)

run, being trailed along the runway with the aircraft in tail-down attitude.

This Manchester aircraft subsequently was used as a 'hack' machine for taxying trials investigating the performance of naval arrester hooks at the heavier end of the range. Engineers will recognise that the stresses in the hook installation of an aircraft of 50,000lb {22,500kg} all-up-weight being arrested under a deceleration rate of ½g are equivalent to those of a machine at 20,000lb {9,000kg} AUW being subjected to a retardation of 1¼g. Hence the hook system for an aircraft of the lesser weight (of which none was available for naval flying in the early 1940s) was adequately proofed by the Manchester installation. On the dynamics side, an aircraft of 50,000lb {22,500kg} AUW taxying into an arrester gear installation at 60mph {27m/sec} would be equivalent to a machine at 20,000lb {9,000kg} AUW landing into the gear at an entry speed of 95mph {42m/sec}. This expedient provided valuable data for subsequent installations.

By the end of World War 2 an appraisal of aircraft arresting procedures was prepared by Willis (#313). This was evolved from the examination of the vast amount of footage of 35mm ciné film used for recording deck landings into shipboard and shore-based equipment throughout the war years. Willis begins:

'The arrester hook installation consists of a specially designed hook mounted on the lower end of an arm extending downwards from the fuselage. The arm is pivoted to the fuselage at its upper end and is stowed horizontally within the fuselage when not in use. All hook installations are divided into two classes, the 'V'-frame type and the articulated type. In the 'V'-frame type the arm consists of two struts in the form of a 'V', pivoted at the open end so that the hook is free to move in a vertical plane only. The arm of the articulated type is a single shaft attached to the fuselage through a Hooke's joint (no pun intended!), permitting it to move in both vertical and horizontal planes. A 'V'-frame installation is usually, but not necessarily, fitted with a snap gear which secures the hook against the underside of the fuselage after an arrester

wire has been engaged. The hook in this case incorporates an ejector that prevents it entering the snap gear (if a wire has not been engaged). A snap gear cannot be used with an articulated suspension. The suspension may be situated at any point on the underside of the fuselage compatible with satisfactory stability of the aircraft during arrestment. In general a snap gear is fitted only to installations forward of the tail wheel.'

It is obvious that this argument was based upon the general use of 'V'-frame installations at that time.

The development of the sting-type hook (which began with the later marks of the Supermarine Seafire) and nose wheel undercarriages (as the Bell Airacobra) were projects yet much in the future. Willis (#ibid.) goes on to say:

'The principal factors governing the position of the arrester hook installation are as follows:

1. prevention of excessive pitching and tail slam
2. prevention of ballooning after engaging an arrester rope
3. strength under side load
4. weight distribution in the aircraft
5. length of the aircraft from nose to hook.'

Willis then proceeds with discussing the geometry of hook installations and the moments of forces experienced by the aircraft during the arresting process. The recommended, sometimes mandatory, conditions were promulgated via the famous, continually updated, Air Ministry Publication, AP.970 –*Design Requirements for Aeroplanes* – together with the various SISs (*Standard Instruction Sheets*) issued by the Central Designs Office at RAE. These design parameters now are of limited, historical interest and are not repeated here.

An interesting diversion in this subject is seen by a photograph of a Japanese arrester hook. This adopted a hinged throat, ostensibly a spilling device for rapid disengagement of the arrester wire after landing. It is significant that little, if any, attention had been given to the

Hawker Siddeley Sea Vixen FAW.2 (XP924) with arrester hook deployed. (Author)

Sting type arrester hook fitted to a Hawker Siddeley Buccaneer. Also the retractable tail skid for use in bridle launching is seen deployed. (Author)

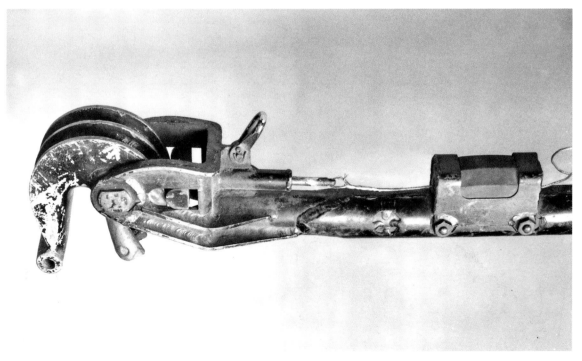

A Japanese arrester hook portraying the release system of the 'spilling' throat to disengage the arrester wire after landing. It is significant that no attempt seems to have been made to accommodate the bight of the rope as it passed across the throat. The bending stresses in the arrester wire must have been extremely high. July 1947. (© Crown Copyright/MOD)

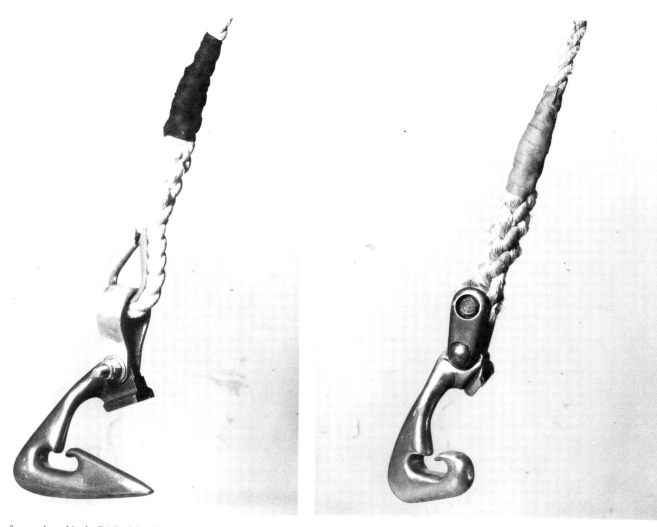

Two types of grapnel used in the RAE trials of in-flight man recovery. The low-flying aircraft would trail this grapnel on the end of a line which would snatch at the loop of a lanyard (supported on goal-posts) attached to the flying harness and parachute pack of the intrepid person hoping to be picked up. November 1945. (© Crown Copyright/MOD)

throat profile at the hook/wire interface. NAD found that nothing of significance was to be learned from these specimens.

A patent application for a British design of a spilling arrester hook was lodged by Lt(E) Jeremy Sidgwick in 1946. It is thought that this proposal was not followed up by RAE.

SNATCH HOOKS

Although these were not arrester hooks, their use for snatching ropes in airborne recovery was similar in concept. Advice on these was sought from NAD staff but the Department was not otherwise involved in these experiments.

Army co-operation aircraft in use with the RAF mostly were of the Hawker family, specifically Audax, Hardy and Hector. All these in turn were seen at Farnborough, particularly when No 4 or No 53 Squadron was in residence. Each of these aircraft flew with a small hook attached to a swinging arm from the undercarriage cross brace, for the purpose of message-retrieval without the need for landing. The article to be retrieved was placed in a pouch lying on the ground, connected to a long lanyard which in turn was supported as a loop between a pair of small

pylons. The aircraft, on approach to the pick-up point, would lower its hook, trail it through the loop in the lanyard and, hopefully, snatch the missive from the ground. It was not always successful but it did put ideas into other people's minds.

The Air Ministry envisaged this snatch-and-grab principle being extended to that of man-recovery by air of 'agents' from enemy territory without the need for landing at the site. A similar set-up would be mounted but instead of a leather pouch to be snatched, there was to be a live person at the end of the lanyard. An Avro Anson X (probably the prototype, NK352) was adapted to carry the hook attached to the end of a nylon rope, to trail beneath the fuselage. This rope was attached to a windlass within the aircraft. Trial runs were conducted, firstly with dummy loads and later with a man-sized rubber torso fitted with an accelerometer to measure snatch loads. Final trials were conducted using human subjects, the volunteers being doctors from the RAE Physiological Department (later to become renowned as the RAF Institute of Aviation Medicine). The candidates were Capt P. L. Lee-Warner, Wing Commander Roland Winfield and Group Captain 'Bill' Stewart. All three were successfully 'snatched' in May 1945 but Dr Winfield went through a 'hairy' experience in a repeat performance – the 53rd snatch. Turnill (*#101 p.72*) recounts this in some detail. In this instance the hook did not engage the lanyard cleanly and the assistant operating the winch aboard the aircraft had to use a boathook device to ensure a safe recovery of Dr Winfield into the aircraft. While the system appeared practical, the risk was high and, with the end of the war in sight, the operational requirement was receding. A total of 97 snatches, mostly with dummies – each known as 'Major Manila', the name indicative of its inner being – was achieved before the project was terminated.

In the immediate post-war years the principle of snatch towing was applied to the picking up, or retrieval, of gliders. The prototype Vickers Valetta C.I aircraft (VL249) and the first production model (VL262) were both fitted (1948) with glider towing hooks. The latter machine was further adapted through the fitting of a retrieval hook and cable winch for snatch pick-up of an Airspeed Horsa II glider (TK994). The trials took place during 1950 but the project was abandoned as no longer being a service requirement. The two Valetta aircraft remained on Farnborough's strength, being used for other trials for more than a decade. (Previous experimental work used an Armstrong Whitworth Whitley as a tug. One snatch pick-up of an Airspeed Horsa glider (DP749) was conducted in May 1944.) NAD was not involved in any of this work, the design of the hook and its quick-release slip mechanism was the responsibility of others. For a decade, most of the research in this field was the responsibility of a relatively small set-up known as the Airborne Forces Experimental Establishment (AFEE), generally based on airfields in the New Forest and later becoming a component unit of A&AEE Boscombe Down.

WIRE ROPES

deck rope which, in turn, was known as the main reeving or main purchase.

ABOVE-DECK ROPES – THE AIRCRAFT INTERFACE

LAUNCHING STROPS OR BRIDLES?

Early in the 1940s a new operational feature appeared with the introduction into the Fleet Air Arm of Lend/Lease aircraft from the USA. Reference is made to this elsewhere when the system of launching carrier-borne aircraft in the Royal Navy was developed from the use of a collapsible trolley to the adoption of US Navy practice of towing the aircraft on its undercarriage, in a tail-down attitude along the flight deck. The Fleet Air Arm referred to this new system as 'strop' launching. The aircraft was connected to the launcher shuttle by a strop made of steel wire rope. This was attached to one or a pair of hooks affixed to the underside of the aircraft and released itself as the aircraft exited the accelerator. The USN used a catcher device on its carriers, hopefully to retrieve the strop as it flew out of the deck shuttle at the end of the launch. This was not always successful and the strop fell into the 'oggin, never to be retrieved. Before long, a new name appeared for this launching strop. The US Navy called it a 'bridle' and thus it became and continued to be throughout the life of Royal Navy carriers. During the author's time at Farnborough this bit of wire rope was known as a strop but it will be referred to from hereon as a bridle.

For the first of the Lend/Lease aircraft, the Grumman Martlet/Wildcat, a single bridle was used. This was formed as a grommet, suit-

What is there of interest in a length of steel wire rope to command the attention of a group of research engineers? Perhaps such a mundane feature should be a matter of contempt rather than demanding time and energy to be devoted to what normally is a common, workaday feature. Is there something new to learn about such a subject as wire ropes? The answer must be in the affirmative when life and limb are dependent upon its satisfactory functioning – virtually hanging on a thread, albeit a steel one!

From the earliest days, the functioning of flight deck machinery in aircraft carriers invariably has been dependent upon the use of steel wire ropes. The towing of a launching trolley (or, subsequently, a shuttle) has employed under-deck wire ropes. Every system of aircraft arrester gear, in addition to such ropes for its under-deck machinery, uses wire ropes as the across-deck elements (centre spans) which the aircraft arrester hook is designed to engage. These ropes need to be highly flexible and of adequate strength to accommodate the accelerating and decelerating loads imposed upon them. Over the years, functional requirements with regard to the weights of aircraft and their speeds have increased, so the design parameters of the under-deck machinery have grown in proportion. Within the decade of the 1940s the performance requirements of flight deck machinery increased four-fold. The 1940 Fairey Albacore and Fulmar aircraft required an applied force of 11·8 tons to launch at 2·4g while the 1950 Supermarine Scimitar needed 42·4 tons to launch at 3·8g, all to be applied from the under-deck gear through steel wire ropes.

During World War 2, the ropes used in aircraft accelerators and arrester gear were generally made up of six strands, each strand – or lay – comprising a number (usually 19 or 37) of wires, the strands being dispersed around a hemp core. The terminology used in wire rope practice was rather loose. The flexible steel wire rope was referred to as cable, rope or wire. That length of the rope which, in an arrester gear, was picked up by the aircraft hook was generally known as the centre span, but occasionally the term 'deck pendant' was used – mostly in the US Navy. This centre span was subject to rigorous working conditions and (when worn or damaged) was made readily replaceable through the adoption of swivel shackles at flight deck level connecting to the under-

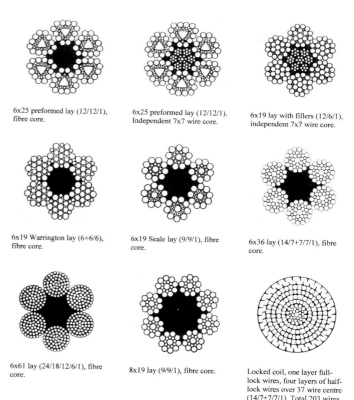

6x25 preformed lay (12/12/1), fibre core.

6x25 preformed lay (12/12/1). Independent 7x7 wire core.

6x19 lay with fillers (12/6/1), independent 7x7 wire core.

6x19 Warrington lay (6+6/6), fibre core.

6x19 Seale lay (9/9/1), fibre core.

6x36 lay (14/7+7/7/1), fibre core.

6x61 lay (24/18/12/6/1), fibre core.

8x19 lay (9/9/1), fibre core.

Locked coil, one layer full-lock wires, four layers of half-lock wires over 37 wire centre (14/7+7/7/1). Total 203 wires.

A representative sample of wire ropes tested by RAE for use in flight deck machinery. Generally these were of 3¾in {95mm} circumference (1⅛in diameter) but some samples of 4¼in {108mm} *circa* (1⅜in diameter) were included in the trials programme. (Author)

Samples of a variety of wire ropes used in accelerators and arrester gears. February 1950. (© Crown Copyright/MOD)

An 'Illustrious' Class carrier berthed alongside South Railway Jetty in Portsmouth Dockyard. The arrester wires are rigged with the bow springs raised but not tensioned. October 1947. (© Crown Copyright/MOD)

Supermarine Seafire F.XVII (SX314) is aligned on the BH.5 Catapult on Jersey Brow. The towing hooks may be seen below the leading edges of the wings, each side of the engine air intake, with the bridle lying on the deck. April 1947. (© Crown Copyright/MOD)

ably lashed and served, to connect the single hook on the aircraft to the deck shuttle. The C.I accelerator on Jersey Brow was the first British unit to be adapted to use this system and the BH.III accelerators on board ship were modified to accommodate bridle launching while retaining the detachable trolley system for the contemporary FAA aircraft. With the advent of the Grumman Tarpon/Avenger aircraft, and for subsequent machines, a twin bridle was used. This was a single length of wire rope with an eye at each end for attachment to the aircraft towing hooks, the bight at mid-point being served for engaging the deck shuttle.

Naval Aircraft Department gave a great deal of attention to the format of the wire ropes used in these bridles and the research staff became authors of a number of technical reports on their suitability, behaviour, efficiency and life-expectancy or overall 'fitness-for-purpose'. One feature to be addressed was the behaviour of the bridle at the moment of release from the aircraft at the end of the launch. A particular point was to examine the tendency of the bridle to strike any

part of the aircraft or its external 'stores' as it left the accelerator. The paths of the flailing ends of the bridle were closely examined by high speed ciné film. The designs of bridle catcher devices also had to be investigated for this aspect. This was difficult to resolve and there was a tendency to allow the bridle to be lost overboard each time an aircraft was launched. However, as a bridle retained in the launcher shuttle could damage the aircraft, it was felt that the bridle must be released from the shuttle, whether it was to be retrieved by a catcher or lost overboard. Various devices were tried as bridle catchers and eventually a system was developed with a high rate of retrieval.

Coincident with the adoption of bridle launching, the need arose for the retention of the aircraft on its launcher during the duration of loading and for holding it against its engine thrust during run-up immediately prior to its launch. With the trolley system, the restraining of the aircraft was achieved by the mechanical design of the front and rear hooks and their attachments. With bridle launching there was no such constraint so a device known as a holdback was provided. Basically, this comprised a hinged hook that was attached to a strong-point at the rear of the aircraft fuselage. The hook was restrained by a frangible link which was adequate to hold the aircraft against its own engine power but would break as soon as the extra load imposed by the accelerator was applied. The deck end of the holdback was spring-loaded and engaged a deck fitting. At first this was a sliding rack used for pre-tensioning the aircraft on the accelerator but in later applications the holdback position was a fixed point with the pre-tensioning achieved by inching forward the shuttle itself.

Beard (#206) provides details of bridles used in the late 1940s, showing that of 13 aircraft, two (Sea Fury & Sea Hawk) were using the single-point grommet bridle, the remainder using the vee type. The sizes of wire used in the bridle construction ranged from 2in to 3⅛in circumference {50mm, 79mm}. The grommet for the Hawker Sea Hawk comprised a 3⅛in circumference wire of 7x37 format, 6ft finished length, to fit a 1¾in diameter aircraft hook and a 6½in diameter shuttle {79mm, 2·4m, 44mm, 165mm}. This was proof-tested to 27·5 tons. Similarly, a vee bridle for the Supermarine Attacker aircraft comprised a 2⅛in circumference wire, again of 7x37 format, 26ft in length, to fit a pair of 1¾in aircraft hooks {54mm, 7·9m, 35mm}. The eye-to-eye lengths of vee bridles ranged from 14ft to 34ft {4·2m, 10·3m}. Beard's table also shows the weights of bridles employed over the range of these aircraft to be 20lb (Seafire) to 130lb (Scimitar) {9kg – 59kg}, only half of these being under 50lb {<23kg}. The weight of the ultimate complex bridle with van Zelm retrieval lanyards for the Phantom aircraft is understood to have been in excess of 200lb {>91kg}.

Early bridles were formed by cut-and-tuck long splices while later ones used either the Brit-loc or Talurit type of cold-swaged terminations. This development enabled any competent craftsman, using a simple machine tool, to make satisfactory cable terminations without the necessity of resorting to the practical and time-consuming craft of a skilled rigger. This proved to be a great advantage for shipborne operations as new bridles could be made up readily on demand, from basic wire stock.

It can be seen that a fair amount of man-power was required to load the aircraft on to the accelerator: deck hands were needed at the shuttle, the aircraft launching hooks and the holdback positions. Those in the front of the aircraft needed to be in the vicinity of propellers or jet intakes; most unsavoury, even dangerous, positions to occupy. It was also apparent that, in spite of mechanical positioning of an aircraft on to the accelerator, the attachment of the bridle was difficult and time-consuming, inhibiting the objective of launching aircraft at 30-second intervals.

The man-handling of bridles in a fairly cramped position beneath an aircraft, with engine running and half a gale blowing down the deck, was no easy matter. NAD addressed this and post-war developments satisfied this requirement. In another paper, Beard (#207) outlined a system in which much of the exercise of loading an aircraft on to the accelerator could be automated. He suggested that some of the preparatory work could be undertaken in the ship's hangar, his proposal being that prior to moving into the 'hold' position on the flight deck, both the bridle and the holdback could be attached to the aircraft while still in the hangar. The holdback and the bridle would be carried by the aircraft from hangar to flight deck with the free ends of both suspended by thin lanyards from fuselage attachment points. The aircraft would then be ranged and taxied into the loading position on the accelerator, passing forward over the mechanical positioner, where it would be centred and brought to rest against the loading wheel chocks. Here the after end of the holdback would be released with it falling into the deck fitting, thereby restraining the aircraft from further forward movement. The chocks would be lowered and the aircraft would move slightly forward until the holdback took up the tension against engine thrust. At this point, the bight of the launching bridle was released from the aircraft and fell across the track in the way of the bridle-engaging unit (sometimes latterly known as the jury loading gear). This, in turn, moved forward so that the bight of the bridle was made to engage with the deck shuttle after which the shuttle was inched forward to tension up the whole system ready for launch. All this happened in less time than it takes to describe the operation.

The most obvious advantage gained by this automatic loading cycle was the removal of operating personnel from the danger areas at both holdback and bridle positions. Each stage of the loading exercise would have been in full view of both the Catapult Directing Officer and the operator at the launching console.

Beard (#*ibid.*) investigated the proposal for a solid link to replace the bridle. One design was for a hinged rod to be provided either on the aircraft or the shuttle, to be retracted as soon as the aircraft was free of the launcher. An adaptation of this principle is now current practice in the US Navy. (See 'Tail-Down Launching – Nose-wheel Tow' *ante*.)

CENTRE SPANS OR PENDANTS ?

Moving aft, our attention is drawn to the wire ropes spanning the flight deck, forming the arresting area for the recovery of aircraft. Of the Fleet and Light Fleet carriers of World War 2 some 32 percent of the flight deck area was taken up with the arrester wires. The first wire was rigged some 40–75ft {12–23m} forward of the after round-down. Ahead of these, stretched athwartships, were (in general, average terms) ten wires, spaced at intervals of 26ft {8m}, covering the next 231ft {70m} of flight deck. The visible cross-deck lengths of these wires were known as centre spans ('deck pendants' in the United States Navy) and formed the wires, one of which was snatched by the aircraft arrester hook on landing. The ends of each centre span were terminated by capels and attached to a wire rope that disappeared below deck by way of pulleys and fairleads towards the arrester-gear machinery. The connection was kept simple to permit rapid rigging and unshipping the centre spans for replacement purposes or simply for clearing the deck when flying was not in progress. A post-war development was to replace the capels with swaged swivel connectors.

For recovery of aircraft, each centre span was raised some 6in {150mm} above the deck. When the arrester gear was in operation, these supports could be lowered to the surface of the deck by pneumatic, or hydraulic, actuators, only being raised on the imminent approach of an aircraft. This enabled the obstructions on the flight deck to be minimised when deck handling parties were actively engaged in ranging or striking down aircraft. In pre-World War 2 ships the centre spans were supported on hinged, cantilevered arms which collapsed to deck level when not in use. The World War 2 fleet carriers, beginning

Grumman Martlet aircraft loaded onto the American AH.II Accelerator aboard HMS *Biter*, showing the single 'grommet' bridle connecting the aircraft launching hook to the deck shuttle. This was the introduction of bridle launching to the FAA. September 1942. (© Crown Copyright/MOD)

with HMS *Illustrious*[40], adopted bow-spring supports which were also made to collapse to deck level by small hydraulic rams. This became the standard pattern for all subsequent ships.

These centre spans had a heavy duty to perform; they were required to be as flexible as possible while undertaking intensive braking loads. The entire braking force to bring the aircraft to rest was taken through this length of steel wire rope. The wire was required to withstand the impact of the aircraft at speed and immediately to bend around the throat of the arrester hook on the aircraft while the braking force was applied. At times it was felt that the limit of strength commensurate with flexibility was being approached but, as from the earliest days, there seemed to be no alternative method which could be adopted.

To the observer watching an aircraft land aboard ship, the centre span – on immediate release from the aircraft hook – was seen to coil up into a spring resembling a 'coiled-coil'. This was due to the heavy loading imposed on the wire during arrestation causing the rope to stretch, and possibly to be subjected to some torsion. When the rope was drawn back and re-tensioned for the next landing this phenomenon disappeared. However, it was a pointer for the flight deck engineer to monitor and to arrange for the replacement of an over-strained centre span when its integrity became doubtful.

A guideline, in 1944, given in one manufacturer's instruction manual for arrester gear, advised:

'that the rope forming the centre span be discarded if the number of broken wires in a length of lay exceeds 10 percent of the number of wires in the rope (ie for 6x37, 2¾in circumference rope, the rope should be discarded if more than twenty wires are broken in a length of 5½in). High wires, or loops of wire standing out from the rope, should always be counted as broken wires.

'For the main reeve through the under-deck gear, no criterion had been established for the main ropes, which are not subjected to such severe wear and bending stresses as the arresting wires. It would seem, however, that a main rope could be used for 1,500–2,000 landings.'

UNDER-DECK CABLING

Many of the preceding comments also apply to the wire ropes reeved into the under-deck machinery forming both accelerator (catapult) and arrester-gear units. In some respects the duties were less severe than for the centre span ropes in that they did not have to withstand the actual impact of the aircraft and inevitable abrasion problems. In other respects these duties were increased through the ropes having to pass through a series of pulleys and fairleads, involving many reversals of direction and bending, under load and at high speeds, often in excess of 100 knots {>51m/sec}. Even here, through space restrictions, the recommended ratio of pulley diameter to rope seldom could be met.

ACCELERATORS

During the decade into the immediate post-war era the under-deck machinery for Accelerators/Catapults underwent a drastic change. The well-established hydro-pneumatic jigger unit was superseded by the slotted tube catapult, generally known as the BS.4 Steam Catapult. The development of catapults for carrier use during World War 2 matched the service requirement relating to the weight of the aircraft to be accommodated. Essentially the BH.III Accelerator was with us for the duration of the war. Particulars are given elsewhere but details of the under-deck reeving are included here. In the case of the single-track launcher, two ropes were fitted to both the towing (leading) and braking (trailing) ends of the trolley assembly. These were Extra Special Flexible 'Beacon Pre-lay' ropes, each of 6x37 wires with a fibre main core, 4¼in {105mm} circumference, having a breaking load of 64·5 tons. The total length of each accelerating (leading) rope was 733ft while each retardation (trailing) rope was 522ft {223m, 159m}.

A point of difference arises here regarding the re-reeving of a rope through the accelerator system. Brown Brothers, in their instruction manual for the installation in HMS *Indomitable*, state that, with the 8:1 reeving, the quantity of wire rope required was three runs each of 459 feet {140m} – two runs for the accelerating rope and a single run for the retarding rope. To renew a rope, one end of the old rope would be disconnected from its attachment to the above-deck gear and four strands would be cut away for a short distance. The remaining two strands would be formed into a becket. The new rope would be similarly treated, the beckets being looped within each other to form a continuous rope. The old rope would be gradually withdrawn from the system with the new rope following its predecessor through all the fairleads and sheaves until the temporary splice appeared at the end of the run. The old rope would then be cut away and the new one connected to the above-deck gear. (This principle of splicing a replacement rope to the old one and threading them through the system also applied to each of the under-deck arrester gear installations.)

However, in the post-war years the launching of aircraft up to 30,000lb {13,600kg} AUW was envisaged and the BH.V accelerator was designed to achieve this. Progress with the steam catapult, however, ensured its pre-eminence as the new generation of launchers, coming into its own by the mid-1950s. This new system immediately made the hydro-pneumatic machinery obsolete. The steam catapult, with its free-moving piston, no longer depended upon steel cables for propelling the shuttle along the deck. With the employment of the BH.V Accelerator in 1946 and its use at Farnborough over the next decade, there was little more to say about the use of wire ropes in ships' catapults; they had reached their peak of usefulness. Ropes were used in the BS.4 units only for pre-tensioning and as trailing cables for hauling the shuttle back to the start position after a launch. The shuttle was brought to rest by

Blackburn Firebrand TF.IV (EK741) carrying its launching bridle and holdback for mechanical loading onto the Jersey Brow BH.5 Catapult. September 1950. (© Crown Copyright/MOD)

This rule-of-thumb assessment of the rope condition appears to the author to be rather generous. He would have preferred to have had somewhat fewer breaks in any particular length. In addition, the flight deck engineer was required to look for any evidence of 'burning' (friction-generated heat) if the wire had been fleeting through the throat of the aircraft arrester hook, particularly if the aircraft, when engaging the wire, was well off-centre.

During World War 2, as mentioned elsewhere, the normal centre spans in RN ships were made of ⅞in {22mm} diameter wire. Those in American ships were nearer ¹¹⁄₁₆in {17mm} – some 21 percent smaller. The requirement was for maximum flexibility combined with adequate strength and robustness for the duty it had to perform. Some of the aircraft arrester hooks were very narrow in width with a small radius curve within the throat itself. In fact, for aircraft of up to 7,000lb {3,200kg} weight the radius of the hook was less than one inch {<25mm}. This dimension was increased to 1¼in {32mm} for aircraft up to 10,500lb {4,800kg} AUW. The need for improvement during the war years became urgent but, even by 1945, the maximum throat radius was increased only to some 2½in {63mm}. Research into wire rope sizes by 1949 had proposed their increase in diameter to 1in and even 1¼in {25mm, 32mm} (Thomlinson (#298)). It may also be seen by the geometry of the centre span at maximum pull-out that a wrap-around of the hook of some 330° would be attained. All these elements combined to produce a highly demanding specification.

Blackburn NA.39 (Buccaneer prototype XK488). The port towing hook is seen with its launching bridle attached. (Author)

means of a buffer under the deck, the arresting load being taken through the ship's structure instead of a braking wire back to the power unit.

ARRESTERS

In the case of the arrester gear the story is vastly different. In 1939, the performance of a carrier's arrester gear was generously set for accepting a weight of 15,000lb at 60 knots entry speed {6,800kg, 31m/sec}. Within 15 years the rating was increased to 35,000lb at 103 knots {15,900kg, 53m/sec}, a four-fold increase. During this period, the design of the under-deck gear had passed through the development stages from Mark 4 to Mark 13. While the general principle of the hydro-pneumatic jigger system – component-wise – remained virtually unchanged, some significant developments had occurred over this period. One particular feature was that the wire pull-out at flight deck level had increased by one-third, from 157 feet to 220 feet {48m, 67m}.

The development of the arrester-gear machinery and its improved performance obviously had an influence upon the wire ropes and their associated ironmongery. All aspects of wire rope technology were addressed and the results of a great deal of experimental work featured as subjects of a variety of technical reports. To reproduce in detail the results of these investigations again would be tiresome but a list of some of the major explorations would contain:

a) Assessment of different make-up and lay of ropes
b) Fatigue of ropes subject to high speed travel and stress reversal
c) Inertia of ropes and their behaviour in passing through sheave pulleys
d) Lubrication of ropes
e) Behaviour when fleeting through arrester hook throats
f) Corrosion effects of ropes under stress
g) Determining load stresses in ropes moving at high speed
h) Advantages of using non-rotating ropes
i) Locked coil ropes
j) Performance when passing through sheaves of relatively small diameter
k) Suppression of tension waves in a rope system.

All of the foregoing relate to the passage of ropes through a system at a speed of the order of 120ft/sec {37m/sec}.

All the ropes were to the Extra Special Plough Steel specification

The bridle of a van Zelm recovery system is seen with its nylon restraining cables. The bridle itself is a single wire rope with ferrules swaged near the bight which engages the deck-launching shuttle. January 1968. (© Crown Copyright/MOD)

having an ultimate stress figure of 110/120 tons/sq in {17,250/18,820kgf/cm²}. The sizes under investigation ranged from 2¾in c (⅞in d) {70mm}; through 3⅛in c (1in d) {89mm}; 3¾in c (1¼in d) {95mm}; to 4¼in c (1⅜in d) {108mm} – the first figures (c) give the circumference measurement, the equivalent diameter (d) being in brackets. The construction generally fell within either 6x19, 6x25 or 6x37, some with fibre core, others with independent wire rope core, employing standard lay or Lang's lay. Some were laid up with normal stranding, following the standard Seale or Warrington construction patterns (of which those that were named had the advantage of greater flexibility). Other construction included non-rotating (18x4 over 3x24), locked-coil ropes to 'Paragon' specification. Pre-formed construction in 6x37 lay also was adopted in some later cases. Coated (galvanised) and uncoated samples were tested from the aspect of corrosion and effect upon performance.

From the kinetic standpoint of the movement of ropes through sheaves, the diameter of the sheave normally (D) is related to the diameter (d) of the wire. Those wires in our field of investigation would command sheave diameters of D=45d, viz 37in to 62in {0·94m – 1·57m}. (A contrasting model is to recall the sizes of the rope wheels surmounting a colliery pit-head gear relative to the size of rope passing over them.) It was not feasible for flight deck sheaves to match these standard requirements so a compromise had to be made. In practice, deck sheaves would be nearer 24in {610mm} diameter. It was necessary to evaluate rope performance through sheaves of this reduced diameter; in fact sheaves of ratios 22·7, 26·5 and 30·0 to 1 were employed for these tests. The crosshead sheaves on the Mark 13 gear – the last of the jigger units – were of the order of 22in {560mm} diameter. Howie (#260) indicates the sheave/wire performance produced a rule-of-thumb evaluation of fatigue life as equivalent to 800 deck landings into the penultimate Mark 12 arrester gear. At that point, the under-deck system would need to be re-reeved with new wire ropes. The situation worsened considerably if locked-coil wire ropes were used. These proved to be very inflexible and required a sheave/rope ratio of 100 for effective performance.

With the advent of jet aircraft and high landing speeds, beginning with the first deck landings of the de Havilland Vampire on HMS *Ocean*[45] in December 1945, a new phenomenon appeared. This was the occurrence of a tension wave in the arrester gear rope system. Willis (#314) states that theoretical investigation, together with limited experimental evidence, has shown that tension waves are induced in an arresting gear rope immediately following engagement and that these waves, travelling at the speed of sound, may be multiplied by reflection. This results in instantaneous rope tensions considerably greater than is indicated by the standard method of estimating arresting gear characteristics (in which an inextensible rope is assumed). The magnitude and speed of these tension waves become of significance at high entry speeds. Research into this unanticipated feature demonstrated that resilient rope anchorages were needed to provide a means for suppressing the reflected tension waves. Both hydraulic and pneumatic shock-absorbing devices were tried out, with the pneumatic unit being adopted. Willis (#315) records this research to have been conducted on the RAE High Speed Track using both standard and purpose-made liquid and air-spring units.

A means of determining the stress in a wire rope while running was also addressed. NAD developed a five-pulley rope-rider, through which the rope passed when running. The five pulleys slightly bent the rope as it passed through the rider and the displacement – against a spring – of the central pulley relative to its two immediate neighbours indicated the stress which the rope was experiencing; the unit was calibrated to produce definitive figures.

Perhaps the reader now will agree to there being more than meets the eye to a length of steel wire rope!

AIRCRAFT ADAPTATIONS

While Appendix 3 provides a list of the FAA aircraft in service during World War 2, only a proportion of these aircraft were operational at any one time; relevant dates are given but these should be regarded as approximate. Some aircraft remained in second line service after the operational squadrons had been either disbanded or re-equipped with later marques. The first list in Appendix 3 indicates which aircraft were in squadron service during the period 1939–45 and those aircraft which began their life – their first flight having taken place – within the same period. This covers both wartime operational and development aspects of Navy flying. The second list shows the development sequence of post-war aircraft, while helicopters appear separately as a third list.

The aircraft selected in this chapter represent those upon which work was required to prepare them for shipboard operation. Mostly this was provisional adaptations of the machines to accommodate either or both accelerator points (spools or hooks) and/or arrester hooks. Introduction of new aircraft types in the normal progress of acquisition of FAA aircraft is not included. Each of the machines treated were 'specials' in the first instance and it is for that reason they are highlighted here.

CONTEMPORARY MARQUES – 1939

In the Introduction, reference is made to the nine types of aircraft equipping front line FAA squadrons serving on aircraft carriers at the outbreak of World War 2 (1939). The list is repeated here:

Blackburn	Skua (1937–41)
	Roc (1938–43)
Fairey	Swordfish (1934–45)
	Albacore (1938–43)
Gloster	Sea Gladiator (1938–41)
Hawker	Nimrod (1931–41)
	Osprey (1931–40)
Supermarine	Walrus (1933–45)
	Sea Otter (1938–46)

All these had been specified and ordered by the Admiralty subsequent to the installation of flight deck machinery on the Navy's carrier fleet, only a matter of a few years earlier. They probably represented the second generation of aircraft fitted for mechanical assistance to shipboard launching and landing. As a consequence, all were equipped with catapult spools – for four-point trolley launching – and all, apart from most of the Supermarine amphibians, had arrester hooks.

Carrier-based aircraft were limited in weight and dimensions, the controlling features being the width of the flight deck, size of lifts, hangar space, strengths of lifts, performance of accelerators and arrester gear. In September 1939 (#4211) the following were applicable: Absolute weight limit 12,000lb {5,440kg}; Span 50ft {15·2m}; Width, folded 18ft {4·5m}; Length 40ft {12·2m}; Height 13ft–6in {4·1m}. The aircraft was required to achieve a free take-off within 300ft {91m} against a wind speed of 20kts {10m/sec} its stalling speed was limited to 58kts {29m/sec}. (Aircraft for catapulting from battleships and cruisers were similar, apart from the length being extended to 44ft {13·4m} and the height to 15ft-6in {4·7m}. These relaxations in the specification made allowances for the floats and hulls of seaplanes and amphibians respectively.)

The Walrus and Sea Otter aircraft were never launched from a carrier deck by the accelerator; their catapult spools were used for launching from other capital warships. Also, the former – affectionately known throughout the Navy as the 'Shagbat' – did not require mechanical assistance when landing on a carrier deck; it was said to settle as gracefully as an elegant, crinoline-clad Victorian lady.

The Sea Otter possessed an enigmatic aura; some pictures show it without an arrester hook, others show it with a 'sting' type hook, reminiscent of that on the later marks of Seafire. Like the Seafire, the hook installation occupied a space immediately below the rudder and deprived this aerofoil of some 20 percent of its control surface. Perhaps with the AUW of the Sea Otter being some 25 percent greater than that of its renowned amphibian ancestor, their Lordships decided that a hook was necessary. Thetford (#97) makes no comment; Hobbs (#43) and Dickson (#27) both show Sea Otters (JN185 & RD872 respectively) in flight with the hooks prominently displayed, but no explanatory comment is given. A photograph shows prototype K8854 on the Farnborough Mark II catapult sporting what appears to be a 'V'-frame arrester hook; this cannot be confirmed but may have been a handling or lashing point. Another (JM987) is sporting an experimental 'jury rig', 'V'-frame arrester hook. This was converted later into a Seafire-type, 'sting' hook assembly on RD876.

NEW MARQUES

Specific new aircraft were already under pre-war development for the Fleet Air Arm; these included the Barracuda and Fulmar; both from the Fairey stable (being to 1937 and 1938 specifications respectively). The Blackburn Firebrand and Fairey Firefly came into the picture in 1940 specifications, followed by the Hawker Sea Fury and Fairey Spearfish, both in 1943. But this is not the end of the story.

Additionally, under the Lend/Lease agreement with the USA a number of US naval aircraft were made available. Four of these, the Grumman Wildcat, Avenger and Hellcat, and the Chance Vought Corsair were issued to service squadrons for embarked carrier operations. The FAA adopted other US marques but, not having been used on shipboard operations, they are not treated here. Under the conditions of Lend/Lease, all material emanating from the USA had to be returned to its source, or destroyed, on the cessation of hostilities. Some American aircraft had been purchased – particularly from the collapse of France in 1940 – and these could be retained. During the Cold War period when the NATO countries needed equipping with military supplies, the Mutual Defense Assistance Program (MDAP) was instituted, from which the FAA benefited.

Perhaps it is apposite to refer here to the Fairey Fulmar. This aircraft may have been somewhat maligned for its not having been in the same class as the Hawker Hurricane and Supermarine Spitfire, but for a year or more it was the only aircraft the FAA possessed with eight-gun fighter capability. Its lack of speed may be attributed to the fact that it was a two-seater aircraft, the need for a navigator being paramount. At the time of its introduction there was little in the way of navigational aids for a single-seater fighter to be able to find its mother ship on return from a sortie. This aircraft went through two marks only, very little difference separating them. Of the 602 aircraft produced for the FAA a total of 112 enemy aircraft were accounted for over the period September 1940–August 1942. This represents one third of all FAA victories during World War 2 (Thetford *#97 p.158ff*).

Of these new aircraft types, only those that possessed unique attributes or featured in some major adaptation are included in the following paragraphs. All marques were the subjects of frequent trials on land and at sea, mostly for their own commissioning as shipborne aircraft, or for commissioning or development of flight deck machinery. The particular machines selected for trials were mostly allocated to either the RAE itself or to the FAA trials unit, No 778 Squadron located during World War 2 at HMS *Condor* (RNAS Arbroath). Those required for trials purposes often were fitted out with temporary instrumentation; some carried distinguishing markings for photographic purposes.

The machines identified below are those for which significant adaptations were required, many of them originating from NAD with much of the initial work being undertaken in RAE workshops.

AIRSPEED AS.39

This aircraft did not feature under adaptations as such but it is included here because there are so few references elsewhere to this machine. It was specified and built as a 'Night Shadower', becoming redefined later as a 'Fleet Shadower', still reflecting the thinking and policy of providing FAA aircraft for the purposes of gunnery spotting for capital ships of the Royal Navy. It was a high-wing monoplane of slightly more than 50ft {15m} span, powered by four Pobjoy Niagara V engines of about 150hp {110kW} each. Its stalling speed was around 30kts {15m/sec}. A feature was its folding wings, enabling it to be struck down into a carrier hangar. One prototype (N1323) was built which arrived for assessment trials at Farnborough in 1940. It featured a stalk-like undercarriage, with a tail wheel well forward of the tailplane, to provide a tail-up attitude for take-off and landing. There is no record of its having been fitted with an arrester hook and the author is not aware of its having undergone trials at sea. It did not see service with the FAA, appearing to have been a 'dead duck', with aerodynamic problems, from the outset. Taylor (*#94 p.96*) states that this machine – like its non-identical twin the GAL.38 – was ruined by its not being able to attain the near-impossible requirements of Air Ministry Specification, S.22/37, against which it was designed and constructed.

BELL AIRACOBRA

This aircraft is selected for special treatment in this record because of a number of unique features. The Airacobra aircraft was built by the Bell Aircraft Corporation as their model P-400. It went into production for the US Army Air Corps as the P-39D after a French order for a large number of these machines was made. On the fall of France in 1940, the order for 675 machines was taken over by the British Direct Purchase Commission at which time the name 'Airacobra' was adopted. After receipt of the first 80 machines in the UK, and resulting from their inconspicuous performance in combat, the order for the RAF was cancelled and the balance diverted to the US Army and to Russia, with Spitfires re-equipping the RAF squadrons.

The aircraft was typical of the Hurricane/Spitfire vintage, being a low wing, single engine monoplane with retractable undercarriage. Its special characteristics were the adoption of a tricycle (nose-wheel) fully retractable, undercarriage and the installation of the water-cooled engine behind the pilot, driving the tractor propeller through a geared layshaft.

A cannon fired through the boss of the propeller. The cockpit became cluttered with layshaft and magazine for the cannon as well as the normal flying controls. Also, it was not generous in space. Entrance to the cockpit was through a forward-hinged, car type door. In order to give adequate propeller tip clearance the nose-wheel oleo strut was some 54in {1·3m} long, producing a very spindly appearance at the front end.

During the late 1930s RAE was investigating the airflow pattern over a variety of wing sections, particularly laminar flow conditions. Many Farnborough aircraft at that time were to be seen fitted with 'cuffs' around the wing surfaces to provide new aerofoil forms. These were often covered with wool tufts so that the pattern of airflow could be photographed while in flight. To supplement this programme two aircraft were purchased, one from France, the other from America. The French machine was a Dewoitine D.501 (L4670) single-seat monoplane of all-metal construction with a fixed undercarriage with the wheels enclosed in spats. It had a very thin wing – a low thickness/chord ratio – flush-riveted and untreated in finish. The trailing edge of the wing consisted of some five thicknesses of light alloy riveted along the edge. There was no attempt to provide a smooth fillet.

The second aircraft was the American Northrop 2E bomber, registered K5053 (contemporaneous with the Spitfire prototype K5054). This too was a low wing monoplane of all metal construction. It had a large radial engine and was fitted with a fixed undercarriage completely shrouded in 'trousers'. This aircraft had a very thick – high thickness/chord ratio – gull-type wing and every rivet over wings and fuselage was finished as a nicely shaped hemisphere, all pop-headed. (Carbuncles everywhere!) Both aircraft were used for limited research and it was thought that, aerodynamics-wise, very little was to be learned from either of them.

It was in this particular field that, a couple of years later, the Airacobra was thought to be of interest. One of the first machines to arrive in the UK (AH574) was diverted to Farnborough for this purpose and was used extensively with a variety of aerofoil section 'cuffs' added to the wings. With the aircraft quickly going out of favour with the RAF, there was no need to return this particular machine to the squadrons and it was transferred to the RAE inventory. With the advent of jet aircraft – and tricycle undercarriages becoming the norm – it was decided that the Airacobra should be adapted as a trials aircraft for investigating its performance in deck landing and catapulting. A pair of towing hooks was fitted to the front spar of the wing for 'tail down' (*sic*) launching trials.

A Martlet I aircraft had been damaged during an early launch on the Mark II catapult. Its arrester hook was salvable so it was purloined and

Bristol Brigand aircraft at the October 1945 Farnborough display of enemy aircraft. Although not destined for ship-board operation, a Brigand I (RH748) was fitted with towing hooks for use by NAD in demonstrating the BH.5 Accelerator at conditions of maximum performance weight. October 1945. (© Crown Copyright/MOD)

fitted to the Airacobra. This latter installation was unique in that the hook was pivoted at the extreme end of the fuselage – almost a 'sting' type feature. Counter to normal practice, it was stowed by swinging the hook forward into a release mechanism rather than being latched in a trailing position; the nose-wheel undercarriage made this a relatively simple exercise.

The aircraft with these adaptations is depicted in this text. The trials were carried out by Lt/Cdr Brown, the RAE senior naval test pilot, whose stature was commensurate with the small cockpit. Brown, of course, was looking to the future and felt that the introduction of nose wheels would not be long delayed, particularly in the case of jet aircraft (especially the de Havilland Sea Vampire prototype which was being prepared for trials).

A few practice landings (ADDLs) into the RAE dummy deck indicated that any anxiety over the intended procedure could be discarded. These trials were followed, on 5 April 1945, by simulated 'touch-and-go' approaches to an aircraft carrier deck, including one actual landing under 'emergency' (predicted and unspecified!) conditions onto HMS *Pretoria Castle*. The Captain, Caspar John, himself an experienced naval pilot of high repute, exercised discretion by not enquiring into the nature of this indeterminate, sudden emergency. This happened to be, historically, the first British carrier landing and take-off of an aircraft with a nose-wheel undercarriage (Brown, #16 p.25ff). Subsequent accelerator and landing trials were conducted at Farnborough and valuable experience was gained from them. The anxiety generated by the thought of the long-stroking nose-wheel assembly encountering the complex of arrester wires on shipboard proved to be unfounded. No problem occurred during the period of trials at Farnborough. This boded well for the arrival of the impending jet aircraft.

The Airacobra also was used latterly by Lt/Cdr Brown in familiarising himself with low level runs over the experimental carpet deck at Farnborough prior to actual landing with the modified Sea Vampire aircraft. This aircraft (with the Fieseler Storch) virtually became 'Winkle's' personal aircraft as he appeared to be using it a great deal for shore-based deck landing and catapulting activities. The American manufacturers became aware of this application and sent one of their staff to Farnborough to examine the machine. It is said that the visitor was appalled that an aircraft in such a state of dereliction should still be flying and he condemned it outright. (Bell's did not offer a replacement!) AH574 was the only example of this type that was modified by RAE for experimental use; it was a valuable acquisition.

BRISTOL BRIGAND

A one-off adaptation was this machine, to provide an aircraft of 30,000lb {13,600kg} AUW for commissioning and calibrating accelerators under development. An early version of a Brigand (RH748) was fitted with a pair of accelerator towing hooks and used in commissioning the BH.V accelerator on Jersey Brow. This machine was used in flying condition, becoming airborne at every launch. Brigands were never embarked in carriers.

DE HAVILLAND SEA MOSQUITO & SEA HORNET

The Sea Mosquito originated from an Admiralty specification of 1944 for a twin-engined aircraft capable of operation from a carrier. The prototype was a converted Mosquito VI (LR359) which was fitted with

an improvised arrester hook – possibly ex-Barracuda. The first deck landing of a high performance twin-engined aircraft took place with this machine on HMS *Indefatigable* on 25 March 1944 with Lt/Cdr E. M. 'Winkle' Brown as pilot. On a subsequent attempted landing the arrester hook claw snapped, forcing the pilot to divert ashore.

The RAF Mosquitos possessed compressed rubber shock-absorber systems for the main undercarriage units; this was felt to be inadequate for shipborne operation so early in the production run these were changed for the normal pattern of Lockheed oleo legs. At the same time wing folding was introduced.

The Sea Mosquito was not equipped for accelerator launching. A number of RATOG experiments at 22,500lb {10,200kg} AUW were conducted against deck operations. It is thought that no Sea Mosquito squadron operated from a carrier although early in the evaluation programme special trials for a 'High-Ball' weapons mission involving the Pacific Air Force were conducted.

The Sea Hornet became the first twin-engined single-seat fighter to equip FAA squadrons for shipboard operations. The prototype (PX212), a converted Hornet, first flew in April 1945 and the first deck landing again was by Lt/Cdr Brown on HMS *Ocean*[45] on 10 August 1945. It was equipped with arrester hook, tail-down accelerator hooks and power-folding wings. The first accelerator trials were carried out at Farnborough with (PX214) in January 1946. The service squadrons equipped with this marque operated from HMSs *Indomitable* and *Implacable*, with the later night fighter version being embarked in HMSs *Vengeance* and *Eagle*[51], all in post-war commissions. The Sea Hornet was the first FAA aircraft marque to go into service fitted with Martin-Baker ejector seats.

According to Lewis (#60 p.83) in 1949 a dozen time-expired RAF Hornet aircraft were allocated to RAE for safety barrier trials. These aircraft, bereft of their outer wing sections, were used for taxying into various configurations and designs of barrier net. The wreckage after impact was often quite spectacular.

DE HAVILLAND SEA VAMPIRE

The first deck landing trials of a jet aircraft are well documented, both by the pilot, Lt/Cdr Brown, in his autobiographical records (#16 p.60ff) and others, including Boddington (#212) in his formal report on this achievement. The procedure was first mooted in January 1944 when Brown, newly arrived at Farnborough, was asked to investigate the feasibility of operating jet aircraft from HM Aircraft Carriers. Among the available prototypes at Farnborough were the first Gloster jet aircraft, the E28/39 (W4041) and the initial Meteor variants, together with a Bell Airacomet XP-59A (US serial 42–108773) on exchange from the USA. Shortly afterwards – February 1944 – the de Havilland Spider Crab (LZ549) arrived and Brown concentrated on making his assessments on this machine. Before long, arrangements were made for one machine of this last marque to be adapted for deck landing; in the meantime its name had been changed formally to Vampire.

The machine allocated for deck landing trials – from October 1945 – was one of the prototype Vampires, LZ551/G. Of the problems to be solved, the first was to decide upon the location of the arrester-hook attachment. With the aircraft possessing a pair of tail booms there was no rear fuselage to provide an anchor point. The only place was at the rear of the nacelle containing the cockpit and the engine, with the whole of the trailing section being occupied by the jet pipe. The only solution was to find – or provide – a strong-point in this area. The normal type of latched fitting, common at that time, was required (why? may be the obvious question!) and the only points were athwart the jet efflux pipe. A cantilever was provided above the jet pipe to house the latching mechanism and the 'V'-frame hook attachment points were located at the wing roots each side of the jet pipe. This meant that the hook would drop through the hot exhaust gases when being deployed. Trials were conducted on this feature and there was felt to be no problem for the short period during which these arrester elements were in the region of high temperatures.

After a period of shore-based ADDLs, during which forty test runs on Farnborough's runway were satisfactorily completed (with Lt/Cdr Jimmy Pratt operating as DLCO), the first shipboard landing was programmed. This significant event occurred at 11.28am on 3 December 1945, when LZ551/G was landed on HMS *Ocean*[45], piloted by Lt/Cdr E. M. Brown. The aircraft picked up No.1 wire which pulled out to 106 feet {32m}, recording a maximum retardation of 1·05g; the wind down the deck was 42 knots and the aircraft approach speed was 41·4 knots {both, 21m/sec}. A total of twelve landings were achieved during two days of flying. However, three years were to pass before the FAA received its first operational jets. Subsequent authors have been highly critical of this delay.

The pilot, Lt/Cdr Brown, in his report of these trials covered all aspects of shipboard flying operations with jets. Of all people, he should be the most authoritative in expressing a viewpoint. He writes (*#15 p.106*):

'We had succeeded, but the Vampire never went into squadron service with the Navy. There were two reasons for this. First was the dangerously slow acceleration pick-up of her jet engine. The Vampire was in all other respects a perfect deck-landing plane, but we felt at this stage no jet could be trusted entirely in an emergency requiring a sudden increase of speed. When a piston-engined machine was opened up suddenly on a wave-off the wings got an immediate increase in lift from the slipstream of the propeller. With a pure jet there was, of course, none of this. A pilot relied on getting his extra lift entirely by increasing the aircraft speed. The jet engine's acceleration response was very sluggish.

'And there was another important snag ... the Vampire's fuel capacity was too low for carrier work. Its radius of action would be far too limited for the long and uncertain hauls over the sea which naval flying often entailed, especially with the possibility of having to circle the ship and make several approaches to the deck before finally getting down. The decision not to put it into squadron service meant that the Americans had jets operating from carriers before we did. But our caution turned out to be justified when the Korean War developed.'

On a different tack, one of the Sea Vampire aircraft from the flexi-deck trials was later adapted for use as a lightweight dummy for land-based investigation into the performance of arrester gears. The aircraft originally was at 9,000lb {4,100kg} but by the time that it had passed through the Thurleigh workshops – who removed all fairings, doors etc and the wings outboard of the tail booms – the AUW was reduced to 6,000lb {2,700kg}.

ENGLISH ELECTRIC CANBERRA

This was a one-off application in which a redundant Canberra B.I (WJ995) – one of the first four prototypes, not an FAA machine – was adapted for land-based development trials of arrester gears designed to accept aircraft of 25,000–40,000lb {11,300–18,100kg} AUW. At the time of its introduction there was no FAA machine capable of operating at this loading; a bonus was that it had a tricycle (nose-wheel) undercarriage. This aircraft was fitted with various reinforcements to

An English Electric Canberra B.2 (WJ995), reinforced at the rear transport joint and bereft of wing tips, is seen fitted with an arrester hook. The Canberra was a convenient high speed aircraft used for arrester gear trials in the 30,000lb AUW bracket. This machine was used at RAE Bedford in the development of the Mark 13 and DAX arrester systems until 1967. July 1958. (© Crown Copyright/MOD)

strengthen the airframe and an arrester hook (recovered from a de Havilland Sea Vixen) was fitted to the after transport joint. The outer wings were removed as the aircraft was to be used only for taxying trials. Its main purpose was to function as a test vehicle for the development of the Direct Acting (Water Spray) Arrester Gear (DAAG) at RAE Bedford.

Canberras issued to the Royal Navy were used as shore-based target tugs, or targets themselves; they were never embarked in carriers.

FAIREY BARRACUDA

Although the Barracuda was designed and built to a 1937 specification and normally would not be considered here for special treatment, there are certain features which make this machine a candidate for additional comment. This machine could be described as grotesque when 'dirtied-up' for carrier retrieval, or seen standing on its grasshopper-like undercarriage.

Thetford (*#97 p.164*) refers to the machine as:

'a beast of burden if ever there was one: few aeroplanes in the course of their career can have been cluttered up with such a remarkable variety of extraneous equipment. Radomes, radar masts, rockets, bombs, mines, torpedoes, lifeboats, even containers below the wings for dropping secret agents in France; all were to be seen festooned about the Barracuda at one time or another. Add to this propensity such intrinsic features as the shoulder wing with its Fairey-Youngman flaps below the trailing edge, the curious undercarriage design and the high-mounted tailplane, and it will be evident why the sight of the Barracuda caused many a raised eyebrow among spectators new to the experience. Perhaps the most amazing sight of it all was that of a Barracuda using rocket-assisted take-off gear being shot off the deck of a carrier amidst a cloud of smoke.'

To all of which, and more, the author was witness!

The wide track of the undercarriage permitted easy access to the underside of the fuselage where, originally, a torpedo would be slung.

Fairey Albacore I (N4380) with practice 18in torpedo fitted with dummy wing and a tailplane. November 41. (© Crown Copyright/MOD)

Fairey Barracuda II (LS760) with a supplies dropping canister fitted in the torpedo bay. July 1944. (© Crown Copyright/MOD)

Fairey Firefly FR.5 (WD848) carrying experimental aerials for a VHF homing system. This aircraft eventually became ground instruction airframe A2378 at Yeovilton. February 1951. (© Crown Copyright/MOD)

This location, as well as under-wing access, enabled a wide range of 'stores' to be carried. In addition to those listed by Thetford (#*ibid.*) the author recalls seeing a large container being slung below the fuselage to enable a Royal Marine Commando unit to be supplied from the air. This required a great deal of research into the engineering facet and the parachute design to meet the operational requirements of the facility. This particular project is illustrated in Pudney (#*80 p.24ff*). Much of the superficial equipment fitted to the 'Barra' was subject to accelerator and arrested landing tests, most of which were conducted first at Farnborough.

The containers (personnel carriers) fitted beneath the wings for dropping secret agents – to which Thetford refers – were known as 'Cuda-Floats'. With one fitted under each wing, each was a capsule to accommodate two parachutists – under extremely cramped conditions – for transport to a dropping zone. The equipment accompanying them was dropped in a container from the torpedo rack at the same time. The drop was controlled by the pilot in a programmed sequence simply by opening doors in the floor of the capsule and allowing gravity (and parachute) to do the rest. The victims had a series of signal lights warning them of their imminent departure. For obvious reasons, instead of the carrier's accelerator, RATOG with its lower 'g' forces would be used for launching the aircraft on such a mission.

Friedman (#*34 p.213ff*), in his detailed appraisal of the Barracuda, suggests that limitations in its design prevented it from being used effectively as a torpedo aircraft. Its usefulness in operational sorties was as a dive-bomber carrying a heavy 1,600lb {730kg} armour-piercing bomb. It was considered to be much superior to the Grumman Avenger, particularly in its navigational facilities, and the Admiralty claimed that it was much easier for night flying operations.

Friedman (#*ibid. p.109*) also records that HMS *Furious*, having a relatively short flight deck, was fitted, in 1944, with a ramp – in effect an early form of ski-jump – to assist heavily laden Barracuda dive bombers to become airborne. HMS *Furious*, the reader will recall, was not fitted with an accelerator. The NAD at Farnborough was not involved in this feature.

The Barracuda, in its later marks, was adapted for tail-down accelerator launching, an early occurrence of this being on 3 April 1945. This aircraft was one of the few World War 2 marques which had undergone both trolley and bridle systems of launching.

FAIREY SPEARFISH

The Spearfish aircraft is included in this list not because of any particular adaptation required of it but through its use by NAD on subsequent trials at Farnborough, for which an amount of special instrumentation would have been installed. This aircraft was an attempt by Fairey Aviation to improve upon its earlier Barracuda marque, taking advantages of some of the features found in the Grumman Avenger. The first prototype flew in July 1945 and a total of four aircraft were built: these were given serial numbers RA356, RA360, RA363 & RN241. The aircraft arrived too late to see operational service in World War 2.

After a series of trials RA360 was retained at Farnborough and used in commissioning one of the new arrester-gear installations on Runway 04/22. Lt/Cdr Brown (#*16 p.101*) was scathing in his remarks about the plane and made a comparison with 'a realistic, aged, female, bovine' (real old cow). Like many another prototype, NAD found a use for this machine; it was operated at around 20,000lb AUW {9,100kg} for arrester-gear trials on Farnborough airfield. Another machine was retained at RNAS Ford as a component of the Carrier Trials Unit. The anticipated place of the Spearfish in squadron service was probably taken by the 1947 re-issue of the Barracuda, which, in turn, was super-

Fairey Swordfish II (HS671) has been loaded with a dummy 18in torpedo fitted with a monoplane tail. September 1943. (© Crown Copyright/MOD)

Fairey Swordfish II (probably NE954 which was at Farnborough at that time for RATOG trials). The attachment in the torpedo bay is an air drop supplies canister. July 1944. (© Crown Copyright/MOD)

A rather cold Fairey Spearfish, possibly the first prototype (RA356). This machine was used to a great extent in proving trials for various arrester gear systems at Farnborough, being a useful machine operating at around 20,000lb AUW. April 1950. (© Crown Copyright/MOD)

seded by the re-issued Grumman Avenger AS.4 (USN – TBM-3E) for a couple of years from 1953 until that was replaced by the Fairey Gannet.

GENERAL AIRCRAFT GAL.38

Another Fleet Shadower aircraft, this was a machine similar to the Airspeed AS.39, mounting four Pobjoy engines in a high wing. Whereas the AS.39 had a normal tail wheel, the GAL.38 had a twin nose-wheel undercarriage. (This firm was engaged in research with their twin Pobjoy-engined Monospar ST.25U aircraft, selected for early trials with nose-wheel undercarriages, a grotesquely rigged prototype being at Farnborough at that time. See Gunston (*#38 pp.16 & 98*).) Its stable of origin was not difficult to detect; the aircraft had all the appearance of a half-size Hamilcar glider – including its large fin and rudder – with four engines. (It was not the powered Hamilcar GAL.58 'glider' of slightly later vintage.) One early design illustration shows the aircraft with twin fins, and both nose and tail wheels as well as a main undercarriage, the last fitted to a lower, stub wing. A liberally glazed 'office' graced the front end of the fuselage, probably offering lavish accommodation for the observation team who would be airborne for some eight hours at a time. No further particulars are available on this machine. It is thought not to have landed on shipboard and it did not enter service with the FAA.

This aircraft, together with the AS.39 (*ante*), were two of an assortment of small four-engined variants at RAE around the late 1930s/early 1940s. The Short Scion Senior comes to mind, this being a half-scale model for the wing and engine mounts for the Sunderland flying boat and the Stirling. All these were powered by Pobjoy Niagara engines.

GLOSTER METEOR

Of the wartime jets there were only two marques which were used for carrier research. The first landing of a jet aircraft on to a carrier was that of the de Havilland modified Vampire LZ551/G, piloted by Lt/Cdr Brown to HMS *Ocean*[45] on 3 December 1945 (see *ante*).

The test of the Gloster Meteor was comparatively straightforward. The first instance of a Meteor on board ship was when an unidentified machine was lightered to HMS *Pretoria Castle* for tests to determine the extent of heating of the flight deck plating and the slipstream to be expected from the efflux of the jet engines. The aircraft was held stationary for a period and then allowed to taxi to various spots where temperatures were measured. The author cannot recall the result but there appear to have been no problems. Advantage was also taken to conduct deck-handling trials and shipboard taxying of a jet aircraft.

As a post-war investigation, two Meteor III aircraft, EE337 and EE387, were fitted with arrester hooks (transferred from redundant de Havilland Sea Hornet aircraft) for flying trials when Lt/Cdr Brown landed the former aircraft on to HMS *Implacable* on 8 June 1948, the first British twin-engined jet to land on a carrier. Apart from 32 shipboard landings, no record of the extent of these trials appeared until four years later (Lean (*#265*)) when their performance was compared with four other jet aircraft. Both Meteor aircraft continued on shore-based projects for some four years, before being scrapped in 1955. Suffice it to say that the Meteor did not become an FAA aircraft. By that time the Hawker Sea Hawk was nudging itself into the picture as a machine with prospects for Navy flying.

GRUMMAN WILDCAT, AVENGER, HELLCAT & CHANCE VOUGHT CORSAIR

The principle of adapting whatever was readily available was not to cease with the arrival of these aircraft. The American generosity in the process of Lend/Lease extended to the provision of aircraft of all types, and the FAA was not to be deprived of its share. The two first types of naval aircraft to be offered were both from the Grumman stable. These were the F4F-3 fighter, to be known as the Martlet (1940), and the TBF-1 torpedo-bomber, labelled the Tarpon (1943). These machines were offered on the basis of 'take-them-as-they-are'; in other words, the suppliers were not interested in making changes to their design which involved adjustment to their manufacturing processes. As a result, these aircraft could not be catapulted/accelerated from British ships as the US Navy exclusively used the tail-down method of launching.

One of the major adaptations was the need to provide catapult spools on the fuselage of each aircraft. This was undertaken on a number of Martlet I aircraft and the first of these (AL247) was proof-tested on the Mark II catapult on Jersey Brow in 1941. The single towing hook on the Martlet/Wildcat was retained and when tail-down launching was available on British carriers this was used in preference to the four-point trolley launching system previously employed. On the early deliveries of the Martlet, the towing hook appeared to be exceptionally small for its duties; later marks were fitted with a more robust design. Later the arrival of the Tarpon provided another problem with its towing hooks. There were two of these, one each at the top of the undercarriage legs. Again, these appeared to be undersized and located very close to the undercarriage fairing. A lengthy series of trials was carried out at Farnborough as the bridles for this aircraft were damaged after almost every tail-down launch. This problem was partly overcome by adding a wire 'serving' to the loops at the ends of the bridles – the points of contact with the towing hooks. It was thought by those who were involved in these procedures that there had been little investigative research undertaken in the USA into these specific problems.

It was the arrival of these first two Grumman types in the UK which precipitated NAD's investigation into the adaptation of the BH.III accelerator to accommodate these aircraft on Royal Navy Fleet Carriers, an expedient which was quickly and easily put into effect. (It was one of the author's jobs to determine and specify the sizes of the frangible rings for use in the aircraft holdback unit when tail-down launching was to be used on the BH.III. On one occasion, to install this holdback at the accelerator position, the author volunteered to drill and tap the threaded holes in the flight deck plating of one of the fleet carriers while the ship's engineering department went to dinner. It was ready for the arrival of the first aircraft shortly afterwards.) The author cannot recall whether the Tarpon/Avenger was treated similarly; the tail-down practice in FAA flying was well established by the time this aircraft appeared in Britain, in 1943.

The arrester hooks fitted to these American aircraft generally operated from the trailing, 'sting' position, in contrast to the captive 'V'-frame system used by the Royal Navy. (Having said this, an exception was the early Tarpon/Avenger aircraft in which was fitted a hinged, trailing hook abaft the tail wheel. Later marks carried the power-extensible/retractable, 'sting' unit.) One disturbing feature was that the American carriers used a smaller size of arrester wire on their flight decks, viz. $^{11}/_{16}$in in diameter, compared with the British $^{7}/_{8}$in {17mm, 22mm}. Their plain, forged steel hooks matched their wires, having a smaller throat size, while the FAA hooks were of fabricated construction with automatic latching devices. Extensive deck-landing trials were conducted to assess the effect of this apparent incompatibility and frequent inspection of both the across-deck (centre span) wires on the flight deck and the hook throats on the American aircraft were ongoing

features of these types. It was a little surprising to discover the extent of research work required in the UK into the launching and arresting aspects of these aircraft. Like the instances of the towing hooks, one cannot help but wonder if these features had been adequately studied in the USA.

Few American aircraft were given type names; a combination of letters and numbers seemed to be the normal practice. It was inevitable that once a new type reached British shores it would be given a title. Perhaps the names of aircraft were selected by the same Admiralty board that produced so many strange names for their ships. The small, tubby Grumman fighter was soon named Martlet, a dictionary definition of which reads 'a small, footless bird' while Heraldry matches it with 'a swallow'. The supposed similarity of the natural and the man-made must have been the product of a dry humourist. During 1944 it was agreed that the name Wildcat, belatedly adopted by the US Navy, should be substituted. Likewise, the torpedo bomber was known in the FAA as the Tarpon – 'large game fish, common on the south coast of the US'. Again, the British name was changed to Avenger, to accord with US Navy usage.

A measure of incompatibility persisted in a few cases throughout World War 2. First, US aircraft were mostly fitted with Colt machine guns of 0·5in {12·7mm} bore (which required manual 'cocking' before firing) against the British standard of 0·303in {7·7mm} Brownings. When Britain looked for an increase in calibre for aircraft armament it adopted the 20mm Hispano aircraft cannon. Another apparently attractive feature was that the Tarpon/Avenger could carry its American torpedo internally, within a bomb bay, while no FAA aircraft had this facility. This enclosure prevented the Avenger from being used as a dive-bomber; the best it could achieve was a shallow dive – glide bombing. The problem here also was that the US Navy used a short 20in {500mm} diameter torpedo while the Royal Navy had a series of longer 18in {460mm} weapons (sometimes fitted with a 21in {530mm} warhead). This mis-match was not resolved until the post-war period. Perhaps this was not as important as it would appear; the FAA adopted a system in which airborne torpedoes were fitted with dwarf aerodynamic fitments. These were either, or both, stub wings or monoplane air tails (one type carried the name TORA) to improve their delivery. These stub aerofoils broke away as the torpedo entered the water.

The arrival of yet another Grumman type, the F6F-3 Hellcat, together with the Chance Vought F4U-5 Corsair, both in 1943, presented no flight deck operating problems. Both were able to operate immediately from Royal Navy carriers as bridle-towed, tail-down launching was by then a common practice in British ships. Of the aircraft supplied by the US during the Lend/Lease period (1940–45), only these four types were used in FAA carrier-borne operations.

HAWKER SEA HURRICANE

Early in the war years it was incumbent that the performance of ship-borne aircraft be drastically improved and methods were sought to achieve this with minimum delay. The only apparent way to do this was to adapt other existing aircraft types to operate from a carrier flight deck. The two immediately obvious aircraft to investigate the feasibility for shipboard operation were the ubiquitous Hawker Hurricane and Supermarine Spitfire, already giving valuable service to the RAF Fighter Command. The manufacturers of both machines had previous experience in providing aircraft for the Fleet Air Arm.

The Hawker Hurricane was a straightforward case. The outline of the Hurricane fuselage, with its fabric-coated tubular construction, so closely followed those of its biplane antecedents – the Nimrod, Osprey and their RAF fellows – that the provision of catapult spools (referring to four-point launching by trolley) and 'V'-frame arrester hooks created no new problem. The first prototype adaptation was soon available for trials and was designated the Sea Hurricane, reaching the first squadron in January 1941. This robust aircraft, in its Mark IA version, was selected, together with the Fairey Fulmar, as 'Catafighters' for the short-term CAM-ships project. (Vide 'The 'P' [Projectile/Pyrotechnic] Catapult'.) The Sea Hurricane passed through a series of Mark numbers but every aircraft was a direct conversion of an existing landplane. No new manufacturing processes were set up to produce the Sea Hurricane as a separate entity.

A modification to the above-deck machinery of the BH.III accelerator was necessary to accommodate the Sea Hurricane and the RAE provided the design for a straightforward solution. This aircraft is believed not to have been adapted for tail-down accelerated launching. In fact from the Mark IIC version onwards catapult spools were not fitted. For flight deck manoeuvring of these fixed (non-folding) wing fighters a series of shallow trolleys running between guide strips welded to both the flight deck and hangar deck were provided. In Rawlings (#81 p.157) such trolley rails fitted to HMS Indomitable are clearly seen.

Although Sea Hurricanes were produced in quantity, their service aboard fleet carriers lasted for slightly less than two years; their last major action was in the Malta convoy of August 1942. Their usefulness, however, with the escort carriers remained undisputed until 1944. They were gradually replaced by one or another of the many variants of the Seafire.

HAWKER SEA FURY

The Sea Fury is recorded here as it was an early attempt to provide an aircraft to become the successor to the Seafire and as an interim before the introduction of carrier-borne jet aircraft. Hawker's were preparing the marque as a replacement for their Tornado/Typhoon/Tempest trio and it was a navalised version of the RAF Fury 1 which became the prototype Sea Fury (SR661) making its first flight in February 1945. After much deck landing activity at Farnborough and on shipboard, two further prototypes (SR666 & VB857) were provided with facilities for tail-down accelerator launching. The Sea Fury comes into the picture here as being the first production aircraft in the FAA to be designed with tail-down launching facilities, but it had to wait until after VE Day before this milestone was reached, the first launch at Farnborough being on 19 July 1946. This last of the wartime initiatives reached the service squadrons in August 1947.

HAWKER HUNTER

Readers may have seen photographs of Hunter GA.11 and T.8 aircraft sporting 'sting'-type arrester hooks below the rear fuselage. Thetford (#97) reports that some 80 aircraft were adapted for the FAA with the installation of arrester hooks for 'naval airfield purposes and not for deck-landing'. They were fitted for emergency use in airfield overshoot situations. Investigations have revealed no record of any deck-landing trials at Bedford. It has to be assumed that Hunters were allocated to RN training squadrons ashore and never embarked in carrier activities.

SUPERMARINE SEAFIRE (HOOKED SPITFIRE)

The navalisation of the Spitfire (at first known as the 'Hooked Spit') was a more complex exercise and took rather longer than the Hurricane to implement. Firstly the fuselage was of all-metal, monocoque construction which required a considerable amount of redesign work on the rear fuselage to accommodate the catapult spools and arrester hook. However, December 1941 saw the first Seafire – as the modified aircraft became known – undertaking deck-landing trials at Arbroath. These were conducted by Cdr H. P. Bramwell DSO, DSC, the Commanding Officer of 778 Squadron, the Royal Navy trials unit resident at HMS Condor (RNAS Arbroath). The first trials at sea, again with Cdr Bramwell, were held in early 1942 in HMS Illustrious[40], newly returned from extensive refit in the USA after her near disastrous, post-Taranto pounding by the Luftwaffe in the Mediterranean during January 1941.

The undercarriage was built for lightness and, with its narrow track, was to cause many deck landing problems later, including the need for a Seafire squadron to carry a liberal quantity of spare oleo legs, wheels and tyres when this aircraft was embarked in a carrier. Even in the late summer of 1945 one of the NAD reports (Cooper, #228) records a series of trials of the Seafire 45 involving some 210 landings into the arrester gear on HMS Pretoria Castle. During this series of trials thirteen replacements to some part or another on main undercarriage units were necessary. Even by that late date all the bugs had not been sorted out.

Popham (#76 p.138), himself a Spitfire pilot, writing of the 1942 North African operation 'Torch' states that this:

'was the debut for the Spitfire as a naval fighter; and it demonstrated at once, what was to be confirmed over and over again during the next two-and-a-half-years, that a land aircraft, however good, did not become a naval aircraft simply by giving it a hook and painting Royal Navy on its fuselage. In the air, regarded simply as a flying, fighting machine, the Spitfire/Seafire was a jewel. As a carrier-borne fighter, with its low endurance (1½ hours), its awkward forward visibility, and its delicate little undercarriage, it was a constant headache to its pilots, to the Flight Deck Party (who had to remove the pieces), to Commanders (Flying), and to everybody else.'

In an attempt to do away with the fitting of catapult spools, one Spitfire at Farnborough was fitted with a 'jury rig' comprising a pair of towing hooks fixed to a steel structure projecting forward of the wings. (To those of the author's vintage who recall the many strange sights to be seen at Farnborough during World War 2, this fitting was as grotesque and unwieldy as the first attempts at providing a nose-wheel to the undercarriage of the General Aircraft Monospar aircraft.) Obviously this experiment was successful in that later variants of the Seafire dispensed with catapult spools and embodied towing hooks for tail-down launching. (The Seafire, with the Fairey Barracuda, were the only marques to which, firstly, catapult spools and, latterly, towing hooks for tail-down launching had been fitted during their period of service.) Additionally, 'sting' type arrester hooks eventually replaced the 'V'-frame units, and more robust undercarriage units were fitted. Many of these aircraft were equipped with RATOG (qv) for carrier operation. The first shipborne, tail-down accelerated launch of a Seafire in this latest series of the marque was conducted during the summer of 1945. The early marks of Seafire did not possess the robustness of the Sea Hurricane and many casualties to undercarriages and airframes occurred during deck recovery. Through development, the later marks of the 'Hooked Spit' became machines worthy of upholding the reputation of Navy flying, the initial disappointments being translated into unqualified successes. They equipped many embarked and land-based FAA squadrons. The shipborne life of the Seafire lasted well into the

Prototype Supermarine Seafire XV (NS487) fitted with both standard 'V'-frame and 'sting' type arrester hooks, with the latter deployed. September 1944. (© Crown Copyright/MOD)

A pared-down version of Supermarine type 529 (VX136), rigged for jet efflux deflector trials on the port BH.5 catapult in HMS *Bulwark*. February 1956. (© Crown Copyright/MOD)

post-war period, the final models bearing only an outline resemblance to their predecessors.

For those who wish to know, the Spitfire progressed through 24 mark numbers, while its seaborne sibling – the Seafire – boasted eight. Many authors have covered the complex history of this aircraft. Two such – Bowyer (*#10*) and Darman (*#25*) – are included in the Bibliography at Appendix 15.

SUPERMARINE 'DUMBO' & SEAGULL

One of the visitors to Farnborough in 1942 was a Supermarine aircraft (R1810) produced to Specification S24/37 for the experimental development of a variable incidence wing. This was a shoulder wing, fixed undercarriage monoplane, powered by a single liquid-cooled engine; something akin to the Fairey Barracuda in appearance and size. Its look of podginess must have inspired the name of 'Dumbo', the Walt Disney flying elephant. This machine was not navalised but it was flown by FAA pilots for carrier evaluation. The intention was for the wing to be used in a new amphibian aircraft to supersede the Walrus and Sea Otter biplanes from the same stable. Penrose (*#71 p.275*) states that a devel-

opment of 'Dumbo' was considered to be a replacement for the Fairey Albacore, though this was not pursued. Its competitor to the same Air Ministry specification – the Fairey Barracuda – was preferred. An account of the test flying of this aircraft is given by Brown (*#16 p.120ff*).

The variable incidence wing must have been a success as it was incorporated in a new monoplane: a parasol type, folding wing, twin tail amphibian of very elegant appearance. It was powered by a single, liquid cooled, engine – probably a Rolls-Royce Griffon – driving contra-props. The illustration in Dickson (*#27 p.68*) shows a Supermarine Seagull (PA143) with 'Royal Navy' designation but apparently possessing neither arrester hook nor overtly apparent accelerator points. However, on 22 September 1949, a Seagull aircraft was logged as being at Farnborough for arrester-gear trials. It is assumed that neither 'Dumbo' nor the Seagull prototypes were used for shipboard trials. Neither aircraft went into production.

SUPERMARINE SEAFANG

This aircraft was a logical development of, and anticipated successor to, the Seafire. Its progenitor was the land-based Supermarine Spiteful which was a virtual redesign of the Spitfire, hence its being renamed. Of the sixteen aircraft built, part were with single propellers and fixed wings, the remainder being with contra-props and folding wings. Flying trials were conducted in 1946 with VG471 but the marque was not allocated to squadrons. It was thought that the development of the Supermarine Attacker was sufficiently far advanced – it first flew in July 1946 – for it to become rightful successor to the Seafire marque without introducing a new interim, Griffon-powered machine.

SUPERMARINE SCIMITAR

The type 508 (VX129) aircraft, a precursor of the Scimitar prototypes – with straight wings and swallow tail – no longer capable of flight, was used at its design weight of 20,000lb {9,100kg} as a dummy aircraft for land-based development trials of arrester gears. In its active life it had taken part in flying trials in HMS *Eagle*[51] during May 1952.

Supermarine Scimitar F.1 (WT859) on the barrier proving base at Bedford. The rear dolly at the arrester hook attachment point was for an unspecified experiment. November 1959. (© Crown Copyright/MOD)

TAYLORCRAFT AUSTER

Three aircraft of this marque were of interest to NAD; of these, two had mini arrester hooks. It is not known whose responsibility it was to design and fit these hooks but it quickly became obvious that the exercise was completely superfluous. The first of these lightweight aircraft (2,160lb {960kg} AUW) was a Mark I aircraft (LB384) which is recorded as having been fitted with a hook on 1 August 1945 and allocated to No.778 (Trials) RNAS for ferry duties in HMS *Pretoria Castle*. From its first shipboard landing it was apparent that arrester wires would never be required. The aircraft simply floated across the after round-down and gently settled on to the flight deck. Shades of the very first landings on board ship in the 1920s (HMS *Furious*) would be recalled. (Brown (*#15 p.104*) records his shipboard landing of a similar aircraft, the Fieseler Storch, spotting it directly on the after lift without it running forward off the lift.)

The second hooked Auster was a Mark III (MS935) machine, seconded to serve A&AEE and HMS *Condor* (RNAS Arbroath), possibly undertaking similar duties to its sibling. It is thought that neither of these aircraft visited RAE for deck-arresting trials.

The third machine, a Mark V (TJ537), was allocated to RAE for NAD to experiment with a method of directional control for flying from an assault vessel, such as a Tank Landing Craft (LCT). This project is treated more fully in the chapter, 'Direction Controlled Take-Off (DCTO) – Auster'. No adaptation was required for this aircraft; only the launcher hardware needed to be designed and manufactured. Even the holdback unit, to prevent premature release, was latched to the diminutive tail wheel trolley.

WESTLAND WYVERN

An early Wyvern TF.1 (VR133), with the Rolls-Royce Eagle 22 piston engine, was used for sea trials in HMS *Eagle*[51] in 1952. These trials terminated in the aircraft being towed into the safety barrier through a rig powered by the BH.V accelerator. The following year it was at Farnborough being used again in towed trials with new designs of safety barrier nets and later for a similar purpose, viz the behaviour of barrier nets when impacted by contra-prop aircraft. It later became an 'iron bird', simulating – by plywood attachments – a Fairey Gannet aircraft for use again in barrier trials.

HELICOPTERS

Although RAE had been conducting research into 'rotating wing' aircraft during the 1930s, progress had been limited. Raoul Hafner – of Austrian nationality – had developed a series of autogyros under the name of Gyroplane. These were flown mostly at Farnborough. Mishaps seemed to occur rather frequently, many of them while Hafner himself was piloting his machines. The AR.III gyroplane (DG670) seemed to offer the greatest promise and was retained at Farnborough until its final accident in April 1942. It was being evaluated for naval use but little progress in this role was made, development in the field of rotaries in the UK having been terminated at the outbreak of World War 2.

By 1943 the Aeronautical Research Committee resumed its interest in rotary lift aircraft, their paper (*ARC 6,459 'Note on Helicopters for Naval Use, February 1943'*) stating that two American upgraded Sikorsky VS-300 helicopters were due for delivery in the UK by May

Sikorsky Hoverfly I helicopter (KL109 – first known as the Gadfly when introduced into the UK) during a test, using a smoke generator unit, for examining the down-wash from the rotor and stability. Lt Cdr 'Winkle' Brown was the pilot. August 1947. (© Crown Copyright/MOD)

Westland Whirlwind HAR.1 (XA865) – wearing the NAD/RAE logo – approaching touchdown on the Bedford helicopter experimental rolling platform. November 1959. (© Crown Copyright/MOD)

1943 for evaluation. The RAF had no requirement for these machines and, although some doubts were expressed regarding their effectiveness, the paper stated that NPL opinion was 'cautiously optimistic'. From then on, it was all up to Farnborough and the Fleet Air Arm!

With the arrival of Lt/Cdr Brown at RAE in 1944 as senior naval test pilot, among his manifold duties he was required to investigate the possibility for deployment of helicopters within the Fleet Air Arm. His first flight trial at Farnborough was with the Sikorsky VS-316, (KL109), known at that time as the Gadfly. Later the appellation reverted to its American original, Hoverfly I. Before long 'Winkle' became the RAE research test pilot for rotating wing aircraft and undertook the major part in the development of this type. Much of this work was conducted during the latter years of World War 2 and the first shipboard landing of a Hoverfly I, fitted with pontoon floats, took place aboard HMS *Indomitable* in December 1945. McCart (*#111, p.137*) records that it stayed on board for ten minutes, taking off for HMS *Daedalus* (RNAS Lee-on-Solent). From 1947, shore-based FAA 'chopper' aircrew were trained in the winch recovery of unfortunates from the sea or shipping, including emergency dinghies.

The first intensive shipboard trials of a helicopter were deferred until 1949 when a navalised version of the Westland Dragonfly landed on HMS *Vengeance*. On the successful completion of these trials, the FAA introduced the helicopter into a new, post-war role. Records indicate that it was not until 1951 that HMS *Indomitable* was the first carrier to embark the Westland Dragonfly (Sikorsky S-51, manufactured in the UK) for search and rescue (SAR) duties.

Previously, whenever a carrier was at Flying Stations it was accompanied by a destroyer or corvette maintaining station on the carrier's port quarter. Its purpose was to act as 'plane guard', speedily coming to the assistance of any aircraft or ship's crew who happened respectively to ditch or fall overboard into the 'oggin. It was quickly seen that there would be many advantages if this standby ship were to be replaced by a helicopter, particularly in areas where the escort was not required for other duties, such as searching for enemy submarines! As soon as these aircraft became available to the squadrons, every carrier included one or more of these 'choppers' as part of its aviation complement.

Following the completion of the transfer to Bedford, NAD was required to take over helicopter research, particularly for the Navy which, it was assumed, was likely to be the main military client. To extend research in this field, RAE designed and built a rolling platform to investigate the landing, handling and securing of helicopters onto a pitching and heaving deck of a typical frigate. This platform was mounted on a railway track which extended into a hangar for a limited series of indoor experiments as well as for servicing the platform itself. The principle of deck recovery was developed here, using a cable (suspended from a winch within the aircraft) which could be snagged into a deck fitting and the helicopter hauling itself down to the deck. As a result of these experiments, 'choppers' became shipborne maids-of-all-work, being embarked in most RN ships. A Westland Scout (XP191) was one of the aircraft used in this project.

One of the unusual features to be seen at Bedford in the 1960s was a single-deck Leyland bus careering along the main runway with a Fairey Ultra-Light helicopter fixed to its roof. It was not trying to become airborne by using a moving vehicle, nor was it attempting to get the bus aloft, but it was a captive test rig for studying rotor wing-tip behaviour under forward motion. A similar experiment had been conducted on one of Farnborough's runways with a helicopter power unit – with rotors running – mounted on a flat-bed lorry.

Helicopters featured prominently in the 1982 Falkland Islands campaign with the deployment of Boeing Chinook and Westland Sea King aircraft. Also the 'whirlybirds' have become a frequent sight at seaside resorts as they patrol the coasts, often providing emergency rescue to so many whose names do not appear in the Navy List.

HOVERCRAFT

In the early days of Christopher Cockerell's invention of the air-cushion vehicle, the Navy considered there could be an operational requirement for use over water. The investigatory project was given to NAD to run parallel with helicopter research in the Department. A Britten-Norman CC2 was delivered to Thurleigh early in 1962. Service trials of a number of machines were conducted at Lee-on-Solent but the research team was dismantled and dispersed in 1970. A number of significant innovations had been introduced, making the air-cushion vehicle principle into a reasonably efficient machine for both civil and military applications. Apart from a few instances, the hovercraft has not really caught on and has never been a feature of flying Navy.

'ALSO-RANS'

Other aircraft of interest to the Naval Aircraft Department are listed here, mostly due to their not being directly associated with Navy flying but requiring a little of the expertise of departmental staff in the implementation of adaptations.

The first of these was the Miles M.38B, 'Libellula'. This was the second canard (tail-first) aircraft to emerge from the Miles stable, the first not having been a great success. It was thought by the builders that a canard arrangement would offer good visibility for deck landing and, with the front wing larger than the normal tailplane, would offer advantages in under-deck storage. This aircraft (SR392) was twin engined (a pair of de Havilland Gipsy Kings) with tractor propellers, mounted in nacelles in the rear wing, all on a tricycle, retractable undercarriage. Triple fins were provided which gave the machine a decidedly tail-heavy appearance. The machine had not been prepared for deck landing and catapulting. Brown (*#15 p.157ff*) writes of his experiences in flight-testing this aircraft. The project proceeded no further than trials flights at Farnborough.

Another aircraft from the Miles stable was subjected to naval evaluation trials without seeing a carrier flight deck. This was the M.19 Master aircraft, a hybrid between the Mark I and Mark II models. However, a third Miles machine, the M.38, Messenger, was involved in deck-landing trials aboard the escort carrier, HMS *Hunter*, in 1943. These trials were conducted without the aircraft having been fitted with an arrester hook. Lt/Cdr Brown in conducting these trials says that in landing he felt he was never going to catch up with the carrier. He continues (*#17 p.134*):

> 'The take-off was just as ridiculous. I ran the aircraft up to full power on the brakes and then released them, holding the stick well back, and the M.38 ran about its own length and then went up like a lift. ... this was the idiot's ideal deck-landing aircraft.'

An aircraft seen at Farnborough for naval evaluation trials was the Boulton Paul Defiant TT.I. Although a number of these machines flew in Royal Navy colours, they were used only for target towing duties. It was thought that they could be embarked and used for other operational purposes but the assessment for these suggested they be retained only for their intended shore-based activities.

One surprising aircraft used for carrier assessment trials at Farnborough was the North American B.25, Mitchell, twin-engined bomber. (This aircraft marque achieved fame on 18 April 1942 in being the first to bomb Tokyo, an operation led by Lt/Col James Doolittle USAAC in a one-way sortie from USS *Hornet*.) With the successful carrier-borne trials on a de Havilland Mosquito in March 1944 behind him, Lt/Cdr Brown, as pilot on that occasion, was invited to comment on the suitability of the Mitchell to do the same. The US Navy carried out similar trials in USS *Shangri-La* in November 1944 but no further progress was made on this project. Brown records this in one of his books (*#17 p.138ff*).

Further instances of aircraft modifications involve a Handley Page Heyford and an Avro Manchester aircraft, neither of which were closely connected to operation from the flight deck of an aircraft carrier. Both types were involved in the 'Direction Controlled Take-Off (DCTO)' system. The latter aircraft, together with an Avro Lancaster, a Handley Page Halifax and a Short Stirling, were adapted for experimental use with the 'Brake Drum Arrester Gear (BDAG)' installation. These projects are treated at length in their own chapters.

Frequently, full-scale experiments were required to evaluate certain equipment or procedures. To use 'real' aircraft could prove to have been a costly exercise, particularly if some measure of damage was to be expected in the conduct of such trials, which often was the case. For shore-based catapult trials in the late 1930s a cylindrical form of dead-weight, built up from a series of large discs mounted on a sturdy steel bar, was used. Subsequent trials were followed by the use of obsolete airframes, ballasted to suit the required weight. Samples of these, seen by the author, were Fairey IIID and IIIF, Fairey Seal and Blackburn Shark aircraft; upwards of a half dozen of these last marques – one of which was K4364 launched on 31 July 1938 – passed through RAE workshops for adaptation as catapult dummies. On almost every occasion the engine had been removed from the aircraft and a box filled with concrete substituted. Inevitably the trial was a one-way trip, there being very little of salvageable material remaining from a launch. Some attempts at recovering the airframe were made – the attachment of a tail rope from the aircraft to a ballasted trolley as a brake formed one

system. Seldom were these successful. Similar trials using obsolete airframes were undertaken at intervals throughout the history of the department, but there were limited numbers of suitable specimens.

The alternative was to adopt the use of dummies. Reference already has been made to 'HMS Flying Flossie', and others of her ilk, in the shipboard proof-testing of accelerators. An entire family of floating dead-loads was provided for shipboard commissioning of the BS.4 steam catapults. In the following list, the first eight were ballasted skids while the last two were fitted with wheels. The name 'Flossie' recurs here; this was a different Flossie from the original HMS Flying Flossie, dating from the sea trials of the BH.III accelerators. (See 'Sea Trials'.)

Flora	39,340lb	{17,860kg}
Flossie	37,828lb	{17,170kg}
Ida	11,984lb	{ 5,440kg}
Ivan	12,152lb	{ 5,520kg}
Nellie	16,072lb	{ 7,300kg}
Noah	15,764lb	{ 7,160kg}
Thomas	28,812lb	{13,080kg}
Wendy	19,012lb	{ 8,630kg}
Eric	24,500lb	{11,120kg}
Jack	14,000lb	{ 6,360kg}

Subsequent photographic records show these dummies (none of which was identified as His Majesty's Ship!) to be quietly rusting away in forgotten corners of various south coast dockyards.

For some of the immediate post-World War 2 land trials of both accelerators and arrester gears a suitably ballasted dummy, each appropriate to the relevant test procedure and representative of an aircraft, had to be provided. Much ingenuity was brought into play, often resulting in grotesque forms, in the provision of these devices. These, of course, were never intended to become airborne – hence the name, 'iron birds'. The aim was to provide a typical space-model suited to the purpose of a particular experiment. Examples come to mind for the early stages in a wide range of safety barrier trials. Another was the launching of the Hawker Sea Hawk from the slotted-tube catapult associated with the flexible-deck project. A third group involved the testing of new arrester gear systems at Bedford.

The safety barrier project, for example, passed through a series of development stages. The first was to determine the effect of the standard (wartime) barrier on twin-engined aircraft where the pilot was not protected by an engine mass between him and the barrier net. A second

The Blackburn Buccaneer Mark 1 'iron-bird' enmeshed in a later development of a barrier curtain employing multiple nylon tapes. May 1961.
(© Crown Copyright/MOD)

The McDonnell Douglas Phantom 'iron bird' with original short-stroke nose-wheel assembly. The dummy is rigged on the cordite catapult for safety barrier trials at Thurleigh. c.1966. (© Crown Copyright/MOD)

BAC TSR.2 'iron bird' undergoing barrier trials at Thurleigh. Some development work would have been necessary as the wrap-around of barrier net over the sharply swept-back wings appears to have been inadequate. September 1962. (© Crown Copyright/MOD)

An Avion Marcel Dassault Étendard 'iron bird' undergoing barrier trials into a nylon strap barrier curtain. This installation was on the Safety Barrier and Arresting Gear Proving Base at Thurleigh. March 1961.
(© Crown Copyright/MOD)

high 'g' loading. A de Havilland Hornet hack was used in the first of these experiments. This was towed into the experimental barrier set-up on RAE Runway 04/22 by a long cable attached, through pulleys, to one of the Jersey Brow accelerators. This principle also was adopted for early seaborne trials, in two instances the flight decks of HMSs *Implacable* and *Eagle*[51] being temporarily adapted for the purpose.

Trials for investigating contra-prop problems at model scale in the NAD laboratory were inconclusive. Full-scale trials, now becoming essential, were conducted on the Meadow Gate site, the location of the initial stages in the flexi-deck project. The same catapult and arrester gear were used but a steel deck was substituted as the landing area. In this case, three redundant Supermarine Seafire aircraft were obtained with five sets of contra-props. No specific mark of Seafire was used, each was simply a mishmash of bits and pieces (a Seafire lookalike!) This allowed five tests – live, with engine running – to be undertaken. On each occasion the aircraft finished in an impossible attitude, its front end having sustained some spectacular damage. It was a convincing demonstration of the need for full-scale dummy runs in trials of this nature.

Another of the early 'iron birds' for barrier experiments was a dummy to examine the effects of a barrier impact by a Fairey Gannet aircraft. This aircraft was powered by an Armstrong-Siddeley Double Mamba prop-jet engine driving contra-props. Fears were expressed as to the safety of the aircrew – accommodated in a cockpit situated well forward – from flailing, severed wire ropes and disintegrating contra-props in a barrier engagement. As no Gannet was available for trials, the time-expired Westland Wyvern (VR133) was fitted with a false canopy

phase was the effect of a single-engined aircraft with counter-rotating propellers (contra-props) introducing a scissors effect as they impacted the barrier net. A further phase was the investigation into the arrestment of jet aircraft at high impact speeds without imposing an undesirably

A de Havilland Sea Vixen 'iron bird', with the bridle attached and the holdback tensioned up, is ready for catapulting into the barrier net on the Thurleigh proving base. December 1959. (© Crown Copyright/MOD)

to represent the cockpit. To replicate the Gannet nose-wheel assembly a rather spindly undercarriage strut, mounting a pair of Spitfire tail wheels, was fixed to the front of the engine nacelle. In this format the hermaphrodite Gannetised-Wyvern was towed into the RAE barrier system.

One of the earliest experiments into barrier contacts by jet aircraft was undertaken with a denuded Supermarine Attacker (WA483) – minus its empennage – forming the specimen. This was used for investigatory work into the use of a single across-deck wire as safety barrier, rigged to impact with the upper part of the aircraft undercarriage. (The reader will recall that this was a tail-wheeled aircraft, known locally as a tail-dragger.) It quickly became apparent that this offered only a limited solution to the problem.

Further research into barrier entries by jet aircraft included early trials with iron replicas of Blackburn Buccaneer, de Havilland Sea Venom, Hawker Sea Hawk and Supermarine Attacker aircraft. Martin (#274) states that these were correct copies with respect to weight, centre of gravity, position, nose profile, leading-edge form, height and undercarriage position. They were of very rugged steel construction, so as to be capable of withstanding an indefinite number of barrier engagements without damage. Facilities were provided on the dummies for mounting instrumentation and ciné cameras. Again, these dummies were towed into the experimental barrier systems by one of the Jersey Brow accelerators.

The proofing of the slotted tube catapult, prepared for launching the Hawker Sea Hawk aircraft minus undercarriage, was undertaken with two 'iron birds'. The first was a cylinder with conical ends. It could be ballasted to a range of dead-weights and carried a battery of three rocket motors to represent the jet thrust. It was fitted with a pair of stub wings solely to engage a primitive drag-chain arrester unit for bringing the dummy to rest after a trial ballistic launch. A later dummy was formed with a profiled under surface to match the underbelly of the aircraft. This unit carried a scorpion-like hook to engage a high level wire of a modified arrester unit as it exited the launching track.

A Supermarine Scimitar dummy was provided for use at Bedford in 1958 for the development of the Mark 7 safety barrier. A further 'iron bird' was a dummy McDonnell Douglas Phantom. This again was of steel construction with the front fuselage, engine intakes and leading edges of wings faithfully reproduced, but without the extended nose-wheel unit. Instead of a rear empennage an open structure tail unit was provided with an overhead device, looking very much like the tail of a scorpion (the arachnid, not the liquid-fuel rocket), forming an arrester-hook assembly. This was used at Bedford in 1969 for trials on the Mark 8 safety barrier. Even the TSR.2 aircraft had its representative 'iron bird' replica.

Of 'iron birds', perhaps one of the strangest was a Farmall 'Rowcrop' tractor: a standard agricultural tractor with its front wheels spaced close together, presumably for working between furrows. To someone in NAD, this appeared to be an ideal machine for experimental work in direct towing of aircraft – in lieu of a bridle – for launching. The configuration of the tractor wheels was not far remote from the footprint of a tricycle undercarriage with twin nose-wheels. The trials proved successful but development was brought to a halt when the projected carrier CVA-01 was cancelled. The US Navy has adopted this system almost exclusively in either towing from, or pushing at, the aircraft nose-wheel undercarriage leg.

160

LAND TRIALS

The concept of determining the suitability of a new aircraft or a new installation of flight deck machinery, prior to its being put to general use, must be accepted as being highly desirable, if only for the purpose of discovering the existence of any operational problems and their early rectification. This applies particularly to the assisted launching of an aircraft – by whatever means – and its recovery to deck. While the *ab initio* flight testing of any new aircraft, or new mark, was undertaken by A&AEE, Boscombe Down, once the preliminary assessment had been made the machine was sent to RAE for its evaluation within the naval operational environment.

Such shore-based flight trials generally began with launches from one of the RAE accelerator units on Jersey Brow. These normally would be with the aircraft at light load, the purpose to determine how readily the aircraft could be loaded on to the accelerator and how cleanly the machine exited the flight deck gear. Accelerometers would be fitted to the aircraft for these trials. Similarly for deck landings, these would be undertaken with the arrester-gear units at either HMS *Condor* (RNAS Arbroath), or latterly at Farnborough. Some of the shore-based arrested landing trials would involve the taxying of the aircraft into the arrester wires. With this, the behaviour of the aircraft when entering the wire at some distance offset from the centre line of the gear – with its tendency to yaw – could be assessed, an activity which obviously could not be undertaken on board ship without potentially disastrous results. Careful attention would be given to the tendency for the aircraft to pitch and yaw during the arrested run and high speed ciné film of the trials would be closely examined.

The following paragraphs relate both to proof-testing of shore-based, flight deck machinery and the use of such machinery for trials both of new marques of aircraft, and the testing of new configurations, attachments, performance characteristics and so on of existing aircraft prior to their shipboard trials. (The major experimental facilities used by NAD, such as the Mark III Catapult, Brake Drum Arrester Gear [BDAG], Direction Controlled Take-Off System [DCTO] and the High Speed Track [HST], as well as the permanent installations on Jersey Brow and the arrester area, are treated in the relevant chapters.) Separation of the two complementary aspects of research into both fields of machinery and aircraft has not been attempted in the following paragraphs as interaction generally was inevitable. Specific instances, where either one investigatory field or the other can readily be identified, will be apparent from the context.

Again, while much of the following text treats with the arrestment of aircraft, the launching aspects have also been considered. The general policy of shore-based trials preceding those aboard ship are common to both aspects of Flying Navy – take-off and recovery – with techniques and practices operating as parallel features.

HYDRAULIC ARRESTER GEAR

In the foregoing pages frequent reference is made to trials both at sea and ashore. For many years catapult equipment had been available on Jersey Brow and this was constantly put to good use. This in-house equipment facilitated the development of new processes and procedures and much useful research work was done with one or another of the installations, old and new. In particular, the early concept of the carrier-

borne Assisted Take-Off Gear, which developed into the BH.III Accelerator, originated in an experimental rig mounted upon the RAE Mark II Catapult.

On the other hand, and of necessity, early experiments involving arrester gear were undertaken only at sea. From 1940 much of this experimental work was able to be undertaken ashore at HMS *Condor* (RNAS Arbroath) which was commissioned as a Fleet Air Arm airfield (Royal Naval Air Station) in June 1940 and was the first to be equipped with an arrester-gear installation on a short length of runway dedicated for the purposes of trials and training. Dummy decks – for deck-landing training purposes, known in the Service as ADDLs (Airfield Dummy Deck Landings) – were provided subsequently at other FAA airfields, the earliest being at HMS *Heron*, (RNAS Yeovilton [Henstridge]) and HMS *Jackdaw* (RNAS Crail). ('Touch and go' training for both pilots and Deck Landing Control Officers [DLCOs – or 'Bats'] could be undertaken without the availability of deck-arrester gear.) The aircraft used in these merry-go-round exercises were fondly referred to, by those involved in the flying activity, as 'clockwork mice'.

It must be realised that such gear in a shore installation could be used only on certain occasions when the wind speed and direction were favourable. It has to be borne in mind that the arrested landing of an aircraft into a strong head wind – as on shipboard – creates a much lower retardation load on the airframe and ship's underdeck gear than landing ashore with reduced head wind. Relative wind speeds and air speeds become significant constituent parts of the equation. It was the development of the Hawker Sea Hurricane and the Supermarine Seafire for deck landing which brought into prominence the need for an arrester installation at Farnborough.

It was in early 1942, when much of the above experimental work was being undertaken, that NAD (when it was the Catapult Section) was provided with its own dummy deck facility at Farnborough. The airfield Runway 04/22 – virtually (and fortunately) a continuation of the alignment of the catapult installations on Jersey Brow and already monopolised by the Department's flying programme – would become redundant when the newer Runway 00/18 (re-designated later as Runway 18/36) was completed. The superseded runway in the centre of the airfield would be used as a testing ground for new arrester-gear installations as well as aircraft trials. This local facility was a distinct advantage in that all the special high speed photographic equipment and instrumentation was available within the Establishment and obviated the long-haul of all this paraphernalia to a quay-side on the Scottish

Second prototype Hawker Sea Fury (SR666) is loaded on the C.I Accelerator. The RAE design of holdback attachment was a step towards automatic loading of an aircraft onto the accelerator. July 1946.

(© Crown Copyright/MOD)

coast whenever a trials programme was required. An example of the intensity of usage of an experimental installation is shown in one of the NAD reports, possibly the last time the equipment at RNAS Arbroath was used by Farnborough. On a single day in March 1942 a total of 52 landings were made into the dummy deck arrester system with a Fairey Albacore (12 runs) and a Hawker Hurricane aircraft (40 runs) providing a mass of data at each entry into the wire, all of it manually recorded and laboriously, yet meticulously, analysed.

The first installation at RAE was one of the Admiralty pattern Mark 4 arrester gear units, similar to those installed in the Fleet carriers of that time – viz HMS *Illustrious*[40]. It was installed below ground with the machinery pit on the west side of the runway. This unit was the standard hydraulic jigger with multiple rope sheaves at each end. It was double-reeved to enable two arrester wires to be installed above deck. The unit comprised a cast-iron cylinder with a sliding piston on a shaft protruding at the open end through a gland. Multiple rope sheaves were fitted to both the back end of the cylinder and to a crosshead on the free end of the piston. Around these the main cable was rigged with the free ends brought up to deck level by a system of sheaves and pulleys. Each end was fitted with a socket (capel) to which the across-deck cable (centre span) was connected.

In operation, the jigger unit was set to its fully extended position and when the aircraft engaged with the centre span deck wire the unit closed up under the action of the paying-out of the arrester rope. The hydraulic fluid (a glycerine/water mixture) at the head of the piston was forced through a varying annular orifice. Part of this fluid was transferred to the other side of the piston while the remainder was forced through the bore of a control valve and cut-off unit into an hydraulic accumulator operating against a working pressure of 300psi {20bar}. It was this pressure which resisted the movement of the jigger and thereby exerted the retarding force on the aircraft via its arrester hook. The mechanics of the design of the hydraulic unit also ensured that in all cases, apart from the landing of very lightweight aircraft, the installation worked to almost its full travel. The geometry of the deck installation ensured that, at the instant of an aircraft picking up the arrester wire, its movement was at its minimum, hence impact loading and the overcoming of the inertia forces of the machinery were also at their minimum.

The distance between centres of the sheaves on the under-deck jigger, with the ram extended, was 23ft {7m} with the working stroke as 10ft {3m}. The pay-out of the cable along the axis of the deck depended upon the athwartships centre distance of the deck sheaves. (vide Pythagoras!). On board ship the maximum pull-out, due to the deck sheaves being only about 80ft {24m} apart, was 156ft {47m}. On naval airfields ashore, in order to maintain an obstruction-free runway pavement, the distance between deck sheaves was of the order of 110ft {33m}. This permitted a pull-out of some 167ft {51m}. The installation on Runway 04/22 at Farnborough was designed to represent a flight deck environment with regard to the geometry and methodology of operation. Although the runway was some 150ft {45m} in width, only the central 80ft {24m} or so were used as a deck-landing area to replicate that on board ship. The deck sheaves were installed in the unused spaces on the flanks of the reduced width runway.

This installation became a valuable research tool when investigation into the problems of off-centre deck landings was being undertaken. With a shore installation an aircraft could be brought into the landing zone by as much as 40ft {12m} off centre line without hazarding the pilot, plane or equipment. This, of course, was not possible aboard ship. A moderate off-centre landing to port would be too close to the deck edge of the ship, with the risk of the plane going over the side and that to starboard would endanger the aircraft crashing into the island superstructure. With the advent of the angled deck, the probability of an off-centre landing could be tolerated more readily. This dummy deck facility was employed in assessing the slewing effects imposed on the aircraft during an off-centre landing and to determine the characteristics

of wear and abrasion at the hook/wire interface where the wire would be sliding (fleeting) through the throat of the hook.

This deck landing area at Farnborough subsequently became the home of five arrester-gear installations. The original gear was recalibrated to accept aircraft of up to 16,500lb {7,400kg} weight – its original design feature being for 11,000lb {5,000kg} aircraft only – having a speed of entry of 60kts {31m/sec} with a maximum retardation force on the aircraft equivalent to 2½g. One development in the area of deck-arresting was the installation of the 'Thirty Thousand' arrester gear system, a prototype for the new HMS *Ark Royal*[55]. This requirement was to accept an aircraft of 30,000lb AUW at an entry speed relative to the deck of 75kts {13,600kg, 39m/sec}. The design of this 'Thirty Thousand' gear was being undertaken by the Admiralty in conjunction with Naval Aircraft Department and was destined for the new 'Audacious' Class carriers, particularly HMS *Ark Royal*[55], in which the under-deck installations would be mounted vertically. The prototype installation at RAE likewise was to install the machinery vertically so that development work in this new configuration would be undertaken

Blackburn Firebrand TF.4 (EK670), carrying a torpedo equipped with an air tail, is rigged ready for launch from the C.I Accelerator. The new RAE pattern of holdback is fitted to the arrester hook anchor point. December 1946. (© Crown Copyright/MOD)

Westland Wyvern TF.1 (TS380). The separated front portion is squatting on its torpedo after arrestment on Runway 04/22. August 1949.
(© Crown Copyright/MOD)

– and, hopefully, bugs removed – before the ship was ready for sea. Details of all these installations are given in the chapter 'Jersey Brow – The Arrester Deck'.

This site, and others in the vicinity, also became the locations for initial trials of an entirely new concept of aircraft without undercarriage landing on to a flexible deck. It is thought also that the first of the water-spray retarding units for carrier work, together with the first experiments on the development of the nylon net safety barrier units, were undertaken here.

These trials were at low key for the start in order that the performance envelope of both aircraft and machinery were not exceeded. Much of this depended upon the speed and direction of the wind relative to the alignment of the airfield deck equipment. With the initial stages completed, the aircraft would be embarked at sea for its full assessment in a further trials programme.

DRAG CHAIN ARRESTER GEAR (DCAG)

In the launching of a newly constructed ship, its progress along the builder's slipway must be carefully controlled to prevent its running away through its eagerness to be floating in its intended element. The standard method adopted for this control is through the use of a number of old, generally very rusty, anchor chains. These chains are placed in an orderly manner, in measured piles, along the length of the slipway with their forward ends attached to the ship's hull. As the ship proceeds down the slipway these chains are dragged along by the hull, progressively being paid out to greater lengths so that the moving weight and the friction against the slipway floor continue to increase with the distance travelled, thereby providing the braking effect to the whole procedure.

In its search for ideas for shipborne – and airfield – arrester gear this principle was not lost to Naval Aircraft Department. Deference must be given to the preacher (*Koheleth*) of three millennia past when, in the Bible, he declared there to be 'no new thing under the sun' (Eccl: 1:9 etc) in that a drag chain had been used already in the field of aviation. Penrose (*#70 p.551*) records that Cody with his Military Trials machine dragged a heavy chain as a braking feature. It is thought that NAD was unaware of this earlier embryonic activity when embarking upon the project with the Supermarine Seafire.

An attempt to replicate this methodology was made on the Farnborough aerodrome during the summer of 1943 under the formal title of 'Portable Inertia Arrester Gear' but more popularly known as the Drag Chain Arrester Gear (DCAG). The installation comprised a single across-deck wire, each end of which was attached to a length of (again, rusty) anchor chain, this latter being of reduced proportions in comparison with those for ship launching. The location for the experiments was on the grass area alongside the eastern edge of Runway 04/22. It is thought that the chains used were recovered from the Laffan's Plain end of the redundant Direction Controlled Take-Off system having been installed for use as its carriage braking system the previous summer.

A series of trials took place, firstly, according to Fawell (*#241*), using two 180ft {55m} lengths of chain of 18½lb/ft {28kg/m} weight (link size 1⅜in {35mm}, each link being 6⁷⁄₁₆in {164mm} in length at 3⁹⁄₁₆in {90mm} pitch) attached to a centre span of arrester wire of ⅞in {22mm} diameter supported 18in {450mm} above the grass surface. The supports were spaced 80ft {24m} apart. This gave a dead-weight inertial mass of 3 tons. These experiments produced promising results and the test equipment was quickly upgraded. Heavier chains of 27lb/ft {40kg/m}, (link size 1⅝in {41mm}, each link being 7¾in {197mm} in length at 4¼in {108mm}pitch) each again of 180ft {55m} in length were attached to the centre span, giving a total mass of 4½ tons. Further developments used a smaller diameter of ⅝in {16mm} for the centre span wire, the reduction in sag of this wire between the supports – reduced to 65ft {20m} – permitted the suspension height to be reduced to 12in {300mm}.

Various configurations for 'nesting' the chain in its stowed position were adopted, in every case the intention was to allow the chain to pay out progressively. The initial arrangement followed the shipyard pattern where each of a pair of chains was stacked alongside the across-deck wire and was progressively pulled out as the aircraft was brought to rest. One alternative – which became the finally preferred option – was the case in which each chain was laid parallel to the line of flight in the forward direction, allowing the chain to turn back on itself as it was drawn out. This provided a graduated braking force depending upon the distance of travel, resulting in the maximum length of chain resisting

A 'Gannetised' Westland Wyvern TF.1 (VR133) on Farnborough's Runway 04/22. This Eagle-engined Wyvern had already undertaken flying trials in HMS *Eagle*[51] and here, as an aircraft with contra-props, it was undergoing trials with a variety of safety barrier nets. Advantage was taken to adapt the aircraft to simulate the front end of a Fairey Gannet, a similar contra-prop aircraft with its cockpit right up at the front. A profiled 'canopy' was fitted ahead of the cockpit and a stylised nose-wheel assembly was formed by a strut carrying a pair of Spitfire tail wheels. 1954. (National Archives)

Hawker P1127 (XP980) experimental vehicle at the end of its barrier arrestment. The problem now would be the disentanglement of the aircraft from the strands of the net. *c.*1976. (© Crown Copyright/MOD)

the motion, thereby providing the arrestment through the increase of mass in motion, inertia and friction forces.

Three Seafire aircraft (MB125, MB299 & NX983) together with the hooked Airacobra (AH574) were allocated for these experiments, operating over the range of 6,000/7,500lb {2,700/3,400kg} AUW. The landing run at maximum weight and entry speed of 85kts {44m/sec} was about 400ft {122m} with a maximum deceleration of ½g. During experimentation it was found that for optimum performance the chains should each be about one half of the length of the calculated pull-out. This meant that, after the chains had been fully extended, they were dragged for a further distance approximately equivalent to their own length. Of course, the condition of the ground over which the chains were to travel had some effect on their performance. Coefficients of friction for the chains (μ=0·24 to 0·36) were evaluated respectively over the range for wet grass to dry concrete.

Bullen *(#221)* recorded that after the initial impact of the aircraft with the cross-deck wire the geometry of the system remained reasonably constant until all of the chain had been paid-out. When the chain was in motion it did not maintain continuous contact with the ground but tended to bounce along (porpoising) and the higher the speed the more severe did this bouncing become. While this tended to reduce the coefficient of friction, the total pull-out distance was not greatly affected as maximum retardation occurred early in the arrested run. Bullen (*#ibid.*) stated that the weight of chain used should be selected to suit each marque of aircraft. He provided formulae to enable this to be calculated, based upon the use of 300ft {91m} of chain each side giving a total pull-out of 508ft {155m}.

The centre span of the cross-deck wire was supported above the runway by a series of rubber discs of about 8in {200mm} diameter, partially split to allow them to be threaded on to the wire. Each end of the wire was attached to one length of the drag chain by a shackle, the chain running parallel to the line of the runway. The main problem, of course, was the resetting of the system after the arrested landing of an aircraft. The entire length of both chains had to be recovered and nested – or laid – into the stowed, ready-for-use position. Although involving a pair of tractors, inevitably, this process was time-consuming. Tensioning was achieved by the tractors hauling the shackled end of the wire away from the runway until the wire was held taught across the flight path, the friction of the chain holding the system in its state of readiness. A resetting time approaching the four minutes ideal for service use was never achieved; the recovery exercise often took ten times as long. In spite of its overall simplicity, the distance of the retarding run and the lengthy and laborious recovery operation could not be tolerated under active service conditions; consequently the development for shipboard use of a viable, retrievable system was not proceeded with and the project was abandoned.

On conclusion of these trials the installation was eventually removed from Farnborough airfield with the normal range of adjacent hydraulic arrester units resuming priority. For a while, the DCAG system was retained in use at Farnborough where the extended time for resetting was of little consequence. This application was adopted as the sled/trolley braking system for an intermediate stage in the use of the High Speed Track (qv). However, the expertise gained by Farnborough in these drag chain arrester trials was put to good use in a couple of post-war projects. One of these was the installation at the RAE airfield, Llanbedr, for the arrestment of Fairey Firefly target drones. The other was the arrestment of the 'iron bird' dummy aircraft used in the development of the slotted tube catapult for launching a Hawker Sea Hawk in the flexi-deck project. Both of these studies are treated in their own chapters.

The DCAG principle was one of two mechanical arrester projects researched at Farnborough during World War 2 which did not involve the sciences of pneumatics and/or hydrodynamics. It was basic in concept and required little sophistication in its application. The other system was the Brake Drum Arrester Gear (BDAG) which possessed a greater engineering content than the DCAG. A separate chapter is devoted to this project.

THE 'SPIKE-AND-BARREL' ARRESTER GEAR

This was not the official title for this system, but one of the author's choosing! During the war years there were two hydraulic arrester systems, both described as 'ram and cylinder' units. They were different in concept, performance and rating. Normally the terms 'ram' and 'cylinder' would refer to a long, cast-iron (or forged steel) cylinder, of relatively small diameter, internally machined to present a smooth bore. The ram would be equally long, fully machined with a piston which closely fitted the bore and projecting from the open end through a machined and packed gland. In most applications for arrester gear and accelerator usage each end of this unit would carry a crosshead upon which a number of rope sheaves would be mounted to form a multiple-reeved, hydraulic 'jigger'. In the case of an arrester gear, the unit, in a set-to-work situation, would be in the extended position and the pick-up of an arrester wire by a landing aircraft would close the unit against a controlled back pressure, paying out the arrester wire accordingly. For the accelerator, the unit would start off in the closed position and be extended by the application of hydraulic pressure on the back of the piston, thus pulling the launching trolley or shuttle along the deck to launch the aircraft. The author's preference is to retain the name 'ram and cylinder' for this type of machinery.

The second system, now to be discussed, still consisted of a moving ram within a cylinder, but embodying a different principle. In this case the cylinder ('barrel') would be shorter than the previous type, of relatively larger diameter and greater wall thickness. The only machining required would be at the open end of the cylinder to accommodate a nozzle (choke ring) and sealing ring assembly to hold a frangible disc. Only the nozzle itself would be accurately bored and profiled to engineering tolerances. The ram ('spike') was completely detached from the barrel and fully machined to a slight taper, again within fine tolerances.

The barrel would be mounted horizontally at the retarding end of a braking system while the spike would be fitted to the moving vehicle – be it trolley, shuttle or whatever. The barrel would be filled with water, retained by a thin disc closing the nozzle end. This disc would be punctured by the arrival of the spike, the displacement of the water through the annulus between nozzle and spike bringing the trolley to rest. Of course, the arrangement could be the opposite way round with the spike fixed and the barrel forming the base of the vehicle. It is obvious that this system could apply only to arrester units, or the bringing to rest of the jigger at the end of an accelerated launch.

The first application of this principle was on the P.I Catapult, designed for CAM-ships in January 1941. It subsequently was adopted for a variety of applications, including the C.I and K.I Accelerators on Jersey Brow. The details of some of these applications are now lost in history but Chisman *(#224)* describes a typical system that was adopted for the first, experimental slotted tube catapult.

'The retarding cylinder was a steel forging 6ft long and 13½in external diameter, the bore being 7in diameter. One end of the cylinder contains the choke ring and sealing disc assembly, the other end the venting and charging valves for filling with water. The water was contained in the cylinder by a ⅟₁₆in thick Hallite disc, held in position in front of the choke ring by a quick release clamp ring. The choke ring was machined to an exact diameter, namely 5·0215in, to provide the orifice through which the retardation ram enters after having pierced the sealing disc. The retardation ram was of tapering profile in diameter over a length of 65¾in ranging from 4·662in diameter to 5·019in diameter over the last few inches of parallel portion. As can be readily seen from these figures this gradually reduced the annular orifice with penetration of the ram into the cylinder, thereby giving a retarding force as the water is forced out of the cylinder. The water charging valve fitted to the end of the cylinder incorporates a non-return valve to protect the water main against the working pressure inside the cylinder of approximately 12,000psi.'

The diameter and length (3in {75mm} and 30in {750mm}) of the spike-and-barrel units for the C.I and K.I/K.II Accelerators, having been designed for lower performances, were about one-half of the dimensions given by Chisman.

It is obvious that fine tolerances only could be permitted between the spike and the barrel orifice. In fact, the above figures show that the clearance between the two moving surfaces at the end of the stroke was slightly over one-thou (0·001in {0·025mm}). Inevitably, after a number of operations, wear would increase this clearance but the system could be restored by electro-deposition of new metal on to the spike surface and replacement of the barrel choke ring. The system described by Chisman was devised to bring to rest the launching piston from a terminal speed of 120kts {61m/sec} within a distance of five feet {1·5m}, thus imposing a retardation rate of some 127g.

In addition to the prototype slotted tube catapult, this system was adopted for the experimental stages in the catapulting of the dummy Horsa gliders of the flexible-deck project and, later, for the catapulting of Seafire test specimens in safety barrier experiments. It was used also on the High Speed Track when high 'g' loadings were required for experimental use. The system – at that time – was used only once on shipboard, the C.II Accelerator in HMS *Pretoria Castle*, where a pair of spike and barrel units were located on deck to arrest the trolley/shuttle and a similar installation below deck to bring to rest the moving crosshead of the accelerator jigger. This was a repeat of the C.I Accelerator installation at Farnborough except that in the ship the under-deck jigger and its barrel were mounted vertically.

The foregoing was the picture in 1945. Inevitably, a system which shows promise at an experimental stage will be developed to the extent that new features and higher levels of performance may be attained. The spike-and-barrel braking system is a case in point. With the introduction of the slotted tube accelerator (steam catapult) and its capacity for expanding the performance envelope required in launching aircraft, the abandonment of some tried and tested operating techniques was inevitable. This simple braking system was examined and in the post-war years was subjected to experimentation and development processes.

One unanticipated feature was the ability to dispense with the frangible disc retaining the water in the barrel. *The Engineer* of 28 January 1955 reports that J. R. C. Waterston, a design engineer with Brown Brothers & Co Ltd of Edinburgh (designers and manufacturers of flight deck equipment), produced this system. A water supply, introduced through a series of jets at the nozzle end and directed inwards, created an annular vortex along the walls of the barrel, the combined jet returning as a plug of water along the axis of the annulus to vent to atmosphere. This phenomenon caused the barrel to remain full of water without the need for a blanking plate at its end. Unavoidably, a lot of water would be slopping about the area – which could be dealt with effectively by recirculation – but the installation did not require an attendant having to replace the blank and manually to refill the barrel after each launch.

Another refinement was that the back end of the spike was fitted with an annular deflector having a profile section similar to the bucket of a Pelton turbine wheel. The water forcibly displaced from the barrel along the axis of the spike impacted with this ring and was deflected through 180°. This diversion of the momentum of the escaping water added significantly to the total braking effect when the spike was the moving element, a simple expedient with an advantageous bonus.

It goes without saying that, due to its high 'g' deceleration rates, this equipment was never used for direct arrestment of aircraft; its intrinsic value was in absorbing the kinetic energy of the shuttle/trolley assembly on release of the aircraft at the end of a catapult launch.

PROVING TRIALS FOR CATAPULTS

The proof-testing of accelerator and catapult equipment was always a necessary and vital expedient before permitting it to be used by humans piloting expensive aircraft. With the Mark I and Mark II catapults on Jersey Brow, and others of that ilk, it was a simple matter to provide a series of test weights to be loaded into the trolley and to carry out typical launch procedures.

A steel shaft, with ends formed in the shape of a standard catapult spool configuration, carried a selection of circular plates to represent the dead-load of an aircraft. This was mounted across the front hooks of the trolley and secured by the hook detents. The launch was initiated and the trolley trundled along its track, releasing the dummy at the end of its stroke. No attempt was made to retard the dummy after release; it was allowed to drop to the ground and continue its progress down the slope from Jersey Brow until it came to rest of its own volition. To the sight-seer, the prospect of a steel cylinder careering along the grass of the airfield was an occasion not to be missed. The dummy was recovered by an airfield vehicle towing it back to the launching area – as a roller – probably for a repeat run.

In later years a dummy aircraft or two would be used for a more elegant presentation. In the author's time at Farnborough, a number of redundant Blackburn Shark aircraft were fitted with 'concrete engines' in the RAE workshops. These 'engines' were wooden boxes ballasted with concrete and attached as replacements to the engine mountings on the airframe, the intention was to replicate the weight and centre of gravity position of a normal flight. Each dummy aircraft would be loaded on to the catapult and launched as for a real flight. In some cases the aircraft was restrained by a rope attached to the tail of the aircraft but mostly they were allowed to travel, howbeit haphazardly, along their projected path, possibly to be recovered for a further trial but mostly ending up as a heap of junk. This procedure was extended for the Jersey Brow proofing trials on the prototype trolley-fitted Assisted Take-Off Gear, later known as the BH.III Accelerator.

With the FAA adopting the tail-down, strop (bridle) launching of aircraft, the proofing of accelerators took a new turn. A special trolley was provided for this purpose, which would be trundled along the path of the accelerator. Smith (*#286*) describes a trolley of his design used for calibrating the accelerators at Bedford at the time of their commissioning. This was a robust platform supported on three pairs of wheels, 18in {450mm} in diameter, running along a track alongside the accelerator slot. (The whole – to those who recall the days of steam railways – became a realistic replica of the underframe to the tender of a railway locomotive.) To this platform were bolted rectangular steel plates, their number being adjusted to suit the amount of ballast the trolley was required to carry. After launch, this trolley was brought to rest by travelling (freewheeling) along a water-filled channel, a scoop at its head picking up the water and functioning as the bucket in a Pelton water turbine; another NAD spectacular!

The dummy aircraft facilities are described elsewhere under the sub-heading of 'Iron Birds'. Proof-testing of shipboard accelerators at sea is treated in 'Sea Trials' (qv).

DATA ACQUISITION AND PROCESSING

To the current reader the state of the art relating to instrumentation at the time of World War 2 would appear to belong to industrial archaeology. Almost all data acquisition would be analogue, some direct-mechanical but mostly electro-mechanical, with photo-electric systems available for a few procedures. Much of the information would be recorded photographically for subsequent analysis; this was the only playback facility available. There was very little television; the cathode ray oscilloscope at that period would have a 4in screen producing a phosphorescent green display. The length of the electron gun providing the signal would be about ten inches, all powered by thermionic valve circuitry. Lasers, microchips, instant playback and digital techniques were all future.

In the proofing of flight deck machinery much of the data acquisition process could be achieved with standard engineering techniques using commercial instruments and gauges. This applied to both accelerator and arrester gear installations. When aircraft became involved in the trials a greater sophistication in instrumentation was necessary.

For trials with a new aircraft – as distinct from those of flight deck machinery – the parameters to be recorded during an arrested run (on shipboard or ashore) would be some or all of the following:

a) Wind speed and direction over the flight deck.
b) Speed of entry of aircraft into the arrester wire.
c) Retardation of the aircraft during arrestment.
d) Attitude and pitching of the aircraft during arrestment.
e) Identification of the wire engaged.
f) Extent of pull-out of the wire engaged.
g) Speed of contact of undercarriage with deck, vertical and horizontal.
h) Sequence of contact of undercarriage units and arrester hook with deck.
i) Weight of aircraft at each landing.
j) Identification of the aircraft (often the name of the pilot was recorded as approach technique could vary between 'drivers').

Of the above, each element would be recorded or calculated respectively:

a) by a hand-held, spinning cup type anemometer, or a deflecting-vane type, directional 'Velometer' for side wind vector measurements;
b), d), g) & h) from a high speed ciné film record;
c) as a continuous trace by an accelerometer; a mean value also could be assessed from the high speed film record;
e), f) & j) visually (Mark 1 eyeball!);
i) manually, at each change of 'stores' carried, or each refuelling stage, with intervals interpolated.

Most of the above records would apply also to the accelerator launching activities. This long-standing recording procedure was described in detail by Sharwood (#285) in a paper produced in 1952.

It was a general policy to record on ciné film as much of the trials programme as possible. Good use was made of Vinten and Newman Sinclair cameras with facilities to record on 35mm film at rates of up to 200 frames per second. These cameras were fitted (by RAE) with an electric spark system – generated by an electro-magnetically pulsed tuning fork – that would record pulsed signals on the film at 50Hz. Such records would be analysed for assessment of attitude, distances, speeds, angular movement, rates of acceleration/deceleration and so on. Post-activity film analysis was a tedious process, of no value where an instant appreciation of circumstances was desirable. The film negative would be developed and projected, frame by frame, onto a large plotting

The RAE recording accelerometer. The trace was made by a brass needle onto a waxed paper wrapped around the recording drum. The instrument is seen fixed, inverted, to an access panel of an unidentified aircraft, thought to be a de Havilland Sea Vixen. This is not an ideal location. December 1959. (© Crown Copyright/MOD)

table from which a step-by-step analysis of the above parameters could be laboriously obtained. Often three such cameras would be set up for a series of trials, requiring the handling of some six or seven pieces of luggage for each camera.

Other cameras would be used for general photographic recording and for overall viewing of attitudes of aircraft undergoing accelerator or arrester trials, including situations of particular interest, or even the occasional mishap!

The accelerometers, to RAE design, were made in the Farnborough workshops. Each comprised a lightweight drum driven by a low voltage (12V dc) electric motor to run on the aircraft circuitry. A tinted, waxed paper strip was attached to the cylindrical surface of the drum. A brass needle at the end of a swinging arm produced a trace by scratching the wax surface. The sensing element which powered the arm was a damped, sliding mass on a rotating spindle operating against a spring which responded to changes in gravitational forces in one direction. Two control springs were used, one each for accelerator and arrester tests, calibrated to read up to 6g and 2½g respectively. The drum rotated at about ten seconds per revolution; this gave an instantly readable analogue trace record of 'g'/time, about two inches {50mm} in length; in other words one fifth of the length of the sensitised paper. The instrument frequently was installed in the aircraft in a spare radio equipment tray, with easy access for the trials staff – the sensitised paper needed to be renewed after each record. All accelerometers were subjected to regular calibration, using the small centrifuge in the RAE Structures Department for the purpose.

In those pre-electronic days the types of instruments available were limited. The Cambridge Instrument Company did produce an accelerometer which was put to occasional use. This employed an electrically damped aluminium vane as sensing element, producing a mechanically impressed trace on a narrow transparent film. For reading the record a light-box was required to project the trace – often barely visible – magnified, on to a frosted screen. Although this light-box, carrying its own battery and projector lamp, was portable it was still cumbersome for flight deck use. Also it was necessary to identify and trace the records on to paper if they were to be reproduced in a report. Although the early RAE design of accelerometer still had its supporters, post-World War 2 development in electronics enabled Farnborough/Bedford to produce up-to-date designs for such instrumentation.

Sometimes the movement of the arrester-hook arm was of interest and a Selsyn/Dessyn telemetering system would be installed. In the case of the measurement of aircraft engine thrust against a holdback unit for tail-down launching, a rudimentary (by modern standards) wire-wound strain gauge system would be employed. Airborne ciné cameras were fitted when they had the advantage over the normal bulky, high speed equipment.

In almost every case the aircraft to be used in any set of trials was prepared at RAE, particularly if special instrumentation was to be fitted; this normally involved accelerometers for recording launching and/or arresting forces. A feature to be found on many of these aircraft was a couple of targets painted on the side of the fuselage. These either took the form of a capital letter 'I', or a circle quartered in black and white, and were of fixed dimensions, orientation and spacing. These were vital elements in the detailed, post-trials analysis of photographic records.

Post-World War 2 efforts were concentrated on the development of more specialist instrumentation for use both ashore and on shipboard trials. The major field for improved techniques was the provision of a direct indication of speed of entry, both horizontally and vertically, in the landing-on process. The system provided a pair of narrow parallel planes of light, arranged vertically at a set distance apart. These would be interrupted by the passage of the aircraft and, from the time between the interruptions, the speed could be deduced with considerable accuracy. For the measurement of horizontal speed the planes were set vertically; natural daylight onto photocells was the sensing element. The trials aircraft – generally for land-based trials and occasionally for sea-borne trials – was fitted with a rectangular plate (with its long edge vertical) to its nose. This 'spade' produced a sharp-edged trigger to the light ray, sufficient to activate the system.

For determining vertical speeds of landing a similar system was adopted with the light planes lying horizontally, one above the other. In this case the aircraft undercarriage was fitted with an optical corner reflector. The system was pulsed at 2,500Hz. As the reflector passed through each optical fan, a signal comprising a group of these pulses triggered the start and stop of an electronic timing circuit, from which the rate of descent was quickly displayed.

The immediate post-war decade, and by the time NAD was in process of removal to Bedford, saw a significant extension in the use of instrumentation of greater sophistication than that which was previously available. Post-Farnborough instrumentation employed, progressively and extensively, the more modern, familiar techniques based upon state-of-the-art electronic systems with strain gauges and transducers to the fore. Ward (#307), in his 1956 report on the commissioning of the first BS.4 steam catapults in HMS *Ark Royal*[55], outlines these new features. During a catapult launch cycle of operations it was desired to record the transient values of pressure, temperature, displacement and acceleration. By that time RAE (Instrumentation Department) had designed 12-channel recording cameras with their associated amplifiers. Each

The double dead-load trolley used at Thurleigh for ballast testing of experimental launcher and arrester-gear machinery. This could be loaded to 60,000lb (27 tons) weight. The three-thousandth run of this test vehicle was a cause for celebration by NAD. February 1961. (© Crown Copyright/MOD)

camera contained 12 quick-response string galvanometers and two event markers. Each galvanometer projected a spot of light onto a continuously moving roll of photographic paper, recording the amplitude change of each variable. The electrical elements contained within each transducer recording these changes were either resistive or inductive, depending upon the type used.

Fluid pressures – steam, water or hydraulic fluid – were measured by barrel transducers. These were tubes of beryllium-copper with its wall thickness reduced at one area as the sensing element. This was wound with a continuous helix of resistance wire bonded to the surface. When the tube was subjected to an internal pressure the thin walled portion increased in diameter thereby increasing the resistance of the winding – as in a strain gauge – this increase being recorded in the camera via a bridge amplifier.

Movement of valves and other small mechanisms were recorded by inductive type displacement transducers all designed and produced by RAE. These were sufficiently robust to tolerate environments of oil, steam and water to withstand vibration and to require the minimum of attention after initial calibration and fitting. For the larger displacements along the length of the catapult track a series of brittle pin switches (BPS) was adopted. A micro switch served as a start-of-launch marker with the brittle pin switches placed as distance markers along the track; one instance spaced these as 10, 25, 50, 100 & 150ft {3·05, 7·62, 15·24, 30·48, 45·72m} markers from the launch position. These BPSs, which were mounted to one side of the deck slot, comprised insulated holders which carried brittle steel pins of ½in {0·8mm} diameter and connected electrically to the recorder. These pins were broken, during a launch, by a striker pin attached to the side of the catapult shuttle unit, and needed to be replaced for each launch. A later system of measuring catapult speed employed a series of reed switches along the track with a magnetic trigger attached to the shuttle.

The measurement of end speed of the launch was achieved by an electronic counter which was started and stopped by a pair of BPSs. These were spaced at 20¼in {0·514m} centres (20·2667in is the distance travelled in one second at a speed of one knot; the discrepancy is less than 0·1 percent). This enabled the end speed in knots to be read instantly, converting the displayed figure from a reciprocal chart.

Temperature readings into the recording cameras were made via quick-response resistance thermometers. Other direct-reading high temperature indication – particularly steam temperatures – was by means of platinum resistance wire thermometers reading into a direct receiver. These latter units were slow in response and of no value for measuring transients.

Special accelerometers were used in which the sensing element was a transducer comprising a small ferro-magnetic armature supported on leaf springs between a pair of coil units. These were used mostly for dead-load shots and for these the sensing heads were fixed either to the shuttle itself – where it was subject to damage through the high retarding 'g' forces – or to the dead-load. In the latter case, at the end of the launch stroke the trailing wire was broken and the sensing head, with the dead-load, ended up in the 'oggin. For live launches the robust NAD electro-mechanical recording accelerometer (described earlier) with its recognised deficiencies was generally adopted. The performance of these instruments was monitored and frequently checked against calculated parameters – as in the old days!

SEA TRIALS

Many of the foregoing premises and activities in land trials relate also to trials undertaken at sea. This was where the 'real' test of experimental work was undertaken under practical operating conditions. Also, when problems associated with in-service flying were brought to the attention of RAE, the Naval Aircraft Department trials team would examine the problem ashore and then take the trials to sea. Some of these were associated with arrester gear on the American Escort Carriers (CVEs) and the determining of control settings for receiving British aircraft. The Seafire was a particular problem case here.

While much of the work was undertaken on aircraft carriers already in commission – including proof trials of the flight deck machinery of all ships from across the Atlantic – development work was essential and at times was restricted through the non-availability of a carrier for trials purposes. A solution was quickly provided in that RMS *Pretoria Castle* – a Union Castle liner recommissioned in August 1943 for naval service and allowed to retain her name in the title of HMS *Pretoria Castle* – was fitted with a full-length flight deck, arrester gear (limited in scope and range) and an accelerator (developed at RAE as the BC.II) and made specifically available for trials purposes. It also was used for training Fleet Air Arm personnel.

One of the intractable problems with trials at sea is that the required specific weather conditions cannot be obtained on demand, a facility much to be desired when the behaviour of aircraft at low wind speeds is being investigated. Frequently, experimental work was required with low wind speeds down the deck, often coupled with a fair amount of cross-wind, for investigation into take-off characteristics. Flying from a carrier is best conducted when the wind speed down the deck is fairly high and from directly ahead. But these ideal requirements are not always obtainable, particularly under service conditions. For trials to be meaningful, the required conditions must be attained by whatever means are available; one of these is by controlling the speed of the ship into wind. For low speed take-off trials, if the natural wind speed was too high, the ship often would be required to steam astern for take-off, then to steam ahead for the landing-on. With multiple screw ships of the 'Illustrious' Class, the ship's engineers would have one of the screws trailing 'astern' while the others were running 'ahead'. To change the situation would involve the speeding up of one set of shafts while the other sets were slowed down, a fairly straightforward procedure. However, when this was tried out with HMS *Pretoria Castle* on the first occasion, the Engine Room Department almost created a riot.

This ship, being a converted Union Castle liner, was powered by two enormous diesel engines – each of 8,000hp {6MW} – both of which were required to be used whether moving ahead or astern. The Chief Engineer was shattered when he found that he was required to reverse these massive engines every few minutes or so. To say the least, this was a most involved exercise requiring the engines to be stopped and a multiple of levers moved before restart in the opposite direction could be achieved. As the chief explained, in a normal voyage to South Africa he would be expected to reverse once in leaving harbour and then once or twice on reaching his destination some fourteen days later. To have to submit to three dozen or so similar movements within a few hours was completely beyond his experience. In due course he became used to this manner of operating his beloved machinery and he and the complete engine-room staff subsequently were always fully co-operative.

In the proving test of an accelerator installation on board a new ship the normal pattern would be for a few dummy runs to be made before committing an aircraft to the system. For this purpose, in the early days

of the arrival of CVEs from America, a ballistic tank on wheels was provided by the Admiralty. It was a roughly boat-shaped water tank, sitting on four small wheels. This would be ballasted up to a weight of some five tons with water but still sufficiently buoyant to float. This angular chunk of infra-basic technology would be trundled along the deck – as in a tail-down launch of an aircraft – and released over the bows to splash into the sea below. A sea-boat was available in the vicinity; its crew would recover this equipment for it to be towed to and hoisted by the ship's seaplane crane and delivered back to the flight deck for a repeat performance if required. It did not take long for His Majesty's matelots to christen this monstrosity 'HMS Flying Flossie'. The 'Goofers' (the non-participating audience) of the ship's company were always delighted to see something splash into the water at speed from the height of the flight deck.

A whimsical sequence to this appears in Ward et al (*#307*) when in 1956 he lists, as an appendix, the names of a series of both skid-mounted and wheeled floating deadloads as: Ivan, Ida, Nellie, Noah, Flora, Flossie (not the original), Wendy, Thomas, Jack and Eric, the last two being the wheeled type. These were pre-loaded at weights covering the range 11,984lb to 39,340lb {5,440kg, 17,850kg} and were used in the dockside commissioning of the BS.4 catapults in HMS *Ark Royal*[55] during her 1966–70 refit.

HMS *Eagle*[51] at speed with uncluttered flight deck. The angled deck can clearly be seen as well as spotting points for eight helicopters. Four arrester wires are rigged forward of the after lift. Port and starboard projector landing sights are visible and adjacent to them the angled deck spotting points for the nylon safety barrier packs. The bow and waist catapult tracks on the port side are readily discernible. (Author)

Proof-testing of the arrester and barrier gear was even more basic than the foregoing. Generally 'Jumbo', the Ransomes & Rapier flight deck mobile crane, was deployed to grab the centre span of one of the across-deck arrester wires and travel along the deck with the rope paying out as it moved, taking care not to become unstable and turn over in the process. Both this and the commissioning trials of the accelerator were carried out with the ship at anchor. With the performance of the arrester gear needing, from time to time, to be rapidly updated, this elementary method of testing had to be improved. A progression of land trials was conducted at Farnborough and eventually its sophistication was carried across to trials at sea using 'iron birds' – full-size space models of various aircraft marques – and time-expired dummy aircraft.

A post-war Admiralty appointment was that of Arrester Gear Overseer. His task was to carry out tests and inspections of arrester gear in

Model of HMS *Fencer*, one of the Attacker class of American CVEs seen in the FAA Museum. The track of the AH.II Accelerator may be seen along the port side of the flight deck. (Author via FAAM)

The author working on the for'ard camera boom, one of three installed for trials purposes, on the starboard side in HMS *Pretoria Castle*. 1945. (Author)

manufacture and commissioning. Working with Farnborough on the installations there, a dynamic system for testing was developed. At Farnborough, one or another of the accelerator installations was used to tow a dead-weight (trolley, or 'gash' aircraft) into the arrester gear located further down the runway. This principle was adopted for shipboard gear first in HMS *Eagle*[51] (*ARC 15,233 'A Note on Recent Developments in Naval Arresting Gears, October 1952'*).

Pulley sheaves were welded to the flight deck abaft the port accelerator and on the centre line. Cables were run from the accelerator shuttle via these pulleys to a dead-load trolley engaging the centre span of the arrester wire on which the test was to be conducted. The 96ft {29m} accelerating stroke caused the trolley to make a similar pull-out of the arrester wire at a realistic speed of impact from an aircraft. With a 2:1 reeving the pull-out would be doubled, encompassing the full capacity of the arrester unit.

For recording by ciné film, RAE designed a series of three retractable camera booms which were installed at flight deck level along the starboard side of HMS *Pretoria Castle*. Each boom was a Warren type girder, about 60ft {18m} long, fitted with a rail track along its top. A trolley carrying the bulky Vinten camera, itself mounted on a turntable, ran on this track to the outboard position. Remote start, stop, aperture and time-base control of each camera was undertaken from a conning position, located at flight deck level on a deckside sponson at each boom. Panning of the camera was achieved by a series of wires and pulleys, sighting being through a convex mirror on the camera turntable and a telescope at the control position. The operator inevitably had his back to the flight deck while taking his pictures, viewing the procedure in a mirror reflection. (One operator – Jim Hampson – was badly injured when a fragment from a breaking propeller struck him in the back. He was too intent on recording the incident to see this, but the film recorded it in detail.) The camera trolley had to be hauled back to the deckside control position for lens changing and film magazine replacement. The camera booms were located, one for'ard and two abaft the island. From the for'ard position the entire length of the accelerator stroke (as it was at that time) could be monitored while from the two abaft booms the length of an arrester run could be recorded. The booms were stowed alongside the deck edge and deployed only when required. A team of the flight deck party would do the rope-hauling in the rigging and stowing activities. On the decommissioning of HMS *Pretoria Castle* in 1946, these booms were removed and transferred across to HMS *Illustrious*[40] then operating as the replacement trials carrier.

The trials aircraft, having been previously subjected to a series of dummy deck landings ashore, normally would be flown on in a lightweight condition. It would be loaded to the desired AUW on board with the appropriate fuel and 'stores' for the conditions required. The arrester gear would be set to the standard rating for the weight of the aircraft and the trials would begin. Accelerometer and ciné records would be taken, the pull-out of the wire engaged would be measured and a visual impression obtained. Adjustments would be made to the flight deck machinery as necessary and the tests repeated. The aircraft AUW would also be adjusted to suit the trials programme; attempts would be made by the DLCO to bring the pilot to pick up the whole range of wires in turn. The forward deck park would be clear of aircraft and safety barriers normally would remain in the lowered position, or left unrigged, for these trials.

The accelerator trials would begin with the aircraft in low AUW condition. The deck handling and loading procedures would be examined first, then the actual launching programme would proceed. Firstly the conditions would be obtained with a high wind speed straight down the deck and the launch made. This too would be photographed so that the actual speed at take-off could be determined and the tendency to 'squat' (with tail-down launching) could be seen. Trials would continue with the conditions being made progressively more severe with regard to AUW and imposed acceleration forces – the 'g' effect. Tests would

One of the ciné camera booms removed from HMS *Pretoria Castle* and re-rigged in HMS *Illustrious*[40] when she took over the post-war function of trials and training carrier. The Vinten high speed camera is seen fitted to the panning turntable in the loading position. (Fleet Air Arm Museum)

Boulton Paul Sea Balliol prototype (VR599) picking up an early arrester wire in HMS *Illustrious* during trials in June 1953. The high speed Vinten camera is rigged on the after camera boom and the photographer is shown panning the camera by viewing through a telescope and mirror geared to the camera platform. February 1955. (© Crown Copyright/MOD)

HMS 'Flying Flossie' aboard HMS *Indefatigable* during commissioning trials. This was a ballistic, dummy load used for commissioning shipboard accelerators (at anchor). It was a buoyant hull, mounted on four wheels, fitted with towing hooks and holdback attachment point. It could be ballasted (with sea water) to any weight within the range 13,020–20,820lb for towing by the accelerator shuttle. On leaving the deck, 'Flossie' described a graceful arc before making an almighty splash as she hit the 'oggin. March 1944. (© Crown Copyright/MOD)

continue with a measure of cross-wind over the deck, achieved by the ship steaming slightly abeam to the natural wind direction. Finally trials would be made with the aircraft at maximum AUW; this inevitably meant a flight ashore for the heavy load (torpedo, bombs, depth charges and so on) to be removed before the aircraft could be landed-on again. Seldom was an aircraft brought back on deck with military 'stores' retained – although in these trials 'dummies' were always used.

For the case of cross-wind launches, Dr J. Thomlinson – a senior colleague on the NAD staff – had designed a pocket-sized ready reckoner, basically a circular slide rule. By this, the Flight Deck Engineer could quickly assess the safety limits within which cross-wind launches could be undertaken.

At times the author became sensibly aware of the responsibility which rested on his shoulders. Often he – in his early twenties – was the senior officer in charge of the trials and his decision as to the progress of a research project carried weight. There was one occasion when he was required to appear before the ship's Captain – Caspar John, (HMS *Pretoria Castle*) who was a highly experienced pilot in his own right and boss of the ship's company (the author's equivalent rank would have been Lieutenant) – to explain why a certain procedure could not be carried out and what action was necessary to rectify the situation.

The author's experience of sea trials over a period of four years 1942–5 included the conducting of trials on a total of 31 different aircraft carriers, amounting to 46 visits, covering 134 days at sea. These included visits to Clydeside, Rosyth, Scapa Flow, Belfast Lough, Liverpool Docks and the Solent area of the English Channel, as well as short visits to various airfields. These trials at times included his being a passenger in an aircraft catapulted from the deck and landing-on again into the arrester gear. On one occasion he had been ashore to service some of the test instruments and was being conveyed back to the ship in a Fairey Barracuda (P9667). The ship was not ready to receive the aircraft so, while waiting for it to turn into wind and increase its speed, the pilot – Lt A. S. Baker-Faulkner, shortly to become the CO of an operational squadron – treated the ship's company to an aerobatic flying display. The passenger was aware of 'g' reactions in a multitude of directions as he was being thrown about in the rear cockpit without the advantage of being secured by straps. This was not so bad, but he was sharing the space with four or five Spitfire undercarriage legs and tyres. At one instance he had to look up through the side window to see the flight deck above him. Inevitably, he was a bright shade of green when at last he alighted from the aircraft. On another occasion the author had landed on board in a Grumman Avenger when none other than the captain of the ship opened the aircraft cabin door. Unknown to the passenger, he was acting as stand-in at a rehearsal for the arrival by air of the local flag officer on the following day.

It was always a fascination for the author to see the activity created by the Bo'sun's Mate's pipe 'Hands to flying stations'. This triggered the maximum of activity throughout the ship but particularly at the hangar and flight deck levels when every seaman of the air branch went helter-skelter to his post. The resemblance to a stirred-up ants' nest is apposite. On one occasion the author suggested that the pipe should be 'Hands to *panic* stations' but this would belie the outcome for within a matter of a few moments all sailors were at their working locations and ready for whatever action was to follow. The term 'panic stations' has now become quite familiar as synonymous with a condition known as 'organised chaos'.

All this added up to an experience which the author was glad not to have missed, even to have seen HMS *Pretoria Castle* on VE Day 'dressed overall' with a gigantic umbrella and bowler hat (the 'uniform' of a city gent) hanging from its mainmast. This ship subsequently was demobilised (March 1946) and 'recivilianised' under the name of RMS *Warwick Castle*, being restored as a commercial liner on the weekly South African run. For a while HMS *Illustrious*[40], rigged with the ex-*Pretoria Castle* camera booms, took over the duty of trials carrier until her own decommissioning and demise in 1956.

During 1948–49, HMS *Warrior* adopted the stance of a special trials carrier when landings without undercarriage by de Havilland Vampire 21 aircraft were made onto a flexible, inflatable rubber deck (vide 'Post-war Developments – Flexible Deck').

HMS *Perseus* was another ship used for a series of sea trials. She was of carrier format, fitted with a full-length flight deck, but was employed as an aircraft service ship. The prototype steam catapult, BXS.I, was superimposed on the for'ard end of the flight deck for the duration of its proving trials over a couple of years, 1950–52.

Perhaps this is the place to refer to HMS *Illustrious*[40] herself, as this ship featured much in sea trials during the war years, at least until HMS *Pretoria Castle* assumed major duties as a trials carrier. Poolman (*#73 p.38*) writes of her building at Barrow:

'The very impact of her name suggested glory, and her crest, inherited from the old *L'Illustre*, captured from the French in the time of Nelson, was three brazen trumpets. The Captain pondered this, then asked his brother, a classical scholar, "To think of a motto involving God and War", because, as he said, "If I don't have a motto linking those trumpets with God and War they'll be connected with blowing hard!" His brother looked in the Scriptures and found a sentence of St. Paul, which ran: "And if the trumpet hath an uncertain sound who shall arm himself for the battle?"

'From this they took the Latin *"Vox Non Incerta"*, "No Uncertain Sound", and made it the motto of *Illustrious*. Her men took great heart from this inspired choice and the new team worked on with greater zest than ever.'

Jones, who served on the lower decks of HMS *Illustrious*[40], writes (*#53 p.36*) that the Captain in question was Denis Boyd who was in command of the ship for the subsequent Taranto exploit. He points out that the Scripture quotation is from I Corinthians 14:8.

ON GOING TO SEA

The preparation and assembling of equipment for trials at sea resembled nothing short of kitting out for an expedition. The basic research equipment comprised mostly accelerometers and camera gear, but there were plenty of both types of instrumentation to be prepared and conveyed to the trials location. Without this equipment the series of trials would

HMS 'Flying Flossie' awash in a maelstrom of her own making. She is about to be recovered by a sea boat after a dummy launch from the BH.III accelerator aboard HMS *Indefatigable*. March 1944.
(© Crown Copyright/MOD)

A 1962 picture of Blackburn Buccaneer S.1 (XK536) about to be launched from the starboard BS.4 steam catapult in HMS *Hermes*[59]. The aircraft is in its typical nose-up attitude, held by the holdback unit against its retractable tail skid.
(© Crown Copyright/MOD)

be almost valueless. Accelerometers would give immediate indication of the stresses involved in both landing and accelerated take-off, while later analysis of exposed photographic film would enable many performance features to be evaluated.

The accelerometers normally were installed into the aircraft at Farnborough, with temporary wiring connected to a control switch in the pilot's cockpit. Generally the radio bay was used as it offered easy access to remove and replace the paper record each time the aircraft was recovered to the flight deck, a necessary feature when the wind down the deck was blowing half a gale and more. The test pilot had to be instructed in the operation of the control switch at the appropriate time, adding to his multifarious duties at both points of take-off and landing-on. Sometimes a trace was missed but generally good results were obtained.

It was the camera gear which constituted the bulk of equipment for transportation to trials on 'site', be they on land or shipboard. For high speed ciné photography, up to 200 frames per second, RAE favoured the Vinten camera. This was a robustly-built camera and therefore fairly bulky for handling. It used interchangeable, matched twin lenses in its optics and carried an external magazine normally loaded with 400ft of high speed (27° or 30° Scheiner) 35mm film. It was electrically powered using a 12 volt aircraft battery.

The optics were adapted by RAE to include a timer unit. This comprised a spark generator, located in the film path, triggered by a magnetically-impulsed tuning fork operating at 50Hz. The spark 'fogged' a spot on the perforated track of the film enabling a relative time to be matched against the recorded picture. For land and sea trials the camera was run at about 120 frames per second, providing a time spot at about 2½ frame intervals. This was adequate for subsequent assessment of rates of acceleration and retardation, distances, speeds, 'g' forces, attitude and pitching movement by meticulous graphical plotting using a manually operated, frame-by-frame projector. (This, incidentally, was one of the tasks involving the author!)

Each Vinten camera comprised the body, an assortment of lens units, tripod, motor, battery, tuning fork unit, cabling and control box. To this was added some half dozen or so loaded film magazines and a host of sealed tins containing further rolls of film. During the progress of the

Hawker Siddeley Buccaneer S.2 aircraft (Code 102, XT288) of 800 Squadron, leaving the waist BS.4 catapult in HMS *Eagle*[51]. The bridle is going overboard on this launch; the bridle catcher device had not been perfected at the time of this picture. (Author)

Hawker Siddeley Buccaneer S.2C (Code 023 of 809 Squadron) towards the end of its launch from the waist BS.4 Catapult in HMS *Ark Royal*[55] during which the bridle will be retained by the van Zelm system.
(© Crown Copyright/MOD)

trials, these extra rolls were loaded into the magazines, replacing the exposed film which in turn was re-sealed in the tins and labelled accordingly. (The ship's darkroom was used for this.) All this adds up to a bulky mound of luggage. When this is multiplied three-fold – and suitable weather-proof clothing (Sidcot flying suits) added – the likeness to an expedition may be justified.

In addition to the Vinten cameras the photographic party would bring along one or two hand-held ciné cameras, mostly the clockwork-driven Newman Sinclair using 35mm film. Also, a twin-lens reflex camera of the Rolleiflex type would be carried for stills photographs (on 9cm square 120 film) where good definition was required. The ship's photographer inevitably would be interested in what was being done and the best of co-operation always ensued between his department and the trials party.

The transport of this equipment from RAE to 'site' was almost a 'royal progress'. Of his 46 visits to ships on sea trials, the author can recall only three occasions when the ship was 'tied up' alongside a dock, and one of these was with the ship in dry-dock. As most trials were conducted on the River Clyde – and points northwards – the equipment had to be conveyed to its destinations. Generally the equipment (and sometimes the personnel) were taken from Farnborough to Euston by RAE transport, occasionally by a 3-ton truck, where it was off-loaded on to a four-wheel porter's trolley (often on a dark, rainy evening) and man-handled into the railway van of the overnight 'Services Special' train to Glasgow. (This train was also known as the 'Ghost Train'; in both directions its existence was not acknowledged by any railway timetable. All seat/sleeper reservations had to be made via the Admiralty.) Thence, at the destination, were three alternatives. The first was the provision of an Admiralty 3-ton lorry to take the equipment and trials party to Greenock, Gourock or Wemyss Bay pier. The second was to transfer all the equipment to another train at Glasgow Central and the third was to cross to Glasgow St Enoch station – pushing the four-wheeled porter's truck – for a train thence, either route to be met at the destination by a further truck to convey the goods to the pierhead. From then on the equipment had to be handled into a drifter for conveyance to the trials ship, generally swinging on a buoy somewhere out at the 'Tail-o'-the-Bank', and then on board. For the last two stages of handling a team of ratings was normally available; a welcome relief to the Farnborough trials party.

Normally the starboard after gangway was used for boarding ship. We would climb this, salute the Officer of the Watch (by raising our hats; we all wore them then) and await the arrival of the gear which would then be taken by the ship's handling party to the aircrew ready room for sorting and deployment. On one occasion the ladder was not rigged for our use. A cargo net was lowered over the side to the drifter's deck, all the equipment and personal baggage was piled into this and it was lifted to the flight deck by crane, the wet weather not helping. We civilians were faced with a 30ft {9m} high scrambling net, hanging over the side of the carrier; one naval officer in the party scrambled up, showing us lesser mortals how this should be done. The lesson could not have been very effective as the rest of us seemed to labour the task, scared out of our wits at the drop, the instability of a large wet mass of steel netting up which we were attempting to make a dignified ascent, the presence of so much sea beneath us, and the rain tippling down from above. We did wonder if His Majesty was aware of the conditions under which his obedient and faithful servants were required to carry out their duties on his behalf!

All trials equipment would have been set up and made ready as the carrier paced its way down the Clyde towards an area between the Isle of Arran and Ailsa Craig – the latter known to every sailor as 'Paddy's Milestone'. The aircraft would be brought on, the pilots briefed as to their duties and the trials programme would begin. On completion, perhaps after a day of busy-ness, or occasionally a few days if the programme was a lengthy one, the trials party would pack their bags and

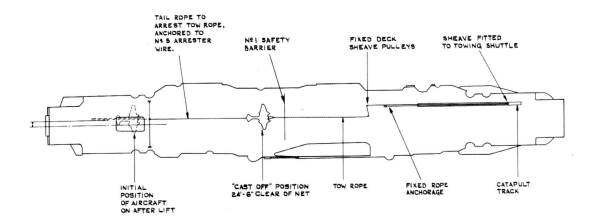

TAIL ROPE TO
ARREST TOW ROPE,
ANCHORED TO
Nº 5 ARRESTER
WIRE.

Nº I SAFETY
BARRIER

FIXED DECK
SHEAVE PULLEYS

SHEAVE FITTED
TO TOWING SHUTTLE

INITIAL
POSITION
OF AIRCRAFT
ON AFTER LIFT

"CAST OFF" POSITION
24'-6" CLEAR OF NET

TOW ROPE

FIXED ROPE
ANCHORAGE

CATAPULT
TRACK

Deck plan for carrier-borne trials in which research aircraft were towed into an experimental safety barrier, using the accelerator as a 'tug', and braking system attached to the ship's arrester gear. This installation was temporarily rigged in turn in HMSs *Implacable*, *Triumph* and *Eagle*[51]. June 1951.
(© Crown Copyright/MOD)

begin their passage home with all the man-handling of gear as on the outward journey. The procedure possibly would need to be repeated just a week or so later on a further set of trials. No matter how intense the pressure, to the author each visit was interesting, perhaps even exciting, providing a host of experiences to treasure for a lifetime.

MARK III CATAPULT

Not all the work undertaken by NAD was related to naval flying. In the early history of the Department – when it was still the Catapult Section – consideration was given to assisted take-off and arrested landing of heavy bombers under development at that time. This was in the years prior to the issue of the specifications that produced the Handley Page Halifax, Avro Manchester/Lancaster, Short Stirling and Vickers Warwick aircraft. The RAE Mark III Catapult, with its design originating in November 1935, was bold in the magnitude of its concept in that it was intended to launch large, multi-engined aircraft in tail-down attitude. This principle of three-point launching was original in that it pre-dated, by some seven years or more, its introduction to carrier-borne Naval aircraft. (To be pedantic, this installation ought to have been designated an 'Accelerator' but from its inception it carried the title of 'Catapult', with its serial number following the sequence of its permanent Farnborough antecedents on Jersey Brow.)

It is the author's impression that this project has never been 'written-up'; references to the installation as to script or illustrations are sparse in the extreme. In fact the only instance, by Bowyer *(#10 p.48)*, owes its origin to a letter written by the author decades ago; he cannot recall having seen any other reference. He therefore feels that this description here must be treated in fair detail, bearing in mind that most of the material content arises from what can be recalled from memory. Apart from a few pictures of the progress of the civil engineering work in constructing the machinery pit, there seems to be very little photographic record remaining of this monster.

Engineering design on this project began in November 1935, the general principle being an upgraded extrapolation of what had become the standard Farnborough type, telescopic tube design. Manufacturing proceeded both at Farnborough and with contractors starting shortly afterwards. Salmon *(#283)* quotes the specified requirements for launching aircraft of 60,000lb at 110mph {27,000kg, 49m/sec} with a

mean acceleration of 2g (in other words, with a launching stroke of 202 feet {61m}) at three-minute intervals. Originally, construction was to have been at Farnborough but geographical limitations overruled and the new RAF airfield at Harwell, opened in March 1937, was chosen for its installation. Site excavation and construction work began towards the end of 1938.

The catapult machinery complex was accommodated as a rotatable turret within a pit 100ft {30m} diameter, about 10ft {3m} in depth, located at the eastern end of the airfield close to the perimeter track alongside Runway 11/29. The whole was roofed over at ground level to allow aircraft to taxi over it without hindrance. Two underground channels ran from this central pit, one each orientated into approximately west and south-west directions. One record is precise in giving the compass bearings as 215° and 275°. These were about 100 yards {90m} in length and carried a slot in the roof which was plated over when not in use.

The entire installation of structural, mechanical and electrical components was on a suspended platform within the turntable structure completely filling the large pit. This would be aligned into either of the two orientations of the channels. The structure below ground was of steel lattice work with the centre space occupied by a system of six fully extendible telescopic rams forming a compound, nested tube hydraulic unit. This ram system started with a machined steel pressure vessel of five feet diameter some 60ft in length {1·5m x 18m}. The second tube was inserted into this with a piston at the back end and the front end passing through a pressure gland. Similarly the remaining four tubes were accommodated. These tubes were produced, on their new 2,000 tons forging press, and machined by the English Steel Corporation at Sheffield. The forward end of the smallest tube was to be attached to the launching trolley and interface component which would connect the catapult with the aircraft above ground. These tubes were to be filled

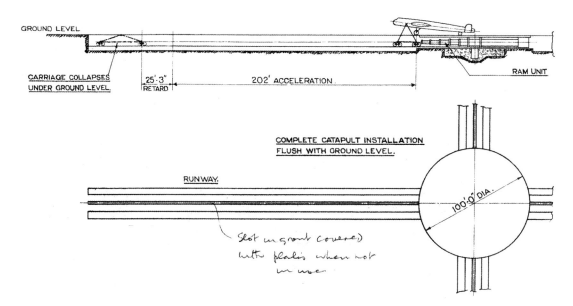

Percy Salmon's ambitious proposals in the mid-1930s are illustrated in a number of his sketches. This one portrays the use of a large hydro-pneumatic, telescopic tube catapult for launching heavy bomber aircraft. This system developed into the RAE Mark III Catapult for installation at RAF Harwell, where two of the four proposed launching tracks were to be provided. The installation was never completed. (© Crown Copyright/MOD)

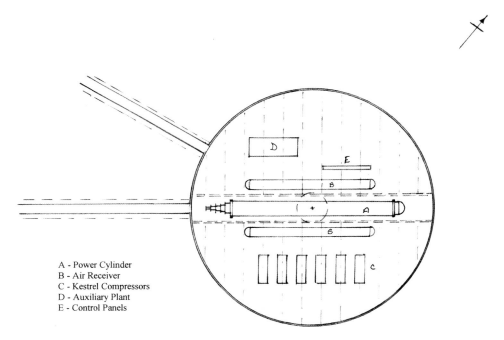

A - Power Cylinder
B - Air Receiver
C - Kestrel Compressors
D - Auxiliary Plant
E - Control Panels

The author's impression of the Mark III Catapult turntable depicting the below-deck-level plant layout. All the machinery was mounted on a platform supported by the central pivot of the turntable. Only the major items of plant are shown here. (Author)

Mark III Catapult at RAF Harwell. The central steel structure within the 100ft {30m} diameter pit is shown. Beneath the centre of the structure was a hydraulic jack system for raising and turning the massive turntable which would be supported by this equipment. The trench on the far side was one of two which would accommodate the rails for carrying the aircraft towing hook, propelled by the hydro-pneumatic telescopic ram system, installed within the turntable structure. April 1939. (© Crown Copyright/MOD)

The main, static cylinder under installation into the turntable of the RAE Mark III Catapult at Harwell. This shows the bore to be about five feet {1·5m} diameter. *c.*1939. (UKAEA)

RAE Mark III Catapult under construction at RAF Harwell. One of the two below-ground launch-ways showing the slot in the roof, as seen from the turntable end. June 1940. (© Crown Copyright/MOD)

with a water/glycerine mixture. Carefully designed interconnecting passages in the tube walls would permit this fluid from the external annulus of each larger tube, downstream of the piston, to pass into the interior, upstream space of the next smaller tube.

Pneumatic pressure (air, stored at 2,000psi {140bar}) would be introduced at the closed, rear end of the largest tube. This would push forward the next tube through displacement of the hydraulic fluid and this in turn would propel forward the next smaller tube at an increased speed. So the process continued until the smallest tube was being moved forward at a speed of the order of 100kts {51m/s}. During this process the front end of the smallest tube – connected to the launching trolley – would have extended some 300ft {90m} from its starting position. (The deceleration of the underground mechanism towards the end of its stroke would be achieved by a set of fluid transfer ports in the telescopic tubes being progressively masked by the tubes passing through the outer glands as in a typical 'F' type catapult.) This extension of the tube system would take place along one of the direction orientated tunnels, the trolley at the front being guided throughout its length by a monorail on the tunnel floor. The roof of each tunnel carried a slot to the ground surface above, through this would protrude the aircraft towing linkage.

The catapult was designed to be fully self-contained with all the equipment, including the power sources, suspended within the turntable structure and slewing around with it. The power plant comprised 12 redundant Rolls-Royce 'Kestrel' 'V'-twelve aero engines, six of them being drastically modified to operate as compressors. These were direct-coupled – without the front end, propeller reduction gear – to the other six that would provide power input to drive the compressors. Those modified as compressors retained their cylinder blocks and pistons, and all rotating machinery below them. The original carburettor, ignition system, camshaft and valve gear were removed and the cylinder head castings adapted for their new use. One cylinder block had packing pieces added to the crowns of the pistons, this unit acting as an air compressor with the output from three (or four) cylinders acting as input to the other three (or two) as a two-stage compressor delivering air at about 200psi {14bar}. The other cylinder block had the cylinder head machined to accept a new heavy steel cylinder block mounted above it. This block had six cylinders of about 1¼in {31mm} bore with poppet-type inlet and delivery valves in its head. The ram-type plungers were connected to the crowns of the engine pistons below, the latter being used as crossheads for the plungers. These plungers

A view along one of the Mark III Catapult runways from the turntable. June 1940. (© Crown Copyright/MOD)

would pump water at up to 2,000psi {140bar} into a pair of large hydraulic accumulators located within the catapult turntable complex.

The intention was for air to be pumped into the storage receivers and accumulators at up to 200psi {14bar} and the pressure subsequently increased by the hydraulic unit to the operational level of 2,000psi

A late mark of a Rolls-Royce 'Kestrel' 'V'-twelve aero engine with modified propeller shaft for driving a hydraulically controllable pitch propeller. Late 1930s. (Rolls-Royce Heritage Trust)

A photomontage from a Rolls-Royce photograph of a 1930s 'Kestrel' aero engine. It is intended to depict the unique set-up in which pairs of time-expired 'Kestrel' aero engines were adapted to act as air compressors to power the RAE Mark III catapult under build at RAF Harwell. The 'Kestrel' engine, to the right, was the power source for its mate, a heavily modified engine adapted to function as a combined air compressor and hydraulic pump with crankshafts connected through a muff coupling. (Author via Rolls-Royce Heritage Trust)

{140bar}. In the author's opinion the concept was faulted by the expedient of attempting to achieve high pressures from devices not designed for such, and at rotational speeds far in excess of those normally found in similar pressurised systems. The six power-plant units were worn out during their development testing.

The same hydraulic power source was to be used for lifting the entire

Sixty years on! The RAE Mark III Catapult pit at Harwell. The pit is 100 feet {30m} in diameter and ten feet {3m} deep. The scale may be seen by comparison with the mechanical digger on the western trench. Following removal of the wartime catapult machinery in the mid-1940s, Harwell scientists used the concrete-lined facility as a temporary waste transit area. It remained buried for the next 50 years when it was decided, in 2002, that in the light of modern environmental considerations the pit should be refilled with freshly quarried materials. The picture shows the pit cleared of waste and backfill material prior to refilling. (UKAEA)

turntable assembly and rotating it between the two directional orientations. This was achieved by floating the central pillar, forming the axis of the turntable, on a large hydraulic footstep bearing. The object was to lift the turntable a matter of a few inches to raise the system above location blocks, to slew it 60° to the alternative orientation tunnel and track, then to lower it into the second set of location blocks. These blocks ensured complete alignment of catapult trolley and track. The author understands these two operations were satisfactorily achieved before the project was abandoned. (These were the component systems in which, design-wise, he had been involved.)

A separate trolley was to be provided in each launcher track. For each change in orientation the trolley in the first track would be disconnected from the ram system and, after the slewing of the turntable, that in the new location be connected. The trolley could not be accommodated in the turntable as there was no provision for a slot in the decking, and there was insufficient space for it to be housed in the collapsed position.

The launching mechanism as envisaged, was to terminate in a linkage embodying a large hook that would fit around a spherical spool located on the aircraft. This hook would be at the top end of a rigid arm pivoted to the front end of the under-deck trolley. As the catapult trolley approached the end of its travel, with the aircraft about to be released, the underground gear would begin to decelerate, the hook would fall away to collapse below ground level and the aircraft would be free for take-off. This hook unit was never completed in design; there were many problems, mainly associated with ensuring its total collapse below ground before the rear end of the aircraft passed over it. Also, the

positioning of the spool attachment on the underside of the aircraft fuselage presented a difficulty as the size of the bomb bays dictated such an attachment to be located either forward or well aft. The preferred position was forward but this would mean the towing bars being lengthy. If the aft position was adopted then difficulties would arise in that the accelerating forces would have to be applied at a point towards the rear of the aircraft thereby reducing its stability in yaw under launching conditions. Another problem was the prevention of the tail wheel from dropping into the shuttle slot. It was with these unsolved considerations in mind that the whole project was abandoned before an aircraft was made available and adapted for its use.

The reader may question why a system using a towing bridle, similar to that employed later with carrier accelerators, was not employed. The short answer was that such a technique had not been established, nor even envisaged, in the late 1930s. If consideration had been given to such a system the weight of the bridle would have been awkward for man-handling when loading the aircraft. The application in theory would have been attractive but much work would have been necessary in the investigation into the use of steel wire towing bridles, including their behaviour during launch and their tendency to flailing immediately after release of the aircraft. This last feature was of importance as the aircraft would still be on the ground with little clearance for the bridle to fall away – not as in seaborne carrier launches.

An aside! At the time when the design of the control system for this catapult was being worked upon in the Main Drawing Office, the draughtsmen involved often resorted to one of the structural columns in the office to check the reach of an operator to turn a wheel or operate a lever at high level on a control panel. (The RSJ columns in this office were painted Midland red up to dado height and white above. A tape measure for use as a height reference was fixed to one of these columns.) Thus in time a lot of pencil marks appeared on the column, the author himself adding to the graffiti. A bit of friendly banter in the office between short and tall over the height of reach was inevitable. At this time an apprentice in the office, Norman Brown – the youngest of three brothers who 'served their time' at RAE – who was well over six feet tall had his opportunity for fame. Rubbing his thumb in the graphite dust from a pencil sharpening machine he reached at full stretch to place his mark some eight inches or more above the rest of the grubby paw marks. This was a 'Guinness Book of Records' achievement in that for many a subsequent re-decoration of the office the thumb-print was never painted over. Now, as for so many others on the 'Factory' site, this building [Q.1] no longer exists, having fallen victim to the subsequent, general demolition of the old 'Factory' site.

The construction work on this project at Harwell was overseen by Alf Crossfield who, for a couple of years, spent much of his time on the site. An interesting write-up appears in Bowyer (#*ibid. pp.48 & 155*), the paragraphs on the former page having been written by the author! Some of the larger plant elements (including the two larger telescopic tubes and some hydraulic control units) were recovered from Harwell in the mid-1940s by the Sir William Arrol company and transported to RAE for use on the experimental 30,000lb accelerator, the penultimate in the series of launching machinery to be constructed on Jersey Brow.

'Harlequin', the AERE Harwell house magazine, in its issue dated Winter 1982/83, carried an article relating to the catapult site. This was based upon some information that the author passed to a curator of the Science Museum in 1976. By the date of its publication, the only evidence to be seen on the airfield was that, after a dry spell, the grass on the two radiating arms and a few smaller areas appeared to turn brown quicker than their surroundings. On a personal visit to the location early in 1997 no evidence of the existence of the Mark III Catapult was visible.

In the late 1940s, following the removal of the catapult machinery, Harwell's scientists used the concrete-lined facility as a temporary waste transit area for some years before backfilling it. At that time the site was still within a high security enclave with restricted access. It remained buried for the next fifty years when it was decided, in 2002, that in the light of modern environmental considerations and the proximity of a nearby junior school and proposed housing estate, the pit should be refilled with freshly quarried materials and grassed over. The evidence of what once was there will be buried again, to be rediscovered by a future archaeologist.

DIRECTION CONTROLLED TAKE-OFF (DCTO) – MANCHESTER

From the early 1920s, the technically inventive workings of the mind of Percy Salmon, head of the Design Department at the RAE, were directed towards mechanical assistance in the launching of aircraft. While most of his efforts were directed towards Navy flying he did not exclude the problems of enabling multi-engined aircraft to become airborne from grass airfields. A short-lived attempt was developed at Farnborough when it is recorded that on 4 June 1930 a Vickers Virginia aircraft was launched from the airfield by means of a surface, rope-hauled system forming what was known as the 'Portable Catapult'.

Not to be frustrated by the abandonment of this project, Salmon proceeded with two further schemes pursuing a similar end. One of these was the sophisticated machinery known as the Mark III Catapult (qv) on which work proceeded for some years at RAF Harwell but, again, was never completed. The latter scheme, conceived in 1938, was the Direction Controlled Take-Off system (DCTO) built on Cove Common. Research in this field was to bear a relationship to an installation which, supposedly, could be transported to an area of rough terrain from which large aircraft could be operated. The Avro Manchester was expected to be the first of a series of aircraft for which this system could be adopted.

In his submission, Salmon (*ante* 'The Catapult Section – Its Heritage') refers to the Courtney system of 'Assisters' in which the aircraft would be mounted on a carriage running on rails, using airscrews (propellers) to drive the carriage. Smith (*#89 p.19*) records that in mid-1937 the question on pro-/anti-runways for airfields was under heated discussion. He writes:

'There was some optimism that the Courtney Take-Off System would solve all the problems. This consisted of a rail track running in any desired direction, regardless of wind, and an auxiliary power unit to launch a carriage on which the aircraft was mounted. The bizarre scheme was never adopted, however, and it appears that it was mainly a red herring to blunt the attacks of the pro-runway lobby!'

Salmon's DCTO project was a workable solution and is treated here in depth; a description of the engineering details is presented as Appendix 9. Previously the author had seen only sparse reference to this installation (for example Caunter (*#222*) and as reported by Cooper (*#22*)) until the publication by Kirby (*#57*) in 1995. This last writer began to make the current author doubt long-held opinions of this project, with the more recent uncovering of Fawell's paper (*#239*) putting him completely right. For the past fifty years it has been thought that a Manchester aircraft was the only machine to have been launched from this system, there having been only one flight, that on 8 September 1942. However, previous dummy runs had been made using a Handley Page Heyford III aircraft (K5184) as motive power and ballast, lashed to the trolley. These were followed by two flights of this aircraft from the system. Fawell (*#ibid.*) describes these events as two separately identifiable stages within the one project.

This provision of a suitable test facility became the second of the non-naval activities of the newly formed Catapult Section. The installation comprised a free-running trolley system operating along a steel

Handley Page Heyford III (K5188) photographed on Jersey Brow at Farnborough with its ventral gun position (dustbin) extended. Its sister (K5184) was adapted later for use on the Direction Controlled Take-Off (DCTO) system. November 1935.
(© Crown Copyright/MOD)

track of nearly a mile in length. The aircraft would ride on the trolley and propel itself and the trolley until flying speed was attained, achieving lift-off vertically from the system. From then onwards the trolley would free-wheel into its braking unit at the end of the track. No external power source was applied to the system. A loading turntable located at the catapult site on Jersey Brow formed the head of this installation while the trolley braking system occupied the final length of track at the other termination on Laffan's Plain. The alignment of the track across Cove Common followed what was originally known to the locals as the 'Mile Straight' – almost parallel with, and to the west of, the later Runway 04/22 – with the first extension of Runway 07/25 crossing the line of the DCTO track at its south-western extremity.

The track consisted of a pair of rails formed from rolled steel channel sections set to the gauge of the main undercarriage wheels of the Manchester aircraft (22ft-9in {7m}). The carriage comprised a pair of bogie trains, one set running in each track rail and cross-braced to form a composite unit. Each train was built up of three bogies, linked together in series. The leading bogies supported the aircraft with its main wheels situated over the bogies. The trailing bogies provided the support for the rear end of the aircraft fuselage. This support comprised a collapsible 'A'-frame structure spanning both train units and terminating at its apex in a spherical cup. A spherical bobbin, to match the cup in the 'A'-frame, was fitted to the underside of the aircraft fuselage abaft the bomb bay doors. The tail-up position was held until a trigger along the track released the tail support, allowing it to collapse away from the aircraft once flying speed had been attained. From this point, the two systems will be discussed separately, the first stage covering what later became known as the RAE trolley.

THE RAE TROLLEY

Stage one, with the RAE trolley, was very much an in-house prototype development. Its progress in manufacture appeared almost to be on a 'trial and error' basis. New ideas were being incorporated in the equipment, and tests were conducted as each stage in construction and assembly was completed. Much depended upon the progress of the laying of the track. In parallel with this, the manufacture of the trolley unit and the development of the hydraulic braking system was proceeding in the RAE workshops. The special features with this trolley were:

    a) the method of collapsing the fuselage support arm and
    b) the braking system.

To achieve the disconnection of the support arm from the aircraft, the trailing bogie-pair was made to separate rearwards from the leading bogie-pair while still being physically connected together. The 'A'-frame assembly was pivoted at its base and as it rotated forward on release, its fall by gravity was checked by dropping into a wire support spanning the track and suspended on the intermediate bogies. The intermediate and trailing bogies of both sides were fitted with diagonal cables for transverse bracing across the track to give a measure of stability. The trailing bogies also were designed to carry the carriage braking system.

As soon as sufficient length of track was available, field tests were conducted with the trolley and its braking system. One rail only was used with a single train unit (forming one-half of the carriage assembly), propelled firstly by being towed by a tractor. Before long, a frame, supporting a Rolls-Royce 'Kestrel' aero engine driving a propeller, was mounted on the leading bogie. A trip device was provided alongside the track to cut the engine prior to the entry of the train into

the braking unit. Problems occurred with the braking system and a number of adjustments had to be made involving a series of as many as 30 test runs. (Engineering details of this system appear in Appendix 9.) The results were so poor that at one time it was thought that the system would have to be abandoned. Separate trials were conducted using a transverse arrester wire, rigged upstream of the original cam-plate system, and connected to a pair of drag chains deployed along the track-side. After a series of five runs in this mode, while mechanically the operation proved to be successful, other undesirable operational features precluded further trials. Eventually, the original hydraulic system was reinstated in a simplified set-up and, after a few test runs, was used for the remainder of this trials programme.

At this stage, realising that this trolley could be considered only as a prototype, a development contract for a new unit, capable of being manufactured in quantity, was placed with Sir William Arrol & Co Ltd. (The fact that, some 50 years previously, a previous generation of this firm had constructed the massive River Forth railway bridge did not imply that the new carriage would have similar bridge-like proportions! That proved to be far from the case.)

With the preliminary trials completed and the full trolley assembly installed in the track, the Heyford aircraft (K5184) adapted for this project was mounted on the trolley and set at take-off attitude to enable the system to be checked. An additional bogie was attached to the rear of each trailing bogie to firmly support the tail of the aircraft when the collapsing of the fuselage support unit was undergoing tests. The aircraft was securely lashed to the trolley for the first series of unmanned dead-load trials. These were to check the free-running properties of the trolley under load, using the aircraft engines as the sole power source. Engine controls were brought down to trackside level (those who were familiar with the ungainly appearance of the Heyford would have seen this not to have been a serious problem) and trip devices were installed alongside the track for cutting the engines prior to entry into the braking system. Following these runs, the next test involved the experimental collapsing of the fuselage support. Minor adjustments were necessary but in all some 18 runs were made at speeds up to 70mph {31m/sec} over distances up to 800ft {240m} to bring these dead-load trials to a satisfactory conclusion.

The next step was to go 'live'. The fourth, temporary, pair of bogies and the aircraft lashings were removed from the trolley, the front wheel chocks were made to collapse at lift-off, and the remote engine controls were removed. The trip device to collapse the fuselage support was set at 300 yards {275m} down the track. On the day of the launch, with the aircraft at about 12,000lb {5,400kg} weight, the pilot ran the engines to take-off power, signalled that all was well and the ground crew triggered the holdback to release the trolley. There was a cross-wind of 5mph {2·2m/sec}. The take-off was uneventful apart from a slight lateral drift of the aircraft on release from the trolley. A second trial took place a few days later under similar conditions in which the drift was more marked but insufficient to cause anxiety. One of these dates is given by Barnes (#4) as 22 August 1940. Records at Farnborough show this aircraft to have been allocated to 'A' Flight and was test flown for 15 minutes on 20 August 1940. This was followed by two launches from the DCTO track, each flight being recorded as five minutes – which sounds right for this experimental project. The first was at 06.30 on 22 August and the second at 10.50 on 10 September 1940. The pilot on all three occasions was Sq/Ldr 'Willie' Wilson. (This aircraft had a special significance to the author: it was the first real 'live' aircraft on which he, as an apprentice, had worked. The machine was in F1E workshop [Building P.67] being fitted with a comprehensive de-icing system, having previously been engaged on early experiments in flight-refuelling.)

The intention was for the Manchester aircraft to follow these trials on the RAE trolley but as the aircraft was delayed in its arrival at Farnborough (possibly through repairs resulting from a previous forced

At the same time as the initial BDAG trials, the prototype Avro Manchester aircraft (L7246) was used at Farnborough on a parallel experiment on launching from a rail-track. The aircraft is poised in the Arrol design launching trolley on the Direction Controlled Take-Off (DCTO) turntable, ready to be slewed into its take-off position. September 1942. (© Crown Copyright/MOD)

landing at A&AEE, Boscombe Down) the substitution of the Arrol trolley went ahead.

THE ARROL TROLLEY

Sir William Arrol & Co Ltd of Dalmarnock Iron Works, Glasgow, were invited to produce a development of the RAE trolley suitable for a possible production run. While this trolley was similar in concept to its predecessor, three main features were new. The first difference in the Arrol trolley was that the wheels carrying the leading bogies were not contained within the depth of the rolled steel channels forming the track but ran on the top surface instead. This enabled larger diameter wheels to be used, thereby operating at lower rotational speeds and reducing problems with bearings. (The photographs of the aircraft sitting on the trolley make these wheels look ridiculously small against the massive size of the aircraft wheels.) The second difference was that the 'A'-frame rear fuselage support was rope-hauled into the collapsed position thereby dispensing with the telescopic linkage between the leading and intermediate bogies. The third difference was that the carriage braking system had given way to an external unit, dispensing with the previous trolley-borne hydraulic system. Overall, these modifications were beneficial in the reduction of the overall weight of the carriage by one-third (3·6 tons to 2·3 tons). A further feature of the Arrol unit was that both leading bogies were rigidly connected athwartships by a streamlined beam spanning the track.

For the development of the new braking system, a wire rope and jigger system – similar to a shipboard arrester gear – was rigged across the track. A rocket-propelled trolley, ballasted to match the weight of the Arrol Trolley, was satisfactorily arrested from entry speeds of up to 127mph {57m/sec}. This rig was re-erected spanning both tracks, replacing the cam-plate system. (Owing to the nature of the surrounding ground, the arrester apparatus could not be positioned below surface.) The Arrol trolley was fitted with a pair of rope pick-ups – similar in shape to, but much larger than, the throat of an aircraft arrester hook – one each mounted on a forward projection of the leading bogies. Resulting from trial runs, the intermediate Arrol bogies were found to be unsatisfactory; they were replaced by those from the now redundant RAE trolley. (It cannot be established, but it is thought that these trials were conducted by using newly available rocket motors as power sources, fixed to the leading bogies. It is this assumption that has given rise to the assertion that the system was rocket powered. This assuredly was not the case.) The new arrester gear performed satisfactorily on all dummy runs of the trolley. A trial loading of the Avro Manchester aircraft (prototype, L7246 – which also was used for the Brake Drum Arrester Gear project) onto the trolley was made on 21 August 1942 but no dummy run with the aircraft mounted on the trolley was conducted.

The flight test with the Manchester aircraft was programmed for early morning of 8 September 1942. The aircraft was loaded on to the trolley in the turntable, swung into alignment with the track, and the pre-flight check carried out. The person who was head of the unit responsible for the flight trials – R. J. Nixon – was expected to be on site that misty morning (it was thought to have been a Sunday) to witness this first 'live' launch, but he failed to appear. (It was found that he had arrived at the RAE Pinehurst Gate [the old location] without his pass and a conscientious security guard would not permit him to proceed through to the airfield. Contact could not be made with any of his colleagues as they all were on the site, so he missed the occasion.) Launching procedure was initiated and the exercise progressed according to plan and design, perhaps to be chalked up as one of the Department's early successes. (On that morning the author was carrying out his patrol duties as a private in the RAE Home Guard. He heard the sound of the take-off but, regrettably, this commitment prevented his being a witness to an interesting activity.)

The pilot again was 'Willie' Wilson who by now had been promoted to Wing Commander. The aircraft weight was about 31,000lb

Prototype Avro Manchester (L7246) on the DCTO Arrol trolley part-way along its track. September 1942. (© Crown Copyright/MOD)

{13,600kg} and take-off took place in ideal weather conditions, with no wind. The trigger to release the fuselage support was placed at 450 yards {410m} from the start point; lift-off from the trolley was achieved within 100 yards {91m} from release, with the aircraft 'floating' off the trolley. The trolley continued freewheeling along the track, trundling to a stop of its own accord some 20 yards {18m} short of the arrester gear! (Why all the hassle over the braking system?) Thus was the project brought to its conclusion.

Contemporary photographs, including those reproduced by Kirby (#ibid. pp.171–172), portray the Manchester aircraft positioned on the Arrol trolley system, supposedly immediately prior to its first and only launch, although it is shown lashed to the track. These pictures show the aircraft set on its main undercarriage wheels which, in turn, are located between front and rear chocks, all of which are mounted upon the leading bogie unit. There has been some doubt as to the date of this launch of the Manchester. Caunter (#ibid. p.116), writing in 1949, gives the date as July 1941 but no subsequent matching reference to this date has been discovered. Kirby (#ibid.) gives dates of 5 and 8 September 1942. Cooper, in researching for his book (#ibid.), has had access to RAE flight records from which he has discovered that Manchester L7246 was flown from the DCTO track by W/Cdr H. J. ('Willie') Wilson on 8 September 1942, take-off at 6·45am, duration of flight 5 minutes; enough time for a half circuit of the airfield, followed by a conventional landing. This date is now accepted as authentic for this particular event. Yet, in so accepting, another problem is posed.

DEVELOPMENTS

Although the system was ripe for development, no further launch was ever undertaken on this installation. The Services requirement for such equipment had ceased to exist with the advent of concrete surface runways to most of the RAF airfields. The DCTO track itself at Farnborough was beginning to appear as a hazard to normal flying activities on the airfield.

Fawell (#ibid.) had been forward-looking in recommending a number of development options. An adaptation for launching flying boats and/or landplanes with their undercarriages retracted was mooted. The fitting of rocket motors to the trolley – to shorten the take-off run – was another proposal. The design of both RAE and Arrol trolleys included prepared facilities for mounting the aircraft in a skewed position to allow for launching in side winds. None of these adaptations was implemented.

Of the DCTO project, perhaps one may be allowed a mental shudder when contemplating the outcome should a pilot need to 'baulk' a launch once it had been initiated. The tail of the aircraft would be without support; the aircraft may even have vacated the launching carriage. The machine could only drop back onto the track with a large chunk of engineering ironmongery careering along beneath it. This would have been catastrophic for the aircraft and its crew. Perhaps the risk was too great to proceed with its further development.

By September 1942 the DCTO track had become an encumbrance to airfield operation. The south-west extremity of the track, with its braking gear, crossed the alignment of the 07/25 runway which was about to undergo its first of three extensions. The truncated track did not immediately go into decay. For many months after the launch of the Manchester aircraft one leg of the track was used as the prototype High Speed Track (HST) along which rocket-propelled sleds were sent speeding with many and varied experimental devices mounted upon them. Eventually, the track intruded into the flight path of proposed extensions to the runway and a new purpose-built HST was constructed close by using, it is believed, much of the recovered DCTO track and its elements re-aligned east-west, roughly parallel with the 07/25 runway and some distance to the south of it. For details see 'The High Speed Track'.

The trolleys and ultimately the ground-level track installation at Farnborough were removed, apart from a few short lengths of track which remain to this day as evidence of occasions long past. Most of the length of this trackway is now grassed over and the runway crossings have been filled in with concrete. At the time of writing – January 1999 – vestiges of the track still may be seen between the bend in Mosquito Way and Runway 11/29 westwards of the crossing of the latter with Runway 04/22 (Ordnance Survey, Landranger 186, Ref SU 860543). The alignment of the track in places still may be recognised, by a barely perceptible discoloration of the grass, in the direction approximately south-westwards of this point. The exact location of the turntable at the head of the track cannot now be determined with any certainty; it would have been a short distance to the north-east of this OS reference point and may now be buried beneath the concrete pavement. An appraisal of the engineering content of the DCTO installation appears as Appendix 9.

Two recent writers refer to this system as the 'Frictionless Take-Off Device'. During the whole of the period in which the author was associated with the project it was always known as the Direction Controlled Take-Off (DCTO) system. The installation may have been relatively frictionless as far as the rolling resistance of aircraft tyres and wheels were concerned, but there were friction and inertia forces of the trolley to be considered. This, by virtue of the evenness of the track contrasted with the grass runway surface alongside, would have offered less friction (hence 'frictionless'?) to the process in its entirety.

A contemporary alternative to this system was proposed by Major F. M. Green – to whom the reader was introduced in the early pages of this book – in his 1940 paper, 'Gravity Assisted Take-Off for Aircraft' (#247). This proposal is treated under 'Launching by Gravity'.

DIRECTION CONTROLLED TAKE-OFF (DCTO) – AUSTER

Whilst bearing the same title as the preceding project, this system – simplicity itself – was vastly different in concept from the Manchester facility. All that was required was to determine if a light aircraft would benefit from some directional control during take-off, particularly under conditions of light cross-winds. The initial intention was to investigate the possibility of launching small observation aircraft from tank landing craft (LCTs) by fitting a track system over the top of the hold. A Taylorcraft Auster V aircraft (TJ537) was allocated for these experiments.

Towards the end of March 1945, a tender specification was prepared for the installation of three rows of low level concrete plinths on Jersey Brow. These were to support three rails to act as runways. Two rolled steel channels were laid on these plinths – flanges uppermost – and spaced at the centre distance of the main undercarriage wheels. The wheels of the aircraft would run in these channels with the flanges acting as constraints during the take-off run. A third rail (RSJ) was laid along the centre line to support the aircraft tail wheel. There was no trolley for the main wheels and therefore no detent retaining system to hold the aircraft with its engine running. It was held by a pair of collapsible wheel chocks that were released on a signal from the pilot. The tail wheel was set into a trolley of elementary design, its track dropped a few inches toward the end of its run to allow the tail to lift off away from the track prior to the main wheels. The system was reminiscent of the earliest shipboard trials in which the aircraft wheels were

guided by a track laid along the flying-off deck. The track was some 167 feet {50m} in length; the concrete plinths raised it about 30 inches {750mm} above ground level.

Take-off was achieved under its own engine power. The engine was 'revved up' and held by its brakes until power was developed; it then taxied along the track until it reached the end of the run and lifted off as in a normal take-off. During the summer and autumn of 1945, a total of 20 runs were accomplished under various weight loading (1,500–1,800lb {680–820kg}) and wind conditions. Brown (#17 p.149ff) provides a detailed account of these trials. In his report, he states that he found the wind speed and aircraft weight affected the take-off run so critically that they had to be correlated to calculate the take-off distance. This was why on one occasion, when he found that 100lb {45kg} of ballast was too little and 200lb {90kg} too much, that the author – at 10½ stone, or 150lb {68kg} – was bundled into the passenger's seat, not as observer but solely as make-weight! With his single, five-minute, flight as passenger on that afternoon of 20 July 1945, he claims to have been the sole civilian observer to have flown from this equipment! This series of trials were assumed to have been successful but the application to LCTs was never adopted. Within a matter of months the track was lifted and scrapped, the project retreating into oblivion.

While treating with Auster aircraft, it is recorded that two others of this marque were fitted with arrester hooks and subjected to carrier landings. Surely this was an anomaly and waste of effort!

British Taylorcraft Auster V (TJ537) ready for take-off from the 'mini' DCTO system, under the shadow of the P.I Catapult on Jersey Brow. The main wheels of the aircraft were guided in steel channels until take-off. December 1945. (© Crown Copyright/MOD)

LAUNCHING BY GRAVITY

Why not? Being a freely available source of power, gravity is waiting to be used by any entrepreneur prepared to harness it. Such availability was exploited from the first in the history of aviation. Barnes (*#4 p.3*) records that José Weiss, an associate of Frederick Handley Page in his younger days, in 1908 used a falling weight within a tripod structure, some 30 feet {9m} in height, to launch a series of gliders from a rail-mounted trolley. These experiments were conducted at Fambridge, Essex.

Shortly before his death through a flying accident in 1910, the Hon C. S. Rolls, the co-founder of Rolls-Royce, had brought a Wright biplane to Farnborough 'for the training of Army officers'. In association with this, Caunter (*#222 p.23*) refers to a Captain P. W. L. Broke-Smith RE erecting a weight and pylon catapult on Jersey Brow – adjacent to one of the airship hangars – for launching this machine. Apart from a photograph of the entrance to the hangar, little appears to be known of this device.

Cronin (*#23 p.71ff*) states that a patent was taken out as early as 1911 by Sueter, Boothby and Paterson for a gravity-operated catapult. The function of the device was described as:

'To start an aeroplane in a confined space, such as from a ship, it is supported by trunnions upon arms, carried by a trolley adapted to run upon widely separated inclined rails. The trolley may be propelled by a falling weight and cable. The weight may instead be suspended from the mast of the ship. Outwardly extending struts may be provided on the trolley for supporting the wings of the machine.'

Cronin further comments:

'Perhaps fortunately, this design was not proceeded with, although the designers' foresight at the birth of naval aviation is remarkable.'

For a couple of decades or more, floatplanes were operated from many of the Royal Navy's capital ships through the expedient of providing platforms, generally atop gun turrets, from which the aircraft would take off without mechanical assistance.

Before the last of these flying platforms were removed in 1935/36 work continued on the design and operation of catapults. Between 1922 and 1927, two destroyers, HMSs *Stronghold* and *Thanet*, were used in experiments on the launching of early guided missiles. RAE had developed a radio-controlled flying bomb the design of which had been prepared a few years earlier in the Royal Aircraft Factory. Its flight trials at that time were inconclusive and abandoned. The project was resurrected in 1922 using four airframes abandoned from the earlier trials, with the production of a further six, all being fitted with more powerful (45hp) engines. For obvious reasons of security, the project was known as the 'RAE 1921 Target'. Cronin (*# ibid. p.76*) continues:

'By mid-1922 attempts to launch the 1921 Target were restarted. Two launches from horizontal rails on HMS *Argus* (1922) ended in failure and both targets were written off after crashing into the sea. It was considered that the transition from horizontal rails to climbing flight was the cause of the problem, so it was decided to launch into the climb from an inclined catapult mounted in the bows of HMS *Stronghold*. The missile was accelerated up the catapult's track by a large bag of water that was allowed to fall into the sea.'

Contemporary photographs of this launcher indicate its angle of elevation to have been of the order of 30°. The final launch from HMS *Stronghold* using this contraption (1925) is recorded as having been completely successful and developed into the subsequent 'Larynx' experimental project. HMS *Stronghold* continued to be used as a trials ship but in a later application the gravity launcher was replaced by the newly developing type of hydraulic, telescopic ram and cylinder catapult.

To the author, it is not clear where HMS *Thanet* comes into the scheme of things. A photograph (Cronin – *#ibid. p.79*) shows the ship rigged with a structure in the bows having a much shallower angle of elevation, possibly around 10°, ostensibly for launching the RAE 1921 Target. A contemporary diagram portrays a similar installation with the towing cable passing over a pulley at the outer end and doubling back beneath the structure, disappearing below the decks into a vertical shaft beneath the launching track. This shaft contained the weight for operating the system attached to the end of the rope through suitable reeving. (Was this a convenient adaptation of the ammunition hoist trunk originally supplying the forward gun?)

Coincident with the development of the Direction Controlled Take-Off project at Farnborough, Major F. M. Green (whom we have met before) produced a 1940 paper (*#247*) for a gravity-operated project for launching heavy aircraft from an airfield. As already seen, the system of adopting a weight, falling under gravity, to launch an aircraft was not new. During his time at Farnborough (1910–16) Green must have been aware of Broke-Smith's contraption rigged to launch Rolls' Wright aircraft. With his continuing close contact with RAE he might also have been aware of the activities on the shipborne, gravity-operated devices carried by HMSs *Stronghold* and *Thanet*.

Green introduced his proposal as 'a method for assisting the take-off of heavily loaded aircraft, the principal advantages of the scheme being that it required very little special apparatus and as little high-class engineering labour as possible'. The basis of the system was the employment of the potential energy of a weight falling down a well to provide, through a system of pulleys and a tow rope, a force to accelerate the aeroplane during its ground run. At the end of the run, retarding the weight at the end of its fall, releasing the aircraft from its tow rope and arresting the free end of the rope after release was thought not to offer problems of any special difficulty. Green calculated that a weight of 30 tons falling in a well of 160ft {49m} depth would accelerate an aircraft weighing 15 tons from rest to 150ft/sec in 400 yards {46m/sec in 366m}. At the completion of a run, the system would be re-cocked by the weight being lifted by a tractor or tank pulling on the free end of the rope. The interval between each successive take-off was stated to be ten minutes.

In a diagram accompanying the report, the proposed 160ft {49m} deep shaft is shown to be 5ft {1·5m} diameter with the 12ft {3·7m} long weight occupying almost the full cross-sectional area. The bottom 12ft {3·7m} of the well is shown filled with water to form the weight-retardation facility. The top of the well is shown covered over and an arrester-wire system (shock absorber cable) is arranged at ground level. In plan the arrester gear is a wire rope set around a hexagon with sides of 50ft {15m} between fairleads and the ends connected to a 'spring loading'. The towing cable disappears down the centre of this set-up, with the intention of the 'shock absorber cable' catching hold of the hook mechanism by which the aircraft is towed. The purpose of the arrester unit would be to maintain a reasonable tension in the tow rope while the weight was undergoing deceleration in the well.

Green's proposal stated that the weight for supplying the energy would be twice the weight of the aircraft. A 30 tons weight would be a cast-iron cylindrical mass about 4½ft {1·4m} diameter by about 11ft {3·3m} long, shaped with an elementary taper at the lower end. This would be guided by vertical rails fixed to the wall of the well. The weight would be assembled in component parts to permit adjustment by adding or subtracting elements as required. Its deceleration by falling into water would produce a retarding force to be controlled by designing the clearance between it and the well. A suitable safety device for holding the weight against a breakage in the rope or slackness would be needed. The driving force would be transmitted through a five-fall pulley system so that the speed of the weight before it was retarded would have been 15ft/sec {4·6m/sec}.

The method of loading is described. With the weight at the bottom of the well, the free end of the tow rope would be held by the surface arrester gear. A tractor (a tank or other aircraft was also proposed) would pick up this free end and haul it to the launching position. The aircraft would be taxied to the launching point and connected to the tow rope with the tractor still holding the tension until the aircraft brakes are full on. The tractor would then be uncoupled, the aircraft engines started and the brakes released. The aircraft then accelerated and took off, freeing itself automatically from the tow rope in the process.

To the author, this appears to be a not-very-plausible proposal; interesting perhaps but naïve to the extreme. With Green's reputation as an engineer, the author is surprised that Green could convince himself (and others?) of the supposed viability and simplicity of such a project. First, the provision of a well some 160ft {49m} in depth can hardly be termed as 'requiring very little special apparatus'. There were problems at Farnborough at this time in producing a pit of only one-quarter of the depth proposed by Green. (Ref: 'Land Trials – Hydraulic Arrester Gear'.) The diameter of 5ft {1·5m} does not permit any access for servicing the bore of the well or for the passage of any pipelines and mains supply to pumps in a sump at the bottom of the well. A service area with heavy handling equipment just below the ground surface would be required. This would be needed to enable adjustment of the driving weight to be carried out to match the AUW of the aircraft to be launched. (The case of 'Force = Mass x Acceleration' still applies here!) How would the aircraft be held against the static towing force induced by the weight (equal to some six tons) prior to engine start-up, and countering this with engine thrust immediately prior to take-off? No holdback unit was proposed. It is not surprising that this proposal was 'placed on the back burner', never again to see daylight.

BRAKE DRUM ARRESTER GEAR

The system known as the Brake Drum Arrester Gear (BDAG) reflected the principle on which it was to function, for which Naval Aircraft Department developed a prototype unit. Although primarily concerned with naval flying, NAD was often involved with other projects, drawing upon the engineering and technical expertise to be found within the Department. The BDAG exercise was the third non-naval project undertaken in the early days of the original Catapult Section and was instigated by Bomber Command in 1940 who required 'an emergency arrester gear for bomber aircraft, to assist in cutting down the number of over-runs of aircraft landing light after raids'. See Caunter (#222 p.125).

This project here is treated in depth, with engineering details given as Appendix 10. Previously the author had seen only sparse references to this installation (Caunter [#ibid.]) until the publication of a short report by Kirby (#57) in 1995. In making contact with Kirby, he put the author on to Blake et al (#8) who again had little information but corrected some long-standing erroneous conceptions of his. Correspondence with Blake's co-authors – particularly Hodgson – triggered the author's response in producing this account in some detail.

A prototype gear was developed for installation at Farnborough from which a workable unit, suitable for quantity production, would follow. It was proposed to equip a number of Bomber Command airfields – originally six, with the possible expansion to 20 – with an installation at each end of their three runways. Of necessity, the equipment was required to be simple in construction and operation, and to be made available quickly. In parallel, the fitting of suitable arrester hooks to the

relevant aircraft would be required. The intention was that, once a bomber aircraft had already landed and the pilot feared an overshoot, its arrester hook would be lowered to engage the arrester rope, bringing the aircraft to rest before reaching the further, upwind end of the runway.

Both entry speed and rate of retardation were to be relatively low, bringing the aircraft to a standstill from 80 knots {41m/sec} within a distance of 450 feet {140m} or from 60 knots {31m/sec} at 400 feet {122m}. The RAE job-file for this project suggests that the first 200 feet {60m} contributed little braking effort; the residual pull-out gave a deceleration of 0·6g. Essentially it was to function as a last-ditch installation. It was never intended to operate in a mode comparable to shipborne arrester gear in which naval aircraft entered the equipment at flying speeds and brought to rest within one-third of this distance.

The proposals submitted towards the end of 1940 envisaged a wire rope stretched across the runway some 200 yards {180m} from its end. This wire would be supported on bow springs (similar to the flight deck of an aircraft carrier) and pre-tensioned to hang above pavement level, running through deck sheaves to a braking system housed in an underground chamber adjacent to the runway position. In fact a single wire was made to cross the runway twice with each end attached to the brake unit, the 'looped' end on the opposite side of the runway passing *en route* through pulleys. The braking system comprised a coupled pair of units, each consisting of a grooved rope drum mounted on a horizontal shaft, around which one end of the rope was fixed and wound as a single layer. Each rope drum was bolted to a brake drum with a pair of heavy duty brake shoes fitted with 'Ferodo' friction pads. The brake shoes

Avro Manchester prototype (L7246) with its arrester hook deployed. (This machine was also used for the Direction Controlled Take-Off [DCTO] experiments. The spherical attachment point for this is seen below the roundel.) This aircraft was engaged in experimental work investigating means of preventing airfield over-run of large aircraft returning from operations. November 1941. (© Crown Copyright/MOD)

Avro Lancaster I (L7528) taxying into No.1 arrester wire on Farnborough airfield during a demonstration run of the BDAG in June 1942. The wire can be seen from the bottom left corner, crossing Runaway 07/25 and supported on three retractable bow-spring elements located on the white line straddling the runway beneath the aircraft fuselage. June 1942. (© Crown Copyright/MOD)

were held against the drums by heavy duty coil springs and the recovery of the rope after pull-out was achieved by a geared motor drive to the drum assembly.

The prototype was developed in conjunction with Sir William Arrol & Co Ltd of Glasgow (we have met this firm before) who also manufactured the original hardware. The entire unit, comprising two rope/brake drum assemblies, its recovery gear, controls and prime mover, was mounted upon a rigid fabricated bedplate. It was transported to site as a complete working unit and lowered into its machinery pit, to be bolted to the floor. This prototype installation was sited adjacent to, and on the south side of, Farnborough's Runway 07/25 with the ropes crossing that runway. The pit to contain the under-deck machinery was excavated about 150ft {46m} from the south side of the newly installed runway and concrete lined, with a heavy duty roof to permit aircraft to taxi over it. The pit was quite large, being 16ft long, 11ft wide and 6ft 6in in depth {4·9m x 3·4m x 2m} with the long side parallel with the runway. The arrester wires left the pit at ground level through a pair of fairleads located immediately below the roof deck plates. The airfield pulley sheaves were of robust construction (accepting a working load of the order of 12 tons, acting horizontally) and were affixed to concrete blocks positioned in the soft verges alongside the runway. The entire system – when in operation or stowed – presented a zero hazard to the normal flying programme at Farnborough.

For this experiment an Avro Manchester (L7246 – the same prototype aircraft used for the Direction Controlled Take-Off experiment, qv *ante*) was fitted with an articulated 'sting' type hook assembly forward of the tail wheel. To this was attached in turn a selection of hook configurations under development by, and manufactured at, RAE. Concurrently with

this project, Naval Aircraft Department was conducting an in-depth investigation into the mechanics of the functioning of arrester hooks for shipborne aircraft and their interaction with deck arrester wires. Various designs of hooks were produced and an extension of this project produced a basic set of hooks of larger scale for testing on the Manchester aircraft. One of these was a hook fabricated from pre-formed steel plates in a welded configuration, probably with a hard-facing of 'Cromaloy' in the throat. Another model featured a large radius beak to assist the cable entry. A third type of hook contained a roller in the throat. Experiments were conducted with this roller as a freely rotating unit, and at other times it was fitted with friction washers to restrict the extent of rotation. It is thought that one of the heavy duty naval hooks, made as a drop forging, was also adapted for the Manchester experiments. Research was also conducted into rudimentary attempts at damping the vertical and lateral movements (bounce and shimmy) of the hook as it trailed along the runway. RAE staff undertook the design and fitting of both the spherical mounting for the DCTO system and the arrester hook installation in the aircraft. The fitting of layers of light alloy reinforcing plates to stiffen the underside of the fuselage immediately abaft the bomb bay to accommodate these highly stressed areas was undertaken in the RAE hangars.

A number of successful taxying runs were completed using a variety of hooks fitted to the aircraft. The Manchester would be ranged at the north-east end of the runway (orientation 25) and taxied into the wire at various approach speeds. (At no time was the aircraft airborne immediately prior to entering the system; the arrester gear was not designed to operate in this mode.) Following these initial tests, the powers-that-be were informed that a demonstration was ready to be presented. A date and time was arranged in August 1941 and Air Marshal Sir Arthur T. ('Bomber') Harris, head of Bomber Command, was among the guests invited. Unfortunately, a few days prior to the date of the demonstration there was heavy rainfall and the arrester machinery pit became flooded; the sump drainage pump was not able to cope with the abnormal quantity of rain. The pit was pumped out by the airfield fire service and the cable recovery/pre-tensioning motor (a Ford V-8 car engine) was quickly made operable but the drums and the rope were not pre-checked for serviceability.

For the demonstration, the aircraft was positioned at the end of the runway and began its taxying run into the experimental area; the plane shuddered as it picked up the across-deck rope, its tail lifted high, the rope stretched and twanged, then parted into two flailing ends whipping across the runway in opposite directions, one end passing uncomfortably close to the Air Marshal. The brake drums had rusted with the rainwater and the increased friction induced by the brake pads caused a partial seizure of the system. Like Queen Victoria, the Air Marshall was not amused and departed forthwith, his dudgeon, appropriately, as high as his rank! The aircraft was taken out of service temporarily for the damage to be repaired by RAE personnel. The project was cancelled. It was conceded that the new airborne radar equipment (H2S) about to be installed in a large bubble housing in a ventral position immediately aft of the bomb doors would preclude the fitting of a hook to this and similar aircraft. (However, Bomber Command had a change of heart shortly after and within a few months issued an instruction to proceed with the manufacture of the equipment and its installation on a number of airfields.)

Unfortunately, there appears to be some conflict of dates with Manchester L7246. Kirby (*#57 p.197*) records the aircraft as having been at Farnborough over the period 18/12/40 to 20/11/42 during which time it was damaged in trials on 1/8/41 and repaired on site. This tallies with the occasion detailed above. The dates of the various earlier trial runs do not appear in RAE records of aircraft movements; taxying runs were not considered as 'flights' and, therefore, were not logged as such. Caunter (*#222 p.126*) in his record, compiled in 1949, refers to a successful demonstration in 1942 (RAE photographs showing Lancaster [L7528] are dated June). This result prompted the Air Ministry to order 120 sets of this equipment for installation at twenty Bomber Command airfields. The production models were developed from the original Arrol equipment and manufactured by Mather & Platt Ltd of Manchester. This latter firm made the unit into a more compact form with all the controls conveniently brought to a panel at the front of the machine. Green (*ARC 11,436 'Note on Arrested Landings, April 1948'*) declares the cost of six units to equip one airfield as £20,000 in 1942 money!

Although further trials on this project were limited to a few sporadic check-outs on some installations located on service airfields, the system was never put to formal use. The adapted Manchester aircraft was retained at Farnborough and continued to be used for arrester hook evaluation trials. The size and weight of the Manchester provided valuable facilities for testing the larger series of arrester hooks under development for Fleet Air Arm aircraft. Manchester L7246 remained at RAE until November 1942 when it was put up for disposal. The last recorded activity for this aircraft was a two-day visit in October 1942 to RAF Woodhall Spa for BDAG trials. This must have been its swan-song. A March 1943 reference records other Bomber Command 'heavies' having been fitted with experimental hooks. These included a Handley Page Halifax (W7650) and a Short Stirling III (R9188), demonstrated at RAE on 23 February 1943. At one time, the prototype Vickers Windsor was considered as a likely candidate for this project but no work was done on this machine.

Prototype Avro Manchester (L7246) for BDAG trials. It is fitted with a welded plate arrester hook containing a roller in its throat. November 1941. (© Crown Copyright/MOD)

Most of the RAF trials party is seen standing around the machinery pit for the demonstration of the prototype BDAG system at Farnborough in June 1942. The runway is behind the viewer. An interesting feature is the wartime blackout masks fitted to the car head-lamps. June 1942. (© Crown Copyright/MOD)

Avro Lancaster (L7528) was one of these adapted aircraft and here a further problem arises. Farnborough's register of photographs records L7528 as being a Manchester, minus the central, fixed tail fin. Other aircraft records, including Kirby (*#57 p.197*) and Thetford (*#96 p.59ff*), indicate that the last Manchester to have been built was serial L7526, and that serial L7527 onwards – in the contract for Manchesters – became four-engined Lancasters, L7528 being the second aircraft in this new series. This problem of identity remains to be solved. Mason (*#64 illn.144*) states that L7528 was fitted with an arrester hook in November 1942 but, apart from a taxying demonstration to VIPs at Farnborough, was never tested.

At the time of writing (1999) few remains of the site location of this short-lived installation on the Farnborough airfield may still be traced. Subsequent civil engineering works in this vicinity are likely to have obliterated vestiges of the pit itself. The shades of a similar system linger on. An airfield arrester installation at each end of the Farnborough Runway 07/25 was installed in the early 1990s as protection against emergency overshoot of the Tornado aircraft.

Reports of installations on Bomber Command airfields have been verified. The production model, by Mather & Platt Limited of Manchester, was designed as a compact assembly; virtually a 'packaged' unit. The existence of the hardware recently uncovered at Bottesford, Woodhall Spa and elsewhere presents a tangible confirmation of such equipment having been installed. The original concept for the installation of the Brake Drum Arrester Gear units was for six airfields to be equipped. Smith (*#88 p.24*) confirms these as Elsham Wolds, Lakenheath, Linton-on-Ouse, Middleton St George, Swinderby and Waterbeach. Caunter (*#222 p.126*), writing in 1949, seven years after the event, suggests that some 120 sets were on contract for installation on 20 Bomber Command airfields. Kirby (*#ibid. p.174*) refers to installations on a number of RAF airfields. The RAE job-file (now in the National Archives as reference AVIA 15/1164) lists these installations as at 20 Bomber Command airfields:

1. Bottesford, Leics	2. Breighton, Yorks
3. Croft, Co Durham	4. Dalton, Yorks
5. Elsham Wolds, Lincs	6. Lakenheath, Suffolk
7. Leconfield, Yorks	8. Leeming, Yorks
9. Linton-on-Ouse, Yorks	10. Marston Moor, Yorks
11. Middleton St George, Co Durham	12. Snaith, Yorks
13. Stradishall, Suffolk	14. Swinderby, Lincs
15. Syerston, Notts	16. Topcliffe, Yorks
17. Waltham, Lincs	18. Waterbeach, Cambs
19. Woodhall Spa, Lincs	20. Woolfox Lodge, Leics

Another record substitutes Ossington, Notts, for Woolfox Lodge. The Farnborough files record progress reports on the installation of these units and it is conjectured that as Ossington came under Flying Training Command in January 1942, arrester units were not installed on its runways.

It is understood that remnants of the underground plant on some of these erstwhile airfields, having been abandoned, were left quietly to rust away. Records of such installations are now coming to light. A photograph shows one of these units in its pristine, newly manufactured condition, taken at the Manchester works of Mather & Platt Ltd. Correspondence with Weir Pumps Limited, successors to Mather & Platt, has drawn a blank through their admission that they no longer hold any records of the aircraft arrester-gear units.

Memories of RAF personnel support the impression that some work had started on the system with the installation of the ground equipment but was quickly stopped when the project was cancelled. One reference appears in Holyoak (*#46 p.45*) stating that his airfield, Bottesford, was to be closed on 15 June (1942) whilst arrester gear was fitted at each end of the three runways. According to Holyoak, the scheme was abandoned in June 1943, the machinery never having been used operationally.

One of these BDAG units, now on exhibit at the Lincolnshire Aviation Heritage Centre, East Kirkby (Silksheen Airfield), had been installed originally at Woodhall Spa airfield during the early 1940s. Since that time it – with its neighbours – had been lying dormant in its underground bunker quietly deteriorating, probably completely immersed in water until discovered by an industrialist (extraction of gravel) who came across the installation when working on the old war-time airfield. This particular unit had been recovered and presented to the Centre in the 'as found' condition by the current owners of the Woodhall Spa airfield. Unfortunately, many of the features of the system are missing from the East Kirkby specimen. It had been vandalised to the extent that all readily accessible components, particularly non-ferrous metalware and the Ford V-8 engine, had been removed.

A second unit has been recovered from Woodhall Spa to a local museum at Thorpe Camp Visitor Centre, located on an accommodation site within the purlieu of the original RAF Woodhall Spa airfield. The machinery was found to be in a comparatively complete, unvandalised condition and has been cleaned and prepared for exhibition, not as a working piece of machinery but as an example of wartime engineering designed against a specific operational need. When the author visited the scene (May 2000) he was given the nostalgic impression that the equipment was almost ready to be driven! All credit must be given to Mike Hodgson and his enthusiastic ground crew of volunteers.

Engineering details of this system are recorded as Appendix 10.

THE HIGH SPEED TRACK (HST)

This installation originally was not an integral part of NAD equipment but over a period of time the Department made extensive use of its facilities. In addition to working on major projects associated with the application of positive and negative 'g' forces to FAA aircraft, NAD staff were looking at improvements to the design of flight deck machinery relating, in particular, to the behaviour of jet aircraft operating at higher speeds and weights. The performance of sections of flight deck machinery, where constituent units were required to move at speeds of 100mph (147ft/sec {45m/sec}) or more, was under consideration. Attention was being given to the cases where highly stressed, heavy duty wire ropes were passing around pulleys and fairleads under accelerating and decelerating loads at these speeds. A variety of test rigs was produced at Farnborough to investigate the situation and to determine if the existing equipment was approaching its limit of safe operation. One such rig involved the use of the High Speed Track (HST) on Laffan's Plain.

The mists of time have shrouded the origins of the High Speed Track, and records seem to be almost non-existent. Its location on Laffan's Plain is recorded as having occupied what was known as the FIDO Strip, a clear area on the northern side of Berkshire Copse. FIDO was the acronym for 'Fog, Intensive Dispersal Of'. In the late 1930s, trials were conducted on Laffan's Plain for dispersal of local ground mist during night flying. A series of paraffin/kerosene flare-path burners were deployed in a pair of parallel lines which, having been lit for a while, tended to disperse the local mist. Further experiments were conducted at Blackbushe airfield, leading to the development of the project for selected Bomber Command airfields in 1942. Post-war research was renewed for a period through 1945–9 when the more up-market title of 'Fog Investigation and Dispersal Operation' was adopted. As Kirby (#57 p.251), Smith (#88 p.138ff) and others have treated this subject in greater detail, it will not be discussed further here.

The first reference to an installation on Laffan's Plain was that of the Exeter Track. The primary purpose of this track was the investigation of effects on aircraft wing sections when impacting defence balloon cables at speed. These trials included the 'Bolas' experiments in which explosive mines were suspended on such cables for defence against enemy bombing raids. The unit conducting these trials, according to Caunter (#222 p.95), was known as the Defence Investigation Department which, in 1939, was out-stationed at University College, Exeter. One recalls the pre-war trials using a Fairey P4/34 (prototype Fulmar) and Battle aircraft, suitably armour-plated along the wing leading edges and the cockpit heavily protected, leaving Farnborough for flying trials and returning some time later with dents in the wings and, occasionally, trailing a length of balloon cable. It was felt that the use of the Exeter Track was a little more conducive to aircrew safety in this trials programme!

Basically, the track comprised a straight length of 'railway' along which a sled or trolley, on which was mounted the research specimen, was propelled at high speed to impact with another, complementary experimental rig. A variation of the experiments was to determine the effects of induced high rates of acceleration upon the test specimen. It is recognised that the adoption of the 3in Unrotating Projectile propulsion element (3in cordite rocket motor) as a power unit for attaining high speeds and rapid acceleration was a trigger for this particular activity. It is known that the original length of track comprised a pair of rolled steel channel sections along which the sled or trolley was propelled. This sled was designed to carry a test rig on a platform and was driven by a battery of rocket motors (similar to the system adopted for the 'P' Catapult).

Initially, high speed experiments took place at the southern end of one length of the (by then) redundant Direction Controlled Take-Off (DCTO) track. The final launch of an aircraft on the DCTO system took place in September 1942 and its subsequent take-over for other trials programmes savours of good-housekeeping by the RAE. Immediately following the launch of the Avro Manchester aircraft the DCTO track was scheduled for partial dismantling. Its southern extremity – with its associated arrester gear – crossed the alignment of Runway 07/25 which was about to undergo its first extension late in 1942. The reduced length of the DCTO track remained in its original position for only a short while but during this time it is thought that some early high speed trials were undertaken on this stretch of track. Regrettably, this cannot be

The High Speed Track on Laffan's Plain. This view shows the extreme end of the track, looking north-east, with the centre rail being installed. The trolley braking system would be installed in this area, the two outer rails stiffening up the system to accommodate the heavy braking loads. August 1953. © Crown Copyright/MOD.

High Speed Track. An experimental trolley (equipped as a fighter cockpit for the Physiological Department/Institute of Aviation Medicine) is mounted on an 'Arrol' type bogie for use on the redundant DCTO track. Rocket motors are fitted to the rear. July 1947. (© Crown Copyright/MOD)

The adapted duplex American Mark 4 arrester gear in its pit alongside the High Speed Track (HST). March 1981. (© Crown Copyright/MOD)

Fl Lt Guinard is seen strapped to an early model of the Martin-Baker pilot ejector seat, mounted on a rocket-propelled trolley on the High Speed Track. March 1958. (© Crown Copyright/MOD)

confirmed. A photograph of a physiological trolley – to carry a human subject – gives the impression of its having been mounted on one of the leading bogies from the redundant Arrol carriage. The adoption of an arrester wire braking system, employing the Drag Chain Arrester Gear unit, also validates this assumption.

Before long, much of the DCTO track was dug up and re-laid on Laffan's Plain solely for use as a high speed track. One piece of evidence may be gleaned from an RAE site plan (#320), dated 1944, some 18 months after the closure of the DCTO track. This shows the High Speed Rail Track running on the south side of Runway 07/25 in the 27 orientation. The start of the track was adjacent to Berkshire Copse Road and terminated some 700 yards {640m} westward, its alignment converging towards that of the end of the final extension to the runway some 350 yards {320m} further on. The alignment of the original Exeter Track (railway) was along the northern edge of the present hard-core road known as the Victoria Track (roadway), which now doubles as a service road to the present installation. The later

(present) installation is known to be aligned 30ft {9m} to the north of the original Exeter Track, and parallel to it.

This basic piece of equipment was used for a wide variety of trials, covering the effects of bird impact with cockpit windscreens, cutters on the leading edges of aircraft wings for severing balloon barrage cables, deployment of parachutes at speed, serviceability of plastic radomes, and so on. Some of these trials – as in the development of aircrew ejector seats – involved tests on human subjects for RAE Physiological Department, a precursor of the RAF Institute of Aviation Medicine. One particular series of trials was conducted for Rolls-Royce in the determination of the behaviour of early jet engines under condition of positive and negative acceleration forces. These experiments were conducted on a general basis, not solely for eventual deck-landing applications.

Remembrances of eye-witnesses to early HST trials indicate that problems occurred with the adaptation of the old RAE DCTO trolley, the running surfaces of the rollers being abraded through skidding. These surfaces frequently required re-facing with a hard-wearing material – 'Stellite' – which was not all that successful. Some deformation of the track also occurred from this high speed operation.

By the time the airfield Runway 07/25 was undergoing its final extension (1946) the Exeter Track was being superseded. It was then that the rolled steel channels of the ex-DCTO track were replaced by 75lb/yard, standard flat-bottom rails obtained from surplus stock held by the Longmoor Military Railway on the Royal Engineers' Camp at Bordon. For a short while both tracks existed alongside each other. Ellis (#30 p.73) refers to this new track as being some half a mile {800m} in length. The anonymous writers of a departmental paper (#317) in 1956 record this track as being 2,000ft {610m} long by 3ft {0·9m} gauge upon which the rocket-propelled trolley could be driven at speeds up to 100kts {51m/sec}. (A contemporary RAE site plan shows the total length of the track to have been 2,053·16ft {625·803m}. With this exactitude, one is tempted to ask at what temperature?) The track was laid with a central, third rail at the same level as the running rails. This was used for guiding the trolley or sled as the running wheels were not flanged. The track was laid on purpose-made, adjustable, steel sleeper elements to permit precision alignment and adjustment as necessary. At the eastern, launching end – abutting Berkshire Copse Road – the track was mounted atop a continuous concrete beam some 24in {600mm} above local ground level. The track rose at a steady gradient of 1:250 for most of its length, ultimately levelling out for the final 100 yards {90m}. The total rise along its length was some seven feet {2·1m}.

This new length of track enabled still higher speeds to be attained. The 'hydraulic' tapered spike-and-barrel – as adopted for the C.I Accelerator – was used as the normal braking system, allowing experiments into the application of high 'g' rates of deceleration to be conducted. In addition to this simple buffer system, a 'standard' wire rope jigger installation – adapted from a redundant Mark 4 American, carrier-borne arrester gear unit – was provided. This was at the instigation of NAD who then used the track as a test rig for examining the behaviour of wire ropes when impacted by arrester hooks at high speeds. This first gear was used for the standard pull-out of 150ft {46m}. Thomlinson and Pierce (#299) state that a need had arisen for an experimental arrester gear to be provided capable of arresting a 12,000lb {5,440kg} aircraft from a speed of 120kts {64m/s} without exceeding a retardation rate of 3·5g. Such an installation was to be provided from existing resources without embarking upon an *ab initio* design. At that time no equipment possessing these capabilities, particularly with such a high speed of engagement, existed. The nearest approach to this specification was obtainable by employing a pair of American Mark 4 units which had been used to arrest a 7,000lb {3,200kg} aircraft at speeds up to 95kts {49m/s}, and here the experience was very limited.

From a critical examination of the specification it appeared that, to maintain a good ratio of maximum/mean retardation, a stopping distance would be not less than 275ft {84m}. This, with a cross-deck sheave spacing of 120ft {37m}, would require 220ft {67m} length of cable to be reeved into each end of the centre span. As this was about twice that normally used on existing types of jigger gear, a workable solution was conceived using two units. With these, the length of reeving per side would be 240ft {73m}, allowing a maximum pull-out of 294ft {90m}, permitting a desirable over-run margin for experimental purposes. It was not advisable for these two units to be installed as separate entities, one at each end of the centre span. As matching the performance characteristics of similar units in such a duplex installation in earlier experiments had been proved to be difficult, the solution of connecting the two units in tandem was explored and found to be practicable. This was the basis upon which the project proceeded.

A pair of redundant American Mark 4 arrester units (probably recovered from the war-damaged CVEs, HMSs *Nabob* or *Thane* which – instead of being returned to USA under the Lend/Lease programme – were decommissioned at Rosyth and Faslane respectively) was available and these were adapted to suit the new situation. This combined, 'tandem', jigger unit was located in a concrete-lined trench, parallel with the line of the High Speed Track and on the port (south) side. The centre span of the rope was supported on cradles each side of the track.

High Speed Track. Experimental aircraft seats loaded with a pair of life-size anthropometric dummies for a simulated crash landing using the high-'g' arrester system. The engineer on the far side of the trolley is Stan Chisman, an ex-NAD apprentice. July 1961. (© Crown Copyright/MOD)

High Speed Track. The trolley is at the point of the beginning of its arrestment. The water spray from the arrester barrel appears at the front while the propulsion rockets are still burning. July 1961. (© Crown Copyright/MOD)

The complete installation was conceived as being used with a single centre span, necessitating the anchoring of the two ends of the main reeving via a pair of air or hydraulic springs. The location of this system was such that the arrester wire spanning the track occurred some 1,439ft {439m} from the start end. Additionally a number of other test rigs, including arrester units, were mounted on this track but the jigger unit was retained on this track until it was dismantled.

One significant use of this track was the development, in conjunction with Messrs MacTaggart Scott, of the prototype Mark 12 shipboard arrester gear system, including the newly developed control unit known as the spline valve. This facility enabled dead-load weights, higher than those found in the then existing aircraft, to be arrested at retardation levels well above those allowed in aircraft at that time.

For this purpose, a special rocket-propelled ballast trolley, of adjustable weight (to represent the mass of an aircraft) was built. The trolley could be ballasted up to a maximum weight of 15,000lb {6,800kg} for operation at speeds up to 120kts {62m/sec}. This was thought to have been the first application of the larger, 5in {125mm} rocket motors. (A test run in early 1951, with the trolley loaded to 12,500lb {5,700kg} at an entry speed of 122kts {63m/sec}, produced a pull-out of 291ft {88m} at a maximum 3·3g rate of retardation.)

Other trolleys were produced in RAE for use on a variety of experiments. When a cabin or seat for physiological trials was used, the firing of the rockets was governed by a series of Strowger distributor switches to ensure progressive sequence control of their ignition. Mostly, in other trials all rockets were fired together to give a rapid build-up of speed. Depending upon the nature of the trials, trolley braking was effected by either the 'spike-and-barrel' unit or the rope-reeved, naval hydro-pneumatic jigger system.

For braking the trolley the normal type of aircraft arrester hook was not used. A large radius shoe on the front of the trolley engaged the centre element of the arrester rope that spanned the track and was supported on a cradle. Control of the speed of engagement of the trolley with the wire was obtained by varying the number of rockets fired and selecting a suitable starting point along the length of the track to produce the required run-in distance. One spot along the length of the remaining track, at the time of writing, still carried a mark to indicate the trolley starting point for a speed run of 150mph {67m/sec} using two rocket motors. Other trolleys to suit various non-naval applications also were provided to various standards of sophistication – or crudity. One of these was a two-part system in which the rocket propulsion unit was braked separately from the arrestment of the experimental trolley, the components separating at an appropriate location along the track allowing the experimental trolley to freewheel for the final stage of its journey. (Records available to the author have not revealed the method by which this separation was achieved.)

The installation achieved all the specification requirements and was used for a variety of research projects. Useful work on research into air springs and liquid springs for rope tensioning was undertaken. Trials on this unit highlighted problems which initiated the preparation of special test rigs for more detailed study of the behaviour of ropes running through pulleys and fairleads at high speeds. During its development the equipment progressively assumed greater sophistication with appropriate instrumentation, and work was still in progress when the Department moved to Bedford in 1955, the rope test equipment being transferred with the Department. NAD found this straightforward experimental facility to be of great value in many of its rope-wise investigations. In fact it was considered to have been an essential piece of equipment, complementary to the conventional catapult and arrester gear. The 'Farnborough Rocket Railway' had justified its existence despite its simplistic concept and basic qualities.

A recent examination by the author of site plans – and the site itself – indicates there to have been a number of arrester systems associated with this track. The first two of these was the hydraulic 'spike-and-

The view from the Eelmoor Gate (south-west) end of Runway 07/25. The diagonal line through the trees on Bridge Hill from the top right is the alignment of the High Speed Track in its permanent location. Test firings were made from the far end towards the viewer with the arrester units in the foreground. September 1987. (© Crown Copyright/MOD)

barrel' unit located at the extreme western end of the track, and the drag chain system used at any convenient space along the length of the track. The selection was made by the nature of the experiment to be conducted, whether the test was orientated for acceleration, or deceleration of the test specimen.

The 'spike-and-barrel' arrester system has been described earlier in this volume. The drag chain system (also described earlier) was cumbersome to operate and time-consuming in resetting after a trial run. This gave way to the third installation, located 615ft {187m} upstream of the terminus. This was the tandem American Mark 4 system to which reference already has been made. By 1953 a fourth arrester system was introduced, another jigger unit, which became known as the Long Stop Arrester Gear. Contemporary drawings indicate that this, firstly, was an extension of the NAD duplex American Mark 4 gear. This system, latterly, was provided with another pit containing the British Mark 4 arrester machinery, with the arrester wire spanning the track at 240ft {73m} from the termination of the track. This had the contemporary standard flight deck pull-out of 160ft {49m}. The installation required three massive caisson foundation units of some 12ft {3·7m} diameter and 12ft depth to support the rope sheaves.

As a final dead-end, a buttress some 12 feet square in plan and about eight feet high, built with sandbag walls, was constructed at the western, terminal end of the track. It is thought that this last-ditch resort was never needed.

In all these cases (in 1972) the total length of the track is shown as 2,100ft {640m}. By 1986 some of the eastern end of the track had been sacrificed to provide space for new laboratory buildings, reducing the total length of track to 1,580ft {480m}. On the author's last visit to the site (February 2000), the track length had been 'topped and tailed' to an overall distance of 1,180ft {360m} with a gravel bed, at the western extremity, forming a drag-trap as the trolley braking system. A vestigial length of ¾in {19mm} diameter wire rope was lying in a trench passing beneath this reduced section of the track, at the location of the third arrester system installed in its history.

For purposes of identification, the track itself was given the building number [T.16] with the terminal structure forming the braking system at the western end as [T.19]. The early NAD arrester system and the latter, 'Tandem' arrester gear were located in a pit numbered [T.17]. The 'Long Stop Arrester Gear' was located at pit [T.58]. The 'Soft Ground Arrester Area' – nothing whatever to do with NAD – for experimental landing of aircraft under overshoot conditions, was located between the HST alignment and Runway 07/25. This was given the identification [T.67].

The grassy mound to the south of the NAD arrester unit – Bridge Hill – with a footpath leading up to building [T20] was used as an observation point for most of the experimental work on the HST. It is now a wooded copse of mature trees, reckoned to be a 'site of special scientific interest'. The ground surrounding the downstream end of this noisy, pyrotechnic device has now been recovered and graded as a brownfield site, probably with a cover of two feet depth over a number of massive concrete blocks, once used as anchorages for pulleys, fairleads and arrester units. In its ignominious end, at the time of writing, the site – although still in occasional use – serves as a (rather remote) car park for the biennial SBAC air display, the Farnborough Air Show. Subsequently, it was learned in February 2001 that the track had been severed to permit the erection of a boundary fence across the site, separating the airfield from the new owner of the adjoining estate.

POST-WAR DEVELOPMENTS

The momentum gained in developments pertaining to Naval aviation during World War 2 continued for a further decade and more after the cessation of hostilities in 1945. NAD continued to be highly active in addressing problems in FAA operation at the time of the removal of the Department to the Thurleigh (Bedford) site during the mid-1950s and with continuing involvement over the next two decades. These studies are outside the author's own personal involvement and he is dependent upon the writings and records of those directly associated with these projects during this period. Many of the immediate post-war activities had their origins in wartime expediencies with their development continuing to the advantage of the FAA. These activities, each of major consequence, are described in the following paragraphs.

FLEXIBLE DECK – (CARPET LANDINGS)

Often, during a trials period on board an aircraft carrier and when flying for the day was finished, discussion in the wardroom (perhaps over a pink gin or two) would ensue between naval officers and their civilian engineer and scientist comrades on various aspects of Navy flying. Some ideas were fanciful, others could be addressed with merit, and a few were genuinely serious. Early considerations of such feasible concepts often developed into the core of an official request for a formal investigation and evaluation; the operation of aircraft without undercarriage was one such. (An earlier instance was the introduction of Seafire aircraft, following a shipboard declaration by a senior official of the Ministry of Aircraft Production to the effect that 'Of course you can have Spitfires, if you can provide the men to fly them'!)

In 1944, RAE began an exploratory investigation into the possibility of landing, handling and launching an aircraft devoid of a normal undercarriage on a carrier flight deck. The weight of an undercarriage was a significant content (6 percent or more) of the all-up weight of an aircraft and it was felt that this reduction in weight would be useful in allowing heavier stores, additional fuel, and so on to be carried. The absence of propellers in jet aircraft enabled research into this idea to proceed and a carrier deck was considered to provide the ideal location in which this research should be conducted. A number of schemes to achieve this were considered. Brown (*#15 p.71*) records a meeting held at RAE on 11 January 1945 at which Major F. M. Green was present. (We have met him already, some three decades earlier when he was in charge of design activities at Farnborough.)

Green proposed a scheme for landing a naval aircraft, bereft of its undercarriage, on a rubber carpet stretched between shock absorbers to give a dead-beat deflection. The aircraft would pitch on to the carpet after picking up an arrester wire at the rear end of the carpet. The meeting agreed that such a proposal should go ahead and the project developed into an inflated, air-cushioned, rubber deck with a single arrester wire rigged athwartships at a height into which the aircraft would fly at a speed marginally above landing speed. The entire development of this project, through six identified stages, was undertaken by NAD at Farnborough.

Stage 1 was to conduct model experiments – to one-eighth scale – in NAD Lab. This was an exploratory investigation to determine the type of flexible material and to examine the method of supporting and tensioning the flexible layer forming the alighting area – the top surface

of the deck. Site work started on the adopted full-scale system – forming Stage 2 – in January 1946. The first experimental rig, a small area of flexible deck 34ft x 30ft {11m x 9m}, was set up on Jersey Brow, located beneath the crane servicing the P.I Catapult. The object was to drop a cylindrical dead-weight from the crane hook onto the experimental rig under various load conditions and tensioning of the carpet using two methods of carpet suspension. Experiments were continued with a ballasted GAL.48, Hotspur glider (BT752) fuselage on both types of carpet suspension, that comprising a trampoline-type of tensioned membrane and that of a tensioned membrane fully supported on an air-cushioned sub-strate. Following these experiments the full-scale ballistic trials proceeded through the next phase, with the development of the pneumatic tube system of carpet support.

Experimental work, beginning in March 1947, comprised a mattress-type sub-strate containing hundreds of inflatable tubes in five layers. This 'air-bed' supported a 'rubbery', flexible top cover (the carpet, a single piece 200ft long and 60ft wide {61m, 18m}) which formed the impact interface with the aircraft. The Meadow Gate area, on the north side of the airfield, remote from the main runways, was selected for this trial installation. Three further Hotspur gliders, complete with modified wings and tail surfaces, were provided for ballistic flight tests. These were suitably strengthened and reinforced on their lower surfaces, and fitted with a 'sting' type arrester hook.

Some confusion exists here as to the identification of the gliders used in this project. The Farnborough aircraft movements log identifies three Hotspur gliders as being set apart for these trials, their serial numbers being recorded as HH175, HH610 and HH889. A photograph exists of glider TJ889 sitting on the carpet; was this previously BT889 and, subsequently, HH889? Although gliders of this marque normally carried BT or TJ identifications, it is possible that for the sake of records these airframes were renumbered to HH serials on their detachment to Farnborough, and ultimate disposal. Such an expedient, although rare, was not unknown.

An arrester wire was rigged athwartships astern of and at some 30 inches {750mm} above the experimental deck surface. A simple rocket-propelled trolley running in a track – which could be adjusted in elevation for angle of approach – was located in line with and abutting the experimental deck to enable the glider to be catapulted into the system. The glider was released from the trolley within a short distance of the arrester wire, its flight path purposely being restricted to a minimum distance. These tests were conducted to examine the impact of the aircraft with the carpet surface and the behaviour of the surface itself; the results after some 34 landings indicated that a full-scale trial with an aircraft was a feasibility. The engineering content of this Stage 2 experimental set-up appears as Appendix 11.

The experimental landing area for flight trials, forming Stage 3, was located alongside and to the west of Runway 04/22. A full-size carpet area was laid, together with its arrester gear, a modified standard hydro-pneumatic unit. A de Havilland Vampire F.1 aircraft (TG285) was adapted for early flight trials but the first arrested landing into this installation at Farnborough was made on 29 December 1947 by Sea Vampire F.21 (TG426). This proved disastrous, the aircraft being wrecked in the process. The pilot, Lt/Cdr Brown, escaped unhurt but the aircraft was a write-off. Some 52 successful landings subsequently were made into this equipment and almost a year after the initial landing at Farnborough, the first shipborne landings on HMS *Warrior* took place.

For sea trials – Stage 4 of the programme – HMS *Warrior*[44] (recently returned after two years' service under the ensign of the Royal Canadian Navy) was adapted to accommodate the Farnborough flexible-deck structure and carpet used in the Stage 2 trials. This was uplifted from the Meadow Gate site and transported to Portsmouth Dockyard where the shipboard installation was undertaken during the summer of 1948. Four de Havilland Sea Vampire F.21 aircraft (from the group containing TG285, TG286, VT795, VT802, VT803 & VT805) were embarked and sea trials began in November 1948. After each shipboard landing the crane abaft the island lifted the aircraft from the 'carpet', its undercarriage was lowered and the aircraft taxied forward for launching by the accelerator for its next attempt. These trials were completed in May 1949 after a total of some 271 landings having been made by ten pilots into this equipment, both at Farnborough and aboard HMS *Warrior*.

The trials indicated that a higher rate of retardation of the aircraft than with a normal carrier arrester gear was to be expected; this was due to the higher speed of entry into the arrester wire, coupled with the friction forces on impact with the rubber deck. The 'g' figures for the flexible deck landings were about double that for the traditional deck-landing systems. On conclusion of these trials, all flexi-deck equipment was removed from HMS *Warrior* and the ship was put into reserve. She was refitted and recommissioned some months later but eventually was sold to the Argentinian Navy in 1958 and renamed *Independencia*. Her service terminated in 1971.

The final development of the flexible deck – forming Stage 5 – was subsequently installed at the Laffan's Plain end of the RAE airfield, adjacent to, and to the north of Runway 07/25. In common with many another project in the defence industry this stage of the flexi-deck activity was given a code name. Alongside Blue Streak, Blue Steel, Black Knight and many other nomenclatures, we meet the term Red Rufus in our application on Laffan's Plain. To a bystander the logic of these titles appears without foundation. Perhaps the intention was to introduce a 'fog factor' to cover any semblance of identity of the real purpose of this, or any, project. This was a new carpet, 400 feet {122m} long and 80 feet {24m} wide, laid in two pieces over a modified form of airbag sub-strate.

In this new location, a further 302 landings were made into the equipment by 23 pilots, flying de Havilland Vampire F.21 aircraft together with a smaller number of landings with a modified Hawker Sea Hawk prototype aircraft (VP413) between April 1952 and January 1955. Thomlinson (*#298*) says that the arrester gear used in this project comprised a pair of redundant American Mark 4 arrester units, connected, head-to-head, to form the 'Tandem' arrester gear which produced a pull-out of some 250ft {76m}. This equipment was a copy of that provided for the High Speed Track (qv) which, a few months earlier, had successfully concluded its commissioning trials on Laffan's Plain. Hotson (*#259*) reports on a series of rapid-handling trials using specially produced equipment for deck-handling the aircraft in its wheels-up condition. The best time interval attained, using three aircraft in sequence, was 44 seconds between landings, a figure expected to be improved upon with the ground crew gaining handling experience.

The closing episode in this project, Stage 6, was the complete experimental cycle of launching and landing an aircraft without undercarriage and its deck handling. The special catapult used for this launching is described elsewhere. (Vide 'Slotted Tube [Ventral] Catapult'). The aircraft adopted for this exercise was the Hawker Sea Hawk prototype (VP413), specially adapted for this trial. The modified deck, used in Stage 5, formed the landing platform. The one and only flight under-

An adapted GA Hotspur II glider (BT889) for the flexible deck trials is mounted on its launching trolley with its 'sting' type arrester hook deployed. July 1947. (© Crown Copyright/MOD)

The Hotspur II glider dummy (BT889) has landed on the flexible deck, its hook having engaged the arrester wire. Often five cameras were deployed for each run, two of them being high speed ciné cameras for free-time frame-by-frame examination of the behaviour of the aircraft in its launch and landing sequence. July 1947. (© Crown Copyright/MOD)

taken by the Sea Hawk from catapult launch to arrested landing was made on 12 November 1953 (five years after the first landing at sea!), the FAA pilot being Lt W. H. Noble DSC. Although this trial was completely successful – all equipment working within specification – it became the 'swan song' for the project. The deck and its equipment were dismantled and removed from the airfield shortly afterwards.

The whole experimental exercise has been written up by various authors, the RAE pilot's records of the initial landings being particularly complete in detail. Unfortunately some subsequent authors seem to wish to discredit this project. At the time of its conception there was a very real, operational, need to obtain maximum performance from shipborne aircraft and every channel to attain this end was investigated. One contention raised was the supposed difficulties in deck handling of aircraft on landing and subsequent relaunching. The aircraft landed within a very restricted area of the deck as only a single arrester wire was used. The installation on board HMS *Warrior* made use of the deck-level seaplane crane for lifting the research aircraft from the carpet – which was specifically located on the flight deck for this purpose.

The subsequent loading on to the accelerator would not present any great problem: the mechanical positioning of aircraft with undercarriages at the launch position of the accelerator was an already established procedure on the Fleet Carriers and the adaptation of this principle would have created no difficulty. It was seen that a similar installation could be established for landing ashore but this may not have been desirable as the crane would present a permanent obstruction to other 'wheeled' aircraft if the installation was made on a normal airfield. With the end of World War 2 there was no operational need for the system to be further developed, not even for the era of the Cold War. At one time the US Marine Corps expressed an interest for their mobile

airfield concept. The project was abandoned some nine years after its conception, having been overtaken by the VSTOL principle.

Marriott (*#63 p.44*), in 1985 writing on the shipboard installation, concludes:

'Although the subsequent sea trials showed that the system could work, it was really a blind alley in terms of development and would have required extensive facilities both aboard ship and at naval air stations in order to support it. Any gains in aircraft performance were more than cancelled out by the complexity of the idea and the cost of its implementation.'

Lt/Cdr Brown, CO of the RAE Aero Flight and experimental test pilot closely involved with this activity, records the history of the project in great detail (*#15 p.71ff*). Let him have the last word (*#ibid. p.84*):

'Just to round the whole thing off, some years after I left the RAE and before the flexible deck installation was uprooted, a catapult take-off and landing was made with a Sea Hawk without the undercarriage wheels ever being lowered. A tidy way to wrap up what was undoubtedly a most spectacular thing to witness. Indeed a flexible deck landing was the standard showpiece for VIPs visiting the RAE during my time there. I enjoyed this trampoline circus act as each landing contributed some more data to our knowledge, and anyway I always found such a landing a most exhilarating experience.'

The only evidence of this field of activity now remaining at Farnborough are three separate areas of concrete paving on the erstwhile Cove Common, containing embedded chunks of ironware and

Modified de Havilland Vampire F.I (TG426) in its first landing onto the flexible deck. The aircraft, piloted by Lt Cdr Brown, damaged its tail on premature impact with the ground and nose-dived into the carpet, badly damaging the aircraft. December 1947. (© Crown Copyright/MOD)

Model of a Fleet Class carrier portraying a tentative layout of a flexible landing area to the flight deck. The 1948 experimental installation in HMS *Warrior*[46] covered a relatively smaller area. February 1946. (© Crown Copyright/MOD)

rusted remains of cropped-off foundation bolts at locations [O.19], [N.6] and [W.5].

Although superficially appearing to some to have been a 'white elephant', a highly significant feature emerged from this project. During arrestment the arrester wire was seen to develop a transient anomaly in the shape of a kink occurring during pull-out, a feature clearly seen in one of the photographs. This triggered an in-depth investigation into a stress-wave transmission phenomenon emanating from high speed engagement and fleeting of arrester wires, first exposed in these activities. This feature is discussed elsewhere in this text.

SLOTTED TUBE (VENTRAL) CATAPULT

Coincident with the development of the Flexible Deck (Carpet Landings), the feature of launching an aircraft without undercarriage was addressed. For much of the following information the author has drawn upon the technical reports by Chisman (#224), Grimes (#250 & 252), Martin (#269) and Thomlinson (#295). This project is treated in some depth as the author is not aware of this experimental equipment having been described in any writings released for public reading.

Experimental work on the launching of aircraft without undercar-

riage required a shore-based catapult – as distinct from an accelerator – with a performance greater than any available at that time. It was therefore decided to construct a purpose-made launcher with the appropriate capacity for this purpose. The desired performance, based primarily upon the requirements for launching a Hawker Sea Hawk aircraft, was for an end-of-launch speed of 120kts {62m/sec} of a mass of 12,000lb {5,400kg} without exceeding an acceleration rate of 6g with a maximum rate of build-up of 10g per second. Grimes (#ibid.) states that a variety of take-off methods were examined before deciding upon this particular type of launcher.

The adoption of a trolley-less launcher was attractive in that the inertia of the trolley itself in the launching mode, and the forces required to brake the trolley at the end of the stroke, could both be neglected. The method of launching was to propel the aircraft, by means of the catapult, along a 'V'-shaped trough without any constraint except from the trough itself. The sides of the trough were to be lined with a resilient cushion along which the aircraft would slide during launching. This would reduce the number of moving parts, external to the aircraft, to as low a mass as practicable. The problem remaining to be addressed was that of the handling of the aircraft prior to the launch and its transfer to the catapult. To solve this, a method was devised to make the transition from a handling trolley (of matching trough-like construction on which the aircraft would rest for ground-handling purposes) to catapult as straightforward as possible. The trolley would be accurately registered with the end of the catapult beam to allow the aircraft to propel itself forward the short distance to engage the towing hook on the catapult and for the holdback to be tensioned. This same trolley would serve the double purpose as conveyance for the aircraft on recovery via the flexible deck.

Ideas were being sought for implementing such a launching device. One of these was for the adaptation of an ex-German slotted tube launcher. Historically, the slotted-tube method of propulsion was tried and discarded in Isambard Kingdom Brunel's day (the early Victorian engineering entrepreneur 1806–59 and his contemporaries) on atmospheric railways and no record can be traced of its *modus operandi* having been adopted since. Relating to the South Devon Railway in the early 1840s, Rolt (#84 p.162ff) describes in detail the problems which faced Clegg (1839), the Samuda brothers, Brunel – together with Mallet in France – at that time, and which seemed to be insurmountable. In Brunel's case the piston was propelled within the slotted tube by atmospheric pressure against a vacuum of some 16in Hg {0·53bar}. The sealing of the slot was achieved by an external hinged leather flap, reinforced by iron strips. This created a plethora of problems. The South Devon Atmospheric Railway lasted less than two years. The installation by Mallet of the St Germain to Nanterre Railway in France was the final system to be abandoned; this was in 1866. The hey-day lasted than 15 years, perhaps confirming the opinions of Brunel's contemporaries, the Stephensons (father and son) and Locke, who dismissed the scheme as 'a great humbug'. Just a century later, in the latter years of World War 2, the Germans seemed to have had some success in this field – with the V.1 weapon, 'doodle-bug' launcher!

A week after D-Day (6 June 1944) the south-east corner of Britain was subjected to an assault by pilotless flying bombs, quickly being dubbed buzz-bombs or 'doodle-bugs'. These missiles were the first of six projected 'V' weapons, known as the V.1. With the Allied Armies progressing through Europe, samples of various pieces of German military material were being retrieved for examination by British engineers and scientists. Amongst the equipment recovered were units of the V.1 flying bomb and its launching gear. These had been delivered to RAE and specimens were displayed in the two aircraft exhibitions held at Farnborough in the autumn of 1945 and the following spring. The German launcher was of immediate interest to Naval Aircraft Department as it demonstrated the successful adoption of the principle of a slotted tube type of catapult for launching these missiles. (Was this

The crew associated with the first successful landing on the Farnborough flexi-deck. They comprise (left to right): J. (Tommy) Thomlinson, R. J. (Nick) Nixon, Lt Cdr E. M. (Winkle) Brown (Pilot), Lewis Boddington, Chas Crowfoot, S. W. (Stan) Chisman, Donald Pierce, E. F. H. (John) Noble, N. W. (Norman) Skelton, Gp/Capt Silyn-Roberts RAF (CO Experimental Flying), Jack Wright (Chargehand Inspector, Flight Sheds), A. A. (Alan) Arnold, Cdr (E) Eric Beard RN. March 1948. (Donald Pierce)

De Havilland Sea Vampire F.21 (VT803) is making a good landing onto HMS *Warrior*[46] in Stage 4 of the flexible deck programme. February 1949. (© Crown Copyright/MOD)

This is a highly significant picture of de Havilland Sea Vampire F.21 (VT802) making a good landing onto the flexible deck aboard HMS *Warrior*[46]. A point to be noticed is the kink in the centre span of the arrester wire. This triggered an in-depth investigation into a wave transmission phenomenon first exposed in these early activities involving high speed engagement of arrester wires. The high speed Vinten ciné camera on its outrigger boom is seen in the distance. February 1949. (© Crown Copyright/MOD)

A Sea Vampire, with undercarriage deployed, is ranged on the twin-track BH.III accelerator aboard HMS *Warrior*[44] for launching. The pilot is clambering aboard for the next launch and landing circuit. The man in the left group in white overalls is the fire-suit man, one of the pair of asbestos-clad firefighters present on deck when flying is in progress. February 1949. (© Crown Copyright/MOD)

an infringement of Mitchell's 1938 patent for such a system?) With a number of sections of this launcher having been retrieved from the previously occupied Low Countries and currently available, it was decided to proceed with an installation using the slotted tube principle. Although the performance specification for the V.1 launcher in its original diabolical service was different from that required for its new application, the dimensions of the V.1 equipment generally suited the new requirements.

Martin (#*ibid.*) had prepared a paper on his investigation into the German project outlining its construction and method of operation. The propellant was an exothermic reaction of two materials known as 'T-stoff' and 'Z-stoff', the reaction being a straightforward decomposition of hydrogen peroxide by means of a liquid sodium permanganate catalyst, in turn producing quantities of superheated steam at a temperature of 505°C. This force acted on the back of a piston-type shuttle which was propelled through the launching tube and carried a spigot which protruded through the slot. The travel of the shuttle was designed to seal the slot behind its progress to maintain gas tightness behind the shuttle. The missile, sitting atop the track, was propelled by the spigot. The shuttle itself, at the end of its stroke, flew free of the 'muzzle', and landed wherever (it was not arrested). It was retrieved for re-use. Clearly, the method of propulsion and its inherently violent

chemical instability could not be adopted at Farnborough; neither could the free ejection of the shuttle at the end of its stroke be tolerated.

By 1950 the principle of adopting the ex-German equipment was agreed upon and work was put in hand to produce the hardware. Chisman (op. cit.) records the design requirements to launch a 12,000lb {5,400kg} aircraft at 125 knots {64m/sec} without exceeding an acceleration of 6g to have been:

Acceleration stroke	150ft	{46m}
Maximum working pressure in cylinder	900psi	{61bar}
Retardation stroke	5ft	{1·5m}
Maximum retardation pressure	12,500psi	{850bar}
Time to maximum acceleration	0·70sec	
Initial rate of acceleration build-up not to exceed	10g/sec	
Bore of slotted tube cylinder	11·70in	{297mm}
Width of slot	0·812in	{21mm}
Weight of launching piston not to exceed	1,000lb	{450kg}

For this, 16 three-metre lengths of V.1 launcher sections were obtained. Their bores were re-machined to provide a precision surface and the method of internally sealing the slot was changed from a tube to a flexible strip. Consideration was given to the alternatives of using compressed air or cordite as propellant, with the preference being for the latter as being more readily adaptable. Wide experience had been gained already by NAD through the use of cordite charges in the C.I Accelerator on Jersey Brow. An explosion chamber (which probably now would be described as a combustion chamber or gas generator) with breech-loading facilities was designed and fitted to the launching end of the catapult. By calculation and subsequent experimentation it was found that a cordite charge of 62lb {28kg} weight would be required to achieve the launch specification.

In travelling the length of the launching tube, the dumb-bell-shaped piston/shuttle would employ its leading edge to pick up the sealing strip from the floor of the bore. This allowed the strip to pass through the piston, and presenting it at high level at the trailing edge to seal the slot in the top of the bore, the pressure of the propelling gases accomplishing the seal. The driving spigot (a 'fid', in naval parlance) protruding through the slot and engaging the aircraft was formed in the leading section of the shuttle.

At the end of its launching stroke the shuttle needed to be brought to rest within the framework of the catapult. Again, a feature much used by NAD was employed here. This was the short, hydraulic cylinder with

The slotted tube catapult on the Ball Hill trials site. The track has been fitted with the rubber-lined 'V'-shaped trough for launching the Hawker Sea Hawk aircraft, but with the dummy 'iron bird' loaded at the firing end. June 1952. (© Crown Copyright/MOD)

The slotted tube catapult transferred from Ball Hill and erected on Jersey Brow. This was for launching the Hawker Sea Hawk (VP413) with undercarriage retracted, in the final trials programme. March 1954. (© Crown Copyright/MOD)

The firing end of the slotted tube catapult on the Ball Hill trials site. The 'iron bird', representing a Hawker Sea Hawk, is loaded on to the track ready for a trials launch. June 1952. (© Crown Copyright/MOD)

the water held by a frangible disc in the front end. The author has already called this the 'spike-and-barrel' system which is described in 'Land Trials'. This combination was of heavy construction; the cylinder was made from forged steel, and was 6ft {1·8m} in length. It was designed to bring a mass of 723lb {328kg} travelling at 120kts {62m/s} to a dead stop within that distance, a retarding force of the order of 200g being imposed in the process, generating an instantaneous hydraulic pressure within the barrel of some 12,500psi {850bar}.

This installation was constructed on the Ball Hill site, away from the immediate flying areas of the Farnborough airfield, occupying a site slightly to the north of [Y.23], the housing for the mini catapult serving the aircraft model ditching tank. (It has been reported that on one occasion the 'iron bird' used for ballistic launches went astray and careered into the ditching tank itself, quickly to be retrieved before too many questions were asked!) A series of 'light' launches was programmed to prove the general operational features and to assess the rates of burn of the cordite charges to produce the required acceleration parameters. On completion of these, trials with a dead-load followed. This dead-load was a horizontal cylinder, with conical ends, mounted upon cast-iron sliders running along the catapult beam above the slotted tube. A towing lug to mate with that protruding from the shuttle was fixed to the bottom of the cylinder, forward, and a holdback device was attached to the rear end to engage a turnbuckle unit for pre-tensioning prior to a launch. The body (fuselage!) of the dead-load was compartmentalised to permit variable loading and to enable the fore-and-aft location of its centre of gravity to be adjusted within fine limits. At the end of the launch, this dead-load was propelled from the track – momentarily becoming a ballistic missile – as the shuttle was retarded within the tube. The dead-load was fitted with stub-wings of substantial proportions, the sole purpose of which was to snatch the suspended centre span cable loop forming the pick-up of its arrester gear.

This arrester system for the dead-load was very basic in concept. The portable inertia arrester-gear principle (forming an abandoned experimental project within NAD in 1943 – vide 'Land Trials – Drag Chain Arrester Gear') was adopted, the retarding effort being supplied by friction from a pair of chains being dragged along the ground. These chains each were 300ft {91m} in length, weighing 5 tons and formed of 2in {50mm} links. The centre span, which was to be picked up on impact by the dead-load, was of 1⅛in {29mm} diameter steel wire rope, 150ft {46m} in length. It spanned the line of launch and was supported as an inverted loop suspended from a pair of stanchions (goal posts). The dead-load was projected into this loop, the bight of which passed beneath its body and impacted by the leading edges of the stub-wings, thus bringing the dead-load to rest. The pull-out of the arrester gear was of the order of 400ft {120m}. At a later stage the dead-load was modified to accommodate a battery of three 5in {125mm} rocket motors to represent the 3,500lb {1,600kg} thrust of the aircraft engine. The trials conducted on this set-up proved satisfactory and the programme advanced to the next stage.

Grimes (#ibid.) takes up the story when the catapult tube was adapted to accommodate an aircraft, in particular the Hawker Sea Hawk. With no undercarriage, and no launching trolley, the catapult runway needed to be provided with a facility to support and directionally guide the aircraft in its own right. Having considered a number of alternatives, with simplicity of operation being the main aim, a scheme was adopted in which a superstructure forming a shallow 'V'-shaped trough with its apex in line with the slot in the tube was fitted atop the tube for its full length. This trough would support the aircraft during its launching run. The problem of launch-induced friction between the aircraft fuselage and the trough needed to be solved. This feature had been addressed in detail and the final solution was found in covering the surface of the trough with a resilient interface. This developed into a series of flexible

Prototype Hawker Sea Hawk (VP413) ready for launch from the slotted tube catapult on Jersey Brow. The machinery was moved from Ball Hill to the Jersey Brow site for this – the first and only – sequential launch and arrestment of a Sea Hawk aircraft as the final phase of the flexible deck project. December 1953. (© Crown Copyright/MOD)

sponge rubber pads retained by a sheet of rubber fabric, much like a conveyor belt, held under tension. Both faces of the trough were lined with this cushion along its entire length, with a thin facing of neoprene as the final rubbing surface. The catapult towing hook would move along the bottom of the trough with the aircraft resting in the cushioned vee. In so doing, the aircraft was supported on two oval-shaped surfaces of contact between its fuselage and the trough lining. As the aircraft was loaded into the catapult with its undercarriage retracted, a matching handling trolley was provided for deck handling. This system was given the title of 'Ventral Catapult', reflecting the reinforcement of the underside of the aircraft fuselage and its propulsion via a detent protruding below the belly of the aircraft.

A second dead-load device (an 'iron bird') was manufactured to prove the system. This had a contoured under-surface to replicate, with reasonable accuracy, the profile of the belly of the Sea Hawk aircraft.

A model of a fleet carrier demonstrating a proposal for an angled flight deck. The model aircraft show the author's 1945 concept of a naval jet – prior to the first landing of a jet aircraft (the de Havilland Sea Vampire) at sea. The model aircraft were produced in the RAE pattern shop and foundry to the author's design. September 1951. (© Crown Copyright/MOD)

An exhibition of the work undertaken by NAD was mounted in the departmental laboratory in 1955, prior to its staged transfer to Thurleigh, Bedford. This scale model of a carrier shows a Sea Vixen on the angled deck with a spread of six arrester wires and the visual landing sight on the port side. The absence of safety barriers is noted. July 1955. (© Crown Copyright/MOD)

Also, instead of a fin and rudder, a sloping cantilever arm with an 'inverted' arrester hook fixed at the top was provided; making the unit resemble the stern of a Norse longship of a past era – or even a scorpion! This hook would connect with the wire of the drag chain arrester gear – now with its centre span supported at high level, and looking like a laundry line bereft of its washing – and bring the dead-load to rest. During the model trials a coefficient of friction of 0·2–0·3 was anticipated. At full scale this figure was generally below these limits. During trials it was found that the friction of the surface was virtually the same whether the surface was wet or dry. The temperature rise of the aircraft fuselage due to friction was expected to reach 100°C; experimentally the figure never exceeded 74°C. Trials, totalling some nine runs of this set-up, proved a success and the entire catapult unit was dismantled and re-erected on Jersey Brow at location 1, alongside the head of Runway 04/22, for its 'live launch' activity.

At the Ball Hill proving base, only the bare essentials of the catapult and trough were erected, with the entire installation being above ground level. When the system was moved across to Jersey Brow certain changes were made, the principal one being the addition of a mechanical device for loading the aircraft onto the catapult and the upwards inclination of the catapult axis by 1½°. To obtain this inclination, the rear end was brought to ground level by setting in a shallow trench with its gas generator chamber below ground level. This facilitated the loading of the aircraft from its transportable trolley and associated engineering equipment at ground level.

After a few dummy, dead-load runs of the catapult in its new location, 12 November 1953 saw the Sea Hawk (VP413) loaded on to its handling trolley and conveyed to the catapult. It was nudged on to the launching track, the pre-flight check carried out, the pilot (Lt W. H. Noble RN) embarked and the aircraft was successfully launched. A chase aircraft flew alongside to check the underside of the Sea Hawk fuselage and absence of damage from the launch. After a short flight, the Sea Hawk landed into the flexible deck system and the project was deemed to have been completed, the exercise never to be repeated. (Thus is history made!) The catapult itself was scheduled to be transported to RAE Bedford to form part of the Naval Aircraft Department equipment on the new site. Being overdue in this transfer it was dismantled quickly and sent on its way. Within a few weeks it was operating again, continuing to give dependable, although more rigorous service as the interim propulsion unit on the 'Safety Barrier and Arrester Gear Proving Base' (qv) at its new home. The performance at Bedford was to propel a test vehicle (perhaps a 'hack' or dummy aircraft) of 12,000lb at 125kts {5,400kg, 64m/sec}. A few months later its replacement was installed. This was another cordite-powered machine, similar to the original, but upgraded to 40,000lb at 150kts {18,100kg, 77m/sec}.

The evaluation of, and experimental work upon, the ventral catapult project bore fruit in the contemporaneous development of the steam catapult which itself adopted the slotted tube principle so recently demonstrated in the technology of the German V.1 launcher. Ward (#306) refers to a shorter length of the Brown Brothers design for the BS series of launchers being erected at the ARDE range at Shoeburyness. This also was cordite-powered and produced valuable data – tests having begun in 1950 – for the development of the Steam Catapult (qv).

ANGLED DECK

A second development with long-standing effect in Flying Navy was the introduction of the angled flight deck. Casualties had been caused to men and machines by aircraft alighting on a deck, failing to pick up a wire and careering into – and sometimes over – the safety barriers. Inevitably, under operational conditions it is desirable to be able to recover to deck as many aircraft as possible within minimum time. This normally was achieved by each aircraft, on landing, being taxied forward of the barriers and the deck handling party would then carry out all post-landing activities and strike down the machine into the hangar below. This deck handling activity was seldom as rapid as the rate of recovery (mostly due to the speed limitations of the lifts) and an accumulation of aircraft would occur at the forward end of the flight deck. It was thought that the problem would be exacerbated by the introduction of the flexible deck, coupled with the anticipated increase in approach speed of future aircraft.

NAD had been looking at this problem and had adapted a model of HMS *Illustrious*[40] on to which a 'skew' flight deck had been painted and a number of jet aircraft models had been disposed. (The author had reason to be rather proud of these models; see below.) To give a realistic impression, this model was floated on a cooling pond and photographed at various angles. Boddington (head of NAD) produced a report (*#217*), dated October 1951, on this principle stating that the take-off and landing speeds of carrier-borne Naval aircraft, particularly fighters, were continuing to increase without any corresponding increase in the length of flight deck. To meet these higher performances, it would be necessary to increase the run-out of arrester wires and barriers, and the launching strokes of catapults. It was becoming increasingly difficult to fit such equipment into flight decks with current layouts. He felt that the landing area was not only growing in length but it was also becoming essential to move it further up the deck to cater for higher closing speeds and rough water conditions.

Boddington continued with the point that in the newer ships coming into commission the size of the landing area and the length of the catapults would leave little or no room for a safety zone. This would mean accepting aircraft barrier engagements when any of the last six wires were engaged and reducing the forward deck park. The alternative would be to have only two aircraft ranged forward for accelerated take-off. It was desirable for consideration to have been given to a deck-operation method where the various functions could be carried out without mutual interference or restriction to any one of them. Alternative suggestions had been made to achieve this without materially altering the size of the carrier. These included the provision of two decks, the upper for landing and the lower for taking-off. (Shades of Fairey Flycatchers emerging from the hangar decks of HMSs *Courageous* and *Glorious*?) Another was the construction of the flight deck in a 'Z'-planform: one arm for landing and the other for taking off. The adoption of either of these solutions would have incurred major alterations to the carrier as a whole. He then described a less drastic solution which to a great extent was expected to satisfy the need. He proposed a 'skew deck' in which the landing area would be aligned to port some 7° or 8° off the centre line, leaving a wedge-shaped area on the starboard bow for take-off and a deck park. This is the first intimation of an angled deck and, although it was within the concept of the flexible deck, the application was appropriate for universal adoption.

Boddington outlined five advantages in developing this scheme, one of which would mean a change in landing technique. Should an aircraft miss a wire it could take off again but it would mean that the aircraft approach would need to be similar to that for the flexible deck. It would have to land with the engine on power. It also implied the doing-away of the safety barrier as a means of protection for the deck park. A barrier could be developed into a means of arresting a damaged aircraft and could be positioned further aft than current practice. It could also be given a long pull-out thereby lessening the danger and damage to both aircrew and machine in its use. (A further point, missed at that time but which has been justified historically, is that fewer arrester wires would need to be rigged across the deck.)

The problem was also addressed by Capt D. R. F. Cambell who submitted a paper (*ARC 14,283 'High Speed Aircraft – the Arresting Problem, October 1951'*) to the ARC in October 1951. During the

HMS *Eagle*[51] with the 8½° angled deck as she appeared to her pilots in April 1965. The bow and waist steam catapults are seen, each with its mechanical centring unit and retractable jet-blast screens. No.1 arrester wire crosses helipad 7, while No.4 wire crosses helipad 6. The wire for the single DAG installation crosses the deck some nine feet abaft No.1 wire, close to the forward edge of the after lift. The hatched areas each side of helipad 6 are the anchorage points of the undrawn nylon energy absorbing packs for the safety barrier. The 24ft high masts for barrier net support are seen stowed alongside, outboard of the anchorages. The projector landing sights are located one immediately abaft the island and the other on a port-side sponson alongside the head of the waist catapult. April 1965. (© Crown Copyright/MOD)

previous summer, Cambell had discussed the problem with Boddington. Cambell's paper stated that there was general agreement that the idea showed considerable promise and that it would not be difficult to implement. From the flying aspect there seemed to be no particular objection to the scheme, with much to be said in its favour. It was felt that a practical investigation into the system should be arranged to obtain the views of pilots in operating and developing the proposed new technique through 'touch and go' trials procedures. An initial programme was quickly prepared and trials were conducted in HMS *Triumph* during February 1952. Grimes' report (*#251*) was submitted to the ARC in March 1952. From this moment it was 'all systems go' and, before long, the angled deck became established as a permanent feature, universally adopted by all navies operating fixed-wing carrier aircraft.

Marriott (*#63 p.42ff*) records that the first trials with an angled deck were conducted aboard the light fleet carrier HMS *Triumph* which, on recommissioning in 1952, was provided with a flight deck centre line painted offside to port at an angle of 10°. None of the flight deck machinery was altered to suit the new configuration and the flying trials comprised 'touch and go' landings only, with the arrester gear unrigged. Sea trials of the de Havilland Sea Venom (XP539) were conducted aboard HMS *Albion* using the angled deck and an early pattern of the mirror-landing sight during August and September 1954. Twenty landings and accelerator launches were made during this period. The first ship to operate with an operational, purpose-built angled deck was HMS *Centaur* (1947–72) from 1955. The centre line of the deck was angled at 5½° to port with the arrester gear re-aligned to conform to this orientation. Subsequent ships were built with varying degrees of offset, the last of these being HMS *Hermes*[59] in her first commission with 6½° angle. HMS *Victorious*, in her new commission from 1958, was the first ship to combine the new steam accelerator with an angled deck, the latter being at the relatively wide angle of 8¾°.

The skewing of the centre line of the flight deck by a few degrees to port would enable an aircraft, if missing the wires, to continue along a clear flight path, by-passing the deck park, and to climb away to re-enter the landing circuit around the ship – in Navy terms, 'doing a bolter'. A further advantage to be gained was that only a single safety barrier need be deployed, and that for emergency use (viz damaged hook, undercarriage failure, and so on). Such an installation could be reeved to provide a 'comfortable pull-out' – a long, gentle arresting force, comparable with that from the normal arrester gear. As a permanent feature, the inevitable modification to a ship would do nothing towards improving the aesthetics of its silhouette but carriers were already an ungainly breed. It was thought that a further element of asymmetry as an operational necessity could be tolerated. For the reader who wishes to pursue this aspect of carrier aviation in greater detail, Friedman (*#34*) provides technical information and progressive history of the ships that were provided with angled decks.

Boddington's paper (*#ibid.*) must have been his last prior to his transfer to Headquarters a month later. His successor, A. W. Hotson, joined NAD from Structures Department. A couple of Hotson's papers continued the thread of the angled deck and in one of them (*#257*) he made a most strange proposition. He advocated a 12ft high steel fence, 200ft long {3·7m, 61m} to be erected alongside the starboard border of the landing-on area of the already angled flight deck. This was to have been a rigid barrier to protect aircraft in the forward deck park. It is difficult to understand his reasoning here as he seemed to have neglected the airflow pattern over the after end of the flight deck as well as the characteristic procedure for recovery and striking down of aircraft. The aerodynamic disturbance was difficult enough for pilots, in

their approach, having to contend with the wake behind the island and the stern of the ship, as well as the hot gases discharge from the funnel. A smokeflow pattern over a wind tunnel model would have proved enlightening.

A couple of months earlier in June 1953, this same author (*#256*) produced a comparative study of carrier deck layouts. Of six diagrams illustrating his points, four are shown with the island structure along the port side of the ship, the other two are traditionally correct. (It is thought that there were only two carriers with islands on the port beam, both were Japanese – IJNs *Akagi* and *Hiryu* – both of which were destroyed at Midway on 4 June 1942.) For some reason Hotson considered a portside island might have contributed some aerodynamic improvement; it could only have produced a mirror-effect as the wake would still have been there, but to the opposite hand!

The author may have contributed to the angled-deck idea, even though the credit has been awarded to others. The author's case rests upon the fact that during 1945 a large model of an 'Audacious' Class carrier – HMSs *Eagle*[51] and *Ark Royal*[55] – occupied a bench alongside his office desk. He was given the task of arranging for a number of jet aircraft models to be produced to the same scale for simulating deck parks and so on. He prepared a drawing of a 'futuristic' jet-plane and this drawing was passed across to the RAE pattern shop and foundry to produce these models as aluminium castings. He had prepared a large, plywood cut-out, copying the deck layout for another purpose, and when the aircraft models arrived they were randomly placed on the model carrier deck. The plywood mock-up was skewed across and the haphazard placing could have triggered the idea of the angled deck in someone else's mind, even though the principle was not adopted until some years later.

Dennis Cambell, to whom the credit is sometimes given for the invention of the angled flight deck, retired from the Royal Navy in 1960 with the flag rank of Rear Admiral. His death in April 2000 at the age of 92 occurred while material for this book was being researched.

VISUAL LANDING AIDS

Coupled with the development of the angled-deck project was the introduction of the Optical Glide Path Indicator – the automatic Mirror Landing Sight. This did away with the need for an experienced pilot acting as a Deck Landing Control Officer or 'Bats'. An operational need for a device of this nature was expressed by the pilots involved in the Carpet Landings project in the late 1940s and it was logical that such a need should be rapidly satisfied. Deck-landing was becoming more hazardous through the increase in closing speed to the ship and touchdown of faster and heavier aircraft. Pilots had a tendency to make their points of shipboard contact progressively further forward up the deck. It was thought that the DLCO/pilot relationship was rapidly reaching the limits of physical practicability.

This path had been trodden before. As a solution to an earlier problem, Hobbs (*#45 p.96*) refers to an Autumn 1936 installation in HMS *Glorious* where:

'Night flying trials (used) new visual landing aids designed on board (these comprised pillar lights and sector lights, both concepts still in use today).'

A decade later, the first intimations that something new in the field of guiding aircraft on to a deck – which is moving forward and possibly pitching and rolling at the same time – appeared in an ARC Paper (*ARC 9,020 'Report of the Naval Aircraft research Committee – 10th October 1945'*). Discussion had been made on the advisability of attempting to automate the deck-landing process, the first proposal being the locking of the aircraft on to a radar beam forming the approach path. At that

An early RAE design of mirror-type visual landing aid used on Farnborough's Runway 04/22, seen here on a transporter at Jersey Brow. February 1954. (© Crown Copyright/MOD)

time such an idea, technically, was too far advanced in concept and would require extensive trials before a system could be adopted with any reliance. The Committee felt that a 'talk-down' system should be developed, with the batsman (DLCO) being still in control, his ability to tell the pilot what to do possibly being better than any form of automatic control, both in advice and response. This Paper also raised the point whether the three-colour 'sector-light' or 'sector-radio' should be dismissed from discussion. The ARC declared its willingness to welcome a report with regard to other forms of automatic or other control which would help in landings, both in rapid recovery of large numbers of aircraft, and in individual recovery under bad visibility with rain having obliterated the radar screen.

The Admiralty, in 1947, set up a sub-committee to advise its Radio Approach and Landing Committee on the problems of blind landings on aircraft carriers. In spite of the multiplicity of radio aids which might have been combined to give automatically controlled landings, the sub-committee could not clearly assess which system would be most suitable. Its recommendations pointed to further investigations having to be undertaken and operational experience gained at sea.

An early example of the RAE design mirror landing sight aboard an unidentified ship. Although the prototype equipment was installed in HMS *Illustrious*[40], this installation probably was in a light fleet carrier, HMS *Bulwark*, which took over duties as trials carrier from HMS *Illustrious*[40] in 1954. (Fleet Air Arm Museum)

A production model of the mirror landing sight. This installation was at RNAS Brawdy for training purposes and shows, in reflection, a Fairey Gannet approaching the landing area. An interesting feature is the anchor chain alongside the runway. This was the braking element of the Drag Chain Arrester Gear researched at Farnborough in 1943. December 1956. (Author)

Duddy and Lean addressed this situation and produced an RAE paper (*#233*) in November 1951, which was also presented to the ARC. In this they stated that it seemed unlikely that a completely automatic landing would be practicable in rough seas but that partial automatic control during the early stages of approach appeared to be worth further investigation. They reduced the technology of the approach and landing process to a number of fairly independent constituents, each element being a step towards the completely automated landing. They identified that control of closing speed and of position in the horizontal plane of approach would be more easily attained than control in the vertical plane, even though this last would represent the greatest single advance towards automatic landing. Complete control of the approach up to the 'cut' was expected to be simpler than determining the actual point of the 'cut' command and the type of manoeuvre following it. They suggested a relatively simple form of control to provide assistance to the pilot in the form of:

a) azimuth control by a beam radiated from the carrier
b) height controlled by a radio-altimeter in the aircraft
c) a 'cut' signal given when the aircraft was a specified distance from the carrier
d) control from the 'cut' to touchdown entirely by the pilot – as at present.

With the advent of aircraft with closing speeds of the order of 100 knots {51m/sec}, the problems of deck landing were becoming more acute. Two such problems were recognised by Cambell and Goodhart in a paper (*ARC 14,467 'The Deck Landing Problem, December 1951'*) presented to ARC in December 1951 (a month after Duddy and Lean's proposals). Their points were:

a) the inability of the batsman to perceive small errors in approach path coupled with his and the pilot's reaction time
b) the shorter length of deck available for landing due to the greater arrester pull-outs necessary for high closing speeds.

These amounted to the fact that, while the difficulty of accurate control of the aircraft during the final approach had increased markedly, the accuracy required also had increased correspondingly.

Attempting to solve the first problem, Cambell and Goodhart proposed, as an initial step, giving the batsman a gyro-stabilised sight so that he became aware of any errors in the approach path. The next step would be to eliminate the batsman and give the pilot direct information (preferably visual) of errors in the approach path. The final proposal was to 'eliminate' the pilot by fitting an autopilot coupled to a radio beam (a re-hash of the proposal of six years earlier?). The authors declared this ultimate solution to be by far the best but considered that a very considerable time would elapse before a workable set-up could be achieved. As the problem was already beginning to obtrude, it was thought that one of the earlier steps should be investigated.

The authors proposed that the ideal line of flight be indicated direct to the pilot by visual means, thus eliminating the batsman – who 'was only a delay mechanism for doing the same job'! This was to be achieved by a mirror located at or near the null point of the ship on which would be painted a horizontal line. Near the stern would be a powerful light source shielded in all directions except towards the mirror. The pilot approaching the device would see the reflected light bisected by the horizontal line. If he was too high he would see the light above the line and vice versa.

By this time Lean (*#266*) himself, in addressing the subject, presented a volte-face by completely abandoning the radio/radar concept, turning his efforts towards a visual, direct line-of-sight system. Within a matter of days he – as all good entrepreneurs – had acquired some items of hardware and was experimenting with them on the Farnborough airfield. For the record, these articles comprised two poles,

The Mark 3 version of the projector landing sight, developed by the General Electric Company, Wembley, installed on the Thurleigh arrester deck. An experimental HiLo display unit is seen alongside. November 1959. (© Crown Copyright/MOD)

An early example of the projector landing sight fitted to the port side in HMS *Ark Royal*[55]. A development of the auxiliary HiLo projector, which subsequently replaced the circular lamps, is seen alongside. March 1961. (© Crown Copyright/MOD)

four sheets of tinplate and sodium lamps from a pair of airfield lighting units. Lean wanted to determine experimentally the workability of his ideas. First, it was not certain that a pilot would be able to appreciate and absorb height-error feedback from a source not precisely on his direct line of flight and, second, it had to be demonstrated that such a new deck-landing technique was feasible.

The hardware for this exercise consisted of four sheets, 6ft x 1½ft {1·8m x 0·5m}, of polished tinplate mounted on two vertical poles 10ft {3m} apart. Its presentation in elevation would be similar to the flag of St George – by the reflective (argent) tinplate sheets as the 'field' and the red (gules) cross simulating the gaps between the plates. The planes of the four plates were inclined at 45° to the horizontal in order to reflect the natural daylight from the sky into the horizontal direction. A gap of four feet width {1·2m} represented the vertical line of the cross while the complementary horizontal gap was only one foot {300mm} in height. The two sodium lamps were located behind the gaps at calculated positions from which they would be seen by the pilot in making his approach at the correct 'angle-of-glide', in this instance 3° to the horizontal. (The technical reason for two lamps is that only one lamp would be visible from the start of the approach – some two miles {3·2km} astern of the 'ship' – with the second lamp coming into view for the last few moments as the pilot approached the round-down.) By March 1952, this jury rig was set up on the starboard side (east) of Runway 04/22 at Farnborough, with a line painted across the runway to represent the after round-down of a carrier deck. Over a period of three months' experimentation some sixty landings were achieved.

At an early stage in these airfield trials the results were sufficiently encouraging for it to be considered appropriate to proceed with the design and construction of equipment more suited to the rigours of service use and for trials at sea. This was put in hand. However, HMS *Illustrious*[40] was available for trials in July 1952 and the temporary rig was conveyed to the ship for the first sea-borne trials. It was rigged for operation with a de Havilland Sea Vampire and orientated so that the aircraft would overfly the after round-down with a clearance of some 8ft {2·4m} and touch down with a vertical velocity of 8ft/sec {2·4m/sec}, 100 feet {30m} forward of the round-down.

The first set of permanent equipment, being a purpose-made polished metal concave mirror (with the lamps positioned on an outrigger abaft the mirror), underwent trials at Farnborough and was transhipped, again to HMS *Illustrious*[40], for trials during October 1952. This again proved highly successful – and the 'Optical Glide Path Indicator', or mirror sight, was born! The principle of this equipment was that, to enable him to get quickly on to the predetermined glide path and to remain near it throughout his approach, the pilot needed to receive a signal which told him, continuously, not only when he was actually on the correct path but also by how much he may have deviated above or below it. With the Optical Glide Path Indicator, this signal came from a large mirror situated on the deck slightly ahead of the required point of touch-down; nominally aimed at No.5 arrester wire. A fixed light on the deck towards the stern of the ship, some 150ft {46m} abaft this mirror, was reflected by it so that, when seen by the pilot during his approach, it appeared in the middle of the mirror when he was on the correct path. The light would be seen to move upwards on the mirror when he was above this path and downwards if he was below it. It was this apparent movement of the reflected light that gave the pilot the progressive indication of his distance above or below the correct glide path. If the pilot was too far above or below the correct path the light disappeared completely and this disappearance may be interpreted as the equivalent of the batsman's wave-off. To achieve the desired approach angle of 3° – to be a direct line, without a final 'flare' – the mirror was tilted from vertical and gyro-stabilised to hold this angle against the ship's pitching motion.

By February 1953 the stabilised mirror system, as the Mark 2 unit, was well advanced experimentally in conjunction with John Curran Limited of Cardiff. Grimes (*#251*) was pressing for its adoption, with

aircraft closing speeds expecting to reach 115 knots {59m/sec} within a short period. In 1953, the GEC Research Laboratories at Wembley were invited to assist in improving this equipment, particularly the source lights and the datum lights. Eventually a smaller mirror was found to be satisfactory and, together with the improved lights, the Mark 1A mirror sight was produced. It was more compact than the original equipment and continued to be in use until its replacement some five years later by the projector sight.

In July 1954, NARC Paper (*#17,022*) reported the Optical Glide Path Indicator as being greatly developed with numerous carrier landings having been successfully achieved. The Paper also pointed out that, although the trials completed so far had established the effectiveness of the system, there was still much that remained to be done to obtain the best results under operational conditions. Resulting from the initial shore and shipborne trials NARC, in the same Paper, expressed the opinion that:

'the method of landing on carrier ships in which the pilot is directed by signals from an Optical Glide Path Indicator, instead of from a Landing Signals Officer, is of such potential value to the future of Naval Aviation as to warrant a considerable effort to bring it quickly into service'.

Friedman (*#34 p.305*), without acknowledging the contribution made by Farnborough, records that Lt/Cdr H. C. N. Goodhart introduced this system from simple principles in 1951, yet most of the original research and development work was carried out by RAE. After initial trials by Aerodynamics Department at Farnborough, the prototype sea-borne installation was installed in HMS *Illustrious*[40] in November 1953. (The US Navy adopted the system in USS *Bennington* in September 1955.) Marriott (*#63 p.20*) states that trials began in October 1952, this being the date relating to the Farnborough prototype. More than 100 landings were made and the installation was approved for general use in 1954 and universally adopted thereafter on all carriers with angled decks. While early experiments were conducted with the mirror sight positioned on the starboard side of the flight deck, pilots quickly expressed their preference for the equipment to be mounted on the port side – the location which the DLCO once occupied. The result was that installations were located on both sides, one functioning as back-up for its mate. Haynes, in August 1955, reported to the ARC (*ARC 17,790 'Note on the Operational Progress of the Deck Landing Mirror Sight Fitted to Angled Deck Carriers, August 1955'*) that from July 1954 four carriers had been equipped with this system, with another following, and that of 3,525 deck landings having been completed using the mirror sight within the first year of use, only six were recorded as 'incidents', unrelated to the functioning of the system. It is thought that the first permanent installation of this type, in conjunction with the angled deck, was in HMS *Albion*. Each succeeding installation progressively became the subject of development in detail.

After a few years of successful use it was realised that the mirror sight had a number of optical and mechanical disadvantages which would make it obsolete and incapable of meeting later requirements. Ferguson, of the GEC Illumination Department, (*#31 p.73*) listed these as:

1   The source lights tended to floodlight part of the ship's side at night and the light spillage around the periphery of the mirror could be seen several miles ahead of the ship. They also had to be mounted on foldable booms over the ship's side, thereby making access for lining-up and servicing rather difficult.

2   The stabilisation was not satisfactory due to the considerable mass of moving parts and to the relatively high accuracy required in stabilising a mirror which doubled the angular errors in the reflected light beam which the pilot had to follow.

3   The azimuth light distribution, though satisfactory by day, tended

to cause dazzle at night when the aircraft approached the ship. If the intensity of the image was adjusted for comfortable viewing at a range of one mile, it became uncomfortably bright at close range.

4 The raise/lower equipment did not allow sufficient travel to deal with the increase in hook/eye-ball distances of the latest aircraft.

A further Mark 3 system was developed by RAE (in conjunction with the GEC Laboratories at Wembley) as a Deck Landing Projector Sight (DLPS) using a Fresnel lens system. Ferguson (#*ibid. p.73*) states that in 1956 GEC, under a Ministry of Supply (RAE) contract, proposed a device which, whilst retaining the same presentation as the mirror sight, promised to eliminate most – if not all – of the current snags. Basically, the proposal was to eliminate the source lights and replace the mirror with a vertical stack of projectors of the same height as the original mirror. Field trials with a crude mock-up were encouraging and in 1958 the company was awarded a development contract.

Lewis (#*60 p.94*) describes this system, and much of the following is the result of his researches. Although the mirror system claimed a measure of success, it was decided to dispense with the remote light source and to replace it with a vertical array of lamps aimed directly along the flight path. This projector sight employed Fresnel lenses to give a narrow and high intensity directional beam of light. The presentation produced two phases: the Outer and Inner Approaches. When the aircraft was still well out on the circuit, the 'Outer' guide lights came into view, projected from boxes known as the 'HiLo' units. A multi-coloured array would be seen, changing from white to red through pink, depending respectively upon the pilot's position above, below or on the correct flight path. The deviation was within ±½° of elevation. As the pilot came to within about 3,000 yards {2,750m} of the carrier, the 'Inner' approach system came into view. This portrayed a yellow spot of light ('Meatball') which moved up and down relative to a horizontal datum bar of light. This again was an indication of a departure from the ideal flight path and, additionally, provided guidance to within ±¾° in azimuth.

The Fresnel system produced a composite beam of light which was wide horizontally but very narrow vertically. When on the correct path, all indication to the pilot was central; deviations became readily apparent by the diversion of the relative light component from its median position. The horizontal spread of the fan of light in total was 40°, being 25° to starboard and 15° to port relative to the fore and aft orientation of the ship. With an angled flight deck of 5° to port, the projected image became central to the flight path. It must be remembered that the approach path to an angled deck would mean that the aircraft some 1,000 yards {900m} astern could be as much as 300–450ft {90–130m} to starboard of the ship's centre line; this situation was covered in the design and installation of the projector sight system.

The first Mark 3 projector sight consisted basically of a boxlike fitting, housing twelve projector lights, flanked on each side by three green datum lights and a red wave-off light. The entire unit could be raised or lowered over a spread of 6ft-6in {2m} to cater for the range of hook/eye distances (the height of the pilot's eyes above the arrester hook) of different aircraft. This full range adjustment was achieved within 20 seconds. The array was positioned 200ft {61m} forward of the deck aiming point. The system was stabilised to cater for pitch and roll of the ship. The complete unit weighed 3½ tons, was 13ft-6in wide and 13ft high {4m x 4m} when in fully raised position. A much heavier and bulkier silhouette than a batsman! As in the days of the DLCO, the red light wave-off was a mandatory order to a pilot on his approach.

The projector box contained twelve lamp systems, each comprising a 150 watts lamp-reflector assembly incorporating a sophisticated lens system – similar in principle to the domestic slide projector with the slide replaced by a calibrated slit. The resultant beam was fan-shaped with a wide horizontal spread. Each projector unit was displaced angu-

A development of the Mark 4 Projector Landing Sight atop a bomb trolley for shore-based trials at Thurleigh. The horizontal base-line lamp arrays contain the HiLo units. June 1963. (© Crown Copyright/MOD)

Supermarine Seafire F.47 (PS948), carrying a variety of stores, ready for launch on the Jersey Brow BH.5 Catapult. October 1947.
(© Crown Copyright/MOD)

larly with respect to its neighbour by 8 minutes of arc, so that the total vertical angular sweep between top and bottom projectors was 88 minutes of arc (approximately 1½° – similar to the range of the mirror sight). The optical centre line of the whole system was canted 4½° upwards – the accepted glide-path angle of approach of the aircraft during recovery to deck.

This equipment was first used in HMS *Victorious* in her 1958 rebuild. It was a practical and logical development of the Mirror Landing Sight, which it eventually replaced. The RAE dummy deck (Runway 04/22) was used extensively for the initial trials of these installations. The first shore-based installation for service training was at HMS *Peregrine*, RNAS Ford. By 1965 the projector sight had been developed in conjunction with the GEC Hirst Research Centre to a sophisticated Mark 4 model and was universally adopted as standard. Rapid vertical adjustment of the unit to suit the variation in height of the pilot's eyeball above the throat of the arrester hook for each aircraft marque, together with the automatic gradation of the light intensity during approach were important features of this equipment. The system suited the pilots as, with the audio air speed signal fed into their headsets, they could best judge their approach to the flight deck and function completely independently of the DLCO ('Bats'). In fact, with the universal adoption of this facility, Bats rapidly became redundant and the post ultimately was abolished, with the sight landing officer replacing him.

Beaver (*#5 p.138*) and Hobbs (*#45 p.19*) both suggest that a Landing Signals Officer (LSO), otherwise termed Mirror Control Officer (MCO), sat alongside the sight, monitoring the aircraft approach through a telescope, bore-sighted to the ideal glide slope. He was able to talk to the pilot and wave him off if necessary by flashing red lights. This Bats replacement remained with the FAA until the operation of fixed-wing aircraft ceased with the paying-off of HMS *Ark Royal*[55] in 1978.

Complementing these developments was an audio system to advise the pilot of his closing speed. This was thought not to have been developed at RAE but it was quickly brought into use with the Farnborough installation of the Mirror – and later Projector – Landing Sights. The editorial in *The Engineer* (4 March 1955) states:

'LANDING AID FOR FLEET AIR ARM – As a further contribution to the safe landing of aircraft on to the flight deck of aircraft carriers, a new sound device for indicating air speed has been developed. Known as Audio, it works on the principle of an electric organ and gives its assurance as to correct speed in the pilot's earphones as a background signal.

'It is made in two separate parts. One of these parts consists of a "sensing" unit to be attached to the pitot head, the other is the sound-producing box, consisting of a "two-note organ", working on the same principle as an electric organ.'

The author comments, 'From the first, deck landing has always been an 'azard!' This point would be understood by an organist: 'Nazard' is the name of an organ stop!

It is thought that, although no Royal Navy ship now possesses flight deck machinery, the four vessels currently operating VSTOL aircraft and helicopters use a projector sight to bring the aircraft to a point off the port beam for recovery to the flight deck. A development of the system is also used for homing helicopters to their non-carrier parent ships

BH.V ACCELERATOR (BH.5 CATAPULT)

With the BH.III Accelerator approaching its limits of mechanical development, attention was given (*c.*1943) to the development of a similar hydro-pneumatic accelerator of increased capacity. This was required to

Grumman Avenger III (KE446) in the process of being loaded onto the Jersey Brow BH.5 Catapult. The holdback is attached and the bridle is held against a pair of loading dogs in readiness for the deck shuttle to be inched forward to engage the bight of the bridle. September 1949. (© Crown Copyright/MOD)

have the capability of launching an aircraft of 30,000lb AUW at 75kts {13,600kg, 39m/sec} (Marriott (#63 p.56)). These were to be fitted to carriers of the 'Albion' and 'Majestic' class which had their keels laid in 1942–4. This proposal, becoming known as the BH.V Accelerator, appeared feasible in concept and, soon after the end of World War 2, designs were put in hand for a cavernous underground chamber to be constructed on Jersey Brow to contain all the machinery for the project.

At this point NAD felt that an interim facility, using some of the large nested tubes from the telescopic ram unit of the abandoned Mark III catapult at RAF Harwell, could be transported to Farnborough for use as a temporary 30,000lb {13,600kg} accelerator. This project was to be located in a pit alongside the C.I Accelerator, with the new BH.V unit in parallel alignment further westwards. The proposed 30,000lb installation became bogged down with problems, mostly from attempts to adapt the untried, ex-Harwell hydraulic control system. A few launches were made but the installation was mostly used to develop systems for bridle catching. It was superseded as soon as the BH.V equipment was ready for use.

Friedman (#34 p.292) states that the design limit for this accelerator was originally set to give an acceleration rate of 5g with the end speed at 120kts {62m/sec}. Since 130kts {67m/sec} speed was required it was agreed to extend the maximum rate to 6g but even in 1945 it was hoped that the figure of 7g would eventually be acceptable. This figure compares with the rate in operation some five years earlier when 3¼g was the norm from the BH.III equipment; the Americans came in shortly afterwards with the figure of 4¼g with their tail-down method of bridle launching on the AH.IV Accelerator.

The development of the machinery for the prototype installation at Farnborough was undertaken by Brown Bros of Edinburgh, an engineering company already well-versed in this field. Dugald Stewart (a life-long designer with Brown Bros, whose death was recorded during

the writing of the closing paragraphs of this volume) recorded the hydro-pneumatic drive for this equipment as comprising a cylinder of 40in {1m} bore with a length of almost 15ft {4·6m} with a rope reeving of 8:1. This produced a working stroke at deck level of 119ft {36m}. This BH.V Accelerator became the largest complex of under-deck machinery located on Jersey Brow.

The installation at Farnborough was completed and brought into use in October 1946. After a series of dummy runs the first aircraft to be launched on the 28th was a Grumman Avenger (JZ298). For many years this accelerator was the regular workhorse for NAD use at RAE, replacing the earlier, wartime equipment on Jersey Brow. All the immediate post-World War 2 generation of FAA aircraft, including the Hawker Sea Hawk, de Havilland Sea Venom, Fairey Gannet and Westland Wyvern, completed their trials on the BH.V Accelerator. The last use of this equipment at Farnborough was on 20 September 1957 when the 3,343rd recorded operation was the launch of Sea Venom FAW.21 (HG656) at about 100 knots, piloted by Lt Hefford RN.

One of the non-FAA aircraft specially adapted for bridle launching was a twin-engined Bristol Brigand (RH748), for no reason other than that it had an all-up weight approximating to 30,000lb, the design limit for this accelerator. There was no indication that the Brigand was intended to be carrier-borne! The first experiments began in November 1949 and continued for some six years.

Another instance of the use of this accelerator was the launching of aircraft with the simultaneous use of RATOG, an activity soon to become known as 'CAT-RAT'.

The BH.V Accelerator was also used as the prime mover for hauling aircraft into the safety barrier systems being developed at Farnborough. This was achieved by the use of about half a mile {0·8km} of rope laid, through pulleys and so on along Runway 04/22 between accelerator and arrester gear. Baines (#205 p.4) states that the energy developed by this

De Havilland Sea Venom FAW.21 (XG656, Code FO 524) of 700 Squadron, RNAS Ford, is seen at the moment of launch from the BH.5 Catapult on Jersey Brow in September 1957. With this, the 3,343rd and final launch of this catapult, all NAD activity at Farnborough was brought to a close. September 1957. (© Crown Copyright/MOD)

unit exceeded 10,000hp {7·5Mw}. On transfer of Naval Aircraft Department to RAE Bedford the installation was dismantled and scrapped.

By 1955 it had been recognised that the Farnborough prototype of the BH.V was likely to be the last in the line of jigger-type launchers. An NAD paper (*#310*) of 1956 records the greatest single shortcoming of the hydro-pneumatic under-deck gear was that of increasing weight and inertia of its moving parts, with the unavoidable increase in demands for power. This, inevitably, led to a state where an unreasonable proportion of energy was absorbed in accelerating these moving parts. Retardation of this moving mass at the end of the launching stroke, and problems associated with wire ropes of ever increasing size and stiffness were becoming obstacles to further development of such machinery. Something new was required – the age of the Steam Catapult had dawned.

The transfer of NAD to Bedford began, according to Bowyer (*#11 p.69*), in 1955. Turnill (*#101 p.128*) gives the date of the transfer as 1958, which is regarded by the author as the completion date of a phased transfer. By this time the family of hydro-pneumatic launching devices was receding into the background through the development of what quickly became known as the Steam Catapult, in reality the

'BS.IV Steam Accelerator' (qv), a significant contribution to the art and science of Navy Flying.

Hobbs (*#45*) records that three carriers commissioned in the early post-war period, viz HMSs *Magnificent* (1948), *Eagle* (1951) and *Albion* (1954), were equipped with the BH.V Accelerator, the first having a slightly reduced capability from the other two. Chapman (*#20*), includes HMS *Bulwark* – replacing HMS *Magnificent* – as the third ship to have been equipped with this accelerator. Garstin (*#242*), who attended the commissioning trials of the two units installed in HMS *Eagle*[51] in 1952, states that the shipborne equipment – although similar with regard to the major items of machinery – differed in many respects from that on Jersey Brow. Benefits gained from the initial trials on the Farnborough catapult (serially numbered by Garstin (*#ibid.*) as BH.5(1)) had been introduced into the shipborne units, particularly in the control system. Advantage was also taken of the availability of greater hydraulic power on board ship. Due to the distance from the plant in the 'ram rooms' to the flight deck, the equivalent weight of the moving element of the ship's catapult was 17,500lb as opposed to Farnborough's 16,400lb {7,900kg, 7,400kg}. HMS *Eagle*'s[51] BH.5(8&9) installations were replaced by two BS.5 steam catapults during the ship's major refit completed in 1964.

POST FARNBOROUGH 1955–1978

THE TRANSFER TO BEDFORD

During the mid-1950s the exercise was put in train for implementing the transfer of facilities and the staff complement from the Naval Aircraft Department at Jersey Brow, Farnborough to the newly developing National Aeronautical Establishment (NAE, later RAE Bedford) on the Thurleigh/Twinwood Farm/Little Staughton complex to the north of Bedford. On a virtually greenfield site (incorporating three disused airfields) the constraints with regard to space in the air and on the ground were few. This was exploited to the advantage of the Department. The first elements of the transfer were set in motion in early 1955.

With various units of new flight deck machinery being installed and brought into service at Thurleigh, the corresponding, redundant equipment on Jersey Brow was being decommissioned, struck off charge and scrapped. However, there was no great urgency for this; in fact on a number of occasions a delay proved to be advantageous. The records of Farnborough aircraft movements show a number of aircraft over this period being virtually on a commuter run. First, for example, the machine might be catapulted at Farnborough, then flown to Bedford for arrester trials and back to Farnborough again for barrier trials or one or another variation of such procedures. These were brought to a conclusion on 20 September 1957 with the final launch of a de Havilland Sea Venom FAW.21 aircraft (HG656) from the BH.V Accelerator on Jersey Brow. With the ultimate dismantling of this equipment the era of NAD at Farnborough was brought to a close. It is conceded that the transfer of the Department from Farnborough to Bedford occupied a roll-over period approaching four years (1955–1958).

Of the three World War 2 airfields (Thurleigh, Twinwood Farm and Little Staughton) forming the Bedford establishment, Thurleigh was developed as the primary centre of flying activities.

Thurleigh – a wartime airfield – came into operation as a bomber station early in 1942 and by September had become the home of the 306th Bomb Group, USAAF. Early in 1946, work began to transform the site into the first stage of the massive new complex forming the National Aeronautical Establishment. The main Runway 09/27 was a completely new construction along the south side of the earlier airfield. A secondary Runway of 06/24 orientation was built, crossing the main runway at about mid-length. The earlier parallel Runway 07/25 was later taken over by NAD as the Arrested Landing Deck experimental area. To the north, and on the same 25 orientation, was the Catapult Site accommodating both the 'Raised' and 'Flush' catapult installations. A portion of the original Runway 18/36, near the intersection with the by then abandoned Runway 12/30, latterly was set aside for the VSTOL pit built for the Short SC.1 trials and the development of the Hawker P1127 aircraft into the Sea Harrier.

In addition to the office and laboratory facilities a dedicated workshop was provided to serve the new NAD site. Outline details of these facilities are given in a publicity brochure (#318) issued c.1970 by an anonymous author. (At the time of publication of this brochure, the wartime differentiation between 'accelerator' and 'catapult' had been abandoned completely in favour of the general term 'catapult'; likewise, the towing strop had become the 'bridle'.) The technical details given below are based upon this brochure.

The brochure summarises the primary functions of the Department as:

a) The determination of mechanical, structural, dynamic and aerodynamic compatibility of aircraft with catapults, arrester gears and other flight deck machinery.
b) The design and development of equipment to assist the pilot in operating from aircraft carriers.
c) The design and development of arresting gears and safety barriers for use on board aircraft carriers and for military use ashore.
d) The design and development of arresting systems for civil aircraft to prevent over-run accident on runways.
e) The design and development of catapults and ancillary equipment for launching aircraft.

Its *raison d'être* was declared in the statement:

'To ensure operational acceptability, aircraft and equipment covered in the foregoing must be subjected to extensive testing and evaluation and the tests conducted at the NAD simulate actual shipboard or operational conditions as close as possible.'

This declaration marginally extended the objective and activities of the department from its World War 2 inception at Farnborough, some three decades earlier.

The Naval Air Department site occupied some 120 acres {49ha}, located in the north-west area of the airfield, with the flying area running parallel with and to the north of Runway 07/25. The site was identified as five specific areas:

a. Catapult Site.
b. Catapult Aircraft Loading Equipment Base (CALE).
c. Arrested Landing Deck (ALD).
d. Safety Barrier and Arresting Gear Proving Base (SBAG).
e. Main Office Block, Workshops and Site Facilities.

CATAPULT SITE

Two steam catapults were installed here, lying parallel with each other on a Runway 25 heading. A paved runway extended 2,500ft {760m} ahead of the catapults and a safety barrier normally was rigged 350ft {107m} from the 07 end of the runway.

a. THE RAISED CATAPULT (BXS.4) was normally employed for the evaluation of aircraft and the proof-testing of airborne stores and equipment. The basic machinery for this installation was that recovered from HMS *Perseus* on completion of the shipboard prototype trials. The operational deck of this catapult was mounted some 6ft {1·8m} above ground level, reputedly to simulate the launching from a carrier. (In real life, the flight deck of a carrier would have some 40ft {12m} of freeboard!) The installation latterly was equipped with a van Zelm bridle arrester which provided controlled arresting of the aircraft bridle after launch. The aircraft loading area was protected by steel plate decking for the operation of aircraft with reheat. The maximum power stroke available was 203ft {62m}, a further 6ft {1·8m} being used in the retardation of the piston and shuttle assembly. Typical performance figures for dead-load weights of 30,000lb and 50,000lb {13,600kg & 22,700kg} would be with end speeds of 155kts and 127kts {80m/sec

& 65m/sec}, with maximum accelerations of 6·0g and 4·2g respectively. (It may be difficult to understand that a length of only six feet {1·8m} is allowed to bring to a stop a shuttle travelling at 155kts {80m/sec}; this means that a braking force imposing a rate of retardation of 177g is applied. To the uninitiated this means that for every 12½lb {5·7kg} of weight in the shuttle an applied force of 1 ton is required to bring it to rest.) The first dead-load launch of this catapult took place in April 1954; the first live launch with an aircraft followed in November 1955.

b. THE FLUSH CATAPULT (BXS.5) was used primarily for catapult component development trials but could also be used to launch aircraft. (It is thought that this installation originally was designated as the BXS.2 and was developed progressively during its life at Thurleigh as the prototype for each change of mark in service installations.) For catapult development trials, aircraft launch conditions were simulated by the use of a dead-load trolley loaded up to 56,000lb {25,400kg}. This trolley ran in rails along the length of the catapult and was arrested by passing through a trough of water 550ft {168m} long, followed by engagement with the centre span of a conventional hydraulic ram and cylinder energy absorber situated 140ft {43m} beyond the end of the water trough. The water trough would be covered with steel decking plates to permit the launching of aircraft from this catapult. The maximum power stroke was 203ft {62m} and, using maximum steam pressure of 750psi {50bar}, typical performance dead-loads for 30,000lb and 50,000lb {13,600kg & 22,700kg} would be with end speeds of 190kts and 154kts {98m/sec & 79m/sec} and with maximum acceleration of 8·8g and 6·0g respectively. This catapult was located with its launching shuttle at ground level with its machinery in capacious underground chambers. The first dead-load launch with this catapult took place in May 1960 with the first aircraft launch following in April 1964.

Steam for these catapults was provided by two boilers, a Foster-Wheeler marine type boiler serving the Raised Catapult with steam at a pressure of 400psi {27bar} and a John Thompson Lamont boiler serving the Flush Catapult with an output of 120x10⁶ BTU/hour at 1,000psi {35MW at 68bar}. The boiler houses were located close by the heads of these catapults.

CATAPULT AIRCRAFT LOADING EQUIPMENT BASE

This location was on a redundant length of perimeter track connecting the northern end of the Catapult Site and Runway 07/25. The site was used for evaluating equipment complementary to the head, the firing end of a catapult. Provision was made here for the complete installation of aircraft-centralising roller systems, power-operated wheel chocks, jet-blast deflectors and deck-cooling panels. Pipework and engineering services and equipment were accommodated within a large vault beneath the test area.

For trials purposes, aircraft were loaded on to the base and full power runs (including re-heat) – with the aircraft tensioned in the launch attitude using standard bridles and holdback units – repeated at operating cycles to evaluate the equipment under test.

Development of equipment for the operation of aircraft up to 60,000lb {27,200kg} AUW and with jet efflux velocities of the order of 800ft/sec {240m/sec} and 1600°F {870°C} respectively have been undertaken on this base. Complete 'production' versions of these installations and systems would have been conducted on this site prior to their installation on shipboard.

ARRESTED LANDING DECK

This site was designed to replicate shipboard conditions. The arrester-gear systems were installed either at ground level or beneath the runway in an underground compartment as required. Access to the underground cavern was alongside the runway, enabling the installation or removal of any one of a number of arrester systems without disturbing the deck or interrupting other trials in progress on an adjacent system. The underground compartment was covered with removable steel-plate decking. A topside grillage permitted the installation of alternative sheave arrangements to accommodate main reeve and centre span configurations of the wire ropes both above and below deck. Up to three arrester systems could be accommodated at any one time. These were located 2,000ft {610m} from the threshold of Runway 25 which is 6,000ft {1,830m} overall. The work on this test site encompassed both shipborne and airfield arrester systems and the related aircraft evaluations.

Three typical installations are listed in the RAE brochure (#318) from which the following technical information is abstracted.

a. DIRECT ACTING ARRESTER GEAR (DAAG) was a constant run-out system using a linear hydraulic energy absorber of the water spray type (Ref: 'Post-war Developments – Water Spray Arrester Gear'). The installation of this prototype equipment was above ground and presented a spectacular display when in operation.

Capacity	30x10⁶ftlb {4·1x10⁶kgfm}
Run-out	265ft {81m}
Deck Edge Sheave Spacing	100ft {30m}
Rope Sizes	4½in {114mm} circumference non-rotating steel wire rope
Performance	40,000lb @ 130kts {18,100kg at 67m/sec}

The first trials with this equipment took place in October 1965.

b. MARK 13 ARRESTER GEAR. This was a conventional hydraulic ram and cylinder – jigger – unit. In effect, this was the final stage in development of the original Mark 3 gear first installed in HMS *Courageous* in 1933. This latest equipment, over a period of four decades, had increased the operational capability to almost a four-fold of the weight of aircraft and doubled the entry speed of the original installation. This Mark 13 gear was used mainly for proofing of aircraft and airborne equipment. It was capable of being double-reeved to produce high 'g' values as decelerating loads.

Capacity	19·6x10⁶ftlb {2·7x10⁶kgfm}
Run-out	220ft {67m}
Deck edge Sheave Spacing	100ft {30m}
Main Reeve	3¼in {83mm} circumference flattened strand steel wire rope
Performance	30,000lb @ 120kts {13,600kg at 62m/sec} 40,000lb @ 104kts {18,100kg at 54m/sec}

c. ROTARY HYDRAULIC GEAR was a constant run-out system expressly designed for airfield use. The energy absorber was a rotary hydraulic unit converting the aircraft's energy into heat by fluid turbulence in a stator/rotor system. In other words, a fluid dynamometer.

Capacity	65x10⁶ftlb {8·9x10⁶kgfm}
Run-out	1200ft {370m}
Runway Edge Sheave Spacing	220ft {67m}

Purchase Tape	1140ft x 6in x 0·25in {350m x 150mm x 6mm} nylon tape per side
Centre Span	⅞in {22mm} diameter, 18x7 non-rotating steel wire rope
Performance	44,500lb at 160kts {19,200kg at 82m/sec}

The main machinery pit for these latter installations was a vast cavern located beneath the runway. This pit was 156ft long and a total width of 98ft with the major area at 17ft 6in depth {47m, 30m, 5·4m}. This cavern was spanned by deep section (7ft depth {2·1m}) lattice structural beams which enabled the arrester machinery to be either floor-mounted or suspended from the roof as in shipboard installations, either supported on bearers or underslung between the lattice beams.

SAFETY BARRIER AND ARRESTING GEAR PROVING BASE

This installation was used for the full-scale development of arresting systems and safety barriers. This proving ground, still on the NAD site, was located away from the main airfield runway system but connected to it via a taxi-track. The system comprised dead-load trolleys running in rails set in the deck and propelled by a cordite-powered slotted tube catapult. (The catapult machinery installation was uplifted from Farnborough on completion of the investigation into the flexible deck trials in which the Hawker Sea Hawk aircraft was launched.) The track was 1,000ft {300m} in length, in a runway orientation of 36.

The specimen arrester system was erected on a prepared base comprising substantial steel grillages to which were fixed deck-edge sheaves, energy absorber units, barrier masts and so on in various configurations. Two arrester systems were installed as 'permanent' fixtures. The first was a duplexed Mark 12 installation while the other was a prototype Mark 13 unit. Either of these could be used to give a linear pull-out of the centre span to a maximum of 250 feet {76m}. The specimen centre span of the arrester wire was supported across the track by the structure installed on the grillages. For development of the arrester machinery, a ballast trolley – loaded to the appropriate weight – was propelled into the system by the catapult and, hopefully, brought to rest by the system. (Provision also was made for the installation of a Friction Arrester Gear in this facility, records of such equipment are lacking.)

For testing a safety barrier, the specimen was rigged on supports at the grillages, spanning the track. Instead of a trolley, a mock-up of the aircraft fuselage (an 'iron bird') – configured and ballasted as required, perhaps an aircraft itself – was provided. This was propelled into the specimen barrier by the catapult shuttle. Should a failure occur in either installation, additional braking systems were installed further down the track to cater for such a runaway emergency. The first trials with the Mark 7 barrier net using a dummy Scimitar took place in August 1958. Those with the Mark 8 barrier net using a Phantom dummy began in December 1969.

In case of an overshoot – in which the arrester unit for one cause or another failed to bring the trolley/'iron bird' to rest – a coke-filled pit on the centre line of the installation was provided and beyond this was an earth bank, constructed of sandbags and faced with straw, as a final barrier. In one instance, even this was not sufficient. The 'iron bird' ploughed through the lot, and then the security fence, ending up in a field outside the Establishment.

The catapult had a maximum end speed of 150kts {77m/sec} with ability to launch a maximum weight of 35,000lb {15,900kg}. Heavier weights could be launched at lower speeds within an energy constraint of 35x10⁶ftlb {4·8x10⁶kgfm}.

MAIN OFFICE BLOCK, WORKSHOPS AND SITE FACILITIES

Naval Air Department was a self-contained unit, fully supported by local site technical and design office accommodation, workshops (for both plant and aircraft servicing), laboratories, hangars and storage facilities. The laboratories included equipment for on-load testing of wire rope systems, including a 50 tons tensile machine for calibrating rope tension riders and testing the performance of nylon ropes and barrier net components. By the time of its demise, with its additional sophistication, the entire area had become a worthy – if somewhat unrecognisable – successor to that to which the author was introduced some four decades earlier.

VALE

With all this new plant and equipment coming into being from 1955, and fully operational by 1958, its working life was destined eventually to be relatively short. Within a decade and a half, Naval Air activities at the Thurleigh base were on a down-turn which lasted for less than a further decade, finally terminating about 1979. During this latter period, helicopter research within RAE became the responsibility of NAD until that work was transferred to Aero Flight, also at Bedford. The Department was also responsible for hovercraft research until, in the 1960s, that was transferred to a new specialist unit.

The final programme of any magnitude involving NAD was the preparation for the entry of the VSTOL Sea Harrier aircraft into naval service in 1978. For this development no flight deck machinery was involved (apart from deck lifts, a shipboard engineering feature with which NAD had never been concerned). By 1980 all RAE research into naval flying had ceased, Naval Air Department was no more, and the heavy engineering equipment at Thurleigh was in process of being dismantled and removed. The NAD office and laboratories were finally cleared by December 1981.

The retraction of Fleet Carriers from Royal Naval service inevitably halted any further development of requirements for flight deck machinery and left no problems to be solved in the operation of fixed wing aircraft in Flying Navy. No more were there to be such aircraft to serve the Fleet Air Arm of the Royal Navy.

DEVELOPMENTS AT NAE/RAE – BEDFORD

BS.4 STEAM CATAPULT

Friedman (*#34 p.303*) records that Colin C. Mitchell, of Messrs MacTaggart, Scott & Co since 1936, patented an invention relating to a slotted tube type of launcher in 1938. In 1941 Mitchell was appointed to the rank of Commander (E) RNVR and joined the E-in-C's Department of the Admiralty – the unit which specialised, among a myriad of other matters, in the engineering procurement of flight deck machinery including catapults and accelerators – and thereby was familiar with the work at Farnborough. On de-mobilisation, Mitchell transferred his allegiance to Brown Brothers where, under his supervision, the steam catapult was developed and manufactured.

By 1945, technological advancements, particularly with the 'Slotted Tube (Ventral) Catapult' (qv) under development at Farnborough, and the hardware retrieved from Germany through the post-war 'Operation Surgeon', had demonstrated that such a system was currently viable and an aircraft accelerator, probably powered direct from the ship's main propulsion steam system, appeared feasible. Experiments with the erstwhile V.1 launcher were conducted at RAE and at the Shoeburyness artillery ranges; these indicated a measure of success sufficient for the principle to be developed within the Department in association with the Admiralty. The prototype steam catapult, although similar in function to, differed in many ways from, the Farnborough installation. The development at Shoeburyness proceeded as a result of experience with the Farnborough activities.

It was quickly realised that the system could be adapted for carrier launching of naval aircraft, obviating the need for complicated hydro-pneumatic machinery for flight deck operations. This would enable a significant reduction to be made in the top weight of aircraft carriers, being matched by the installation of the under-deck arrester gear in a vertical mode rather than as horizontal units suspended below the flight deck. Further advantages would accrue from the implication that the launcher need not be restricted in length, and the braking unit for the launching shuttle could be a simple buffer unit. A feature which gave impetus to this development was the realisation that the operation of wire

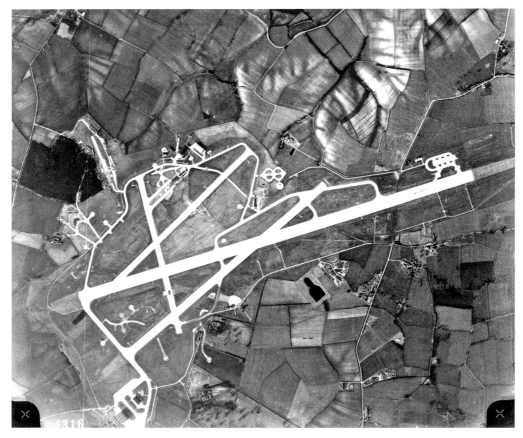

Thurleigh Airfield at RAE Bedford showing the two later runways, south of the original runways, and Naval Air Department's installations on the north side. *c.*1978. (© Crown Copyright/MOD)

ropes and pulleys in hydro-pneumatic jigger systems was reaching its limit when linear speeds in excess of 100kts {52m/sec} were experienced. Dynamically, with a slotted tube propulsion unit, only the weight of the free piston – being the sole mechanical component in motion – needed to be considered.

Ward (#306 p.10) states that the Admiralty:

'decided that an alternative type of catapult, better suited to the high performances required, should be developed, and a contract was placed with Messrs Brown Brothers & Co Ltd (Edinburgh), for the development of a steam operated slotted cylinder catapult, and the construction of three prototypes, one of which should be installed ashore for preliminary tests, and a second in a ship for more extensive trials'.

Ward does not give the location of the third unit, but he continues (#ibid. p.13):

'British activities in the field of slotted cylinder catapults owe much to the work of Cdr(E) C. C. Mitchell, ... a Director of Brown Bros, who undertook the design and production of the BXS.1.'

(Ward, perhaps inadvertently, appears to gloss over all the work undertaken by Naval Aircraft Department at Farnborough during the seven years prior to his report! C. C. Mitchell was a good friend of NAD and occasionally was to be seen on Jersey Brow witnessing some trial or another.)

The prototype (BXS.1) steam catapult was fitted as a temporary installation in the light fleet carrier, HMS *Perseus* (1944–58). Of ten light fleet carriers of the 'Colossus' Class laid down in 1942/43, two – HMSs *Perseus* and *Pioneer* – were never fitted with flight deck machinery. These ships were designed and operated as Aircraft Maintenance Ships. In all other respects they were complete 'flat-tops', each with a full-length flight deck but carrying deck-houses at the stern and midships with other 'top-hamper' constructions scattered about the flight deck.

The rationale of the selection of HMS *Perseus* as an experimental and demonstration carrier for the development of the steam catapult could well be questioned. During its 1949 refit the prototype steam catapult – designated BXS.1 – was installed, forward and atop the existing surface of the flight deck, surrounded by a specially constructed false deck simulating the flush deck of a normal carrier. Trials began in July 1950 and continued until June 1952. The utmost secrecy surrounded these trials, whether conducted at Belfast or Rosyth, with some 1,000

dummy, dead-load launches being achieved, followed by a further 127 launches with piloted aircraft. The astounding thing about these experiments was that for each time an aircraft was to be launched, it needed to be ferried to the ship and lifted aboard by crane to the flight deck before loading on to the catapult. (The ship did carry two aircraft lighters for this purpose!) Once launched, the plane could not return to the ship as there was no deck space for it to land! The man-handling exercise had to be repeated yet again. Perhaps this unaccounted-for procedure – resembling the process of loading the ship for the normal purpose of aircraft ferrying – enabled the operation to remain as covert as considered desirable.

The reader may question why in the case of a steam catapult was there the necessity for a hydraulic jigger to be incorporated within the system? Ward (#ibid. p16) explains that:

'The installation in HMS *Perseus* originally included a steam operated retraction gear. This, however, was found to be unsatis-

Douglas Skyraider (without its 'Guppy' ventral radome), ready for launching from the prototype steam catapult, BSX.I, aboard HMS *Perseus*, while tied up in harbour. 1951. (© Crown Copyright/MOD)

Fairey Firefly FR.4 with bridle and holdback tensioned up ready for launch from the prototype BXS.1 steam catapult in HMS *Perseus*. January 1953. (© Crown Copyright/MOD)

Hawker Sea Fury FB.11 (TF989) is about to be launched from the prototype steam catapult, BSX.1, aboard HMS *Perseus* with the ship in harbour. The location would appear to be the US Navy Yard, Philadelphia. 1952. (© Crown Copyright/MOD)

factory and was not used, the retraction being carried out using the winches on the flight deck.

'Between August and October 1951, a completely different system was fitted in which shuttle retraction, positioning, and bridle tensioning, were accomplished by a hydraulically operated unit with a 16 to 1 wire rope jigger connected to a grab. The jigger was operated after a launch had been made, control being carried out automatically. The grab advanced down the track in the wake of the shuttle assembly and clipped on to the shuttle through a spring loaded attachment. On reversing the gear, the grab with shuttle and piston assembly attached, moved back to the loading position. The grab remained engaged until the next launch when "break-out" occurred. A rope tensioning gear, hydraulically operated, and a mechanical rope stretching gear were provided to reduce any slack in the wire ropes of the jigger system.'

This adaptation is thought to have been general for later shipborne installations.

The trials are reported to have been successful insofar that the ship sailed to the United States. HMS *Perseus* arrived at the US Navy Yard in Philadelphia in February 1952, complete with a flight of Sea Fury aircraft, to demonstrate the prototype British steam catapult. However,

De Havilland Sea Vixen FAW.1 (XJ474 or XJ485) leaving the BXS.4 raised catapult at Bedford. OD May 1964. (© Crown Copyright/MOD)

Fairey Gannet AEW.3 (XL452) – with the ventral 'Guppy' radome – exiting the BXS.4 raised catapult at Thurleigh. March 1964.
(© Crown Copyright/MOD)

the Americans wanted to see how the system handled their own jets. It was fortuitous that Lt/Cdr Brown (*#14 p.173*) was already in the USA having been appointed British Naval Pilot on exchange duty at the Naval Air Test Center at Patuxent River, Maryland. He was invited to demonstrate the equipment by flying the Panther aircraft from the ship – which was to remain alongside in the Navy Yard.

On the day set for the trials the carrier was carrying a full complement of American 'Brass' and VIP officials. The snag was that instead of a head wind being available, a five knots tail wind was blowing along the deck. The Americans shook their heads at the idea of launching a jet with the ship tied up, and a tail wind into the bargain. The huddle of frustration was broken by Cdr Mitchell – the developer of the system – saying that 'Of course we can launch. These are the conditions which this catapult likes'. Without consulting 'Winkle' Brown, the ship's Engineer Commander volunteered, 'We'll risk the British pilot if you'll risk the aircraft'. With this challenge, the Americans agreed for the launch to proceed and all went well. Commander Brown was back in his element and added another 'first' to his flight log-book. Trials were continued in Chesapeake Bay and the end result was that the US Navy quickly adopted the steam catapult for its own ships.

Marriott (*#63 p.47*) prints a photograph showing nine USN jet aircraft ranged on HMS *Perseus*' flight deck for catapult trials. Behind them one of the deck-houses may be seen spanning the width of the flight deck. With what bewilderment does one behold the art of Navy flying! Subsequent to the success of these 'hybrid' experiments – which lasted just two years in HMS *Perseus* – the steam catapult was universally adopted for carrier operation. (Ward et al (*#306*) states that the equipment was operated on 1,560 occasions, 164 of which were piloted flights.) It was said that the demand for steam with these high-powered catapults was so great that often the Engineer Commander had to make the decision whether to use the steam for giving greater speed to the ship or to divert his resources to permit the catapult to be used. At the conclusion of these trials the BXS.1 catapult was recovered from HMS *Perseus* at Rosyth Dockyard and transported to RAE Bedford where it continued to give valuable service in both research and functional modes. The specification for the BXS.2 Slotted Tube Catapult for installation at RAE Bedford was issued in July 1953.

These catapults also introduced a further feature into bridle launching. At least three of the latest naval aircraft types of this era, although fitted with tricycle undercarriages, were launched in a nose-high attitude. (These aircraft were the Supermarine Scimitar and the Blackburn Buccaneer; the American McDonnell Douglas Phantom for the FAA was provided with an extra-long nose-wheel undercarriage for this purpose.) Ray (*#82*), who was closely involved with the development of the BS.4 and BS.5 installations at RAE Bedford, has provided further details of their operational features. The launching strops – bridles – for these aircraft were of sophisticated design and the holdback unit tensioned the aircraft on the catapult against a tail strut; this gave the aircraft a nose-up attitude with the front wheel raised some 2ft {600mm} above the deck level. This launch attitude, it is assumed, gave the aircraft some benefit in lift as it left the catapult; perhaps it was set at the ideal 'angle of attack'.

Hobbs (*#43 p.17*) writes that the steam catapult has seen service with all the world's carriers and is still in use in the US, French and other navies. He suggests that:

'A better name might be "Slotted Cylinder Catapult" since steam is merely a convenient gas; the high pressure gas from slow-burning cordite performs as well. The catapult consisted of a shuttle, to which the strop towing the aircraft was attached. The hook on the shuttle was the only part visible on the flight deck. Two parallel cylinders ran under the deck, which housed pistons fixed to the shuttle and "sliders" which opened and shut slots in the cylinders as the shuttle moved, keeping in the steam pressure.

Hawker Siddeley Buccaneer S.2 (XN975) exiting the raised BXS.4 steam catapult at Thurleigh. August 1968. (© Crown Copyright/MOD)

McDonnell Douglas Phantom FG.1 (XV566, Code 010 of 892 Squadron) at the end of its launch from the waist BS.4 catapult in HMS *Ark Royal*[55]. The bridle is released from the aircraft and is held by the van Zelm retrieval system to be brought to a stop at the end of the sloping track. (© Crown Copyright/MOD)

Steam from the boilers was fed to an accumulator and the actual launch pressure could be varied to take account of different aircraft weights. At launch the aircraft would select full power but be restrained by a holdback fixed near the tail. When the catapult was fired the combined force of engines and steam pressure broke a weak link in the holdback, allowing the aircraft to accelerate rapidly forward, to reach a speed of the order of 135 knots in as little as 150 feet {78m/sec, 48m}. The pistons would be stopped by water rams and the shuttle motored back for the next launch. A good crew would launch an aircraft every two minutes.'

These experiments were successful and during the period 1955–64, when the Royal Navy was equipping with new fleet carriers, eight ships were fitted with BS.4 steam catapults.

Hobbs (#45) – a mine of comprehensive information on carrier specifications and activities – details the development of the BS.4 catapult. He records HMCS *Bonaventure* (ex HMS *Powerful*) as carrying a single catapult operating with aircraft of 40,000lb {18,000kg} AUW at 78kts {40m.sec} with a launch stroke of 103ft {31m}. This was upgraded in HMS *Centaur* to two units each of 40,000lb {18,000kg} AUW at 94kts {48m.sec} in 139ft {42m}. HMS *Victorious*, after her final rebuild completed in 1958, operated two units each with a capability of launching 50,000lb {22,700kg} AUW at 97kts {50m/sec} with a stroke of 145ft {44m}. These ships were all laid down during World War 2 but were adapted during post-war refits to operate steam catapults.

The first carrier to be fitted with the steam catapult from original build was HMS *Ark Royal*[55]. Initially this ship was fitted with two steam catapults forward in the flight deck, both capable of launching an aircraft of 50,000lb {22.700kg} AUW at 91kts {47m/sec} with a stroke of 151ft {46m}. Subsequently, when converted to an angled deck, her catapults were uprated to an 'Improved' BS.4 specification and relo-

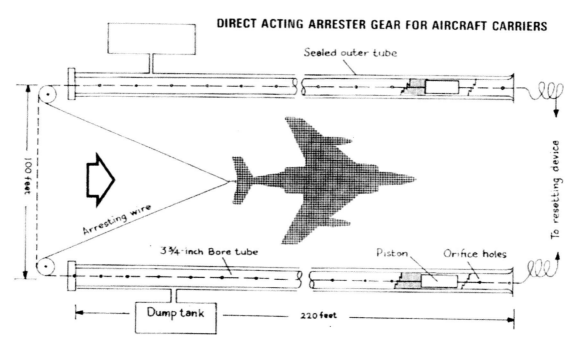

Direct Acting Arrester Gear (DAAG). A diagram showing the basics of the installation in HMSs *Eagle*[51] and *Ark Royal*[55]. (© Crown Copyright/MOD)

cated. These were rated at 50,000lb {22,700kg} AUW at 94kts {48m/sec} in 151ft {46m} stroke and 50,000lb {22,700kg} at 105kts {54m/sec} with 199ft {60m} stroke respectively, one in the port bow and the other in the waist, port-side, on the angled deck. Similar installations were repeated in HMS *Eagle*[51] with the newer, uprated BS.4 catapults during her extended rebuild in 1964.

Most subsequent carriers were equipped during refit with these powerful BS.4 units. Although capable of operating with rates of acceleration within the range of 4½g-5g the aircrews liked the smoothness of build-up to maximum 'g'. This was an improvement upon the rapid build-up to a peak of 2¾g – or more – as experienced with the BH.III and BH.V hydro-pneumatic accelerators. Engineering details of this project are featured as Appendix 13.

In developing this system from their Director's (Colin C. Mitchell) patent, Messrs Brown Bros rapidly gained valuable experience in the manufacture and commissioning of this unique machinery. Although the Royal Navy has not used a steam catapult since 1978, there are British units known still to be operating. British steam catapults now in service can be found installed in the Brazilian Navy's *Minas Gerais* and *São Paulo* (ex *Foch*) and the French Navy's *Clémenceau*. These were built by Brown Bros who continue to have an operational interest in them.

The reader of engineering history sooner or later will come across the name of Dr Dionysius Lardner who, prior to the Victorian era, had the reputation of being egregious in his theoretical pronouncements on matters relating to steam and, according to Rolt (*#84 p,80*), was 'one of the mainstays of that pseudo-science of which he was one of the first and greatest masters'. However, a century and a half later none would challenge his statement in calling the steam engine 'the exclusive offspring of British genius fostered and supported by British capital' as it related to the steam catapult.

DIRECT ACTING (WATER SPRAY) ARRESTER GEAR – (DAAG)

Naval Aircraft Department was ever considering the possibilities of alternative means of mechanically bringing an aircraft to a halt. Reference above has been made to some of its ideas having borne fruit in the arrester-gear installations along Runway 04/22 at Farnborough. One of the proposals was to adopt the use of the watery environment by making use of the sea water so freely available. It was an interesting project, placed on the 'back burner' in the author's days with NAD. The development of this system followed model experiments with a pilot rig on the Thurleigh site at RAE Bedford.

The intention was to retrieve the heavier types of naval aircraft by a simple installation in which the arrester wire was braked by the passage of a hydraulic piston through a tube with graduated perforations along its length. (The principle of resistance of a fluid flowing through a variable orifice was well known to NAD; it had been adopted in all Salmon type, telescopic tube catapults in the 'F' series for bringing the machinery to rest after the launch of an aircraft, and in all hydro-pneumatic arrester gears from Mark 3 onwards.) In this new case, the movement of the piston was restricted by the rate at which the water was permitted to discharge under pressure from the tube through a multitude of small orifices. In 1967 Thomlinson (*#304 p.12*) described such equipment under 'Direct Acting Gear'. It was originally known as the 'Water Spray Gear', which title, rather than the less prosaic substitute, is preferred; it was under that name that the author first met the subject.

The system is described as being extremely simple. The above-deck appearance was no different from that of the standard deck layout. A pair of tube assemblies ran in the fore-and-aft direction just below the flight deck. The arrester wire ran through the full length of these tubes and connected with the renewable centre span across the deck. The rope terminated at each end with a piston sliding within the tube. A gland, in an otherwise solid end plate, allowed the wire to be withdrawn from the tube. Along each tube was a series of holes of designed size and spacing. As the aircraft picked up the centre span the pistons were hauled through the tubes, water was displaced at high speed through the perforations and the induced resistance provided the braking force at the aircraft arrester hook.

The land-based prototype at Thurleigh (RAE Bedford) was tested using an English Electric Canberra B.2 aircraft (WJ995) adapted for the purpose, with the first arrestment into this gear taking place in August 1962. During an arrestment the displaced water, discharging to atmosphere, presented a magnificent spray effect – a vertical curtain, soaring to an impressive height – and running to waste. In subsequent development, for shipboard installation, this spray discharge was contained within an outer tube and re-circulated through the system.

Dr John Thomlinson pointing out a detail on a performance chart of the DAX – Direct Acting (Water Spray) – Arrester Gear. The demonstration model comprises a pair of perforated arresting cylinders with a Buccaneer about to engage the centre span of the arrester unit. Thurleigh Open Day, May 1964. (© Crown Copyright/MOD)

An impressive display of the water discharge from the port-side unit of the experimental DAX.1 arrester gear on the Bedford Safety Barrier and Arresting Gear (SBAG) proving site. The aircraft under arrestment would be travelling from the right, into the picture. OD May 1964. (© Crown Copyright/MOD)

Compared with the complexity of the Mark 13 arrester gear (the acme of the hydraulic jigger arrester system) and its immediate predecessors, this new project was unbelievably simple in concept and less costly to produce and install. Equipment was manufactured by MacTaggart Scott & Co Ltd of Loanhead, Edinburgh, and units were installed in HMS *Eagle*[51] (one wire for experimental purposes in 1967) and HMS *Ark Royal*[55] (all four wires in 1970). Pilots were reported to have preferred engaging the wire connected to this system as the retardation forces appeared to be less severe than those using the normal hydraulic gear. Further installations were made on bases ashore. Besides that at Bedford, one unit is known to have been installed at RNAS Lossiemouth. This project is treated in greater detail as Appendix 14.

One reference has been made to this system having been adopted for the projected – and aborted – super-carrier known as the CVA-01. The Ministry of Defence announced in July 1963 that this ship would be built for completion in 1973 but, within three years (in February 1966), for political reasons, this super-carrier project was cancelled, virtually dismantling the future of flying fixed-wing aircraft within the Fleet Air Arm. Naval flying policy was changing; regrettably, not for the better.

A specially adapted Supermarine Scimitar F.1 (WT859) engaging an early experimental DAX arrester gear at Thurleigh showing the impressive 50ft {15m} high water spray discharge from the nozzles in the braking tubes. May 1962. (© Crown Copyright/MOD)

DAX.1 experimental arrester gear trial at Thurleigh. The trials aircraft, a Supermarine Scimitar (WT859), is shown at the instant of pick-up of the arrester wire. August 1962. (© Crown Copyright/MOD)

Incidentally, for obvious reasons, sea water was never used in this equipment!

AIRFIELD OVERSHOOT ARRESTERS

Two systems, devised to contain an aircraft in danger of overshooting the runway on landing, were developed by NAD during World War 2. These were the Brake Drum Arrester Gear (BDAG) and the Drag Chain Arrester Gear (DCAG), details of which appear in separate chapters elsewhere in this record. First, the BDAG system, having a sizeable engineering content, was not adopted for any subsequent application. A follow-on development of the DCAG system is included in the relevant chapter. Post-war operations, particularly with fast jet aircraft, demanded a fresh look at such systems. These are treated below.

Of the systems which were developed, the primary requirement for emergency use was that each was to be completely reliable, able to be rapidly deployed and easily maintained. Some had a measure of success, others existed for a short time – generally for a specific application – and were then abandoned. Records are sparse, with information not being readily available, but an attempt has been made to correlate data, however piecemeal, into the summaries given below. The systems are identified as:

1. Drag Chain Arrester Gear – 1952 development
2. Friction Rail Arrester Gear – 1952 development
3. Spray Arrester Gear – 1968 development
4. Arresting Gear. Rotary Hydraulic, Mark 21 – 1969
5. Aircraft Arresting Barriers.

1 – Drag Chain Arrester Gear (DCAG)

Following the initial experiments on this equipment at Farnborough in the early 1940s – and its abandonment – a decade was to pass before thoughts were directed once again to this system when two channels of application were addressed. The first of these was in association with the early Flexible Deck/Slotted Tube Catapult experiments at Farnborough in 1953 in which a heavier version of the DCAG principle was adopted for the arresting of the dummy test load in the development of the cordite-powered, slotted tube catapult at its Ball Hill location.

Working model of an airfield DAG over-run barrier system proposed for use with Concorde. The nose-wheel has passed through the barrier net before the arresting strands come into contact with the wings. November 1967. (© Crown Copyright/MOD)

De Havilland Sea Vixen FAW 2 (XN653) at the point of pick-up in an experimental airfield over-run arrester unit at Farnborough. (Author)

Hawker Siddeley Buccaneer S.2B (XW986) at the point of engaging the overshoot arrester unit on Runway 07/25. May 1975. (© Crown Copyright/MOD)

Contemporaneously, in 1953, 34 Fairey Firefly AS.7 aircraft were converted to Mark U.8 machines as pilotless target drones (officially known as RPVs – Remotely Piloted Vehicles) operating from RAE Llanbedr, Merionethshire (Gwynedd) for use on the RAE Aberporth, Cardiganshire (Ceredigion) guided missile range. As these aircraft were already equipped with arrester hooks from their previous FAA service, in 1955 an installation of DCAG units was proposed at Llanbedr for use in recovery of these remotely controlled aircraft. Experimental work was re-started, adopting alternatives to anchor chains as the drag load. A single unit – employing two centre span wires connected to a single pair of braking elements – formed the test installation. This proved successful and a total of six single-wire units were taken into use in 1957 and remained operational until Firefly aircraft were phased out of service in 1960. Vide Thetford (*#97 p.188*) and Turnill (*#101 p.141*)

All six wires were installed on the extended runway used for operating RPVs. Dyer (*#234*) states that landing was a very hazardous operation in that the aircraft – by remote control – had to be brought on to the approach so that its arrester hook would engage one of the first five wires to which – according to Dyer – tons of heavy anchor chain were attached. If these were missed, there was the final reserve wire, acting as a

'long-stop' some hundreds of yards further along the runway, into which the aircraft did not always make a clean entry, with inevitable attention required by the aircraft 'doctor' (Dyer's word).

The units installed at Llanbedr did not all use redundant anchor chains as braking elements. The basic principle was continued but, due to difficulties in obtaining chain, and the high cost of such acquisition, the adoption of alternatives was investigated. Pierce (*#280*) records that in 1953 experimentation was undertaken firstly with cast-iron 'beads' threaded on lengths of wire and, secondly, replacing iron with reinforced concrete!

The iron 'ball and bead' chain comprised a number of balls of 2¾in diameter by 1⅞in long {70mm x 48mm} alternating with beads of 4¾in diameter by 7¼in long {121mm x 184mm}, both with 1in {25mm} diameter bores, threaded onto a ¾in {18mm} diameter wire forming a chain element of some 20ft {6m} length. The weight of this 'chain' was about 40lb per foot run {60kg/m}. Elements were linked together *in situ* to form two composite chains each of 500ft length {150m}. These were laid along the edges of a 150ft {46m} wide runway and linked to the centre span across the runway at the head of the system. These centre spans were pre-tensioned to 500lb {230kg} and held by shear pins in ground anchors. As in the original installation, they were supported across the runway on split rubber discs.

Cast iron, again, was not a cheap material for employment of this kind so the researchers turned to concrete. These mesh-reinforced concrete beads were all identical in size and shape, being 6¼in {160mm} in diameter and 18in {460mm} in length (sizes chosen by experimentation) and similarly threaded onto 20ft lengths {6m} of wire rope. The weight of this chain was 30lb per foot {45kg/m}.

The recorded pull-out distances of both iron and concrete chains were somewhat greater than that of the original anchor chains. A contributory cause of this, apart from the reduced inertia of the chains, was credited to the skewed rolling tendency of the cylindrical bead elements during pull-out, thereby reducing the overall friction of the assembly. This was felt not to be of any significant consequence; the performance of the system remained within workable limits. It was found that the gear could accept aircraft weighing 14,000lb {6,350kg} with entry speeds of up to 90kts {46m/sec} requiring a pull-out of 1,000ft {300m}. By the inherent nature of the installation, there was no requirement for a short time interval in resetting the system after use; a 30-minute cycle was acceptable. In 1964, some six years after the departure of NAD to Bedford, such a scheme was prepared for a DCAG system to be installed at each end of Farnborough's Runway 07/25. Each was to have a pull-out of 1,000ft {300m} employing a pair of chains of 500ft {150m} length. It is not known if this equipment was installed.

Dyer-Smith (*#235*) records that upwards of 100 airfields in the USA were contemporary users of drag chain installations at runway extremities to prevent overshoot. An inherent problem was the time taken for resetting the gear after an engagement.

2 – Friction Rail Arrester Gear

Although friction for the dissipation of energy is used in many instances in engineering practice, the principle – apart from the Brake Drum Arrester Gear – had not been widely adopted for aircraft arrestation. Undoubtedly, the problem of absorbing a large amount of energy in a short period of time without developing prohibitive rises in temperature within the braking elements was difficult to resolve. However, advances in the development of aircraft brakes prompted an investigation into the adoption of a frictional braking system. It was felt originally that the system primarily would be examined for airfield use but it was thought that adaptation of such a system for flight deck operational use was feasible.

Thomlinson (*#297*) outlines the basic feature in this investigation. He proposed that the ends of an across-deck wire (centre span) would be attached to bogies/sliders – equipped with pressure-loaded brake shoes – drawn along a straight rail. This rail could be of normal railway section, rolled steel sections, a flat plate or a cylinder. The design parameters would include a pull-out of 250 feet from an entry speed of 120 knots and all-up weight of 52,000lb. Ambitiously he thought that a landing cycle of 20 seconds per aircraft could be attained.

This project did not even reach the stage of cutting of metal. Attention was rapidly diverted to the development of the Direct Acting Gear (water spray arrester) which quickly established itself as a workable and highly effective system for both shipboard and airfield applications.

3 – Spray Arrester Gear, RN Airfields, Mark 1

At the same time as the Direct Acting Arrester Gear (DAAG) was under development for shipborne installations, Naval Air Department was aware of its suitability for use on airfields as overshoot prevention equipment. This was recommended for, and adopted on, a selection of RN Air Stations where the majority of landings were by FAA aircraft fitted with arrester hooks.

Like the prototype research installation at Thurleigh, this gear was designed for aircraft at 50,000lb {22,700kg} weight entering the system at 130kts {66m/sec} with maximum pull-out of 600ft {180m}. Two centre spans were provided with the rope sheaves each side of a standard 150ft {46m} wide runway. The centre spans were supported above the runway on split rubber discs and connected at each extremity by beckets and purchase cables into a pair of arrester units, one laid along each side of the runway.

Each arrester unit (energy absorber) comprised a pressure tube, 520ft {160m} long, fitted with spray orifices in alignment atop the tube. This pressure tube was located at the invert of a jacket pipe, with the whole buried in the ground. The jacket pipe was half-filled with braking fluid (water containing an anti-freeze), completely submerging and, thereby, flooding the pressure tube through the spray orifices. A piston attached to the centre span by its purchase cable was drawn through the pressure tube and the braking fluid was expelled through the spray orifices into the annular space contained within the jacket pipe. This resistance to fluid flow provided the hydrodynamic effort to arrest the aircraft. Engineering details of this system are included in Appendix 14.

4 – Arresting Gear, Rotary Hydraulic, Mark 21 (PUAG)

This was the long-winded title given to a simple system developed in 1969 which carried the sub-title of 'Purpose Use Arrester Gear', or PUAG as its acronym. (Why Mark 21? Where were the previous models?) It was designed to arrest an aircraft of 56,000lb {25,000kg} weight at an entry speed of 130kts {66m/sec} within a pull-out of 600ft {180m}. The centre span spacing across the runway was adaptable between 170ft and 350ft {52m & 107m}; the wire was supported some 3in {75mm} above the runway by split rubber discs, with an applied pre-tension of 1,200lb {540kg}. The location of the centre span was 850ft {260m} from the end of the runway.

Each end of the centre span was attached to an arrester unit housed in a pit below the ground, in proximity to the runway. The pull-out element of each half of the system comprised a 650ft {200m} length of woven yarn, nylon tape 6in or 8in {150mm or 200mm} in width wound onto a drum on a vertical axis. This was direct-coupled to a water brake, similar to a fluid coupling (Froude dynamometer), the water being cooled by a radiator unit. A main feature of this system was the short resetting time, allowing a six-minute landing interval.

This system was one of two recommended for use at RNASs, being particularly suitable for aircraft fitted with naval arrester hooks. It is thought that HMS *Heron*, RNAS Yeovilton, was provided with this equipment. Similar equipment was installed on certain RAF airfields, particularly those occupied by squadrons devoted to advanced flying training and where arrester hooks were fitted to training aircraft for overshoot situations.

5 – Aircraft Arresting Barriers (AAB)

Again, this was not a Naval feature but NAD was involved in research in this field alongside that of the shipborne installation. The general function of the system was similar to that of shipborne aircraft barriers in which the net arrested the aircraft by constraining the aircraft on impact by contact with, and wrapping around, the wings. These Aircraft Arresting Barriers (AAB) were designed to be installed towards the end of airfield runways to prevent overshoot of jet aircraft with minimum risk to aircraft and crew. Two systems were devised, known respectively as AAB Type A and AAB Type B, the difference being in the number of energy absorbers – braking units – adopted in the system.

Hawker Siddeley Buccaneer S.2B (XW986) at the end of pull-out of the airfield overshoot arrester unit on Runway 07/25. March 1975. (© Crown Copyright/ MOD)

Rolls-Royce captive flight Engine Thrust Measuring Rig, affectionately known as the 'Flying Bedstead', on display at the RAE Golden Jubilee exhibition, July 1955. This 'Flying Machine' carried the official aircraft serial number XJ314. July 1955. (© Crown Copyright/MOD)

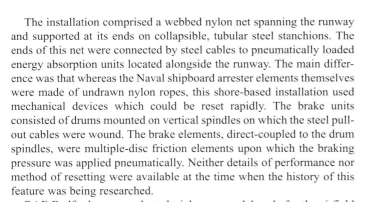

Hawker P1127 (XP980) described as a Ground Experimental Vehicle on the safety barrier proving base at Thurleigh. This was for a radio-controlled barrier entry at 110 knots. *c.*1976. (© Crown Copyright/MOD)

RAE's own two-seat Harrier T.2 (XW175) taking off from the Bedford experimental ski ramp set at 9° angle of elevation. *c.*1978. (© Crown Copyright/MOD)

The installation comprised a webbed nylon net spanning the runway and supported at its ends on collapsible, tubular steel stanchions. The ends of this net were connected by steel cables to pneumatically loaded energy absorption units located alongside the runway. The main difference was that whereas the Naval shipboard arrester elements themselves were made of undrawn nylon ropes, this shore-based installation used mechanical devices which could be reset rapidly. The brake units consisted of drums mounted on vertical spindles on which the steel pull-out cables were wound. The brake elements, direct-coupled to the drum spindles, were multiple-disc friction elements upon which the braking pressure was applied pneumatically. Neither details of performance nor method of resetting were available at the time when the history of this feature was being researched.

RAE Bedford even conducted trials on a model scale for the airfield arrestment of Concorde aircraft. The high angle of sweep-back of the leading edges of the delta wing plan form was expected to produce problems, particularly with the containment of the aircraft within the barrier curtain. It was felt that the delta wing would sweep aside the elements of the barrier curtain and pass through it without having any effect in retarding the aircraft. These trials did not proceed beyond the experimental stage.

One unidentified system was installed on the main runway at Farnborough. With the re-ordering of the airfield for civil aviation usage this runway, now identified as 06/24, was resurfaced and the overshoot arrester installation removed in September 2001.

VERTICAL/SHORT TAKE-OFF AND LANDING – (VSTOL)

Throughout World War 2 the Aeronautical Research Council frequently addressed the subject of take-off and landing of progressively heavier and faster FAA aircraft from carrier decks. Engine design and performance was the main consideration but the employment of aerodynamic enhancement devices also figured in their deliberations. Their attention sometimes was diverted to the adoption of helicopters – and a great deal of investigatory effort was made in this direction – but the use of rotary wing systems at that time appeared to be remote.

A paper presented to the ARC ((*ARC 9,020 'Report of the Naval Aircraft research Committee – 10th October 1945'*)) in October 1945 anticipated new problems in the use of jet-propelled aircraft and the NARC considered that means for obtaining extra lift, either by the use of flaps, or by deflecting the slipstream from the engine, or by using a thrust spoiler, should all be considered. RAE was invited to prepare a note on this. (The author has not been able to trace a record of such a submission of that time.) Some six years later an ARC report (*ARC 14,033 'Report of the NARC of the use of Small Jet Engines, June 1951'*) authorised RAE to undertake a thorough study of the small-engines-for-lift project.

This was followed five months later by another paper (*ARC 14,390 'An Interceptor with Deflecting Jets, November 1951'*) by Dr A. A. Griffith – already well known for his early 1930s RAE research into gas turbines – specifying the form of a typical aircraft of 12,000lb {5,400kg} weight with a narrow delta wing, mounting ten small jet engines, all situated in the fuselage but divided into two groups, one group forward of the centre of gravity and the other well aft of it. These engines were to be fitted with rotatable deflectors for turning the jets. A further report was presented to the ARC (*ARC 14,451 'Report of NARC on Deck Landing of Fighter Aircraft, November 1951'* and *ARC 13,962 'The Use of Small Jet Engines in Naval Aircraft, April 1951'*) which gave an outline of a new type of aircraft which could be operated successfully from the deck of a carrier.

These suggestions submitted by Dr A. A. Griffith drew attention to the advantages inherent in the use of small jet engines, a gospel he had been proclaiming since 1941. Because of the high thrust/weight ratio of small jet engines, the total thrust could be considerably greater than the weight of the aircraft and, by deflecting the jets downwards, the aircraft might be supported and controlled while at rest or at low speeds without having to rely upon lift from the wings. It was felt that the immediacy to meet the anticipated difficulty of landing high performance fighters of normal design could be achieved, with the additional advantage that the wings could be designed solely to give the best performance at high speeds. It was even suggested that a speed of Mach 2 could be attained with a considerable reserve of thrust available for manoeuvre.

NARC thought the project to be a practicable one and recommended that the problem of controlling an aircraft by operating the jets should be investigated experimentally. It was felt that the currently available small engines would be suitable for adaptation and a number could be

RAE Bedford's two-seat Harrier T.2 (XW175). At the time of writing, this aircraft (December 2007) is still flying with QinetiQ and is known as the Vectored Thrust Aircraft for Advanced Flying Control (VAAC) Harrier. *c.*1978. (© Crown Copyright/MOD)

installed in an existing fuselage – without wings – to assess controllability. If these experiments proved successful, the wider application for other aircraft would be obvious but NARC asked for emphasis to be placed on the immediate initiation of experiments on control. At this stage (1951) the project was to be regarded as Top Secret. This must have been the seed germ for what was soon to be known as Vertical Take-Off and Landing (VTOL), subsequently adding the description 'Short' to the title and making it Vertical/Short Take-Off and Landing (VSTOL) or – to the troops – 'Veestol'.

From this point the thread of continuity is momentarily lost but by early 1953 NARC was discussing the relative merits of VTOL *vis-à-vis* the flexible deck project. It suggested that already VTOL could be seen, by the use of deflected jets, to take off and land vertically but, in the interim, progress should continue with the flexible deck installation. It added the comment that in the most favourable circumstances it would still be many years before vertical lift aircraft would come into general service and if this were to happen it would be possible to construct flexible decks from which VTOL machines could operate. It was a little later in the same year that Rolls-Royce introduced their captive flight 'engine thrust measuring rig', probably better remembered as the 'Flying Bedstead' – a most apt title to say the least! (Did it really bear the official aircraft registration of XJ314?) This used two engines with jets discharging through a pair of hover nozzles and a further four mini-jets for lateral stability control. This research aircraft, together with the Short SC.I (XG905), are now on display in the FAA Museum, RNAS Yeovilton.

Through 1955 and 1956 NARC was still pressing for action in 'the energetic and independent pursuit of the long-term developments associated with vertical lift'. In making its point the Committee was still aware that a practical application of VTOL would not be seen in under ten years. One of its reports (*ARC 18,694 'VTO Aircraft in the Navy, September 1956'*) suggested that all the development given over many years in take-off and landing aids had ensured that carrier operation was able to continue where an aircraft could not operate in all conditions under its own power. The use of VTOL aircraft implied the throwing away of the advantages already accrued if these aids were to be abandoned. A subsequent paper (*ARC 19,302 'V.T.O. Aircraft in the Navy, May 1957'*) echoed the point of it being unwise to discard all the advantages which had been built up from development of carrier arresting and launching gear, especially as VTOL unaided would probably impose some penalty on range and payload. The author of the Paper continued in saying that further consideration had tended to confirm the validity of those opinions. The Paper concludes by saying that, apart from any inherent advantages, there would not seem to be any operational gain in the latest VTOL project. Nothing in it seemed to alter the previously expressed opinion that some alleviation in deck operating limitations would be of benefit. To take the solution to its complete extreme would impose more penalties than could be readily accepted.

However, the Short SC.1 (XG905) – with five engines – put in a brief flying appearance at the Farnborough SBAC display in 1959, followed in 1962 by the Hawker P.1127 with a vectored thrust engine. Shipboard trials of this latter machine were conducted in HMS *Ark Royal*[55] during February 1963. In development, this marque soon became known as the Hawker Kestrel F(GA) Mk.1, the world's first strike and reconnaissance fighter capable of VTOL. One of the early pre-production machines (XP984) first flew in February 1964 followed by carrier trials aboard

HMS *Bulwark* in January 1966. At last – after a long period of gestation covering 25 years from the 'Flying Bedstead' concept – the first flight of the prototype navalised Harrier aircraft took place in January 1978. Its progeny, the FAA Sea Harrier, achieved notoriety when a number of these machines acquitted themselves so magnificently with the Royal Naval Air Squadrons in the air support of the Falkland Islands campaign in 1982. (See 'Sharky' Ward [#104]).

It goes without saying that Naval Aircraft Department at Farnborough and subsequently at Bedford was not closely involved with the flight development aspect of VSTOL, but when the time came for its application to shipboard operation, including the development of the ski-jump, then the Department's interest became more formal and practical. This, for many years, ran parallel with application of rotary wing aircraft, mostly helicopters, for carrier operation with both achieving success in their various roles.

THE SKI-JUMP

A tentative concept of the ski-jump was placed before the Aeronautical Research Council (*ARC 15,952 'An Analysis of the Effect of of a Curved Ramp on the Take-Off Performance of Catapult-Launched Airplanes, June 1953*) by Wilmer H. Reed III of the USA Langley Aeronautical Laboratory in June 1953. His proposal was for a ramp of circular-arc profile, 50ft long with a total rise of 1·73ft {15·2m & 527mm} built at the for'ard round-down. (These surprisingly precise dimensions produce an exit angle of elevation some 2° above horizontal.) The start of the ramp occurred at the launch end of the accelerator. Reed stated that:

'the case of the straight deck launchings, the airplanes considered settled from 6ft to 9ft {1·8m to 2·7m} below deck level, whereas for similar launching from the curved ramp there was no tendency to lose altitude'.

(This is a reference to the frequent occurrence of an aircraft disappearing from view below the flight deck level after launching and its scrambling to gain some altitude and avoid contact with the sea.)

As the start of the ramp coincided with the end of the accelerator, the additional flight deck length required for its installation could not be made available in any contemporary carrier. The aircraft would encounter the ski-jump immediately on release from the accelerator, with the 'mini-hump' occurring within a half second of launch.

Since the 1978 paying-off of HMS *Ark Royal*[55], Navy flying has been restricted to helicopters and VSTOL aircraft, the latter in the shape of the British Aerospace Sea Harrier. The prototype aircraft of this marque – a Harrier carrying the 'Royal Navy' designation – first undertook deck landing trials in HMS *Hermes*[59] in February 1977 at which date NAD at Thurleigh was well on its way to oblivion. The first Sea Harrier underwent sea trials, again in HMS *Hermes*[59], in November 1978; with delivery of the first aircraft for Royal Navy squadron use beginning in June 1979.

A couple of decades after Wilmer H. Reed III's proposal, Lt/Cdr D. Taylor produced his idea for assisted take-off by a ski-jump ramp at the for'ard end of a catapult-less flight deck. This was designed to induce an upward directional boost to the aircraft at the end of its free (non-accelerated) take-off run. It was of a greater magnitude and in a different class from that proposed by Reed. Trials were conducted at RAE Bedford and an impressive Bailey Bridge structure was built as a temporary feature at Farnborough – adjacent to Jersey Brow – for the 1980 SBAC air display. Baxter (*#5 p.64ff*), writing in a popular vein, tells of his experiences as passenger in some of the demonstration flights of the company's Harrier (G-VTOL) during the display.

Marriott (*#63 p.99*) treats the subject briefly in a more technical manner and records that the ski-jump constructed in HMS *Hermes*[59] during the refit completed in June 1981 was 150ft {43m} long, 45ft {14m} wide and 16ft {5m} high with the final angle of elevation being 12½°. Similar provision was made in the three 'Invincible' Class ships now forming the FAA 'carrier' complement. The first two ships, HMSs *Invincible* and *Illustrious*[82], were fitted with ramps terminating in an angle of 7°. HMSs *Hermes*[59], in her last refit, and *Ark Royal*[85], from the outset, were fitted with similar ramps but with an angle of 12½°. This steeper angle was adopted by the former two ships in subsequent refits. This system was first used 'in anger' during the 1982 Falklands campaign and continues to this day with a dummy deck for shore training purposes at HMS *Heron* (RNAS Yeovilton).

'FLYING STATIONS' – THE PILOTS

During the early period of World War 2 most of the research flying of Fleet Air Arm aircraft at Farnborough was undertaken generally by the RAF test pilots of the RAE Experimental Flying Department. This contingent (the RAE Squadron) was under the control of a Group Captain as many of his Department held the ranks of Wing Commander and Squadron Leader. The Department operated in 'Flights', the titles of which during the war years included Aero (Aerodynamics), Armament, Engine, High Speed, Instruments, SME (Structural and Mechanical Engineering) and Wireless & Electrical. Some of the pilots' names, taken at random – and written into history – may be recalled as M. J. Adam, R. L. R. Atcherley, A. E. Clouston, W. B. D. S. Davie, A. F. Hards, A. F. Martindale, H. A. Purvis, W. Stewart, F. R. D. Swain, J. R. Tobin, J. Unwin, A. Wheeler and H. J. Wilson. Even W. S. Farren, Director of RAE during the latter war years and one of the very few civilian pilots at Farnborough, used regularly to 'keep his hand in' by taking to the air, often in a noisy North American Harvard. Delivery to and collection of aircraft prepared at RAE for sea trials was generally the responsibility of one of the test pilots from the Royal Naval Trials (778) Squadron at HMS *Condor* – RNAS Arbroath. Pilots from this squadron would undertake the requisite ADDLs at Arbroath and then would fly the aircraft to the carrier in which sea trials were to be conducted, generally by the same pilots.

Coincident with the installation of the first arrester-gear unit at Farnborough, a naval test pilot was appointed to serve alongside his colleagues of the RAF. Lt K. J. ('Robbie') Robertson DSC arrived in May 1942 for attachment to Aero Flight and quickly took over the major part of test flying FAA aircraft. Having had operational experience – flying Spitfires during the Battle of Britain – he quickly adapted to handling the development of the Seafire for shipboard use. Sadly he lost his life when flying a Seafire (MB141) in RATOG trials on board HMS *Chaser* in January 1944.

With the successful placing of a naval pilot at Farnborough, it was felt that a similar appointment should be made to A&AEE, Boscombe Down, Wiltshire, for test flying duties. The posting of Lt E. M. ('Winkle') Brown DSC to A&AEE was made effective in December 1943. He hardly had time to don his Sidcot flying suit at Boscombe Down before he was redirected to Farnborough to replace Lt Robertson. Lt Brown had been one of four trials pilots involved in the commissioning trials of the American-built Lend/Lease CVEs and, by virtue of which, had become a familiar personality to the Farnborough 'Boffins' in NAD. He arrived at Farnborough as replacement within a few days of Robbie's death.

'Winkle' – why this mode of address, the popular name for the genus and species of a crustacean, *Littorina Littorea*, was given to Brown remains an enigma – is a Scot of slightly smaller stature than average. His service *forte* had been flying Martlets from HMS *Audacity* on Convoy Protection Duties. (Perhaps his having survived a couple of ditchings of these aircraft in his earlier days of flying Navy earned him this popular nickname.) He carries a small scar on his chin, having received a mouthful of windshield splinters on one of his sorties involving a military confrontation with a Focke-Wulf Kurier. This does not adversely affect his slight Edinburgh accent. Scotsmen often carry the adjective 'dour' as an indication of their mien. This does not apply to 'Winkle': the word 'canny' would be far more appropriate. This would reflect the meticulous attention he would give to the details of any experimental flight proposal, particularly if something unusual was required of him. As a consequence, he has many 'firsts' to his credit.

Popham (*#76 p.222*), himself a pilot of no mean ability, writes of him as:

'perhaps the FAA's most immaculate pilot'.

Eric Melrose Brown CBE, DSC, AFC, MA, PPRAeS, Captain RN (Retired) arrived at Farnborough on 17 January 1944 as a Lieutenant (A), assuming duties as Senior Naval Test Pilot. Shortly after, by March 1944, he was given his 'half-ring' on appointment to Lieutenant-Commander and it was in this rank that he served most of his time at Farnborough. During this period he was appointed Commanding Officer of the prestigious RAE Aero Flight, accepting responsibility for a number of RAF test pilots within his command. As the Admiralty required him, career-wise, to move on to watch-keeping duties and to take command of a front-line squadron, he left RAE in August 1949. Two years later he was again into test flying, this time at Patuxent River, Maryland, USA. While there he was involved in demonstrating to the US Navy the British Steam Catapult (Accelerator) BXS.I in its experimental installation in HMS *Perseus*. On return to the UK he was awarded his second 'half-ring' on appointment to Commander, eventually retiring from active

Lt Cdr E. M. 'Winkle' Brown, Head of Aero Flight and Senior Naval Test Pilot, dressed typically for duty in between flight assignments. (Capt E. M. Brown)

Aero Flight, posed in front of de Havilland Sea Hornet XX aircraft (TT191). From the left they are Fl Lt C. M. Adams; Fl Lt J. S. R. Muller-Rowland DSO, DFC; Fl Lt R. W. F. Ellis; Lt Cdr E. M. Brown OBE, MBE, DSC, AFC, RN (Senior Naval Test Pilot and Commanding Officer); Fl Lt C. J. Lawrence; Fl Lt G. E. C. Genders DFM; Sqn Ldr R. W. Whittome OBE, DFC; Lt R. B. Mancus RN. May 1948. (© Crown Copyright/MOD)

service duties in 1970 with the rank of Captain RN. As for most of the period when 'Winkle' was at Farnborough his rank was Lieutenant-Commander, it is with this title – Lt/Cdr Brown – that reference is made to him throughout this text.

His technique of taking off from Farnborough in an Avro Lancaster aircraft, as if it were a naval machine, had to be seen to be believed! All carrier flying adopts the left-hand circuit when circling the ship; Farnborough airfield also adopts the left-hand circuit for aircraft approach. (This means that a pilot when circling the ship, or Farnborough, keeps his destination on the port – left-hand – side in his approach.) A good naval pilot, as soon as he was airborne from a carrier, would put his starboard wing down to get out of the way of, and to reduce the slipstream effect on, the aircraft immediately following and then join the left-hand circuit. The first occasion Lt/Cdr Brown did this with a Lancaster was impressive, but in no way foolhardy! *Verb sap*! This technique was subsequently perpetrated at Farnborough by every naval pilot; in the best tradition of the Senior Service honour had to be maintained. 'Winkle' also took on the major part of the development of

the helicopter as a viable Naval aircraft. The earliest of these 'work-horses' was the American Sikorsky Gadfly (KL107) – this was the name given to this aircraft at that time. At the other end of the scale were the new jet aircraft and the requirement in their assessment for deck flying. The work-load in the early post-war period developed to the extent that in January 1946 a second FAA pilot, Lt B. E. Bullivant, was appointed to Farnborough to add his contribution to the interests of Flying Navy.

As Captain E. M. 'Winkle' Brown, his experiences and exploits have been well recorded in his own books. Reference to some of these is made in the Bibliography at Appendix 15. In particular, the latest edition of his autobiography, *Wings on my Sleeve* (2006, Weidenfeld & Nicolson), is of immense value in portraying the activities described in these pages from the pilot's viewpoint. Various biographies by other writers on the lore of Flying Navy are listed but any record from which the name of 'Winkle' Brown is omitted surely cannot be regarded as authentic history. It is a most satisfying privilege to have his approbation of the author's efforts through his ready agreement to writing the Foreword to this work. The author's deepest thanks are due to him.

THE GROUND CREW

The manifold research departments of an experimental establishment, such as RAE, normally would consist of scientists – university graduates – supported by experimental staff. Both Scientific and Experimental grades were recognised as such within the hierarchy of the Civil Service. NAD, with its stable-mate Structural and Mechanical Engineering (SME) Department, were hybrids; regarded (by the scientists) as oddballs in their 'nonconformist' approach and application to their tasks. The complement of both NAD and SME comprised professional engineers of chartered status, supported by engineering technicians. In these cases the Civil Service recognised the posts within the hierarchy of Professional and Technology grades but, nevertheless, in the local research environment such were felt to be 'outsiders'.

Both NAD and SME were large departments, each capable of producing spectacular demonstrations of their expertise. Both possessed and operated equipment comprising great chunks of engineering hardware, sometimes referred to as 'ironmongery'. NAD at times had four catapults on Jersey Brow and a series of five arrester-gear installations along Runway 04/22 with two others elsewhere on the airfield.

SME Department was accommodated within large structural test laboratories, one of which housed its airframe test facility, a massive structural steel enclosure locally identified as the 'cathedral'. Others contained large fatigue test rigs and the undercarriage drop-test facility, while an outdoor feature was the deafening propeller-spinning tower. The 'cathedral' when rigged with a multitude of levers forming a complicated whiffle-tree system for the application of test loads all over a wing surface was an impressive sight. In the case of SME, their trials often were to destruction, ending with a big bang from the collapse of a wing under overload tests or from a tyre bursting during an undercarriage drop test. When an aircraft fuselage started to tear apart, the progressive popping of rivets sounded like a fusillade from a machine gun. On the other hand, NAD prided itself on being able to demonstrate a more positive, constructive approach in its trials – except when things went disastrously wrong!

Throughout the war years, the staff complement of the Naval Aircraft Department eventually amounted to some 30, mostly chartered engineers by qualification or in-training. At the outset, as the Catapult Section, only two members were university graduates. Most of the engineering staff had qualified through the National Certificate procedures. The extensive practical – 'hands-on' – experience acquired during such training was of distinct advantage to the type of work undertaken within the Department.

Percy Salmon, Head of the Main drawing Office – 1916–42 – instigator of many catapult systems and designs. c.1942. (© Crown Copyright/MOD)

Lewis Boddington, the first Head of Catapult Section (later to become Naval Aircraft Department) over the period 1938–51. August 1949.
(© Crown Copyright/MOD)

Departmental staff and associated Designs Office personnel assembled outside NAD offices on Jersey Brow. The occasion was that of Lewis Boddington's departure to Headquarters in November 1951. Those identifiable are (*left to right*): *Back row*: A. K. M. (Alec) Howie, Norman Edwards, W. (Bill) Waldron, Les Maybury, Tom David, Eric Kent, Johnny Johnson, Gilbert Barrett, *, Roy Branch, Tom Dawson (Chargehand). *Second row*: Jonathan (Dave) Davies, W. (Bill) Davis (Foreman), J. M. N. (Willie) Willis, D. Sharwood, Ann Ross, Sid Tonks (Foreman), Betty Spong, Horace Reynolds (Chief Designer), Joyce Holloway, *, *, *, W. Watt, Donald Pierce, N. Holey. *Third row*: Tom Reeves, Alan Arnold, James Hampson (Photographer), Jack Wright (Foreman inspector), M. R. (Mardy) Jones, Norman I. (Nibs) Bullen, W. (Bill) Lyle, S. W. (Stan) Chisman, A. Kerswell, *, Miss M. Perkins, Alf Crossfield, Sid Cox, *, *. *Front row*: Lt Innes RN (Pilot), E. F. H. (John) Noble, Sally Ayres, R. J. (Nick) Nixon, Miss J. Turner, J. (Tommy) Thomlinson, Lewis Boddington, C. W. (Chas) Crowfoot, Miss M. Phillips, P. R. (Pincher) Martin, Arthur Grimes, Les Andrews. November 1951. (Donald Pierce)

From the outset, Catapult Section had its own small design office, mostly for on-site projects, but other departments were used for theoretical and practical assistance on speciality work. This latter field included installation or modification work on aircraft; stress calculations of an involved nature, generally on airframes and their attachments; structural and mechanical testing of assemblies, components and the like; manufacture of major components; civil engineering work on the airfield; all aspects of photography; and provision of instrumentation, often to a demanding specification.

While not realising it during the time, the level of responsibility given to, and expected of, the members of staff in the Department was surprisingly high. As a young (trainee) engineer in his early twenties, the author was required to accept and to undertake responsibility for conducting trials at sea with only a photographer accompanying him. He was expected to attend meetings on board ship with Commander Flying and his staff – and, at times, the Captain himself – to advise whether or not certain procedures could be undertaken. All this was accepted by the author and his colleagues as part of the job, each performing in his own particular, specialist field. Reports had to be written after each series of trials, ashore or at sea, and these were circulated to high authority in both the Admiralty and the Ministry of Aircraft Production/Supply.

Inevitably, as far as project work was concerned, the staff forming the Department tended to be loosely compartmentalised. This was unavoidable as one section would be concerned with testing of shipboard accelerators while its mates would be investigating new systems for accelerators. Similarly, one section would be involved with the design of a new type of arrester gear while another would concentrate on the development of arrester hooks and their attachment to aircraft. There was a policy section, others took care of examining the feasibility of projected schemes, and yet another section took care of standardisation and its effects. In-depth examination of the viability of new aircraft, including preliminary field and flight trials, and related processes occasionally incurred a heavy work-load. While such sections normally operated within closely defined limits, the system was kept sufficiently flexible to permit transfers of staff for short or long periods to be made between the discrete functional units. The whole Department held together very well with a highly satisfactory sense of overall camaraderie, irrespective of age, background, experience or ability. Staff members were able freely to approach each other, whatever their relative grades, and this mutual respect and confidence contributed to the sense of well-being throughout. We could work, laugh – and commiserate (when necessary) – with each other in a well-knit community.

Overseeing these activities was the head of department, Lewis Boddington. He joined the RAE (1936) as technical assistant to Percy Salmon and by 1938 was made Head of the Catapult Section. For 13 years – until his promotion to a headquarters post – he managed, steered and cajoled his crew for the wartime and immediate post-war achievements of his department. He was a product of the Principality of Wales, was happy with his slight Welsh accent, and – seemingly – liked to appoint fellow Celts (six of them) to his 'Empire'. He was approachable and well-liked by his staff. His system of management was to let his people get on with their projects under a loose rein, offering advice when and where it was needed. This tacit recognition of the capabilities and expertise of his staff was respected by each member of the team. For his contribution to the development of Flying Navy he was awarded £1,500 for inventiveness; the American government awarded him the Medal of Freedom; and the British government, in 1956, appointed him CBE.

There was a camaraderie between NAD and SME with the occasional exchange of personnel. In fact, on the departure of Lewis Boddington, the three departmental heirs-apparent to his post (John Thomlinson, R. J. Nixon and P. R. Martin) all preferred to remain where the action was – at the coal-face – rather than take on the responsibility of departmental management. Boddington's successor, A. W. Hotson, was from SME Department, as in turn was his successor L. G. H. Sterne, who oversaw the transfer of NAD to Bedford. John Thomlinson was eventually appointed Head of Department just prior to the time of rundown when the practical application of his mathematical expertise was no longer in demand.

The individuals – the author and his contemporaries – who comprised this Department are listed in Appendix 2 – Personnel. They formed the team to which the renowned Naval test pilot, Captain E. M. 'Winkle' Brown (*#14 p.160*), in 1978 awarded the following accolade. He writes:

'For sheer inventive ingenuity the Naval Aircraft Department team, led by the genius of Lewis Boddington, was supreme. The names of Dr Thomlinson, John Noble, Charles Crowfoot and others of this Department must go on record as those responsible for giving the Royal Navy a technical lead in aircraft carrier equipment that it still holds to this day. These men and women were Civil Servants, but they worked hours, took responsibility, and produced results far beyond what their country paid them for. To me they represent the true measure of Britain's greatness.'

It was an immense pleasure and an outstanding privilege to have been, at that time, a part of this unique team of engineers.

EPILOGUE

For the closing sequence of this book I (the author) will revert to the use of the first-person pronoun as being more apposite to my experiences and observations.

My final involvement of major consequence with the Naval Aircraft Department was the association with, and attendance at, the first deck-landing trials of a jet aircraft, the de Havilland Sea Vampire prototype (LZ551/G), in HMS *Ocean*[45] in December 1945. (I am writing this when the Golden Jubilee of its occurrence has passed.) A fortnight later I was away to pastures new. With the advent of the jet age it was left to my colleagues to continue the task of bringing the Fleet Air Arm into its majority and its most effective era in the operation of superb, high performance aircraft with ships and flight deck machinery to match their service requirements.

During my spell of almost six years with Naval Aircraft Department I had been witness to, and a part of, the development of Navy flying from the aircraft and equipment outlined in the Introduction to the advent of machines of vastly superior performance. These modern aircraft included:

Bell	Airacobra (experimental aircraft)
Blackburn	Firebrand (1945)
de Havilland	Sea Mosquito (1944)
	Sea Hornet (1945)
	Sea Vampire (1945)
Fairey	Fulmar (1940)
	Barracuda (1941)
	Firefly (1941)
	Spearfish (1945)
Gloster	Meteor (experimental aircraft)
Hawker	Sea Hurricane (1941)
	Sea Fury (1945)
Supermarine	Seafire (1942)
	Seafang (1945)

as well as the first naval helicopter

Sikorsky	VS-316 Hoverfly (1945).

Of these, some were more successful than others in satisfying service requirements. I saw the Fairey Swordfish outlast its intended successors. Subsequently, from a detached viewpoint, I saw all of these 'modern' aircraft superseded by machines of even higher performance, particularly in the field of jet flying, culminating in the ultimate of shipborne aircraft, the Blackburn/Hawker Siddeley Buccaneer (1963) and the McDonnell Douglas Phantom (1968). The flight deck machinery was developed in line with, and reached its zenith in, the sophisticated, operational needs of these later aircraft.

It was my privilege also to see the first aircraft with a tricycle (nose-wheel) undercarriage (Bell Airacobra) used on a carrier, also the first twin-engined aircraft (de Havilland Sea Mosquito) and the first jet aircraft (de Havilland Sea Vampire). The helicopter was coming into its own with its initial sea trials (Sikorsky Hoverfly). (Helicopters eventually took over the standby guard duties [Search and Rescue role] in escorting aircraft carriers when at 'flying stations'. They replaced the corvette or frigate maintaining station on the port quarter as patrol for rescuing any aircrew from a ditching, or other of the ship's crew who fell overboard.)

During the same period the Fleet Air Arm benefited from the American Navy aircraft which equipped FAA carrier-embarked squadrons viz:

Chance Vought	F4U-5 Corsair (1942)
Grumman	F4F-3 Martlet/Wildcat (1940)
	TBM-1 Tarpon/Avenger (1943)
	F6F-3 Hellcat (1943).

These all were subject to improvement in performance during this period but no new US Navy types were introduced into carrier operation with the Fleet Air Arm during World War 2.

I must say that it is rather distressing to find that some authors – thankfully few in number – writing of their experiences in the Fleet Air Arm, appear to disparage the efforts and achievements of those who bore the brunt of wartime research and development as if all their activities were of little account. This is doubly surprising when at least one of these authors was a test pilot during a part of his own naval career. According to this author, the Barracuda was wrong, the Firebrand was wrong, the Mosquito was wrong, the Meteor was wrong, the Vampire was wrong; what was there left?

Crosley (*#24 p.21*) – with a particular penchant for ETPS – has a paragraph on:

'the miraculous way the Royal Navy – assisted by ex-ETPS (*Empire Test Pilots' School*) students – transformed the machinery and landing aids on its carriers' flight decks to allow jets to land safely in all weathers and how, by their invention of the steam-powered catapult, made it possible to operate top performance fighter/attack jet aircraft anywhere in the world more easily and more cost-effectively than the same aircraft could hope to operate from an airfield'.

The RAE, and its contribution by Naval Aircraft Department, is dismissed by this author in less than a dozen lines!

Again, how a history of naval aviation during World War 2 can be written without making even a passing reference to the renowned naval pilot who, in addition to his squadron service, was leader of a prestigious team of test pilots, 'Winkle' Brown, baffles me. To put the record straight, none of the civilian engineers in Naval Aircraft Department, nor any of the Navy pilots who flew at RAE during my time had any connection with the Empire Test Pilots' School. (ETPS was formed at Boscombe Down on 10 June 1943, transferred to Cranfield a few months later and did not arrive at Farnborough until August 1947.)

A second writer, Hobbs, (*#45 p.29*) makes the following statement:

'It is interesting to note that all the improvements to carriers that have been made to enable them to keep pace with operations by heavier, faster and more complex aircraft have been made by naval officers, both in the US Navy and the Royal Navy. Scientists seem to have lacked the practical experience necessary to lead the way.'

As one who, as a research engineer, was in the thick of things when it was all happening – and presumed to be among those *'lacking the practical experience'* – I have to wonder whence this writer sourced his history!

One of the features in working within the Civil Service is that credit for any scientific development must always remain with the Department and not with any particular individual within that Department. Exception is made in the case of a spectacular invention and in such a case an application for a patent may be permitted. On the other hand, an officer in the Services is allowed to be given credit – and may even be given an award – for some out-of-normal activity or practice. Perhaps this is the reason why modern authors appear to assume that all achievements were basically due to an entrepreneurial approach, assumed to be found only in serving officers! The classical example of this is in the case of the development of jet engines, or gas turbines. The name of Fl/Lt Frank Whittle is renowned as he was awarded promotion (ultimately to Air Commodore), honours and a knighthood for his part in this work.

Farnborough airfield and RAE, looking NNE from Berkshire Copse area. The sites of most NAD activities may be seen in this picture, including Jersey Brow (with its catapults), Runway 04/22 (with its arrester gears), the flexi-deck landing areas and Ball Hill. The High Speed Track is out of the bottom of the picture. (Peter J. Cooper/Falcon Aviation)

Conversely, the names of Dr A. A. Griffith, Hayne Constant, A. R. Howell and a host of others are seen only as the authors of abstruse technical papers recording their not insignificant contributions to the science of jet propulsion. In this instance, my case rests upon the series of 35 papers delivered to – and published in the 'Proceedings' of – the Institution of Mechanical Engineers (*#109*) on the subject of gas-turbine technology. These were presented in symposia and individual lectures by mostly NGTE Staff, together with the first James Clayton Lecture by A/Cdre F. Whittle, over the period 1945–50. My point is supported by a brief résumé of the genesis of gas turbines which appears in a recent brochure, 'Pyestock – A Celebration of the Gas Turbine Engine' (*#319*), published by DERA (the acronym for the Defence Evaluation and Research Agency, the current all-embracing name given to a consortium of major government research establishments, many of them originally having the prefix 'Royal'). In 1957 the Royal Commission on Awards to Inventors published a title – 'Use of Inventions and Designs by Government Departments' – in which a list of awards for some major wartime inventions was printed. This, at least, was an acknowledgement although pitifully limited in extent.

Closely bonded working associations were established by the Naval Aircraft Department with the relevant branches of the Ministry of Aircraft Production (later, Ministry of Supply) and the Admiralty civilian and service engineering staff direct. The MAP department was 'ADRDN, Assistant Director, Research and Development, Navy'. Those for the Admiralty were DACR, 'Directorate of Airfields and Carrier Requirements'; DNC, 'Directorate of Naval Construction', and E-in-C, 'Engineer-in-Chief's Department'. While both DNC and E-in-C were staffed by engineers (some of them naval officers), and both were involved in carrier flying, DNC looked after the arrester-gear installations while E-in-C was in charge of the accelerator/catapult functions. The wartime developments, inventions and techniques related to Navy flying are well-documented in wide-ranging references and these give the appropriate acknowledgements – where they are due

– to those to whom reference is made elsewhere in this volume, including the paragraphs below. Almost without exception the publications listed in the Bibliography confirm the points which I make. No meanderings by any one or another writer can deny or alter the facts and remove the credit and honour due to those whose dedication was to the art, skill and practice of Flying Navy.

Perhaps I may take heart from Popham (*#76 p.222*) – himself a naval pilot of no mean reputation – who wrote some 25 and 32 years before Crosley and Hobbs respectively, and thereby much closer to the events which he records (and I quote at length):

'On 4 December, 1945, Lt. Commander Eric Brown MBE, DSC (later a Captain, but known throughout the Service as 'Winkle', and perhaps the FAA's most immaculate pilot) brought a suitably modified Vampire on the deck of *Ocean*, before an anxious and slightly breathless crowd of bigwigs. This was the first deliberately-planned jet deck-landing. ... It is difficult to recall now just how dramatic an event this was, and how almost intolerably risky and exciting it seemed to the onlookers; after all, the last Swordfish squadron had only been disbanded six months before. However, there was no fuss and no blood. The fifteen take-offs and landings which Winkle Brown performed during that day and the next went without a hitch; and opened what was to be naval air's most inventive and successful decade. With its role at last fully understood, with the Navy behind it, money available, and a brilliant team of engineers and designers contributing a flow of novel ideas, it ceased being 'the monkey's orphan', and became one of the most professional and efficient flying services in the world.

'It took a little time, of course. There was the post-war run-down and all the adjustments that it entailed. ... For five years the FAA had to make do with aircraft which had been converted or designed for it during the war. The interceptor fighter was still the Seafire,

The remains of the Direction Controlled Take-Off (DCTO) track on Farnborough airfield, north of Runway 12/30. The view is looking south-west, along the alignment of the port rail with its termination in the way of Runway 07/25 in the far distance. (Author)

On a cold January morning the rusty covers to arrester Pit A.5 are seen in the foreground, on the western edge of Runway 04/22, with the other pits receding towards Jersey Brow. The dark patches on the runway are the sites of the fair-lead pulleys for the cross-deck arrester wires. The RAE buildings fronting the airfield are seen in the far distance. (Peter J. Cooper/Falcon Aviation)

Richard Snell (ex-DERA Archive Department) and the author in the cavernous underground plant room which housed the BH.5 Catapult propulsion unit on Jersey Brow. The machinery space was about three times the area seen in the picture. (Peter J. Cooper/Falcon Aviation)

with a steadily improved performance. The Firefly, promised as a fighter in 1941, had finally come into service two years later as a night fighter, reconnaissance, and dive-bomber and remained operational until 1956, finishing its working life as an ASW aircraft. The Firebrand ... made its long-threatened appearance between 1945 and 1953 as a single-seat, torpedo-strike aircraft, with eighteen feet of nose in front of the pilot, and a reputation for unmanageability second only to that of the Corsair Mark I. In addition, for three or four years, one or two squadrons were equipped with the twin piston-engined Sea Hornet. But in 1947 the first post-war naval fighter/strike aircraft, the Sea Fury, though still basically an adaptation of an RAF design, came into service. It was a fine rugged machine and it gave a good account of itself.'

I feel this last author appreciated the true content and context of the wartime situation, having himself been involved at the centre of operations during the period of which he writes.

At the time of my writing – half a century later – all past efforts in technical research and engineering development of exciting processes, innovations and applications of Flying Navy appear to have been of little importance as they fade into insignificance. The exception, of course, is the development and introduction of VSTOL in the form

of the Sea Harrier which superseded the high performance Sea Vixen, Buccaneer and Phantom aircraft of the fleet carrier era.

It is hoped that the US Navy, who still fly 'orthodox', fixed-wing aircraft from their carriers, have reaped the benefit of this intensive war-time research and application in the United Kingdom. One of their authors, Rear Admiral P. T. Gillcrist (#36 pp.137–148), is generous in his acknowledgement and appreciation of this effort. He sums up with this appraisal:

'The key to this dramatic progress in carrier aviation was mainly the legacy of the technical genius of our British comrades in arms – the enhancements to operational capability provided by the angled deck, the steam catapult, and the optical landing system. ... Truly, carrier aviation had come of age.'

Commander Colin C. Mitchell – who we have met earlier in these pages and who latterly was Managing Director of Brown Bros, Edinburgh – cherished an appreciation awarded him by the USN Commander, Air Systems Command, in the form of a 'Shield of Honour'. This bore the legend:

Maurice Shakespeare and the author by the shortened remnant of the High Speed Track on Laffan's Plain. The original Exeter Track ran along the gravel road some 30ft {9m} to the left of the HST. (Peter J. Cooper/Falcon Aviation)

'Presented to Colin C. Mitchell in recognition of his contribution to the United States Navy's carrier strike capability through his invention of the steam catapult, in his generous advice and counsel over the past twenty years which led the way to the modern carrier flight deck. No man has contributed more to the carrier navy.'

Flying in the Royal Navy currently is only a shadow of what it once was. It is well that the Fleet Air Arm Museum within the purlieu of RNAS Yeovilton (HMS *Heron*) continues its yeoman effort to remind us of past greatness, and avoiding its disappearance into oblivion. The period of which I write was an era of adventure in which it was a privilege to serve.

In 1985 Marriott, (*#63 pp.12–13*), was writing in the strain:

'By 1945 the aircraft carrier was firmly established as the capital ship of the fleet, a position which had only slowly been achieved in the Royal Navy, whose last battleship did not commission until 1946. Indeed, it was not until 1948 that all work associated with the projected 'Lion' class battleships finally ground to a halt and only in 1951 did carriers feature ahead of battleships in the influential *Jane's Fighting Ships*. The story of the British carrier force since 1945 brings out perhaps the worst aspects of British defence policies in the period under question, but it is to be hoped that the recent slight resurgence in naval aviation, mainly a result of the Falklands Campaign, will be continued and that the lessons of history will perhaps be finally absorbed by those responsible for ensuring that the British armed forces are provided with the proper weapons necessary to ensure the peace and security of our nation.'

In the immediate post-war period, having been closely associated with RAE during the war years, the Admiralty decided to establish a permanency at Farnborough. A small group of Fleet Air Arm and Engineering Branch officers was formed under Cdr Beard with the title of Carrier Equipment Division. When Naval Aircraft Department began its transfer to RAE Bedford in 1954, the Navy's Carrier Equipment Division went with them and, in the process of removal, the two organisations merged to form the Naval Air Department. The transfer to Bedford was not completed until 1957, the equipment on Jersey Brow being in use throughout the intervening period. In fact, the final launch from the Farnborough BH.V Accelerator was made on 20 September 1957 when a Sea Venom FAW.21 (HG656) was flown by Lt Hefford RN. This equipment was the last to be dismantled and the Jersey Brow site became subject to redevelopment for other purposes. (The airfield fire brigade eventually took up residence there, occupying a fine new

building and using the vast BH.5 catacomb for one or another of its training purposes.) NAD remained effective at Bedford until government policy decided upon the virtual abandonment of fixed-wing naval aviation in the phasing out of the fleet carrier. The Department ceased to be effective from 1973.

By 1978, when HMS *Ark Royal*[55] was finally decommissioned and laid up for scrapping, the era of equipping Fleet Air Arm carriers with flight deck machinery came to an end. The Royal Navy was reduced to a couple of 'Through Deck Cruisers' – with another on 'standby' or undergoing refit – as the essence of Navy flying, operating VSTOL aircraft and helicopters, with the latter now embarked on most ships of the Royal Navy. A fascinating period in British Naval aviation rapidly drew to a close. For a half century and more entrepreneurial engineering had been to the forefront in providing 'Wings over the Navy'. Now, those of us who had a part to play in these exciting activities in the history of naval aviation, quietly reminisce with no little nostalgia over past events and their regrettable, demise. Humble (*#48*) treats at length what I consider to be the regrettable proceedings and political manoeuvrings of this era.

Since writing the foregoing, there has been the launch of the new HMS *Ocean*, virtually a commando carrier. More encouraging is the current activity relating to the provision of two new fleet carriers. These are thought to be for operating high performance fixed-wing aircraft, in addition to VSTOL machines.

It will be fascinating to discover how much of the past expertise will be incorporated in operating both planes and ships. Perhaps new technology will supersede all that has gone before and new methodology will become as familiar to the modern generation as the research and development of now-outdated systems and practices were to its forebears.

Perhaps steam will not be so readily available in new ships for catapulting; encouraging sounds are being heard in the field of magnetic-levitation. *Professional Engineering*, the Journal of the Institution of Mechanical Engineers for 20 October 1999, suggests current research indicates that within a decade Maglev (magnetic levitation) technology will be available for launching space vehicles. Its adaptation for carrier operation is not far removed from this. Burgess (in Nicholls [*#68 P.63*]) states that the US Navy is about to adopt an Electromagnetic Aircraft Launching System (EMALS) now under development to replace steam catapults. EMALS is designed to be lighter in weight than a steam catapult and to require a smaller operating crew; also it is expected to reduce the peak forces applied in launching aircraft.

It is unlikely that a similar advance in deck-landing procedures will be seen; the across-deck arrester wires and nylon barrier nets are unlikely to be superseded. The under-deck braking system may undergo changes; perhaps eddy-current devices will be adopted. Interesting days are ahead for entrepreneurial engineers in a, hopefully to be resuscitated, Naval Air Department.

If, after all the foregoing, a further definition of a catapult is required, the reader is directed to a 21st century explanation by James Dyson, the inventor and manufacturer of the dual cyclone domestic vacuum cleaner – the machine with its entrails unashamedly exposed. In outlining the development of a catapult device for launching projectiles, he writes in the 'History of Great Inventions' – which was distributed with the *Daily Telegraph* of 3 June 2000:

'The catapult had a new lease of life in the 20th century when its elastic energy principles found use in hurling fighter aircraft from sea-borne carriers – and for trapping them when they land. Which just shows that a great idea never becomes obsolete.'

Verb sap!

Buccaneer S.2C (XV344) forms a fitting finale to this book. The Buccaneer typifies the last word in Flying Navy technology applied to British-built, fixed-wing aircraft operating from Fleet Carriers. The Mark S.1 format formed the first operational squadron (RNAS 801) in July 1962 with the Mark S.2C aircraft remaining in service until the demise of the Fleet Carrier in November 1978 (subsequently transferring to the RAF). This particular aircraft (XV344 ex-809 RNAS, embarked in HMS *Ark Royal*[55]) arrived at Farnborough on 12 September 1978 and remained at RAE for the next sixteen years operating with Flight Systems Department. In preservation, this venerable workhorse has been selected to serve as 'gate guard' to RAE's successor, QinetiQ's, headquarters complex on the Cody Technology Park, to the west of the original 'Factory' site. (Peter J. Cooper/Falcon Aviation)

APPENDICES

INTRODUCTION

Seventeen appendices may appear to be bordering upon an overstatement of facts! The purpose, however, is to avoid cluttering up the main text with minutiae, suffixes, footnotes and so on. To some readers these would probably be irrelevant. It was felt preferable to include material of selective or specific interest as appendices rather than to introduce them into the text as footnotes. In fact, some of the subjects are too extensive in content to be considered anywhere other than as separate chapters. Such articles as the treatment of engineering particulars, aircraft identification, performance characteristics, bibliographic references and light-hearted asides are included here. Perhaps the grouping together of relevant data presents a convenient format to the reader for comparative and reference purposes. He/she may turn to these at will.

APPENDIX 1
CHRONICLE OF EVENTS

The following is a schedule of significant dates, leading up to the formation of the special Catapult Section at Farnborough, its progress through to becoming the Naval Aircraft Department, its transfer to Bedford, where it became the Naval Air Department, and its subsequent decade of demise when the Royal Navy, regrettably, was forced to abandon fixed-wing flying. This period amounts to some six decades of investigation, experimentation and development.

1908	José Weiss used a falling weight within a tripod pylon to propel a trolley in launching his gliders at Fambridge, Essex.
1909	Capt P. W. L. Broke-Smith erected a weight and pylon catapult on Jersey Brow, Farnborough.
February 1911	Patent No.3333 granted to M. F. Sueter, F. L. M. Boothby & H. G. Paterson for a gravity-operated catapult, which was never developed.
1912	First successful catapult launch carried out by the US Navy. The aircraft trolley was propelled by a pneumatic ram. After a run of 50 feet, the trolley was lost overboard with each launch.
July 1914	*The Aeroplane* published an artist's impression of the launch of a Scout aeroplane from a warship, using a falling weight, operating within a hollow mast, as the power unit for a catapult.
1916	Percy Salmon was recruited from Brighton railway workshops to Farnborough, as Head of the Main Drawing Office in the Royal Aircraft Factory.
1916	Specifications placed for catapults. Two compressed-air driven units were produced, one by Armstrong Whitworth, the other by Waygood-Otis to a design by R. F. Carey.
1917	Avro 504C aircraft launched from Carey compressed-air catapult at RAF Hendon.

August 1917	First landing of an aircraft on to a ship under way, HMS *Furious* forward flying-off deck.
1917	Avro 504B aircraft flown into first type of arrester gear using fore-and-aft arrester wires.
October 1917	Armstrong Whitworth catapult fitted in HMS *Slinger*.
May 1918	First live catapult launch from HMS *Slinger*.
1918	First safety barrier fitted in HMS *Furious* to protect aircraft from impacting the amidships top-hamper of bridge structure and funnels when alighting on the after landing-on deck.
1918	Fairey IIID and Flycatcher aircraft, wearing Bryer hooks, landed on HMS *Furious* with longitudinal (fore-and-aft) arrester wires spanning a well deck.
1920	First athwartships arrester wires (trailing sandbags) fitted in HMS *Furious*.
1921	HMSs *Stronghold* and *Thanet* fitted with gravity-operated launcher gear for pilotless aircraft (RAE Project 1921).
October 1921	First catapult installed on Jersey Brow site.
1922	Sopwith Snipe (?) aircraft fitted with Bryer arrester hooks.
1922	Designs for two hydro-pneumatic catapults; these resulted in the 'Carey' jigger type catapult and the RAE (Salmon) telescopic ram catapult.
October 1922	Percy Salmon initiated a development programme for catapults.
1922	Longitudinal (fore-and-aft) wire arrester gear removed from HMS *Furious*.
21 July 1925	First catapult launch of Supermarine Seagull (N146) aircraft from RAE Mark I catapult (Type F.I.H?) on Jersey Brow.
October 1925	First catapult launches at sea, from modified Carey Catapult, in HMS *Vindictive*, using Fairey IIID and Flycatcher aircraft.
1926	Carey Catapult, removed from HMS *Vindictive*, installed in HMS *Resolution* for service.
1926	Carrier-borne longitudinal arrester gear abandoned in all ships.
19 May 1926	Jersey Brow catapult converted to cordite operation.
1927	Carey catapult installed in HM Submarine M.2.
27 July 1927	First launch of 'Larynx' radio-controlled target aircraft from HMS *Stronghold*.
May 1928	First cordite-operated catapult launch of manned aircraft at sea, HMS *York*.
1928	Admiralty decision to equip all capital ships and cruisers with catapults.
January 1929	Catapult (F.IV.H) for HMS *Hood* tested on Jersey Brow.
13 June 1930	Vickers Virginia VII (J8326) aircraft launched from portable airfield catapult at RAE.
1930	First launcher fitted in HMS *Ark Royal*[14] (HMS *Pegasus*).
January 1931	Arrester gear, using across-deck wires connected to friction devices, installed in HMS *Courageous*. Trials with Fairey IIIF aircraft.
1931	First arrester-gear installation, using across-deck wires, at RAE; location and details of this system are not known.
1932	Catapults fitted in seven warships.
January 1933	Arrester gear, with multiple across-deck wires reeved into under-deck hydraulic machinery (Mark 3) fitted in HMS *Courageous*, followed by HMS *Glorious*. Trials with Fairey Seal aircraft.

1933	Start of de Havilland Queen Bee launcher trials at RAE.
1934	First accelerators (later known as BH.I) fitted in HMS *Glorious*.
June 1934	Start of development of crash/safety barrier systems.
November 1935	Mark III Catapult concept, start of planning and manufacture of smaller components and power plant at RAE.
1936	BH.I Accelerators fitted in HMS *Courageous*.
1936	Installation of visual landing aid in HMS *Glorious*.
June 1937	First crash/safety barrier trials at Farnborough.
1937	Start on the design of 'Assisted Take-Off Gear', subsequently developed as the BH.III Accelerator.
9 May 1938	Percy Salmon submitted his case to Chief Superintendent for the formation of the Catapult Section at RAE.
1938	BH.I Accelerators fitted in HMS *Argus*.
1938	BH.I Accelerators, and Mark 4 arrester and safety barrier gear fitted in HMS *Ark Royal*[38].
September 1938	Catapult Section, headed by Lewis Boddington, formed at RAE as a unit within the Main Drawing Office under P. Salmon.
1938	Installation of Mark III Catapult began at RAF Harwell.
1938	Work began on the Direction Controlled Take-Off (DCTO) system for large aircraft.
1938	RAE Mark I Catapult dismantled and removed from Jersey Brow.
May 1939	Percy Salmon submitted paper 'A Review of Launching Schemes for Aircraft'.
1939	Catapults had been fitted in 61 warships, mostly battleships and cruisers.
March 1940	The author – Geoff Cooper – joined the Catapult Section for his final year as Engineering Apprentice.
May 1940	First installation of BH.III Accelerator in HMS *Illustrious*[40].
16 August 1940	RAE bombed by Ju.88 aircraft – daylight raid.
22 August 1940	Handley Page Heyford III (K5184) aircraft flown from the DCTO system.
December 1940	Admiralty requested priority investigation into provision of simple catapults for fitting to Merchantmen – CAM-ships.
18 January 1941	P.I Catapult – rocket-propelled – first trials with temporary rig on Jersey Brow.
January 1941	First Sea Hurricane RNAS formed.
15 March 1941	On completion of his apprenticeship, the author was appointed Laboratory Assistant in the Catapult Section, wages 75 shillings (£3.75) per week.
1941	Construction of concrete runways began at Farnborough.
April 1941	First CAM-ship equipped with 'P' (Rocket) Catapult taken into service.
1941	C.I Accelerator (permanent rig) installed on Jersey Brow.
August 1941	Avro Manchester (L7246) aircraft arresting trials into Brake Drum Arrester Gear (BDAG) on Runway 07/25.
December 1941	First 'Hooked Spitfire' trials in HMS *Illustrious*[40].
February 1942	First catapult launch of Seafire.
1 May 1942	The author was promoted to Technical Assistant Grade III at £206 per annum.
June 1942	Avro Lancaster (L7528) arresting trials into BDAG.
1942	First naval arrester gear (Mark 4 – 'Illustrious' type) installation at Farnborough on Runway 04/22.
May 1942	First Naval test pilot, Lt K. J. Robertson DSC, appointed to RAE.
1942	RAE Mark II Catapult dismantled and removed from Jersey Brow.
8 September 1942	First, and only, launch of Avro Manchester (L7246) aircraft from the DCTO system.
September 1942	Percy Salmon retired.
7 January 1943	First catapult launch of Grumman Martlet.
23 January 1943	First catapult launch of Fairey Firefly.
May 1943	First MAC-ship sailed.
19 Aug 1943	Drag Chain Arrester Gear (DCAG) trials at Farnborough with Supermarine Seafire (MB299) aircraft.
July 1943	Brake Drum Arrester Gear (BDAG) project terminated.
1943	Start of experimental trials on High Speed Track (HST) at RAE.
1943	Admiralty decision to remove all catapults from warships, apart from aircraft carriers.
19 January 1944	Lt E. M. 'Winkle' Brown DSC appointed to replace Lt Robertson as Naval test pilot.
1944	Concept of undercarriageless landing on to a flexible deck developed.
March 1944	First carrier landing of twin engine aircraft, de Havilland Mosquito, LR359/LR387 on HMS *Indefatigable*.
26 May 1944	E M 'Winkle' Brown promoted to Lt/Cdr as Chief Naval test pilot at RAE.
1944	Runway installation at Farnborough completed with the third extension to Runway 07/25 and the new fourth Runway 18/36 brought into use, releasing 04/22 for use by Catapult Section.
1944	Tail-down, bridle launching to be standard for all RN carriers; no further 'spooled' aircraft to be built.
April 1945	Catapult Section renamed Naval Aircraft Department (NAD).
1945	Start of model trials on flexi-deck concept (Stage 1) at RAE.
20 July 1945	Taylorcraft Auster (TJ537) flown from 'Mini-DCTO' system on Jersey Brow.
3 December 1945	First world-wide carrier landing of jet aircraft, de Havilland Sea Vampire (LZ551/G) on HMS *Ocean*. Pilot Lt/Cdr E. M. Brown.
1946	Accelerators aboard aircraft carriers adopted the title of 'Catapults'.
1 January 1946	The author departed from NAD for 'pastures new' (MoS Headquarters).
24 January 1946	First catapult launch of twin-engined aircraft, de Havilland Hornet (PX214) aircraft at RAE.
1946	Start of comprehensive development of safety barriers at RAE.
September 1946	Admiralty Carrier Equipment Division formed to augment NAD at RAE.
October 1946	BH.V Accelerator (BH.5 Catapult) brought into service on Jersey Brow.

March 1947	Flexi-deck trials (Stage 2) full-scale ballistic trials at RAE with adapted GAL Hotspur gliders.	November 1953	First and only launch of an 'undercarriage-less' aircraft, Hawker Sea Hawk (VP413), in direct flight from slotted tube catapult to flexi-deck at Farnborough.
December 1947	First attempted full-scale flexi-deck landing (Stage 3) at RAE with Sea Vampire aircraft (TG426). Unsuccessful – aircraft written off.	June 1954	Naval Aircraft Department and Carrier Equipment Division combined to form new Naval Air Department.
March 1948	First successful flexi-deck landing with Sea Vampire aircraft.	July 1954	Mirror landing sight adopted in four carriers.
May 1948	Safety barrier trials with twin-engined aircraft at RAE.	1955	Start of staged transfer of NAD from Farnborough to NAE Bedford (Thurleigh) – later retitled RAE Bedford.
November 1948	Start of sea trials with flexi-deck (Stage 4) in HMS *Warrior*.	1955	First flying trials on angled deck in HMS *Centaur*.
1948	First shipborne BH.5 Catapult installed in HMS *Magnificent*.	1955	First operational mirror landing sight installed in HMS *Centaur*.
May 1949	Flexi-deck trials in HMS *Warrior* terminated.	1955	First operational steam catapult (BS.4) installed in HMS *Ark Royal*[55].
September 1949	Shipborne barrier trials with de Havilland Sea Hornet.	September 1957	Final (and 3,343rd) launch from BH.5 Catapult on Jersey Brow. This brought to a close all NAD operations at Farnborough.
July 1950	First sea trials with the steam catapult (BSX.I) in HMS *Perseus*.	September 1957	L. H. G. Sterne replaced A. W. Hotson as head of NAD.
October 1951	Angled deck project proposed by Lewis Boddington.	1958	Mirror landing sight replaced by Deck Landing Projector Sight (DLPS) embodying a Fresnel lens system in HMS *Illustrious*[40].
November 1951	RAE proposed experiments for visual landing aids – mirror sight.	August 1962	Experimental water spray arrester gear trials began at NAE/RAE Bedford.
November 1951	Lewis Boddington transferred to HQ as AD/RDN; A. W. Hotson appointed successor as head of NAD.	1964	Prototype water spray arrester gear (later named the Direct Acting Arrester Gear – DAAG), DAX.1 installed at RAE Bedford.
February 1952	First angled deck configuration painted on flight deck of HMS *Triumph* for 'touch and go' flight trials.	1967	First shipboard installation of experimental DAAG (one unit – DAX.2) in HMS *Eagle*[51].
February 1952	BXS.I steam catapult in HMS *Perseus* demonstrated to US Navy.	1970	Full installation of DAAG in HMS *Ark Royal*[55].
March 1952	RAE experimental mirror landing sight erected at Farnborough.	1970	Beginning of rundown of Naval Air Department at RAE Bedford.
June 1952	Sea trials with steam catapult (BXS.I) in HMS *Perseus* terminated.	1978	Final project within a greatly depleted NAD was the preparation of the British Aerospace Sea Harrier for entry into Naval service.
July 1952	RAE mirror landing sight transferred to HMS *Illustrious*[40].	December 1981	Demise of Naval Air Department became absolute.
1952	Elevated P.I Catapult dismantled and removed from Jersey Brow.		
1953	Development of nylon nets for safety barriers at RAE.		
July 1953	Specification issued for manufacture of first BXS.2 steam catapult at Bedford.		

APPENDIX 2
PERSONNEL

The RAE Catapult Section was formed as a result of a case Percy Salmon – head of the Main Drawing Office – submitted to the Chief Superintendent, A. H. Hall, on 9 May 1938. In this he proposed that all catapult and kindred work should come under a special section to meet the requirements of technical development, control and responsibility. To cover immediate needs the following staff were proposed:

1 Senior Technical Officer (STO)	to take charge of all technical work
1 Technical Officer (TO[a])	to support STO
2 Technical Assistants (TA)	to support TO(a)
1 Technical Officer (TO[b])	to take charge of installation, tests and experiments
1 Technical Assistant (TA)	to support TO(b) in experimental work
1 Chargehand	to support TO(b) in experimental work

Other help to be recruited from 27 Dept (Fitting Shop) as required.

It is obvious that the proposal was upheld in that the Catapult Section was formed as a separate entity from the Main Drawing Office [04 Department] in September 1938.

Eighteen months later (12 February 1940), Salmon submitted another proposal to increase the staff complement from seven to 12, including Lewis Boddington, the STO head of Section. His staff would be:

1 Technical Officer
1 Technical Assistant, Grade I
1 Assistant Engineer, transferred from Ministry of Transport – equivalent Grade TO
1 Technical Assistant, Grade II
2 Technical Assistants, Grade III
1 Designer
2 Draughtsmen, Grade I
2 Draughtsmen, Grade II

To this complement were added the industrial staff engaged in field work on projects for the Section. These were listed as:

10 Construction work at Harwell
4 Experimental work on the DCTO project
4 Operating the two catapults on Jersey Brow and conducting smaller experimental work.

Salmon, rather plaintively it seems, makes a point that 'Nearly all this work is original and, involving both life and material, represents a heavy responsibility'. He supports his case with a schedule of the projects in hand at that time:

(i) Mark III Catapult under construction at Harwell (approx Cost – £106,000)
(ii) Direction Controlled Take-Off Scheme under construction and test at RAE (approx Cost – £35,000)
(iii) Queen Wasp Catapults designed at RAE and placed to contract to be tested and calibrated by RAE (contract – £36,000)
(iv) Arrester Gear out to contract (contract – £2,500)
(v) Assisted Take-Off Gear for HM Carrier Ships investigation, development and final design. Construction by Chatham, to be fitted to new carrier ships (6 sets). Tests under RAE responsibility (development Cost – £5,000)
(vi) Miscellaneous Catapulting Work
 (a) Proof tests on all FAA aircraft
 (b) SIS (Standard Instruction Sheets) in connection with the launching of FAA aircraft
 (c) Catapulting attachments to RAF machines
 (d) Arrester gear – general problems in carrier ships including tests and redesign of the crash barrier equipment for HMS *Ark Royal*.
(vii) Other Miscellaneous Work not Associated with Catapulting
 (a) Development of 20mm gun
 (b) Development of Smoke Screen

(approx Cost – £5,000)
Total development expenditure – *£189,500*

(The cost figures are given here for comparative purposes. It must be remembered that in 1940 the salary of a scientist or a professional engineer was less than £500 per annum. For a similar sum a three-bedroom house on a quarter acre of land could be purchased, freehold, in Farnborough!)

It was into this environment that I was recruited a month after Salmon's submission of his second case. All these projects, apart from those under section (vii), were then in progress. Of the exceptions, the development of the Liquid Propellant Gun, after a few experiments, quickly fell by the wayside (but was resurrected a couple of decades or so later by the Armaments Research and Development Establishment). Nothing further was heard of the Smoke Screen project.

The following represents the staff complement of the Catapult Section (subsequently Naval Aircraft Department) in late 1945. Of this complement, five, including myself, were ex-apprentices. About one-quarter were young undergraduates, or graduates, directed to RAE from university. Those who wished to complete their studies returned to university during 1946.

Members of the original team, in post at March 1940, are indicated by an asterisk (*).

*Percy Salmon	(Chief Draughtsman)	Head of Main Drawing Office
*Lewis Boddington	Bodd	Senior Technical Officer
*John Thomlinson (Dr)	Tommy	Senior Scientific Officer
*Charles Crowfoot, MBE	Chas (to his face)	Technical Officer
Otherwise, Rootsies (Rook's Tootsies!)		
P. R. Martin	Pincher	Technical Officer
*R. J. Nixon	Nick or Jack	Technical Officer
W. G. Brunton	Bill	Technical Assistant
S. W. Chisman	Stan	Technical Assistant
*G. G. J. Cooper	George	Technical Assistant
*A. R. Crossfield	Alf	Technical Assistant

*Jonathan Davies	Dave	Technical Assistant
E. R. Farmer	Eric	Technical Assistant
*H. D. Fawell	Fawell	Technical Assistant
(On loan from Ministry of Transport)		
H. D. Griffiths	Griff	Technical Assistant
(?) Loxton-Peacock	L-P	Technical Assistant
J. Mollart-Rogerson	Rodge	Technical Assistant
D. Pierce	Donny	Technical Assistant
W. Sargeantson	Bill	Technical Assistant
*N. J. Skelton	Norman	Technical Assistant
A. Barker	Alan	Scientific Officer
E. Rees	Eric	Scientific Officer
J. Salter	Jeff	Scientific Officer
J. M. N. Willis	Willie	Scientific Officer
P. J. Boyce	Paul	Junior Scientific Officer
*C. J. Matthews	Matt	Designer
*W. G. Barnes	George	Senior Draughtsman
*A. A. Arnold	Alan	Leading Draughtsman
A. K. M. Howie	Alec	Leading Draughtsman
A. Cope	Arthur	Draughtsman
E. Hall	Eric	Draughtsman
K. Kirtley	Ken	Draughtsman
K. Rogers	Ken	Draughtsman
A. Waller	Alec	Draughtsman
Miss V. M. Daws	Vera	Department Clerk
*S. Vincer	Sid	Chargehand
*W. Davis	Bill	Chargehand
*Workshop Personnel	4 Mechanics/labourers	

The following were not an integral part of the staff of Naval Aircraft Department but their services were always available as required:

*H. Ranks	Harry	Head of Photographic
B. W. Chalk (Sgt RAF)	Chalky	Photographer
*J. A. Hampson	James	Photographer
*A. Sherliker	Sherlie	Photographer
L. Andrews	Les (Tich)	Stress Office
*M. R. Jones	Mardy	Stress Office
*E. F. H. Noble	John	Stress Office

FLEET AIR ARM AIRCRAFT IN WORLD WAR 2 SERVICE

ACTIVE & PROJECTED, 1939–45

Name	First Flight	Squadron Service
Fairey IIIF	1926	1929–40
Fairey Seal	1930	1933–42
Hawker Osprey	1930	1932–9
Hawker Nimrod	1931	1932–41
Blackburn Shark	1933	1935–48
Supermarine Walrus ('Shagbat')	1933	1935–44
Fairey Swordfish ('Stringbag')	1933	1936–45
Blackburn Skua	1937	1938–41
Grumman Martlet I & II (USN – F4F-3) – L/L	1937	1940–6
Grumman Wildcat IV – VI (USN – F4F-4) – L/L	1942–6	
Supermarine Sea Otter	1938	1944–50
Gloster Sea Gladiator	1938	1938–41
Blackburn Roc	1938	1940–3
Fairey Albacore	1938	1940–4
Fairey Fulmar	1940	1940–5
Fairey Barracuda I – III	1940	1943–5
Fairey Barracuda V	1944	1947–54
Hawker Sea Hurricane	1940	1941–4
Bell Airacobra (USAAF P-39D)[1]	1941–6	
Fairey Firefly	1941	1943–6
Grumman Avenger I-III (USN–TBM-1)-L/L	1941	1943–6
Grumman Avenger AS.IV (USN–TBM-3E)	–	1953–5
Blackburn Firebrand I	1942	
Grumman Hellcat (USN – F6F-3) – L/L	1942	1943–6
Supermarine Seafire I.B, II.C & III	1942	1942–6
Chance Vought Corsair (USN-F4U-5)-L/L	1943	1943–6
Blackburn Firebrand II & III	1943	
Blackburn Firebrand 4 & 5	"	1945–53
de Havilland Sea Mosquito	1944	1946–7
Fairey Barracuda V	1944	1945–7
Fairey Firefly 4 – 6	1944	1947–56
Supermarine Seafire XV & XVII	1944	1945–54
Supermarine Seafire 45, 46 & 47	"	1946–52
de Havilland Sea Hornet F.20	1945	1947–51
de Havilland Sea Hornet NF.21	"	1949–54
de Havilland Sea Vampire F.20	1945	1948–53
Fairey Spearfish[2]	1945–52	-
Hawker Sea Fury	1945	1947–57
Supermarine Seafang	1945	-

POST-WAR AIRCRAFT

Name	First Flight	Squadron Service
Short Sturgeon[3]	1946	1950–7
Westland Wyvern TF.1 (Eagle)	1946	-
Supermarine Attacker	1947	1951–7
Gloster Meteor III[4]	1948	1948–52
Hawker Sea Hawk	1948	1953–8
Fairey Gannet	1949	1954–60
Fairey Gannet 3	"	1960–78
Westland Wyvern TF.2 (Clyde)	1949	-
Westland Wyvern TF.4/S.4 (Python)	1949	1953–8
de Havilland Sea Venom	1951	1954–9

de Havilland Sea Vixen I	1951	1959–64
Douglas Skyraider (USN–AD-4W)–MDAP	1951	1953–60
Boulton Paul Sea Balliol T.21[5]	1952	1954–7
Supermarine Scimitar	1954	1958–69
Blackburn Buccaneer	1958	1963–78
Hawker Siddeley Sea Vixen	1962	1963–72
Hawker Siddeley Buccaneer S.2	1963	1965–78
McDonnell Douglas Phantom (USN – F-4K)	1966	1968–78
Sepecat Jaguar[6]	1969	–
British Aerospace Sea Harrier	1977	1979–Date

HELICOPTERS

Sikorsky VS-316 Hoverfly I (USAAF-R-4)[7]-L/L	1945	1945–7
Sikorsky Hoverfly II (USAA-R-6A)	–	1947–50
Westland Dragonfly	1949	1950–61
Westland Whirlwind	1953	1957–77
Westland Wessex	1957	1961–79
Westland Wessex HAS.3	"	1967–71
Westland Wessex HU.5	1963	1964–81
Westland Sea King	1969	1970–Date

Notes:
1. One research aircraft, AH574.
2. Four research aircraft, RA356, RA360, RA363, RN241
3. Twenty-three aircraft; most were shore-based for target towing duties.
4. Two research aircraft, EE337, EE387 for shipboard trials in June 1948.
5. Thirty aircraft, shore-based trainers, a few fitted with arrester hooks.
6. One research aircraft for accelerator trials at RAE Bedford.
7. Seven aircraft to FAA for assessment, first known in UK as Gadfly.

General:
L/L – Aircraft received from the USA during World War 2 through the Lend/Lease arrangements.
Under the terms of the agreement these were either returned to the USA or scrapped/ditched during 1945–6.
MDAP – Aircraft supplied by the USA under its 'Mutual Defense Assistance Program'.
(USAAF) – US Army Air Force designation.
(USN) – US Navy designation.

AIRCRAFT CARRIERS IN WORLD WAR 2 SERVICE – HM SHIPS

FLEET, LIGHT FLEET AND ESCORT CARRIERS

These ships comprised the Capital Ships of the Fleet Air Arm, capable of service far from home waters for sustained periods of time, operating an aircraft complement of three or four squadrons in all fields of defence and attack. They were self-supporting with their attendant escort and supply facilities. These ships were equipped with relevant flight deck machinery of varying sophistication in accordance with their period of service and operational needs.

Five fleet carriers were lost to enemy action early in World War 2. Two British and two American (Lend/Lease) escort carriers were lost, one of the latter through an unfortunate accident in the Clyde estuary. A further two of the American ships were badly damaged through enemy action and were not repaired. They were formally handed back to the US Navy at the end of the war but in the 'as seen' condition in the Clyde and Forth respectively and were disposed of there. These 'hulks' were valuable sources of spare parts; even some of the flight deck machinery was recovered for experimental use elsewhere.

British Fleet and Light Fleet Carriers	British Escort Carriers	American Escort Carriers (Contd)
Ark Royal – 1938 (*14.11.41)	Activity – 1942	Hunter – 1943
Colossus – 1944	Argus – 1918	Khedive – 1943
Courageous – 1928 (*17.09.39)	Audacity – 1941 (*21.12.41)	Nabob – 1943 (§ '44)
Eagle – 1924 (*11.08.42)	Campania # – 1944	Patroller – 1943 - {2}
Formidable – 1940	Nairana # – 1943	Premier – 1943 - {2}
Furious – 1917	Vindex # – 1943	Puncher – 1944 - {1}
Glorious – 1930 (*08.06.40)	Pegasus # – 1914+ (ex Ark Royal[14])	Pursuer – 1943 - {1}
Glory – 1945 - {2}	Springbank – 1941+ (*27.09.41)	Queen – 1943 - {1}
Hermes # – 1924 (*09.04.42)		Rajah – 1944 - {1}
Illustrious – 1940 - {4}		Ranee – 1943 - {1}
Implacable – 1944 - {1}	**American Escort Carriers**	Ravager – 1943
Indefatigable – 1944	Ameer – 1943	Reaper – 1944 - {1}
Indomitable – 1941 - {1}	Arbiter – 1943 - {2}	Ruler – 1943
Ocean – 1945 - {1}	Archer – 1941	Searcher – 1943
Perseus – 1945	Atheling – 1943	Shah – 1943 - {1}
Pioneer – 1945	Attacker – 1942	Slinger – 1943 - {1}
Pretoria Castle – 1943 - {7}	Avenger – 1942 (*15.11.42)	Smiter – 1944 - {2}
Unicorn – 1943 - {2}	Battler – 1942 - {1}	Speaker – 1943 - {1}
Venerable – 1945 - {1}	Begum – 1943 - {1}	Stalker – 1942 - {1}
Vengeance – 1945	Biter – 1942	Striker – 1943
Victorious – 1941 - {1}	Chaser – 1943	Thane – 1943 (§ '45)
Warrior – 1946	Dasher – 1942 - {1} (*27.03.43)	Tracker – 1943 - {1}
– – – –	Emperor – 1943 - {1}	Trouncer – 1944 - {2}
	Empress – 1943 - {1}	Trumpeter – 1943
	Fencer – 1943 - {1}	– – – –

* = Lost in Action or on Active Service.
§ = Damaged and taken out of service, not repaired. See Hobbs, (#43 pp.138, 178).
+ = Seaplane/Catapult trials ship.
# = The second ship to bear this name; earlier ships were World War 1 vintage.
HMSs *Perseus* and *Pioneer* were aircraft maintenance ships equipped with flight decks but not used for flying activities; *Perseus* was adapted for initial shipborne trials of the BXS.1 steam catapult.
HMSs *Pegasus* and *Springbank* were catapult ships.
Figures in brackets {*} indicate the number of occasions on which the author visited this particular ship.

AUXILIARY CARRIERS – MERCHANT CONVERSIONS

These ships were merchant auxiliaries of about 8,000 tons, adapted to carry a single fighter aircraft (CAM-ships), or a flight of three or four Swordfish equipped with depth charges (MAC-ships). They were used extensively on convoy escort work during the period known as the Battle of the Atlantic. The ships were manned by the Merchant Service but the aircraft and flight deck machinery were the responsibility of service personnel of the RAF or Fleet Air Arm.

CAM-SHIPS

Catapult Aircraft Merchantmen – CAM-ships – were freighters of the Merchant Marine, each carrying a rocket (P.I) catapult in the bows above the port-side anchor winch. The sole Fairey Fulmar or Hawker Hurricane aircraft, with which the ship was equipped, was used only once. It was loaded on to the catapult in harbour and the rocket motors fitted before sailing. The aircraft would not be recovered from a sortie unless it was within flying distance of a shore establishment. The pilot would have the option of either ditching, or abandoning the aircraft and parachuting near to one of the ships in the convoy.

The first CAM-ship was commissioned in April 1941 and the last saw action in June 1943. The first three of these ships were commissioned as Naval vessels. Those succeeding remained as Merchant Vessels and, as such, were the pride of their crews, taking delight in the knowledge that they had the ability to hit back at any undesirable interloper. Of 36 ships adapted for this service, 10 were lost to enemy action. The following is a list of these ships; the losses by enemy action are indicated by asterisks (*).

CAM-ships – (April 1941–June 1943)

HMS *Ariguani*, HMS *Maplin*, HMS *Patia* (*27.04.41) and subsequently HMS *Springbank* (*27.09.41) and HMS *Pegasus* were Auxiliary Fighter Catapult Ships.

MVs: *Daghestan, Eastern City, Empire Burton, Empire Clive, Empire Darwin, Empire Dell*, Empire Eve, Empire Faith, Empire Flame, Empire Foam, Empire Franklin, Empire Gale, Empire Heath, Empire Hudson, Empire Lawrence* (*15.05.42), Empire Moon, Empire Morn, Empire Ocean, Empire Rainbow*, Empire Ray, Empire Rowan, Empire Shackleton*, Empire Spray, Empire Spring, Empire Stanley, Empire Sun*, Empire Tide, Empire Wave*, Helencrest, Kafiristan, Michael E* (*01.06.41), Novelist, Primrose Hill** (36 ships).

It is interesting that Commander Colin C. Mitchell RNVR (who, during the war years, was on the staff of the Admiralty Engineer-in-Chief filling the post of 'Officer Commanding' Flight Deck Machinery) compiled a list of CAM-Ships which omits HMS *Patia* and MV *Michael E*, both of which were lost to enemy action in April and June 1941 respectively. His schedule probably was compiled after the loss of these two ships. He has substituted MV *Dalton Hall*, which the author has not seen in any other list, and is not included above. The Auxiliary Fighter Catapult ships, HMSs *Pegasus* and *Springbank*, also are excluded, being somewhat of a special character. Mitchell also states that two catapult units were sent to RAF Speke, presumably held against a further requirement. It is understood that each MV carried a single Hawker Hurricane aircraft while the two 'Catafighter' ships (HMSs *Pegasus* and *Springbank*) carried more than one Fairey Fulmar each.

MAC-SHIPS

Merchant Aircraft Carriers – MAC-ships – were either tankers or bulk grain carriers (Container Ships had not been invented at that time) fitted with a full-length flight deck above the normal weather deck. None of these was fitted with a catapult. There were two types of MAC-ship. The earlier vessels carried a complement of three Fairey Swordfish aircraft ranged on the flight deck. In this configuration four arrester wires were provided with a capacity to arrest 15,000lb at 55kts. A single safety barrier installation was rigged forward of the landing-on area. The below deck gear comprised the standard hydro-pneumatic jigger units as used by the fleet carriers. (As three under-deck systems would be provided, common sense would have suggested that the 'spare' way would have been used as a fifth wire. Some aerial photographs indicate that such a wire did exist.) During operation, the aircraft remaining on deck would be man-handled along the deck and parked, forward of the barrier, until the machine airborne had been recovered. The aircraft then would be ranged aft again until the subsequent machine had taken off and the exercise repeated.

Each MAC-ship originating as a bulk carrier, and thereby possessing a weather deck reasonably clear of obstructions, was adapted to provide a small hangar below the flight deck. In these instances four arrester wires were provided, the additional safety barrier not being necessary. The aircraft could be serviced out of the weather and each machine ranged as necessary without a deck-handling party being required, a decided advantage in North Atlantic weather. Conversely, where a tanker formed the basic hull, having the clutter in the way of pipe-work, manifolds and valves on the weather deck, provision could not be made for an aircraft hangar. In such instances a safety barrier was installed in addition to the four arrester wires. Of the 19 ships adapted, six were provided with hangars.

The first MAC-ship was commissioned in April 1943 and the last just a year later. There appears to be no record of any of these ships having been lost through enemy action. The adaptation of these ships reduced their bulk carrying capacity by some 20 percent. This was thought to be a small price to pay for the benefit of their usefulness in convoy-protection duties. The crews of these ships were very jealous of their role in operating these aircraft; in some cases they asked for the identification of the Merchant Service to be displayed on the Swordfish which flew from their ships. The aircraft for this service comprised RNAS836, each complement for a ship belonging to a flight with a distinguishing service letter, thus preserving a sense of 'belonging' to a particular squadron and enjoying its camaraderie. The following is a list of these ships. (Poolman (#71 p.117) adds the name of MV *Medula*. The author has not seen this reference elsewhere and it may be a misprint for '*Miralda*'.)

MAC-ships – (April 1943–March 1944)

MVs: *Acavus, Adula, Alexia, Amastra, Ancylus, Empire MacAlpine* (h), *Empire MacAndrew* (h), *Empire MacCabe, Empire MacCallum* (h), *Empire MacColl, Empire MacDermott* (h), *Empire Mackay, Empire MacKendrick* (h), *Empire MacMahon, Empire MacRae* (h), *Gadila, Macoma, Miralda, Rapana* (19 ships). (Ships marked [h] were dry cargo [grain] ships, equipped with hangar decks.)

APPENDIX 5

CATAPULTS, PRE-1939 – SHIPBOARD & ASHORE

The following is a rather sketchy summary of the catapults mounted in battleships and cruisers during the inter-war period. They were used mostly for launching floatplanes and, later, amphibian aircraft for reconnaissance and gun-spotting purposes. Only brief references may be made, mainly through detailed information not having been located by the writer. Dimensions have been rounded to the nearest foot. Not all prefixes and suffixes can be identified but it is thought that 'C' refers to units built at Chatham Dockyard, 'D' indicates a double-acting, athwartships installation and 'E' describes an extending structure for the catapult beam. (To add to the confusion, some of the 'D' series were built on extending beams – similar in these respects to the 'E' Series. The exceptions were the D.II.H and D.III.H machines which had the trolley track, non-extendible, fixed direct to the deck.) 'F' refers to a Farnborough design, 'S' describes the slider type of unit and 'V' is thought to refer to one or more machines built by one of the Vickers consortium. The central Roman numerals indicate Mark numbers, while the sub-scripts 'H', 'L' and 'T' identify 'Heavy', 'Light' and 'Turret-mounted' equipment respectively. Originally 'Heavy' referred to aircraft of weight up to 8,000lb and 'Light' up to 6,000lb launched at about 50kts. These figures were uprated as the development of catapults progressed.

The identities of manufacturers, where known, are given at the end of each entry:

(BB) = Brown Brothers & Co, Edinburgh.
(MS) = MacTaggart Scott & Co, Edinburgh.
(R&R) = Ransomes and Rapier Limited, Ipswich.

The 'F' series of catapults were designed and, apart from some of the larger components, were mostly manufactured at RAE, Farnborough. A large, flat bed, centre lathe was installed in the RAE Machine Shop [Building Q.27], immediately inside the east door, for work on these telescopic tubes. The tubes themselves were obtained from the English Steel Corporation, Sheffield. Occasionally the industrial resources of HM Dockyard, Chatham (sometimes Portsmouth) were used for the manufacture of other heavy ironware.

To avoid the proliferation of numbers in this summary, metric equivalents of dimensions have not been provided. The figures given for the stroke shows the accelerating distance first, followed by the distance for retarding the mechanism.

C.IV.H No specific information for this marque is available. It is thought that this identity relates to catapults constructed in Chatham Dockyard to the Carey pattern. It is known that in addition to the proto-type fitted in HMS *Vindictive*, two catapults were built to a 'modified' Carey design. These were installed in HMS *Resolution* (1926) and HM Submarine M.2 (1927). They were operated by compressed air to launch 7,000lb at 45kts at 2g. Accelerating stroke 34ft with 3:1 reeving. A Carey type catapult, to RAE design, was built by MacTaggart Scott (1927) for installation at Farnborough. This unit is probably that described as being in use in 1930. It is not known for how long this installation remained at Farnborough. It probably worked alongside the Salmon type F.I.H. on Jersey Brow for a while and was then removed for operation elsewhere.

D.I.H (15 units) Double-acting, athwartships unit. Design performance 8,000lb at 56kts (upgraded to 12,000lb at 56kts) at 3g. Length of beam with double extension 90ft, stowed 50ft; with 8:1 reeving. Stroke 61/19ft. Cordite-operated. Weight 34 tons (BB).

D.I.L (2 units) Similar to D.I.H unit. Design performance 5,500lb at 53kts. Length extended 82ft, stowed 46ft; with 8:1 reeving. Stroke 55/17ft. Cordite-operated. Weight 32 tons (BB).

D.II.H (3 units) Length of trolley runway fixed to deck 90ft, to launch 8,000lb (upgraded to 12,000lb) at 56kts; with 8:1 reeving. Cordite operated. (MS)

D.III.H (8 units) Design performance (upgraded) 15,000lb at 70kts (MS).

The D.II.H and D.III.H machines by MacTaggart Scott were installed with the track fixed direct to the deck. A pair of rolled steel channel sections formed the track and was set to a gauge of 4ft 6in, standing 15in proud of the deck. The machinery was installed below deck, occupying a space 7ft deep and running athwartships by some 80–100ft, this distance depending upon the beam of the ship.

D.IV.H (17 units) Similar to D.I.H unit, Design performance 12,000lb (upgraded to 15,000lb) at 60kts (BB).

E.I.H (9 units) Double extension, pivot mounted unit. Design performance 8,000lb at 50kts. Length of beam extended 76ft, stowed 46ft; with 4:1 reeving. Stroke 51/14ft. Cordite-operated. Weight 24 tons (MS).

E.I.T Single extension, gun turret-mounted unit. Design performance 7,000lb at 50kts. Length of beam extended 71ft, stowed 52ft; with 6:1 reeving. Stroke 50/12ft. Cordite-operated. Weight 21 tons (MS).

E.II.H (8 units) Double extension, pivot-mounted unit. Dimensions as for E.I.H unit. First unit installed for trials in HMS *Ark Royal*[14] in 1931 (MS).

E.II.T Double extension, gun turret-mounted unit. Design performance 8,000lb at 52kts, at 2¼g. Length of beam extended (hydraulically) 75ft, stowed 45ft; with 6:1 reeving. Stroke 53/13ft. Cordite-operated. Weight 26 tons. First unit installed for trials in HMS *Ark Royal*[14] in 1938 (MS).

E.III.H (10 units) Double extension, pivot-mounted unit. Design performance 8,000lb at 57kts. Length of beam extended 90ft, stowed 53ft; with 6:1 reeving. Stroke 63/18ft. Cordite-operated. Weight (units 1–6) 27 tons (units 7–10) 30 tons. First unit installed for trials in HMS *Ark Royal*[14] in 1933 (MS).

E.III.T (2 units) Single extension, gun turret-mounted unit. Design performance; 8,000lb (upgraded to 12,000lb) at 52kts at 2¼g. Length of beam extended 75ft, stowed 53ft; with 6:1 reeving. Stroke 53/13ft. Cordite powered. Weight 26 tons (MS).

E.IV.H (3 units) These units are recorded as having been developed from the E.II.H series. The first and third units were manufactured by MacTaggart Scott, Glasgow, while the second unit is recorded as having been manufactured by Stothert & Pitt, Bristol. Only the first of these units was a shipboard installation, in HMS *Effingham*. The second was despatched for installation in HMAS *Perth* but was returned to store in UK. The third was installed firstly at Woolwich Arsenal for ballistic trials, then transferred (1943) to RNAS *East Haven* – satellite airfield to HMS *Condor* (RNAS Arbroath) – but was never used; ships' catapults were about to be phased out of service.

F.I.H Located at RAE in May 1924. (Latterly, and locally, known as the Mark I unit on Jersey Brow.) Design performance 7,000lb at 39kts at 2¼g. Four unit telescopic ram machine, originally powered by compressed air but modified for cordite operation. It had completed 500 operations by October 1926. Similar units of improved performance were installed in HMSs *Barham*, *Resolution* and *Royal Sovereign* (MS).

F.I.L Located at Watchet, Somerset, for 'Queen Bee' (unpiloted, radio-controlled) experiments. Performance; 5,000lb at 54kts at 2¼g, or at 49kts @ 2¾g. Stroke 51/8ft, overall length 83ft. This was a catapult in which the forward length of the beam could be power or manually folded back on itself, reducing the stowed length to 44ft. The beam was mounted on a circular track of 24ft diameter. It was a telescopic tube system with four extensions, originally powered by compressed air at a pressure of 1,800psi, but subsequently was cordite-operated. The retrieval pressure at the end of the launch stroke was 350psi. Weight 12

An 'E' type extending catapult nearing completion at the works of Messrs MacTaggart Scott, Edinburgh. The catapult beam is in its fully retracted configuration. The triangular structure mounted on the trolley base is the temporary rig for launching a dummy weight when conducting acceptance trials for the equipment. (MacTaggart Scott)

A 1930s scene with an 'E' type extending catapult loaded onto a railway flat-bed bogie wagon in the goods yard of Loanhead railway station. After a trial erection in their works, MacTaggart Scott dismantled the catapult and then re-assembled it on the railway wagon. On delivery to site, a dockyard crane would off-load the entire unit for shipboard installation. (MacTaggart Scott)

tons. One of these units was installed for a while in 1927 on Jersey Brow.

F.II.H A three-ram telescopic tube unit, built in February 1926. Performance 7,000lb at 39kts at 2g. Stroke 33/3ft. Cordite-operated. This probably was the 'Larynx' catapult in HMS *Stronghold*.

F.II.L (4 units) – As F.I.L, designated for launching of Queen Bee aircraft.

F.III.H No information available. One speculation is that this unit was the RAE Mark II Catapult on Jersey Brow, powered by compressed air.

F.IV.H Located at RNAS Gosport (HMS *Daedalus*). A folding catapult, mechanically similar to the F.I.L. Design performance 8,000lb at 55kts at 2½g. The first unit installed for trials in HMS *Ark Royal*[14] in 1929.

R.III.L (2 units) One tested at RAE, scheduled for Singapore; the other at Weybourne (?). (No record of this machine has come to light; speculation would suggest that it is a misprint for F.III.L.)

S.I.H Slider type catapult, pivot-mounted unit. Design performance; 8,000lb at 49kts or 6,000lb at 54kts at 2¼g to 2¾g for launching Fairey IIIF, Hawker Nimrod & Osprey aircraft. Length of beam 46ft with overall travel envelope of the slider 65ft; with 3:1 & 2:1 reeving. Stroke 48/9ft. Cordite-operated. Weight 20 tons. The first unit was installed for trials in HMS *Ark Royal*[14] in 1932 (R&R).

S.I.L (1 unit) As for S.I.H, cordite-operated, with 2:1 & 2:1 reeving. First unit installed for trials in HMS *Ark Royal*[14] in 1930 (R&R).

S.I.T (1 unit) As for S.I.H unit but mounted atop a gun turret. Design performance 7,000lb at 50kts at 2¼g. Length of beam 47ft with overall travel envelope of the slider 66ft; Length of fixed beam 47ft with overall trolley travel 67ft. with 3:1 & 2:1 reeving. Cordite-operated. Weight 16 tons (R&R).

S.II.T (1 unit) (R&R).

S.II.L (4 units) One destined for Malta. Slider type catapult, pivot mounted. Design performance; 5,500lb at 50kts at 2¼g. Length of beam 46ft, overall travel envelope of slider 83ft; with 2:1 & 2:1 rope reeving. Stroke 4812ft. Cordite-operated. Weight 16 tons. The first unit installed for trials in HMS *Ark Royal*[14] in 1931 (R&R).

S.III.L (1 unit) Located at Manorbier, Pembrokeshire. Slider type catapult, pivot mounted. Design performance; 5,500lb at 54kts at 2¼g for launching Queen Bee aircraft. Length of beam 46ft; overall travel envelope of slider 83ft; with 2:1 & 2:1 rope reeving. Stroke 58/17ft. Cordite-operated. Weight 17 tons (R&R).

V.I.H One unit located at RAF Leuchars, Fifeshire. This mark also could be the machine to which reference is made by Friedman (#32) as destined for HMS *Hood*. Performance 7,000lb at 55kts with a folding catapult beam of 74ft extended length.

Crowfoot, in the issue of *RAE NEWS* (May 1953, p.17), refers to the first Navy catapults having been built at RAE as well as the RAE Mark I (F.I.H.?), F.IV.H and F.I.L machines. Friedman (#32 p.179) gives the weights of various launchers as:

Heavy catapult (12,000lb at 55kts)   23–60 tons
Light catapult (5,500lb at 50kts) 16–20 tons

APPENDIX 6
SHIPBOARD ACCELERATORS – 1935–1978

INSTALLATIONS IN UK SHIPS

BH.I ACCELERATOR

HMSs

Glorious (2) (P/t)	1935	(Refit)
Courageous (2)	1936	(Refit)
Argus	1938	(Refit)
Ark Royal[38] (2)	1938	
(ex-*Audacious*)		
Magnificent (P/t) 1948 Maj		

BH.V (BH.5) ACCELERATOR

HMSs

Albion (2)	1954	Cen
Bulwark (2)	1954	Cen
Centaur (2)	1953	Cen
Eagle[51] (2)	1951	Aud

BH.III ACCELERATOR

HMSs

Illustrious[40] (P/t)	1940	Ill
Colossus	1944	Col
Formidable	1940	Ill
Glory	1945	Col
Implacable	1944	Imp
Indefatigable	1944	Imp
Indomitable	1941	Ill
Ocean[45]	1945	Col
Sydney	1949	Maj:
(ex-*Terrible*)		
Theseus	1946	Col
Triumph	1946	Col
Unicorn	1943	A/c repair ship
Venerable	1945	Col
Vengeance	1945	Col
Victorious	1941	Ill
Warrior	1946	Col

BS.IV (BS.4) STEAM CATAPULT

HMSs

Ark Royal[55] (2)	1955	Aud
Bonaventure	1957	Maj
(ex-*Powerful*)		
Centaur (2)	1958	(Refit)
Eagle[51] (2)	1964	(Refit)
(ex-*Audacious*)		
Hermes[59] (2)	1959	Cen
(ex-*Elephant*)		
Melbourne	1955	Maj
(ex-*Majestic*)		
Perseus (P/t)	1945	Col
Victorious (2)	1958	(Refit)
Vikrant	1961	Maj
(ex-*Hercules*)		

BC.II ACCELERATOR

HMS

Pretoria Castle (P/t)	1943

Notes:

1. The keels of all the above wartime and post-war ships were laid before the end of World War 2; completion of some was deferred until post-war requirements had been formulated.

2. Figures '1945' and so on depict the year of completion of the ship or its relevant refit.

3. Abbreviations:
   Aud: 'Audacious' Class   Ill: 'Illustrious' Class
   Cen: 'Centaur' Class   Imp: 'Implacable' Class
   Col: 'Colossus' Class   Maj: 'Majestic' Class

4. (2) refers to two accelerators having been fitted; all other ships carried one accelerator.

5. (P/t) refers to prototype shipboard installation of accelerator.

The following carriers were/are not equipped with accelerator gear: HMSs *Activity*, 1942; *Ark Royal*, 1985; *Audacity*, 1941; *Campania*, 1944; *Eagle*, 1924; *Furious*, 1930; *Hermes*, 1924; *Illustrious*, 1982; *Invincible*, 1980; *Leviathan*, (0); *Nairana*, 1943; *Ocean*, 1998; *Pioneer*, 1945; *Vindex*, 1943.

Equipment removed from: HMSs *Albion*, 1962; *Bulwark*, 1960 (Commando carriers).

INSTALLATIONS IN US LEND/LEASE SHIPS

AH.II ACCELERATOR		AH.IV ACCELERATOR	
'Archer', 'Attacker' &		'Ruler' Class	
'Avenger' Classes			
HMSs		HMSs	
Archer	1941	*Ameer*	1943
Attacker	1942	*Arbiter*	1943
Avenger	1942	*Atheling*	1943
Battler	1942	*Begum*	1943
Biter	1942	*Emperor*	1943
Chaser	1943	*Empress*	1943
Dasher	1942	*Khedive*	1943
Fencer	1943	*Nabob*	1943
Hunter	1943	*Patroller*	1943
Pursuer	1943	*Premier*	1943
Ravager	1943	*Puncher*	1944
Searcher	1943	*Queen*	1943
Stalker	1942	*Rajah*	1944
Striker	1943	*Ranee*	1943
Tracker	1943	*Reaper*	1944
Ruler	1943		
Shah	1943		
Slinger	1943		
Smiter	1944		
Speaker	1943		
Thane	1943		
Trouncer	1944		
Trumpeter	1943		

All ships in the 'Archer', 'Attacker' and 'Avenger' Classes were equipped with the AH.II Accelerator. These were ships completed between November 1941 (HMS *Archer*) and June 1943 (HMS *Pursuer*). Subsequent ships were all of the 'Ruler' Class and were equipped with the updated AH.IV Accelerators. All 'Ruler' Class ships were constructed by the Seattle-Tacoma Shipbuilding Corporation, Seattle, the last two of the series (HMSs *Puncher* and *Reaper*) being completed in July 1944.

Of these, HMS *Avenger* was lost to enemy action in November 1942 and HMS *Dasher* by an explosion in the Clyde in March 1943. HMS *Nabob* was torpedoed in Norwegian waters in August 1944 and HMS *Thane*, similarly, in the Clyde approaches in January 1945. Both were able to reach home ports where they were decommissioned and laid-up.

Being operated under the Anglo-American Lend/Lease agreement, at the end of World War 2, the surviving ships were formally returned to the USA. The damaged HMSs *Nabob* and *Thane*, were retained in the UK and their flight deck machinery installations removed, some of which were adopted for research purposes. *Nabob* was bought by a company which restored the ship for commercial use, retaining its wartime name in its civilian role.

All the remaining ships, apart from 11 retained for ferrying and trooping duties, crossed the Atlantic on their return journeys, departing the UK between November 1945 and February 1946. Of those remaining, HMSs *Chaser*, *Reaper*, *Smiter* and *Trumpeter* were returned within the next four months and the final ships, HMSs *Atheling*, *Fencer*, *Patroller*, *Queen*, *Rajah*, *Ranee* and *Speaker* had all returned home by the end of 1946.

ACCELERATOR SPECIFICATIONS

The following information has been obtained from a variety of sources, the principal ones being Brown (*#13*), Friedman (*#32*) and Hobbs (*#43*). It is not possible to give precise dates for some of the upgrading exercises. These generally were undertaken during a refit, particularly if a new marque of aircraft was to be embarked during the subsequent re-commission.

BH.I – BRITISH HYDRAULIC ACCELERATOR
(HMSs *Glorious*, *Courageous*, *Argus* and *Ark Royal*[38]
First fitted in HMS *Glorious*, 1936.

Prototype (1935)	8,000lb at 56kts	{3,640kg at 29m/sec}	Trolley
upgraded to	10,000lb at 52kts	{4,550kg at 27m/sec}	Trolley
upgraded to (1938)	12,000lb at 56kts	{5,450kg at 29m/sec}	Trolley

AH.II – AMERICAN HYDRAULIC ACCELERATOR
('Archer', 'Avenger' & 'Attacker' Classes)
First fitted in HMS *Archer*, 1941.

Prototype (1941)	7,000lb at 61kts	{3,180kg at 31m/sec}	Bridle
upgraded to (1941)	7,000lb at 70kts	{3,180kg at 36m/sec}	Bridle

BH.III – BRITISH HYDRAULIC ACCELERATOR
('Illustrious', 'Colossus' and early 'Majestic' Classes)
First fitted in HMS *Illustrious*[40], 1940.

Prototype (1940)	11,000lb at 66kts	{5,000kg at 34m/sec}	Trolley
upgraded to	12,500lb at 66kts	{5,680kg at 34m/sec}	Trolley
or	14,000lb at 66kts	{6,350kg at 34m/sec}	Bridle
upgraded to	16,000lb at 66kts	{7,260kg at 34m/sec}	Trolley
or	20,000lb at 56kts	{9,090kg at 29m/sec}	Bridle
upgraded to (1949)	20,000lb at 66kts	{9,090kg at 34m/sec}	Bridle

AH.IV – AMERICAN HYDRAULIC ACCELERATOR
('Ruler' Class)
First fitted in HMS *Ruler*, 1943.

First installations	11,000lb at 74kts	{5,000kg at 38m/sec}	Bridle
All ships (1943–44)	16,000lb at 74kts	{7,260kg at 38m/sec}	Bridle

BC.II – BRITISH CORDITE ACCELERATOR
(Prototype BC.I installed on Jersey Brow, Farnborough, 1941.)
Sole shipboard installation fitted in HMS *Pretoria Castle*, 1943.

Installation (1943)	14,000lb at 66kts	{6,350kg at 34m/sec}	(Trolley and upgraded to
	16,000lb at 66kts	{7,260kg at 34m/sec}	Bridle)

BH.V – BRITISH HYDRAULIC ACCELERATOR
(**BH.5** – CATAPULT)
('Majestic' and Fleet Carrier Classes)
First fitted in HMS *Centaur*, 1948.

Prototype (1948)	20,000lb at 56kts	{9,100kg at 29m/sec}	Bridle
upgraded (1954)	30,000lb at 75kts	{13,600kg at 39m/sec}	Bridle

BS.IV – BRITISH STEAM CATAPULT
(**BS.4** – CATAPULT)
('Majestic' and Fleet Carrier Classes)
Prototype installation (BXS.1) fitted 1950 in HMS *Perseus* (1945) for trials purposes subsequently installed at RAE Bedford.

First installation (1953)	40,000lb at 78kts	{18,150kg at 40m/sec}	Bridle
upgraded to	50,000lb at 94kts	{22,700kg at 48m/sec}	Bridle
upgraded to	50,000lb at 105kts	{22,700kg at 54m/sec}	Bridle

BS.4 Catapults were fitted in the following ships:

HMS *Ark Royal*[55]	2 x 151ft {46m} stroke – upgraded later
HMCS *Bonaventure*	1 x 103ft {31m}
(ex-*Powerful*)	
HMS *Centaur*	2 x 139ft {42m}

HMS *Eagle*[51] (ex-*Audacious*)	1 x 151ft & 1 x 199ft {46m & 61m}
HMS *Hermes*[59] (ex-*Elephant*)	1 x 151ft & 1 x 175ft {46m & 53m}
HMAS *Melbourne* (ex-*Majestic*)	1 x 103ft {31m}
HMS *Perseus*	1 x 200ft {61m} – BXS.1 prototype
HMS *Victorious*	2 x 145ft {44m}
INS *Vikrant* (ex-*Hercules*)	1 x 103ft {31m}

Brazilian Navy ship *Minas Gerais* (ex-HMS *Vengeance*) is believed to be fitted with a BS.4 installation.

Shore-based:	
RAE Bedford	1 x 200ft {61m} – BSX.1, ex-HMS *Perseus*
"	1 x 200ft {61m} – BSX.2 experimental, developed through to Mark 4 (production) model, then to prototype Mark 5.
ARDE Shoeburyness	1 x 48ft {15m} – BSX.3 (Cordite-operated, short stroke model for catapult component testing.)

Friedman (*#32 p.331)* gives the weight of the BS.4 installation as 275 tons. This was stated to be similar in weight to that of a carrier lift, and occupied about 1,300sqft {120m^2} of equivalent hangar space.

The table below is based upon Lewis (*#58 p.42*) showing the calculated power generated by each of these accelerators. The imperial units are in foot-pounds (force) and horsepower while the metric equivalents are given as kilogrammes-metres (force) and kilowatts respectively.

Name	ftlb	kgm	hp	kW
BH.I	600,000	{83,000}	550	{410}
BH.I	1·5M	{307,000}	1,360	{1,020}
BH.III	2M-	{277,000}	1,820	{1,360}
"	4M	{553,000}	3,640	{2,710}
BH.V	8M-	{1·1M}	7,270	{5,420}
"	10M	{1·4M}	9,090	{6,780}
BS.4	17M	{2·4M}	15,460	{11,530}
BS.5	30M	{4·2M}	27,270	{20,350}

The horsepower/kilowatt ratings are calculated on the basis of the power being generated for two seconds' duration of launch.

SHIPBOARD ARRESTER GEARS

The tables below depict the performance characteristics of the various arrester gear installations. The data have been extracted from Chapman (*#19*) by permission of the Royal Institution of Naval Architects. Metric dimensional equivalents have not been included.

The following is an historical summary of these machines.

Mark 1 – This, in January 1931, was the first attempt at mechanical arresting of aircraft on shipboard with HMS *Courageous* being the 'guinea pig'. The equipment was manufactured by Portsmouth Dockyard to RAE design. A wire was stretched athwartships, across the deck, with each end terminating in a winch unit fitted with a drum type, dry friction brake. The pull-out was designed to be 200ft {61m}. A Fairey IIIF aircraft (S1781) was fitted with an arrester hook and trials were carried out; these were not too successful. It was not possible to equalise the forces on the two winches and the aircraft, on landing, tended to slew across the deck, sometimes violently. The applied braking load was constant throughout the arrest and the aircraft was subjected to tail slam in the process. As aircraft at that time were fitted with tail skids instead of tail wheels, damage to the aircraft often ensued. Some improvements had been incorporated by the end of the year to the effect that further units were put into production. However, by the end of 1932 a replacement system had been prepared. The whole procedure in fact, although unsatisfactory from many aspects, was regarded as establishing a principle.

Mark 2 – Is thought to have been the logical follow-through of the Mark 1 design, being the production version of its prototype. The brakes were to be cross-coupled through a differential gearbox. Friedman (*#32*) regards this to have been an abortive friction-brake production model and was never installed.

Mark 3 – This was the first 'jigger' type unit adopting the hydro-pneumatic principle, introduced again in HMS *Courageous* during her 1933 refit. It had an improved performance with a reduced pull-out of 140ft {42m}. This would suggest that the 12:1 rope reeving was adopted at this point, to be continued as standard for many years. It is thought that this mark was fitted subsequently to the remaining pre-World War 2 carriers during their respective 1933–8 refits.

Mark 3* – An updated version of the Mark 3 unit; fitted, 1938, in HMS *Ark Royal*[38].

Mark 4 – Although developed from the Mark 3 unit this was a new version and, from 1940, was adopted for HMSs *Illustrious*[40], *Formidable* and *Victorious*. This gave a deck pull-out of 150ft {46m}. Farnborough came into the picture here when one of the Mark 4 units became the first equipment to have been installed alongside the newly completed Runway 04/22 as a research facility for the Catapult Section (later, Naval Aircraft Department).

Mark 5 – Supposedly designed, but not fitted in any ship.

Mark 6 – This was an improved version of the Mark 4 unit, dating from 1941, and was installed in HMSs *Indomitable*, *Implacable* and

TYPE	DATE	WEIGHT (lb)	SPEED (knots)	MAX 'g'	PULL-OUT (feet)	SHIPS FITTED
Mk.1	1931	6,000	60	1·0	200	*Courageous*
Mk.3	1933	8,000	60	1·0	140	*Courageous*
Mk.3*	1938	8,000	60	1·0/1·5	140	*Ark Royal*[38]
Mk.4	1940	7,000/20,000	60	2·0	150	*Illustrious*[40] *Victorious* *Formidable*
Mk.6	1941	7,000/20,000	60	2·0	150	*Indomitable,* *Implacable,* *Indefatigable*
Mk.6*	1950	7,000/20,000	62/81	2·0/3·4	150	*Illustrious*[40] *Indomitable* *Implacable* *Unicorn*
Mk.6**	1950	15,000/20,000	68/77	2·2/2·4	145	*Illustrious*[40], *Implacable*
Mk.7&8	1945	7,000/15,000	60	2·0	150	*Magnificent, Warrior*
Mk.8*	1949	10,000/16,000	61/74	1·7/2·4	150	'Colossus' Class
Mk.10	1952	12,500/30,000	75/88	2·3/3·7	160	*Eagle*[51]
Mk.11	1955	12,500/30,000	75/88	2·3/3·7	160	*Albion* *Bulwark* *Centaur*
Mk.10*	1955	12,500/30,000	75/100	2·0/4·0	160	*Eagle*[51]
Mk.11*	1956	12,500/30,000	75/100	2·0/4·0	160	*Albion* *Bulwark* *Centaur*
Mk.12	1953	10,000/20,000	85+	–	180	*Bonaventure, Majestic* *Warrior*
Mk.13	1955	10,000/30,000	85+	–	220	*Ark Royal*[55], *Centaur* *Eagle*[51] *Hermes*[59], *Victorious*

'*' indicates an updating modification to have been incorporated in the basic unit.
Designs were prepared for Mk.2, Mk.5 & Mk.9 equipment but were not fitted in any ship.

Indefatigable. By 1947, one unit was installed at Farnborough alongside Runway 04/22.

Mark 6* – A 1950 uprating of the Mark 6 gear installed in HMSs *Illustrious*[40], *Implacable* and *Indefatigable*.

Mark 6** – Yet another 1950 upgrading of the Mark 6 machinery, used in ships of the 'Illustrious' Class.

Marks 7 & 8 – The difference between these two marks is not known. It is thought that the Mark 7 was a modified, and simpler, development of the Mark 6 equipment destined for the British Escort Carriers, exemplified in HMS *Audacity*. The Mark 8 was a further development to suit the British Light Fleet Carriers fitted, 1945, in HMSs *Magnificent* and *Warrior*, possibly other ships in these categories as well.

Mark 8* – Updated versions of the Mark 8 fitted from 1949 in all ships of 'Colossus' Class, HMS *Magnificent* & HMAS *Sydney*.

Mark 9 – No record of equipment to this mark has been found.

At this stage it was generally considered that both the Mark 6 and Mark 8 systems had reached their limits with their performance envelopes no longer capable of being extended. Further attempts at upgrading were felt not to be worthwhile. All future experimental effort, it was proposed, should be devoted to the development of an entirely new arrester system. This may be the reason why the Mark 9 system does not appear in any schedule of arrester equipment.

Mark 10 – This was an entirely new concept of the jigger installation with its origin in 1946. Its prototype was installed at Farnborough for evaluation and development. The departure from the norm was due to its being mounted vertically in a deep pit alongside Runway 04/22. Shipboard arrester gear installations of earlier marks invariably were constructed with the under-deck machinery installed horizontally below the flight deck, in the hangar roof, or in adjacent machinery flats. All this heavy ironmongery at high level in the ship's hull caused headaches to naval architects and an investigation was being made to evaluate what would happen if the machinery were mounted vertically, thereby lowering the centre of gravity of these heavy masses of ironwork. Such an installation would affect the performance of the aircraft and the equipment in that the lifting of the dead-weight of the moving mass in the arrester gear would need to be added to the inertia and pressure forces already acting within the system. The intention was for this equipment to be installed in HMS *Ark Royal*[55] but as construction of this ship had been delayed, HMS *Eagle*[51] was programmed to operate this project from her first commission in 1951.

Whipp (*ARC 15,233 'A Note on Recent Developments in Naval Arresting Gears, October 1952')*, provides an illustration which shows this ship as having 16 arrester wires and six safety barriers; it is to be presumed from this that 11 (or more) sets of under-deck machinery were provided. The installation at Farnborough was used to determine and evaluate new methods of fluid transfer and pressure control within the jigger unit to meet the enhanced requirements imposed by new aircraft with increased weight and performance characteristics.

Mark 10* – Progress had been made through trials both at Farnborough and on shipboard which enabled the equipment to be upgraded in 1955.

Mark 11 – The Mark 10 equipment was installed (1955) reverting to the previously standard horizontal format, slung from the deckhead in HMSs *Albion*, *Bulwark* and *Centaur*. This would suggest that the advantages of installing the units in vertical trunks were not as great as anticipated. (Additionally, the presence of such vertical trunks may well have compromised the damage control integrity of the ship should the hull be punctured through mishap or combat action.)

Mark 11* – The Mark 10* uprating in HMS *Eagle*[51] was carried across to the other ships mounting the Mark 11 equipment.

Mark 12 – This was new equipment of conventional design but with increased stroke providing a pull-out of 185ft {56m}, with 14:1 reeving. Its prototype was installed and tested at Farnborough in 1955 and was installed in the modified 'Colossus' and 'Majestic' Classes, including HMSs *Bonaventure*, *Warrior* and *Majestic*.

Mark 13 – A completely new development, again of conventional design with further upgrading, but providing a mean pull-out of 220ft {67m}, with a higher rope reeving ratio of 16:1. (The Instruction Manual gives the pull-out range as 224–35ft, relevant to the ship's installation.) The prototype was developed on the High Speed Catapult site at P&EE, Shoeburyness, using the BSX.3 slotted tube catapult for propelling dead-loads into the system. Some 270 trial engagements with dead-loads were completed here. Commissioning and proof-testing with dummy and 'live' aircraft were conducted at the new Proving Base on the RAE Bedford site, where, by March 1955, more than one thousand trial runs had been completed. During the development phase the original specification requirements of work-load were increased by some 30 percent, thereby doubling the performance of the Mark 10 unit. The equipment was installed in HMSs *Ark Royal*[55], *Centaur*, *Eagle*[51], *Hermes*[59] and the modernised *Victorious*.

At this stage, with the adoption of the angled deck and the ability to accommodate a much longer pull-out of the arrester wires, the number of wires was reduced to four, five or six. HMS *Ark Royal*[55] carried six wires but numbers one and six generally were left unrigged, the centre span of the former being in the way of access to the after lift.

Mark 14 – This was intended to be a replica of the Mark 12 equipment with increased capacity for installation in the modernised ships of the 'Colossus' Class. It is thought to have been fitted in HMS *Eagle*[51] in 1967 for what was to be her final commission. The equipment was rated to accept an aircraft of 40,000lb weight at an entry speed of 125kts {18,000kg, 64m/sec}.

Griffith (*#248 & 149*) reported in 1955 on trials with both Mark 12 and Mark 13 arrester units, recording the following comparisons:

	Mark 12	*Mark 13*
Reeving ratio	14:1	16:1
Maximum stroke	12ft-10in {3·9m}	12ft-3in {3·7m}
Ram external diameter	12in {300mm}	14½in {368mm}
Cylinder external diameter	15in {381mm}	19¼in {489mm}
Sheave (groove) diameter	19in {483mm}	22in {559mm}
Wire rope circumference	2¾in {70mm}	3¾in {95mm}

Loynes (*#59*) in 1968 states that the Mark 13 equipment came into service in 1955 and was fitted to all 'fixed wing' carriers, except HMS *Eagle*[51]. Its capacity was:

30,000lb at 112kts -to- 37,000lb at 100kts.

In 1960, based on service and trials experience, it was uprated to:

30,000 at 120kts -to- 30,000/40,000lb up to total energy of 19·2 x 10[6]ftlb.

Loynes (*#ibid.*) continues with the statement that, following trials at RAE Bedford, no further development of the Mark 13 gear was planned. Another source suggests that equipment of this last highest rating was installed in HMS *Eagle*[51] during the 1967 refit, and named the Mark 14 system. Its rating was said to be:

40,000lb at 125kts.

In comparison, the rating of the water spray, Direct Acting Arrester Gear (DAAG) developed to replace the Marks 13 and 14 hydraulic equipment, was:

15,000lb at 110kts and 4·2g max to 40,000lb at 125kts, with a deck pull-out of 270ft.

One experimental unit (DAX.2) was fitted in HMS *Eagle*[51] in 1967, while HMS *Ark Royal*[55] in 1970 was fully equipped with DA.2 units.

From the early 1930s, when both the US and Royal Navies began their investigation into the use of hydraulic systems for shipboard arrester gears the development of the under-deck ironmongery proceeded in slightly different directions.

The American gear became known as the 'constant pressure' system. The fluid displaced by the jigger ram passed directly through a pair of control valves, located in parallel, into an hydraulic accumulator. These controls were identified as (a) the limited lift valve and (b) the constant pressure valve. The ideal settings of both for each marque of aircraft were obtained mostly by experimentation, adjustments being made for changes in weight of the aircraft and the anticipated speed of its engagement.

The British gear from the Mark 3 through to the Mark 9 equipment, although similar in concept, adopted the 'constant pull-out' system of control. This allowed the arrester wire, and its jigger energy absorber, to use the full travel stroke irrespective of the speed of engagement and weight of the aircraft. The configuration of the under-deck machinery was that the piston displaced the hydraulic fluid through a variable orifice in which the annular clearance progressively decreased in size in accordance with the extent of pay-out of the arrester wire. This principle of graduated choking was well known to RAE as, in slightly modified form, from the earliest catapults it had been adopted to bring hydraulic machinery to rest at the end of the launching stroke.

In the case of the arrester gear, the orifice was formed by a choke ring in the piston head passing along a tapered ram which occupied the full length of the jigger cylinder. The cut-off profile of the ram followed an approximate parabolic curve with the diameter at the start of the run being about 4·7in {119mm} increasing to within a shade of 5·0in {<127mm} at the end of a stroke of 10ft {3m}. With the bore of the choke ring being 5·0in, the final clearance was a matter of only a few 'thous' (to the uninitiated – thousandths of an inch {fraction of a millimetre}). Volumetric displacement of fluid through this orifice passed into a pressurised accumulator through a control valve, graduated – for example, the Mark 4 gear – from fully open at 7,000lb AUW to closed at 16,000lb {3,175kg, 7,250kg}.

With the Mark 10 vertical installation at Jersey Brow, a departure was made and a system similar to that of the US Navy using the constant pressure principle was adopted. This was found not to be successful and with the reversion to horizontal units, as Mark 11, the control also reverted to constant pull-out. Instead of using the tapered rod variable orifice as in previous installations a new control feature was adopted; this was known as the spline valve.

The spline valve comprised a variable choke orifice mechanism with, as before, the size of the aperture being dependent upon the location of the jigger piston throughout its stroke. Lewis (#58 p.69) best describes its features as comprising:

'a cylindrical drum, or "barrel", with a series of longitudinal splines cut into it all round the surface. The depth of the splines was varied along their length in such a way as to vary the annular area between them and a close-fitting choke ring. Thus as the barrel advanced along its cylinder it varied the pressure in the main cylinder, and hence the characteristic of the arrest, by throttling the outflow rate. For a range of aircraft weight, matched pairs of splines, radially opposite each other, worked in parallel as chokes in the two passages carrying fluid expelled from the main cylinder.

'By rotating the barrel to indexed positions, different pairs of splines could be brought into play as required by different aircraft types and this could be done either at the gear position or remotely, by telemetry, from a control centre. In effect, the spline valve was a whole series of calibrated cut-off rods with a simple way of switching from one to another.

'An axial drive was necessary for the spline barrel and was provided by static fluid pressure, generated from a small diameter ram connected to the moving crosshead, and fed to the large diameter piston formed by the rear face of the barrel. A ratio of 12 to 1 was chosen for this, giving a spline movement of about 15 inches. This arrangement greatly simplified the arresting gear control design, was relatively easy to install, modify and service and, of course, retained the constant pull-out feature. In all respects these units were similar to the earlier marks. An air pressure of 650psi was used to give adequate wire tension at reset.'

Perhaps the nearest parallel sample to this system of cut-off would be the sleeve-valve application as in Bristol Centaurus and Napier Sabre aero engines or, in effect, a three-dimensional cam.

The layout of the across-deck wires enabled the wires to be at an average pitch of 21 to 27 feet with the number of wires dictating the actual spread of the arrester area. The table gives details.

Class of Ship	Spread of wires – feet	No. of Wires	Mean Spacing – feet
Fleet *Illustrious*	211	9	26·4
Fleet *Implacable*	220	9	27·5
CVE *Smiter*	195	9	24·4
Light Fleet *Colossus*	237	10	26·3
Escort *Pretoria Castle*	101	6	20·2
Angled Deck *Eagle*	78	4	26·0

The short spread and the low mean spacing of the arrester area in HMS *Pretoria Castle* created a few problems, mostly with barrier entries. This situation did not occur with the later, angled-deck ships as the visual projector sight enabled a greater precision in approach and landing to be achieved.

SHIPBOARD SAFETY BARRIERS

The wire net and hydro-pneumatic jigger safety barrier installations on board any particular ship normally matched the accompanying arrester gear systems, although the rope reeving in all probability would have been different. This commonality of machinery simplified the servicing, spares and maintenance aspects of their operation. The information in the table below shows this similarity when compared with the details in the previous table.

In the cases below, up to Mark 11, the barrier nets were constructed of three horizontal steel wires interlaced with vertical and cross wires, reeved to hydro-pneumatic under-deck gear. In the Mark 13 the barrier nets comprised a series of vertical nylon ribbons laced to a pair of horizontal wires linked to the under-deck arrester unit. Each of the ships fitted with this system – having an angled deck – carried only one operational barrier which was rigged as required. Ample spares were carried as each aircraft engagement, in effect, destroyed the nylon components of the system.

Ultimately, HMSs *Ark Royal*[55] and *Eagle*[51], while still fitted with nylon barrier nets, were both equipped with deck packs of undrawn nylon cords, replacing the under-deck arresting units, for their barriers.

TYPE	DATE	WEIGHT (lb)	SPEED (knots)	MAX 'g'	PULL-OUT (feet)	SHIPS FITTED
Mk.3	1938	8,000	40	1·0/1·5	–	*Ark Royal*[38]
Mk.4	1940	7,000/20,000	40	2·0	62	*Illustrious*[40]
						Victorious, Formidable
Mk.5	1943	7,000/20,000	40	2·0	40	*Unicorn*
Mk.6	1944	7,000/24,000	57/81	3·0/6·0	40	*Indomitable*
						Implacable
						Indefatigable
Mk.7/8	1945	7,000/15,000	40	4·0	40/88	'Colossus' Class
Mk.10	1952	12,500/30,000	75/88	2·3/3·7	100/150	*Eagle*[51]
Mk.11	1951	12,500/30,000	75/88	2·3/3·7	100/138	*Albion*
						Centaur
Mk.13	1956	12,500/30,000	85	2·0	150	*Albion*
						Ark Royal[55], *Centaur*
						Eagle[51]
						Hermes[59]
						Victorious

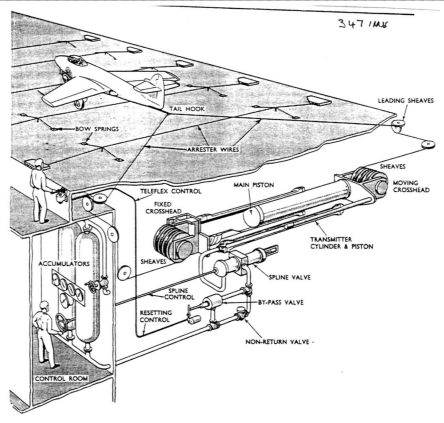

Diagrammatic layout of a shipborne arrester gear system. The above deck gear is self-explanatory but each under-deck gear would serve two arrester wires. The wire reeving around the jigger is only representative; it was more complicated than that shown here. (Institution of Mechanical Engineers)

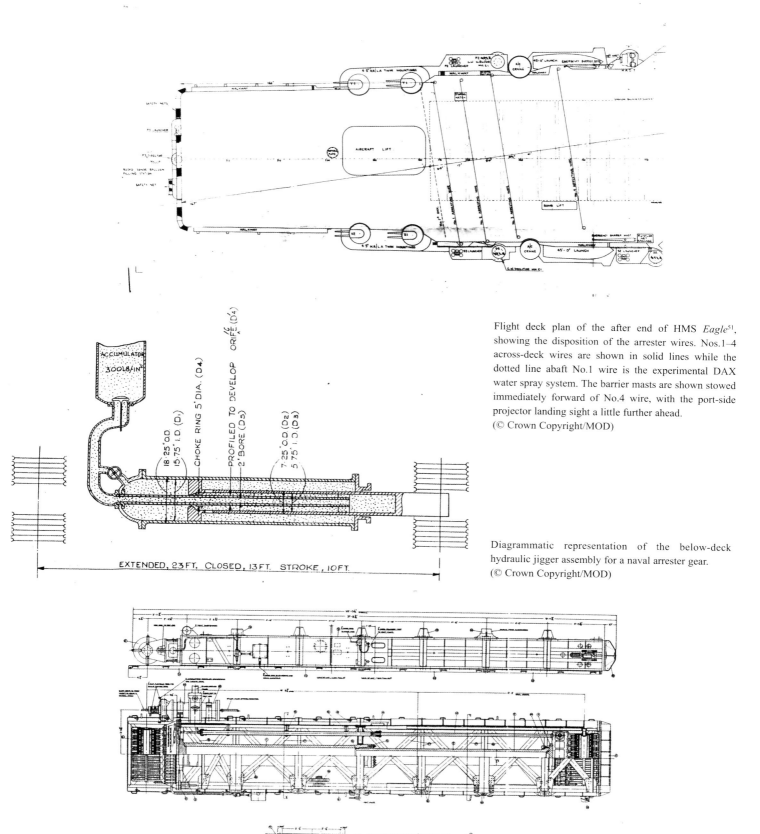

Flight deck plan of the after end of HMS *Eagle*[51], showing the disposition of the arrester wires. Nos.1–4 across-deck wires are shown in solid lines while the dotted line abaft No.1 wire is the experimental DAX water spray system. The barrier masts are shown stowed immediately forward of No.4 wire, with the port-side projector landing sight a little further ahead.
(© Crown Copyright/MOD)

Diagrammatic representation of the below-deck hydraulic jigger assembly for a naval arrester gear.
(© Crown Copyright/MOD)

General arrangement of a Mark 13 below-deck arrester unit. Introduced in 1955, this was the 'last word' in Royal Navy carrier installations.
(© Crown Copyright/MOD)

APPENDIX 8
ARRESTER HOOKS – NEW DESIGNS

The range of arrester hooks developed by NAD is summarised below.

Original Design – Vee frame application (The final phase of the original 10,500lb hook)

	LOAD	WEIGHT	LOAD	WEIGHT
	12,000lb	7lb 12oz	{5,443kg	3·5kg}

Welded Plate Design – Vee frame application

W24	6,000lb	4lb 0oz	{2,722kg	1·8kg}
W48	12,000lb	7lb 12oz	{5,443kg	3·5kg}
W60	15,000lb	9lb 8oz	{6,804kg	4·3kg}

Welded Plate Design – Sting application

W32	8,000lb	3lb 8oz	{3,629kg	1·6kg}
W48	12,000lb	6lb 2oz	{5,443kg	2·8kg}
W64	16,000lb	7lb 12oz	{7,257kg	3·5kg}
W80	20,000lb	9lb 8oz	{9,072kg	4·3kg}

Forged Design – Vee frame application

F74	18,500lb	8lb 0oz	{8,391kg	3·6kg}

Forged Design – Sting application

F32	8,000lb	3lb 8oz	{3,629kg	1·6kg}
F48	12,000lb	4lb 0oz	{5,443kg	1·8kg}
F64	16,000lb	6lb 8oz	{7,257kg	2·9kg}
F80	20,000lb	8lb 8oz	{9,072kg	3·9kg}

A few statistical points may be seen in the design development of these hooks. In the original pattern, a hook fabricated from built-up plates with the detachable nose made from a forging, the mean ratio – load/weight-of-hook – across the group, is seen to be 1557:1. The welded plate, 'V'-frame, design has a comparable ratio of 1551:1, while the identical form for sting application shows 2099:1, an improvement of 35 percent. The forged hook, also for sting application, offers 2618:1, a further improvement of 25 percent.

To achieve the above, an extended series of experimentation in profiles and related trials was put in train by NAD. Brunton, whose project this was, records (#219) the range of experiments covering three types of arrester hook. The first was with the 'standard' hook in two widths; the next was with hooks having roller inserts of two diameters within their throats. The final class was with cast-steel hooks of constant shape but with varying hardness of the throat. The early experimental hooks described in the text comprised eleven variants within these three classes. All were of the type attached to a 'V'-frame suspension and designed to be housed within a snatch gear within the aircraft fuselage. The first, designated hook (a), was the standard 10,500lb hook which had been uprated to 11,500lb at 2g. This was followed by a basic fabricated hook (b) carrying a roller as its throat. The latter showed promise in its being 'kinder' to the rope, but produced some swing to the aircraft in off-centre engagements.

Hook (c) was a development of the roller hook carrying a roller of 3in diameter. Hook (d) was similar but of slighter scantlings with a 2½in roller. Hooks (e) and (f) were developments of Hook (c) but fitted with variants of friction washer to provide restricted movement of the roller during fleeting of the arrester wire. All of these were fabricated in RAE workshops from steel plate.

The following hooks indicated a departure from those above in that they were steel castings. The first of these, hook (g), were of cast chromium/molybdenum alloy steel with a brass facing introduced into the throat. In this case the abrasion of the wire through the hook physically removed the lining. Hook (h) was similar but without any lining to the throat. Its inherent hardness in manufacture (DPN 250) was thought to be adequate but the abrasion of the hook surface after each engagement was measurable. Hook (j) again was a cast Cr/Mo alloy but hardened in the throat area to DPN 550. In this case the abrasion still could be measured but was considerably less than the previous hook. Hook (k) was formed of case hardening steel (BSS S.14), hardened in the throat to DPN 620. The final hook (l) was fabricated from high-tensile steel (BSS S.11 – UTS 55ton) and hardened. Fifty arrester engagements across the gamut of hook variables were undertaken, the upshot being that hooks with hardened throats carried the day. The closer the hardness of the throat surface to that of the arrester wire, the more satisfactory the

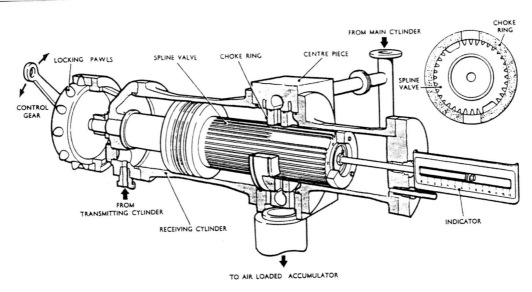

A spline valve control unit for pre-setting an arrester gear to suit the aircraft about to be recovered to deck. The spline profiles varied both along the length and around the periphery of the barrel. The linear and rotary motions of the barrel may be identified. (Institution of Mechanical Engineers)

situation for both hook and wire. From these trials it was thought that hooks made from cast Cr/Mo alloys and hardened, produced the best effects at the hook/wire interface.

While all the foregoing experimentation was in progress, a need was identified for the existing plated hooks to be upgraded with the minimum of both effort and requirement for new materials. During 1943, Davies (*#230*) experimented with adaptations to the standard 7,500lb and 10,500lb hook configurations, together with a variety of welded plate hooks. These latter were designed for intermediate ratings of 8,000lb and 14,000lb, matched respectively to the immediate requirements for Supermarine Seafire and Fairey Barracuda aircraft. Some of these hooks were provided with 4in wide throats and a variety of throat treatments was adopted. In this series of tests the comparative effects of rope wear at the interfaces were also noted. All-in-all, seven ranges of experiments were conducted on the dummy deck installation at HMS *Condor*, RNAS Arbroath. Each range covered rope engagements at both mid-point of the centre span and off-centre pick-ups. All this added up to 93 arrested entries having been achieved with the Seafire and 160 runs with the Barracuda. All of this work showed the advantage of adopting a smooth rope-throat interface, with the welded hooks displaying distinct advantages over the standard hooks. It is interesting to see that these and subsequent welded hooks were constructed of special quality, high-tensile steel plate (to the current aircraft specification, DTD.124A), the plate thicknesses being:

W32	8,000lb	10 swg	(0·128in {3·25mm})
W48	12,000lb	8 swg	(0·160in {4·06mm})
W64	16,000lb	6 swg	(0·192in {4·88mm})
W80	20,000lb	4 swg	(0·232in {5·89mm})

This was not the end of the story. With welded plate hooks becoming the norm for a short period they were soon superseded. Subsequently, further series of forged hooks were produced with the snatch requirement (for the hook to be held captive in its housing during arrestment) being dropped. This opened the door to extensive development of the 'sting' type of hook, following the pattern of, and adding improvements to, the hooks fitted to the Lend/Lease aircraft arriving from the USA.

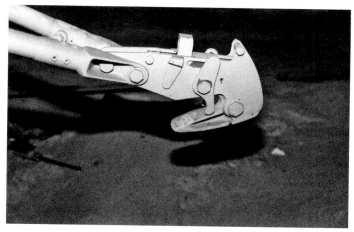

The 7,000lb 'V'-frame arrester hook on a Fairey Swordfish I shown deployed. This hook, from the 1930s, was the standard fitted to FAA aircraft designed to arrest all machines within that weight at 2g deceleration. (Author)

Hawker Sea Hawk FB.3 (WM961) showing the 'sting' type, forged steel arrester hook in the stowed position. This aircraft is now at the Caernarfon Air Museum. (Author)

A cast-steel arrester hook showing abrasion marks in the throat from transverse slippage (fleeting) of the wire across the throat. This hook appears to be the S48 model which was fitted to the standard 'V'-frame suspension using the latch gear (without the detent unit) for stowage. October 1942. (© Crown Copyright/MOD)

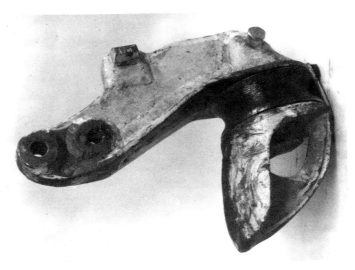

Roller type arrester hook for 'V'-frame installation, operating a 12,000lb aircraft at 2g. This was fabricated in RAE workshops from high-tensile steel plate. It is a well-worn hook having been subjected already to a series of trials. Note the Tecalemit grease nipple for lubricating the roller spindle. October 1942. (© Crown Copyright/MOD)

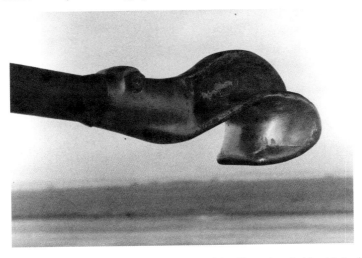

A forged arrester hook similar to that fitted to de Havilland Sea Vixen aircraft. A hard lining has been applied to the throat for experimental wear analysis, probably fitted to the NAD Canberra aircraft. November 1960. (© Crown Copyright/MOD)

APPENDIX 9

DIRECTION CONTROLLED TAKE-OFF (DCTO)– MANCHESTER

ENGINEERING DETAILS

This project was to investigate a proposal for launching an aircraft of 64,000lb {29,000kg} AUW at a maximum take-off speed of 120mph {54m/sec} and a maximum side wind condition of 50mph {22m/sec}. The concept was for the aircraft to be mounted on a trolley running along a 'railway' track. The aircraft was to be supported on the trolley by its own undercarriage and at a jacking point on the rear fuselage. The take-off was to be unassisted, the aircraft engines providing the power to accelerate both itself and the trolley to flying speed. In this system the aircraft from the start would be in flying attitude, its tail supported by a collapsible structure until flying speed was attained. At an appropriate point during the launch the aircraft would liftoff vertically from the trolley, with the latter freewheeling into a braking system located at the end of the track.

The track comprised a pair of rails installed to a gauge of 22ft 9in {6·9m}, the centre line spacing of the Manchester (and, eventually, the Lancaster) undercarriage units. Each rail was fabricated as a pair of rolled steel channels, 6in x 3in {150 x 75mm} in section, mounted on flat soleplates as chairs with the flanges inward-facing and forming a continuous gap of some 12in {300mm} between the pair. The carriage system, formally known as a trolley, was mounted on tapered wheels running within these flanges, with projections passing through the gap to which the above-deck installation was attached. The track was prefabricated in lengths of 39ft {12m} with transverse steel sleepers at 3ft {900mm} pitch. The joints between the rolled steel channels (RSCs) were carefully scarfed on top and bottom flanges to maintain a smooth

running surface. The track was laid in a pair of shallow trenches, supported on concrete bases cast into the ground, the top surface of the channel sections being level with the ground adjacent to the track. The total length of track from Jersey Brow reaching into Laffan's Plain was 1,472 yards {1346m} using some 150 tons of RSCs in the process. The track followed a gently falling contour for the first 300 yards {270m} or so, then continued on fairly level ground for the rest of its length. The Jersey Brow OD was about 224·39ft {68·40m} falling to a mean OD of 214·00ft {65·23m}, gently rising to OD 218·35ft {66·56m} at its far termination. The carriage braking system occupied approximately the final 100 yards {90m} of the track.

A turntable, carrying matching rail sections and large enough to accommodate the whole launching trolley, was installed as the loading terminal on Jersey Brow. This enabled the carriage to be installed into the system initially and, when in operation, to be swung into the location from which the aircraft would be loaded rearwards into the system and positioned in its launch attitude. Pre-flight servicing would take place in this location with the whole system being swung around to align with the track and all temporary access ways removed immediately prior to launch.

A site plan of 1938 shows the track to be provided with a second turntable at the Laffan's Plain end, with the distance between the centres of the turntables to be 4,364 feet or some 1,455 yards {1,330m}. This is misleading as a second turntable was never installed.

FIRST STAGE – THE RAE TROLLEY

As explained in the main text, two trolley units were provided. The first, named the RAE trolley was designed and built within RAE. It was used for a series of dead-load trials, culminating in the launching of a Handley Page Heyford aircraft (K5184). The format of the trolley was a pair of bogie trains – one unit running in each of the two-track

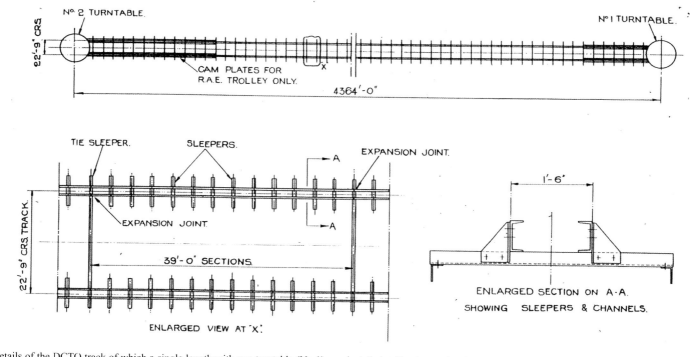

Details of the DCTO track of which a single length with one turntable (No.1) was installed at Farnborough. The cam-plate system for the RAE trolley braking was not too reliable. For the launch of the Avro Manchester aircraft from the Arrol trolley a centre span wire was rigged to a drag chain arrester system. (© Crown Copyright/MOD)

263

The DCTO track seen from the turntable at the head of the system on Jersey Brow. Neither of the alternative trolleys has yet been installed. The alignment approximates to the 'Straight Mile' of the RAE early years, looking in the south-west direction. September 1942. (© Crown Copyright/MOD)

elements – each comprising three separate, but linked bogies. The leading bogies, of heavy construction, were designed to carry the weight of the aircraft on its main undercarriage. The port and starboard bogies were interlinked by wire cross-bracing. This contributed a marginal measure of rigidity to the whole carriage as a unit but imposed side loads. Each of these bogies was provided with pairs of wheels, fitted with large diameter ball-race bearings, running within the rail flanges. The above-deck gear was bolted to projections on the bogies rising through the gaps between the rail flanges. Transverse loads were taken by side rollers fitted to each bogie.

The leading bogies ran on four pairs of wheels apiece and supported a small platform on which the aircraft undercarriage wheel rested; a pair of wheel chocks located on each platform held the aircraft in position. The rear chocks were fixed but the front chocks were demountable to permit the aircraft, on loading, to be hauled backwards on to the trolley. These chocks were of substantial size as the front elements, when secured to the leading bogie, were required to hold the aircraft against forward movement with the engines running, and to accommodate the propulsive thrust during the launch itself.

The trailing bogies also ran on four pairs of wheels and were designed to carry the support for the rear fuselage. This support comprised an 'A'-frame, spanning the track and pivoted at the bogies, but joined at the top to hold a hooked, hemispherical cup into which a matching fitment on the underside of the aircraft fuselage would be accommodated. This unit would support the aircraft in flying attitude during the early stage of its launch. The support was designed to collapse at a trigger point alongside the track, releasing the aircraft for take off. (For trials with the Heyford aircraft the trigger point was set at 300 yards {275m} from the start.) This collapse was achieved by the simple means of permitting the trailing and intermediate bogies as a unit to move rearwards, relative to the leading bogies. This separation allowed the 'A'-frame support to fall forwards as it became detached from the aircraft. For the remainder of the

launch run, the collapsed 'A'-frame was supported on a transverse wire rigged between the intermediate bogies of both trains, to prevent inadvertent contact of the 'A'-frame with the ground during its collapse and subsequent travel.

These intermediate bogies ran on two pairs of wheels and were connected to the trailing bogies by solid links within the rail sections. The attachment to the leading bogies was a little more sophisticated. This provided a telescopic extension of the connecting link via a dashpot system which, in the closed format, was secured by latches. Track-side triggers released these latches, permitting the rear sections of the bogie train to move slowly away rearwards from the leading bogies. This extension movement, amounting to 6ft {1·8m}, imposed a gradual rotation of the 'A'-frame support, which disconnected the hemispherical cup at its top and permitted the support eventually to drop away from the fuselage on to the intermediate suspension wire, releasing the aircraft for take-off.

The trailing bogie pair also was designed to accommodate the trolley braking system. This retarding mechanism comprised a pair of independent, closed circuit hydraulic systems, one each on the port and starboard trailing bogies. Each unit comprised 16 pairs of hydraulic rams with the 'outer' (free) ends of the rams facing inboard, horizontally opposed, about 6in {150mm} apart. The outer end of each ram carried a roller which traversed a longitudinal system of ¾in thick {18mm} steel plates laid alongside, and on both sides, of each track – four runs in total. These cam plates constituted the basic element of the retarding mechanism. They were manufactured with their edges forming a series of sinusoidal undulations over which the ram rollers would travel. These plates induced a reciprocative motion to the pistons as the trolley passed over each ridge and trough, causing the fluid in each circuit to be transferred backwards and forwards by way of orifices and a control system between the groups of cylinders. This rapid movement of the hydraulic fluid dissipated the kinetic energy of the trolley through the applied resistance on the rams. This eventually brought the carriage to rest by reason of the squeezing effect between rams and retarding plates – almost adopting a 'peristaltic' principle. The installation resembled sets of 'roller-blades' (the current type of roller skate favoured by youngsters) placed in opposition athwart the cam plates. In order to maintain a constant decelerating force within the system during braking, the pitch, but not the magnitude (some 4in {100mm}), of the undulations in the cam plates was progressively decreased towards the far end of the track. For recovery of the trolley, pressure was taken off the rear of the hydraulic units allowing the pistons to 'idle' over the retarding plates as the trolley was retrieved to its starting point, firstly by manpower, latterly by tractor and rope haulage.

(As an apprentice, the author witnessed the manufacture of these profiled plates in the RAE Fitting Shop, 27 Department. It was one of the earliest instances of the application of a portable, pneumatically driven, template-following, oxy-acetylene gas-cutting machine, later commercially known as a 'Pug'.) The braking system was of robust construction; unfortunately the weight of the dynamic part of the system was carried by the trolley itself. With this amount of sophistication there were problems in attempting to obtain some reasonable consistency during initial trial runs. This scheme was abandoned for a time when a drag chain arrester system was substituted; this, although fairly consistent in results, was not altogether successful. The hydraulic system was tried anew with more satisfactory results; it was then retained and continued to function for the remainder of this series of trials.

The trolley wheels ran within the rolled steel channel sections (RSCs) below ground surface, and here a problem arose. The trolley wheels were of some 5in {130mm} in diameter and were slightly tapered across the width to match the internal contours of the RSCs. They were machined from iron castings and fitted with large diameter ball-races. If it is assumed that the aircraft would need to attain a speed of 90mph {40m/sec} for lift-off, it is seen that these trolley wheels would be spinning at a rate of some 6,200rpm at that point. Coupled with this, they, to all accounts, may have been reversing in rotation from time to time through the mechanics of the installation, mostly running on the lower flanges of the RSCs, at other instances making contact with the upper flanges. It can well be imagined that these conditions were most severe and it will not come as a surprise to learn that, after a very few runs, some of these wheels were damaged with significant 'flats' having been burned into their treads. The application of a hard-facing material – 'Stellite' – did nothing to improve the situation.

SECOND STAGE – THE ARROL TROLLEY

Although similar in concept to the RAE trolley, which was considered to be an experimental prototype, the Arrol trolley was designed with a view towards quantity production. The main differences were in mechanical details affecting the bogies, and the detachment of the braking system from the trolley itself.

The front bogies were designed to run on the top surface of the track, enabling larger size wheels to be adopted with corresponding benefits from lower rotational speeds. Also, both leading bogies were spanned by a streamlined beam which, in turn, made a more rigid assembly than with its predecessor. Each intermediate bogie was solidly connected (as distinct from a telescopic strut) with its leading bogie in the train and the trailing bogies did not carry the trolley braking system.

The fuselage support, although similar to the RAE Trolley, was physically retracted away from the aircraft by a pneumatically operated rope system, again triggered by a trackside trigger as previously.

A new wire and hydro-pneumatic jigger arrester gear, similar to that found on aircraft carriers, was proof-tested on the track. For these trials a purpose-built rocket propelled trolley was used. A few dead-load tests were carried out on the new trolley. For these the trolley was powered by a cluster of eighteen, 3in cordite rocket motors. (At no time was the system rocket-powered with an aircraft on the trolley.) A forward extension to each of the leading bogies was provided with a wire-catcher, somewhat similar in shape to the throat of an aircraft arrester hook. This would impact the across-deck wire and the trolley would be brought to rest. A result of these tests was that the intermediate bogies were found to be insufficiently stable in performance and they were replaced by the now-redundant units from the RAE trolley. The first, and only, live test was conducted with the Avro Manchester aircraft (L7246) at 31,000lb {13,600kg} AUW. The point of release was set at 450yds {410m} from the start. Actual lift-off occurred smoothly within 100yds {91m} of the release point. The trolley continued its course and, through friction forces, came to rest of its own volition some 20yds {18m} short of the arrester wire!

One decided advantage with the Arrol gear was that the trolley was considerably lighter than its predecessor. The total mass of the RAE trolley to be accelerated – and arrested – was 8,500lb {3,800kg} whereas its successor was 5,250lb {2,400kg}, 36 percent lighter.

The loading of the aircraft on to the DCTO trolley was a time-consuming process. The procedure involved the following sequence. The loading turntable containing the trolley was slewed around 90° from the launching alignment into the loading position. Demountable decking was laid along the length of the turntable to allow the aircraft tail wheel to cross the structure to the rear and for the main wheels to roll to the pair of rear chocks attached to the trolley beam. The aircraft was towed, tail first, to the hard-standing apron leading to the loading position. At this point a cable connected the tail wheel across the turntable to a winch on the far side. This was a manual unit requiring two men to operate. The aircraft was then hauled, tail first, on to the trolley until its main wheels were hard against the pair of rear chocks. The front chocks were securely fixed to the trolley to retain the aircraft in its launching position.

By this time the tail wheel would be sitting on a small platform atop a 'handraulic' jack. This jack was one of a number of 'Skyhi' long-stroking, manually operated hydraulic units to be found within the hangars. With the front chocks secure, the jack under the tail wheel was extended to bring the aircraft into flying attitude. At this stage the demountable decking was removed, the haulage cable detached and stowed, and all temporary encumbrances removed from the area. The 'A'-frame support now could be manually raised until the cup at the apex engaged the support spool on the underside of the fuselage; the jack would then be lowered a fraction until the aircraft was partly supported on its spool and 'A'-frame. The aircraft was then secured to both the jack and the 'A'-frame by temporary lashings. In this position, the aircraft would be prepared for launching. The flight crew would embark and conduct their pre-flight checks. When ready, the temporary lashings would be removed and the tail jack fully lowered to its stowage position. The turntable could then be rotated into and secured in the flight path orientation. Finally, with the trolley held by its release gear in the holdback unit, the safety retaining pins would be removed and the launch could proceed.

APPENDIX 10
BRAKE DRUM ARRESTER GEAR (BDAG)

ENGINEERING DETAILS

This is an appraisal of equipment developed and installed on a selection of Bomber Command airfields during the period 1940–42. The engineering details have not been published before and it is felt that a description of this unique plant may be of interest. The system was designed to reduce the number of aircraft (and aircrew) casualties caused by aircraft overshooting the runway on return from operational sorties. It comprised a simple installation in which an arrester hook fitted to the underside of the aircraft would engage one of a pair of wires stretched across the end of the runway which, in turn, was connected to an underground braking system to bring the aircraft to rest. The installations were completed on a number of airfields but the project was abandoned early in its history. The development of longer runways was a contributory factor but the fitting of arrester hooks to aircraft became a near impossibility, through the provision of additional radar equipment against operational requirements assuming a greater priority.

This arrester gear was required to be simple in both construction and operation, and to be made available quickly. Both entry speed and rate of retardation were to be relatively low, matched to an aircraft towards the end of its landing run. This was interpreted into working parameters related to the acceptance of an aircraft of 60,000lb {27,200kg} entering the system at a maximum approach speed of 70kts (36m/s) and bringing it to a standstill within a distance of 450ft {130m} at a peak rate of deceleration not exceeding ½g. Essentially it was to function as a last-ditch installation. (It was never intended to operate in a continuous mode as in the case of a shipborne arrester gear in which case naval aircraft entered the equipment at flying speeds and were brought to rest within 156ft {47m} at a maximum rate of retardation of 2g).

By this time, the width of an airfield runway had been standardised as 150ft {45m}. As a result, the span of the tranverse wires was set at 400ft with the spacing along the runway between the two wires at 100ft {120m & 30m}. This was seen as a single length of steel wire rope issuing from one half of the brake unit through a system of sheaves and fairleads, traversing the runway, then running parallel on the remote side until its next crossing, returning to the other half of the brake unit. When an aircraft entered the system the rope impacted would pay out simultaneously from both brake units. Because of this feature, the rope would be continuous with no replaceable across-deck wires (as distinct from carrier-borne arrester equipment). This principle is known as the 'fleeting wire' system of rope rigging; in other words when one rope is being paid out under load, the corresponding rope in the idle leg will also be moving at speed through its deck sheaves. Again, this ran counter to shipboard practice but, in this application, it was thought to present neither operational nor safety problems.

The brake unit comprised a pair of grooved rope drums, each bolted to a companion brake drum to which a pair of heavy duty brake shoes – lined with 'Ferodo' friction pads – was fitted. These bore on to the external cylindrical surface across a horizontal diameter, acting inwardly as a clamping device over the braking surface. The across-deck wires were held in tension by heavy duty coil springs forming the braking force. This had a quick-release facility. The recovery of the rope after pull-out was achieved by a geared motor drive to the drum assembly. Problems encountered in the early history of shipboard mechanical arrester gear, having an independent brake unit at each end of the arrester wire, were recalled. Differential pull-out was avoided through connecting both brake drum units by a simple dog clutch.

The entire unit, of both rope/brake drum assemblies, its recovery gear, controls and prime mover was mounted upon a fabricated bedplate. It was transported to site as a complete working unit and lowered into its machinery pit, to be bolted to the floor. The whole system – both when in service or with the wires stowed – presented a zero hazard to the normal flying programme of the airfield. The proto-type installation was sited towards the south-east corner of Farnborough airfield, adjacent to and on the south side of the Runway 07/25 with the ropes spanning that runway. The airfield pulley sheaves were bolted to

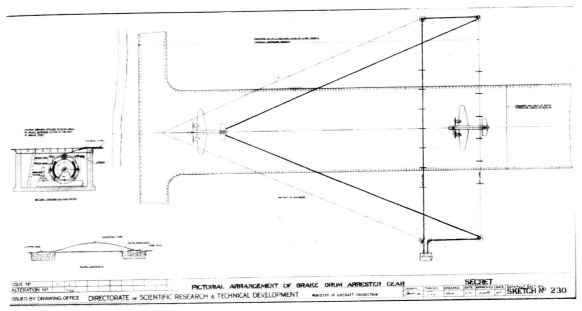

Diagram of the operation of the Brake Drum Arrester Gear. The aircraft is pictured having engaged No.1 arrester wire. The typical location of the machinery pit – always on the port side of the runway – is seen, and the spacing of the fairleads and bow-spring supports for the across-deck wires are indicated. A scrap section of the machinery pit is shown as a thumbnail sketch. January 1942. (© Crown Copyright/MOD)

One of 120 Brake Drum Arrester Gear (BDAG) units supplied for Bomber Command airfield runways ready for despatch from its manufacturer, Messrs Mather & Platt of Manchester in early 1942. July 1942. (© Crown Copyright/MOD)

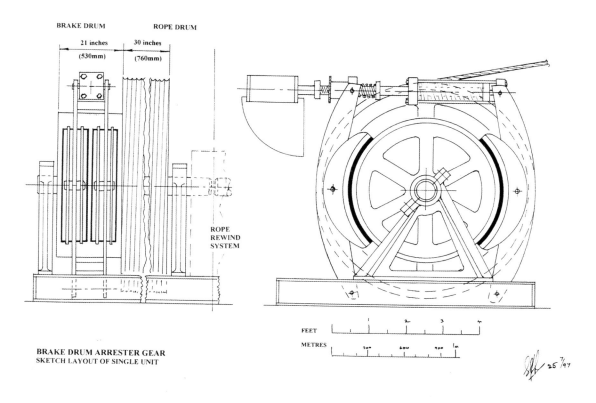

BRAKE DRUM   **ROPE DRUM**

21 inches   30 inches
(530mm)   (760mm)

ROPE
REWIND
SYSTEM

BRAKE DRUM ARRESTER GEAR
SKETCH LAYOUT OF SINGLE UNIT

FEET

METRES

Diagrammatic representation of a Brake Drum Arrester Gear (BDAG) showing the left half of the below-ground machinery. The relationship of the grooved rope drum with its brake drum and brake shoes may be seen. The end elevation shows the brake loading mechanism above the brake drum. (Author)

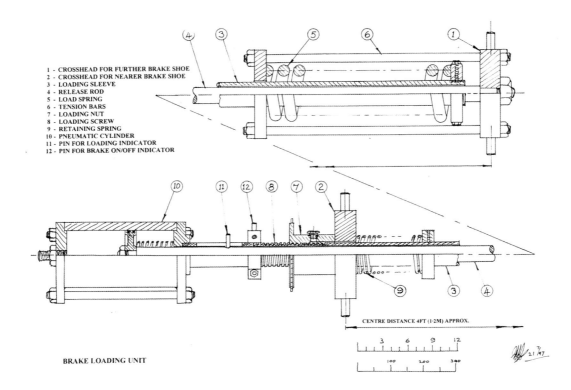

1 - CROSSHEAD FOR FURTHER BRAKE SHOE
2 - CROSSHEAD FOR NEARER BRAKE SHOE
3 - LOADING SLEEVE
4 - RELEASE ROD
5 - LOAD SPRING
6 - TENSION BARS
7 - LOADING NUT
8 - LOADING SCREW
9 - RETAINING SPRING
10 - PNEUMATIC CYLINDER
11 - PIN FOR LOADING INDICATOR
12 - PIN FOR BRAKE ON/OFF INDICATOR

CENTRE DISTANCE 4FT (1·2M) APPROX.

BRAKE LOADING UNIT

Author's sketch showing details of the brake-loading unit mechanism. The brake setting was applied from the control panel through a chain and sprocket wheel. Instant release of the brake was obtained by application of pneumatic or hydraulic pressure to the cylinder at the left. (Author)

four concrete blocks cast in the ground; they were designed to accept a braking reaction of some twelve tons, acting horizontally. The pit to contain the under-deck machinery was excavated and concrete lined, with a heavy duty roof and access cover to permit aircraft to taxi over it.

At the start of a run, both rope drums would be fully wound with the across-deck wires held taut and supported above the runway surface on hydraulic, retractable bow-spring units. The wires would be held in this position by the application of the brake shoes to the brake drums. The brake settings would be adjusted in the machinery pit to suit the weight of the aircraft destined to enter the gear. A calibrated indicator was provided to each unit with the appropriate aircraft type names engraved alongside their relevant settings. The aircraft would taxi into the system, picking up either the first or second wire with its arrester hook and pull out the wire by its impact velocity. The mechanical design of the system was such that the build-up of braking resistance experienced by the aircraft was gradual, the inertia of the plant being overcome and the braking effect being induced by a sequentially hardening procedure. (Resource to a little bit of plane geometry and trigonometry will explain this.) On completion of an arrested run, the brakes would be released, the wires retrieved and re-spooled on to the drums from the control position in the machinery pit.

The system comprised a pair of large drums, each with a spiral groove scored around its circumference to accommodate the wire rope. This drum was 5ft 6in {1·8m} diameter and carried 30 grooves across its width of 30in {760mm}. The wire was ⅞in {22mm} nominal diameter comprising six strands of 37 galvanised steel wires of 23swg (0·024in {0·61mm}) in standard lay around a fibre core. The wire was wound only as a single layer on the drum. The capacity of the drum, allowing for three complete dead turns retained at the attachment point, amounted to some 510ft {155m}, more than adequate for the designed pull-out. (Check your geometry with Pythagoras!) One end of the wire was attached to the first drum, taken out through the system of deck

sheaves, to return to the second drum. The off-takes were from the tops of the drums.

Each rope drum was rigidly attached to its associated brake drum. Both were fabricated from iron castings, firmly bolted to each other on a common shaft. The brake drum was another cylinder of 4ft diameter and 21in width {1·2m x 530mm}. Two pairs of brake shoes were fitted, one pair located each side of the horizontal centre line and wrapped around 90° of arc at the circumference. Each individual shoe was 9in {225mm} wide and was located side by side with its matching unit to form a pair. These dimensions provided a braking surface area for each drum of 9·4sqft {0·87m²}. Each pair of shoes was mounted in a cres-

One of the BDAG machinery pits and fairlead unit as seen in 1995 at ex-RAF Bottesford from the runway side. The arrester wires would appear from the two slots in the pit roof up to the pulleys and then branch each way to the runway. October 1995. (Brian Walker)

The condition of one BDAG installation as revealed in October 1995 when the steel covers to the machinery pit had been removed, after some 50 years of disuse, at ex-RAF Bottesford. It is likely that most of the remaining 120 units are/were found in similar condition, preserved through having been submerged in oily water for half a century. October 1995. (Brian Walker)

Thorpe Camp. An intermediate stage in re-assembly (after a de-rusting exercise) of the cable/brake drum units of a BDAG unit, recovered in 1999 from ex-RAF Woodhall Spa. (Mike Hodgson)

Brake Drum Arrester Gear. The below-deck machinery was recovered from ex-RAF Woodhall Spa and is now exhibited in its refurbished state at Thorpe Camp Museum (TCM). The only component missing from this equipment is the Ford V.8 engine and its starter battery for rewinding the arrester wire onto the drums after an arrested landing pull-out. It gives the appearance of being ready to drive! (Author)

cent-shaped structure, hinged to the framework at the bottom, and at the top were connected by a tensioning unit to its opposite number across the far side of the brake drum. The brake loading was produced by springs in the tension unit and was readily adjustable from the control position.

The wire was retrieved by a Ford V.8 petrol engine connected through clutches to a common gear train. The brakes would be released, sliding gear pinions would be set to engage the respective (generally both) gear train, and the engine would retrieve the cable by spooling it back on to its grooved drums. Care would have to be taken that each drum received its due quota of rewind; there could be problems if the aircraft landed off-centre and more cable was taken from one drum than the other. Fortunately this was not a serious problem although to be avoided if possible. A heavy off-centre landing would cause the aircraft to yaw badly during its arresting run and the wire most likely would be drawn across the face of the hook, causing a fair amount of abrasion. The wire itself would need to be inspected for broken strands after such a landing.

If the system were to be taken temporarily out of commission, the wire would be unshipped from the remote sheaves at the far side of the runway and wound back on to the drums as temporary storage.

To re-rig the system, a tractor would be required to unspool the stowed wire from the drums by dragging it across the runway for it to be relocated in its sheaves.

Consequent upon a visit to the recovered unit at East Kirkby in July 1997, a set of sketches of the system was produced. These are to be regarded as indicative of the *modus operandi*, not as scaled working drawings. The first sketch is a general picture of the brake drum unit, approximately to scale. This shows the relative positions of one rope drum and its associated brake drum, together with the appearance of the brake shoe elements. This portrays one half of the installation; the other part is, generally, the mirror image of that shown.

The second drawing shows the brake loading unit which spanned the upper ends of the crescent levers supporting the brake shoes. This device was used to apply the braking resistance to each half of the system and was readily adjusted to suit the type of aircraft to be accommodated. For re-spooling the wire, the brakes were released and reset as soon as the appropriate tension was obtained in the across-deck wires. The upper ends of the crescent levers were attached to a pair of crossheads, some 4ft {1·2m} apart. Item 1 on the drawing is the crosshead for the brake shoe assembly further from the control panel while Item 2 is that for the nearer brake shoe assembly. These two crossheads supported the brake loading unit which was the active component for applying the braking load. A tubular loading sleeve, 3ins {75mm} diameter – Item 3 – contained a release rod, 2ins {50mm} diameter – Item 4 – which spans the unit.

At the furthermost end the connection between the loading sleeve and its crosshead was made via a load spring – Item 5. This spring was a coiled, compression unit formed of steel bar of 1⅜in {35mm} diameter wound as 14 turns on a pitch diameter of some 6ins {150mm}. Its free length was of the order of 24in {600mm}. The loading was conveyed to the crosshead by a cage formed of four steel tension bars – Item 6. The end of the unit for the nearer brake shoe element consisted of the crosshead – a loose fit on the loading sleeve – which was constrained by a loading nut – Item 7. This nut ran on a square screw thread of ¼in {6mm} pitch cut into the loading sleeve and was driven through an integral sprocket and chain from a hand-wheel on the control panel. The crosshead was kept abutted to this loading nut by a coiled

BDAG at TCM. The left-hand indicator panel. The top pointer showed when the system was 'on'; the small pointer at the edge of the quadrant indicated the setting for the reception of one of four types of aircraft: Stirling, Lancaster, Halifax or Warwick. (Author)

BDAG at TCM. Right-hand side of the machine. The arrester wire would be taken off the top of its drums, peeling off to the right onto the airfield runway. (Author)

Thorpe Camp. Some of the crew responsible for the refurbishment of the BDAG photographed in front of their finished handiwork in May 2000. *Left to right*: Keith Brimson, Derek Mewitt, author, Ian Cooper (no relation), Darren Swinn (rear), Bill Skinner (front), Mike Hodgson (curator). (Mike Hodgson)

retaining spring of fourteen turns of ⁵⁄₁₆in {8mm} bar on a pitch diameter of some 6in {150mm}.

The unit was terminated by a hydraulic cylinder – Item 10 – the casing of which was connected to the loading sleeve while the piston was fixed to the release rod. Pressure on the brake was applied through the control of the loading nut. The brake was released by introducing hydraulic pressure into the cylinder which separated the two crossheads by pushing the release rod and extending the distance between the crossheads.

The across-deck wires were supported at a minimum height of 4ins {100mm} above the runway by a series of bow-spring units. These bows were some 4ft {1200mm} in length and, like their shipboard antecedents (of which they were a direct copy), were made to collapse on to the runway when not in use. The retractors were hydraulic cylinders, contained within the deck unit and operated from compressed-air bottles through an accumulator incorporated in the main machinery frame.

The operation of this feature was through the control panel; indication of the position is shown on a photograph of the brake indicator quadrant. Items 11 and 12 of the previous drawing refer to pins on the brake loading unit which operated a pair of pointers. Item 11 was a pin fixed to the release rod which indicated the extension produced by the loading nut by moving a brake setting pointer along a sector affixed to the quadrant.

This sector was calibrated to suit a number of aircraft (for example, four), each under two weight conditions ('light' and 'heavy'), the operator adjusting the setting to suit the aircraft to be accepted by the system. The second pointer was linked to the near crosshead and indicated the brake 'on/off' position.

The recovery of a second unit, restored by volunteers and displayed at Thorpe Camp – originally an adjunct to Woodhall Spa airfield – has enabled details of the control panel to be reproduced. This panel was constructed of steel plate of substantial thickness to act as a protective screen to the operator should the wire whip or break under operation. It was felt that a fracture of the rope would not affect the coils still wrapped around the drum; the mishap during one Farnborough demonstration confirmed this. However, the existence of this screen provided a psychological assurance to the unfortunate 'erk' in his troglodytic spot, having to spend his working shift in a subterranean environment with aircraft landing almost over his head.

Details of the panel may best be seen from the illustrations. The controls were simple and the operation of the equipment was reasonably straightforward. With the air bottle installed, already charged to a pressure of 1,000–1,650psi {68–112bar}, the Ford V.8 petrol engine

270

would be started to pressurise the hydraulic system. During this procedure the manual setting of the brake pressure – by the chain and sprocket system – would be set to its lowest reading, indicated by the position of the pointers on the quadrants each side of the machine. As soon as hydraulic pressure was available, the brakes would be released and the mechanical drive clutched in to ensure all the arrester wire was wound onto the drums. With this accomplished, the bow-spring units on the runway supporting the centre spans would be raised. The system would then be set to accommodate the aircraft type, again as indicated on the quadrants, and the brakes set by exhausting the hydraulic pressure from the brake release cylinder. In addition to the brake-setting quadrants, the state of readiness was indicated by a dial gauge (hydraulic pressure gauge) showing that the arrester wire supports were raised, and a set of

indicator lights operated by limit switches to show when the brakes were on. The operator, standing by his hand-cranked field telephone, then just waited for things to happen!

An arrestment having been made and with the aircraft clear of the runway, the brakes would be brought to the 'off' position, the wire retracted onto the drums, any adjustment made for the next expected aircraft, and the brakes reset.

Access into the machinery pit was by way of a set of tread-plates fixed to the face of the control panel. The man-cover to the pit was raised, a handrail at the top of the 'ladder' was hauled upwards and secured by locking pins, enabling the operator to descend to his subterranean workplace. The operator would retract the handrail back into the pit to reduce its above-deck protrusion.

APPENDIX 11
FLEXIBLE-DECK – ENGINEERING DETAILS

Whilst the project for experimental development of techniques for landing of FAA aircraft without undercarriage was under consideration before the end of World War 2 it was not until the autumn of 1945 that material and equipment were being brought together for this exploratory activity. Boddington and Nixon (#211) describe the process of acquisition of the hardware (ironmongery – not computer!) – in itself a fascinating story – and much of the following detail has been abstracted from their report. The work initially was divided into four stages:

Stage 1	⅛th-scale model experiments
Stage 2	Full-scale model trials
Stage 3	Airfield flying trials
Stage 4	Shipboard flying trials.

To these, further stages were added:

| Stage 5 | Airfield trials with relocated carpet – project 'Red Rufus'. |
| Stage 6 | Airfield trials with an aircraft without undercarriage launched from a catapult and landing into the flexible deck. |

From early 1945 an experimental system using aircraft models at one-eighth scale was set up in the NAD lab on Jersey Brow. This was known formally as Stage 1 of the project. A dummy section of 'carpet' deck of reinforced rubber was laid out within a framework and tensioned to form a resilient surface. A basic form of catapult, comprising a beam with an underslung trolley, was produced, the model aircraft being suspended beneath this trolley. The trolley was gravity propelled by a falling-weight system, identical to the catapult used elsewhere in RAE for aircraft model ditching trials. On being triggered, the model was propelled towards a scaled-down arrester gear and released in free flight into the arrester wire at the end of its short travel. Boddington and Nixon (#ibid.) admit that, from the first, investigations showed that the motion of the aircraft model was highly sensitive to precise characteristics of the arrester gear and that conclusive results

were not to be obtained from proceeding with these small-scale experiments. They were terminated to await the Stage 2 set-up, but the equipment was retained for purposes of demonstrating the overall technique. The process of acquisition of the hardware for Stage 2 was accelerated in which the experimental work would proceed with full-scale models and equipment. It is the provision of this equipment which has proved to be of absorbing interest.

By virtue of its magnitude, Stage 2 could not be other than an outdoor project, and its location had to be remote from the day-to-day flying activities on the airfield. The site chosen was near to Meadow/Ively Gate on the north side of the airfield, reasonably close to the NAD offices on Jersey Brow. At this stage, the intention was to cover investigation into three activity phases: (a) the launching of the dummy aircraft, (b) the arresting activity, and (c) the resulting landing/impact characteristics of both aircraft and deck. For each of these phases a sizeable amount of specialist equipment would be required, much of it unique in character.

During the summer of 1945, NAD prepared specifications for the machinery and site-work required for full-scale trials. Sir William Arrol was contracted to supply the catapult beam and elevating tower for carrying a rolling load of 10,000lb. The launcher trolley was to be provided by RAE workshops, and G. Percy Trentham was contracted to provide all civil engineering work including machinery foundations. Much of the soil excavated was to be used as backfill to local marshy areas. A duplexed American Mark 4 naval arrester unit was supplied by the Admiralty. The project was costed at £85,000 of which £40,000 was for the civil engineering contract. This price excluded the cost of the carpet and the pneumatic sub-strate, both of which were matters for investigative development.

The full-size aircraft allocated for these trials were four redundant GAL Hotspur gliders, suitably adapted, strengthened and fitted with catapult launcher spigots and arrester hooks. They were capable of being ballasted up to 12,000lb {5,400kg} AUW, operating at entry speeds of up to 95kts {48m/s}. The system comprised the rubber mat (carpet) and its sub-strate, with a pair of arrester wires supported across the entry end and a catapult upstream to launch the Hotspur dummies.

A section of the experimental flexi-deck on Jersey Brow, showing pre-tensioning equipment, for early impact trials (Stage 1 drop tests) onto the 'carpet' surface. January 1949. (© Crown Copyright/MOD)

The trials area [O.15] to the north of the airfield, near Meadow Gate is prepared for the early, Stage 2, 'Hotspur' trials on the flexible deck project. The deck equipped with its tensioned carpet occupies the centre of the picture. The launcher track extends from the left of the picture with the trolley under a tarpaulin at the release end. Two pairs of stanchions in the foreground carry the athwartships arrester wires (in alternative locations) with the jigger arrester unit on the floor to the left. The catapult beam is shown in the stowed position for loading the glider. It would be raised and pitched at an angle to suit the trials programme. June 1947. (© Crown Copyright/MOD)

The rocket-propelled catapult trolley for launching the Hotspur glider in the flexi-deck experiments. Both pairs of front and rear legs tilt forward on release of the aircraft. The pair of retarding spikes can be seen to the left. October 1946.
(© Crown Copyright/MOD)

The flexi-deck carpet is on its transport roller ready for installation over the multi-layer, pneumatic tube substrate, forming the Stage 2 experimental rig at the Meadow Gate site. The team enjoying a photo-call was closely involved in this activity. The first person on the left is Alec Howie; the sixth – hatless – is Jack Nixon, project engineer; while second from the right is Norman Skelton. March 1947.
(© Crown Copyright/MOD)

De Havilland Sea Vampire F.21 (VT795, pilot S Lt Harrison) making a landing on the Stage 5 flexi-deck system. July 1952. (© Crown Copyright/MOD)

THE CARPET

The flexible landing area, from its inception dubbed 'the carpet', comprised a rubberised cord fabric sheet of six-ply construction – approximately ⁵⁄₁₆in {8mm} thickness – made as a single sheet 200ft by 60ft {61m x 18m}. Two schemes were proposed for its structure. The first was for the carpet to be supported along its long edges by clamping to a continuous series of drums of 24in {600mm} diameter, capable of rotation around their horizontal axes for the full length of 200ft {61m}. In this mode the carpet would be tensioned athwartships to a predetermined loading up to 300lb/inch-run {5,440kg/m} throughout its length through the partial rotation of the drums by hydraulic jacks. The upstream edge of the carpet would be fixed to rigid drums while the downstream edge would remain free. There would be no other support; in fact the whole presented itself as a gigantic trampoline!

The alternative scheme – and that which was adopted – was for the carpet to be fully supported across its entire area by a multitude of air bags, but still under a small amount of tension. These air bags were inflatable tubes arranged as a continuous bed, athwartships in three tiers, each tier being separated horizontally with a fabric sheet to preserve the location of the tubes after displacement. This arrangement prevented a 'squabbing-down' effect caused by mutual entrapment of one tube by its neighbours. The tubes were formed from lengths of 7¼in diameter {185mm} 'deluge' hoses, obtained from surplus wartime stock held by the Home Office for the National Fire Service. Of the entire available stock of some 25 miles {40km}, 24 miles {38km} were requi-

sitioned and delivered via Cardington to Farnborough, half of which quantity would be required in each of two projected systems: one at Farnborough and the other on shipboard. Again, the long edges of the carpet would be clamped to drums, but in this instance they would be of 18in {460mm} diameter. The tubes were fitted with steel dished heads at both ends and, in use, were inflated to a pressure of 10–15psi {0·7–1·0bar} from a battery of compressed-air bottles (ex-balloon barrage hydrogen cylinders). The Balloon Development Establishment at Cardington undertook the task of producing some 1,000 individual tubes for the initial installation at Farnborough.

At first, an experimental section of the carpet and its inflatable tube support of some 30ft² {9m x 9m} was set up on a site on Jersey Brow for vertical impact trials. This enabled the principle to be tested and for the mechanics of impact of an aircraft fuselage, over a range of angles of incidence, onto the full-scale carpet system to be examined. First, a weighted steel drum – suspended from a gantry – was dropped on this test element. Later, one of the adapted GAL Hotspur glider fuselages was used. This was suspended from the local crane by a heavy duty bomb-slip unit; ideal for a quick, clean release. A maximum vertical velocity of 12ft/sec {3·7m/sec} at impact was the target. (A site drawing shows the location of this equipment to have been underneath the seaplane crane adjacent to the P.I Catapult on Jersey Brow but eyewitnesses recall it having been under one of the cranes at the Meadow Gate site. Perhaps it occupied both sites, being transferred from one to the other!)

The production of the flexible deck element – the carpet itself – is recorded as having presented a most formidable manufacturing problem. As nothing of this magnitude had been attempted before, a great deal of experimental work with the manufacturers had been necessary before the design could be finalised and manufacturing details settled. A redundant hangar at RAF Greenham Common was made available and adapted for this project. A level granolithic screed to contain the full size of the finished carpet was laid on the floor of the hangar. A pair of rails ran parallel to the long edges; these supported a low, portal gantry spanning the full width of the carpet and traversing the entire construction area. This gantry carried, span-wise, a series of electrically heated platens specially constructed for the curing process. The platens were lowered progressively on to the carpet as the layering exercise proceeded, and operated at a temperature of 200°F {93°C} under a pressure of 80–100psi {5–7bar}, with an electrical loading of 72kVA.

The transhipment of the completed carpet across to the Farnborough landing site was another major undertaking. A tubular steel mandrel, 18in diameter and 68ft long {460mm x 21m}, was provided. This was mounted on trolleys which ran alongside the carpet, allowing the latter to be rolled onto the mandrel from one end to the other. The loaded mandrel plus carpet was 4½ft {1·4m} in diameter at a combined weight of 18 tons. The method of unspooling the carpet at the landing site was the reverse of the rolling-on process. Again the protruding ends of the mandrel were supported on carriages running along a temporary track, allowing the carpet gradually to unroll as the mandrel travelled the length of the deck. While the carpet was being laid on top of the inflatable tube bed, the edges were secured progressively and provision made for final tensioning.

A derrick crane, with cranked jib, was located on the starboard side, at the toe of the carpet, for recovery of the glider after its arrestment.

A second full-size carpet was produced, this was thought to have been made in the hangar [O.15] adjacent to the test site at Meadow Gate. Records are sparse and it is not clear where and when this carpet was installed it is thought to have been that installed alongside Runway 04/22 forming the equipment for Stage 3, the first free-flight, shore-based landing trials at Farnborough. At a slightly later stage, the original carpet at Meadow Gate was prepared for trials aboard HMS *Warrior*. The carpet for the final Stages 5 & 6 trials parallel with Runway 07/25 was larger than the earlier units. This was 400ft x 80ft {122m x 24m} and was constructed in two parts with the split across the flight path about halfway along the landing area. The carpet here was supported on a single layer of large, 30in {760mm} diameter, 'bolster-like' inflated tubes rather than multiple layers of smaller hoses as in the earlier installations. The inflation pressure was as low as 3¾psi {0·26bar}. Lessons learned in the trials at Stages 3 & 4 were incorporated into the new set-up.

On completion of the Stage 3 trials and with the carpet and its tensioning supports removed, the site, retaining its catapult and arrester gear, was used for experiments with Seafire aircraft fitted with contra-props impacting into a variety of safety barrier nets.

ARRESTER GEAR

The under-deck machinery for this section of the project was adapted from one of the standard, American Mark 4 arrester gear units, equipment with which Naval Aircraft Department already was familiar. (It probably was retrieved from one of the two abandoned, war-damaged CVEs which were not returned to the USA under the World War 2 Lend/Lease agreement.) The hydraulic jigger was surface-mounted, in a shallow trench at ground level and was reeved to provide, at the head of the carpet, two arrester wires in a non-fleeting system.

The centre spans were supported on pairs of pylons, capable of height adjustment within the range of 0–3ft {0–900mm} above the carpet surface level; the under-deck return wires were routed at ground level beneath protective coverings. The span between supports was 90ft {27m}. The fore-and-aft spacing between the wires was variable over the range of 10–20ft in increments of 30in {3–6m, 750mm}. For its initial setting, No.2 wire was positioned 10ft {3m} upstream of the head of the carpet, while the longitudinally adjustable, No.1 wire was located at a maximum of 30ft {9m} from the carpet edge. The deck sheaves were mounted on a carriage to permit the foregoing adjustments. In addition, facilities were provided, enabling simulated off-centre landings up to 18ft {6m} to the port side of the catapult centre line to be accommodated.

Early in the trials programme it was found that a single arrester wire was effective and adequate, No.1 wire was removed and the reeving of the system was adjusted to suit the reduced requirement. The remaining wire was rigged to span the deck at a height of 30in {750mm}. This was thought to be the maximum clearance above deck to be permitted without creating a hazard if the plane were to be making its approach too low, with a tendency to slide under the wire.

CATAPULT

The simplicity of the rocket-propelled trolley developed for the P.1 Catapult some five years earlier had its attractions and the principle was adopted for this application. A straightforward open-truss, Warren type girder was built as the main beam. The launching trolley was designed to run on rails on top of the beam, with the aircraft – fitted with catapult spigots (spools, of previous memory!) – supported in mating sockets on the trolley. This trolley was powered by the familiar battery of 24, 3in {75mm} rocket motors. The trolley braking system also was almost identical with the P.1 Catapult, employing 'spike-and-barrel' units fixed respectively to the trolley and the toe of the beam. This was designed to accept a maximum braking load of 180 tons. With the system being a virtual replica of previous work undertaken by NAD, the performance of the launcher in its new application was readily predictable and its operation would have produced no new surprises.

The lattice beam was 136ft {41m} length; of this, the accelerating stroke occupied 86ft {26m} and the retardation stroke was 4½ft {1·4m}. The remaining space was taken up with the length of the trolley and its holdback fitment. The downstream end of the beam (the toe) was located 44ft {13m} upstream of the leading edge of the carpet, at the head of the landing area. The beam was pivoted at the toe; this allowed its head (the loading end) to be progressively raised at 1° increments until a downward slope of 10° was attained. A 40ft {12m} high tower contained the head of the beam and the beam itself was raised and lowered by a winch. The height of the pivot at the toe of the beam was adjustable by means of radius arms. These allowed the pivot axis to be level with the top of the carpet or raised at 6in {150mm} increments to 60in {1·5m} above the carpet level, the angular increments remaining throughout this range. In plan, the centre line of the beam and that of the carpet were in direct alignment.

A trench was provided to accommodate the beam when at its lowest position. This excavation was 160ft long by 8½ft wide and 8ft in depth {49m x 2·6m x 2·4m}. The beam was retracted into the trench for servicing the trolley and for loading the glider. For these purposes a second derrick crane was installed to starboard at the head of the beam.

EXPERIMENTAL AIRCRAFT

The four General Aircraft Limited (GAL) Hotspur gliders allotted to this project were modified to suit the experimental requirements of mechanical launching and arrestation. The fuselage of each airframe was stiffened with an interior tubular framework, and the wooden skin reinforced in the way of the impact surface. The internal framework catered for the imposed catapulting and arresting loads.

Of the first glider, the fuselage only was used for the drop tests on the small specimen carpet. It was ballasted to various operating weights. The second glider, also ballasted, retained its flying surfaces and was used to develop the experimental technique on the Meadow Gate site. Once this was established the machine subsequently was used for demonstration purposes. Neither instrumentation nor controls were fitted to this machine.

The two remaining machines were fully instrumented, with flying controls operable and capable of carrying a human observer. It is not recorded whether this last expedient was exercised. The flying controls were thought to have been rigged to provide only a limited movement thereby inhibiting the presence of a pilot or passenger.

One feature into which research was necessary was the reaction of the hook support arm to its impact with the arrester wire. Previously, all carrier-borne aircraft would pick up a wire suspended 4in or so {100mm} above the deck. This ensured that only the hook would experience the shock loading as it caught the wire. With the carpet deck installation, a new challenge was presented in that the wire could be as high as three feet {900mm} above the deck. This implied that impact could be at any position along the length of the arrester arm, from pivot to hook. This new feature would require attention to be given to the strength and behavioural reaction of the hook installation at the moment of impact and immediately afterwards. One item known to have been addressed in detail was the effectiveness of the hook damping unit.

TRIALS

The trials at Stage 2 disclosed the need for additional cushioning beneath the carpet. A fourth row of inflatable tubes was inserted to permit a maximum depression of 30in {750mm} when accepting a vertical rate of impact at 16ft/sec {5m/sec}. Some 34 landings were made with this installation.

On the successful conclusion of these trials, Stage 3 was introduced with the second carpet being installed alongside Runway 04/22 for flying trials with modified de Havilland Vampire aircraft. The carpet installation – with some adaptations – and the arrester gear were identical to that for Stage 2. In Stage 3 the carpet underlay was increased to five layers of inflatable tubes. The fire-hose source of material had dried up and the additional layer was made from dinghy fabric – the yellow material used for inflatable airborne dinghies. The catapult unit was not used, but the development of the Slotted Tube Catapult for Stage 5 opened a new chapter in the saga of the Carpet Deck project and, eventually, to the whole aspect of Navy flying via the introduction of the steam catapult.

The Stage 3 trials began rather catastrophically on 29 December 1947 when the modified Vampire aircraft (TG426) was irreparably damaged on its first landing. It was thought that the approach flight path across a number of buildings ahead of Jersey Brow was not the ideal. With new Sea Vampire aircraft available, and a revised pattern of approach devised, a series of 40 successful landings began on 17 March 1948. This led to preparations for trials at sea.

The Stage 2 carpet was lifted from the Meadow Gate site and transhipped to HMS *Warrior* – in Portsmouth Dockyard – for the Stage 4 trials. One of the existing shipboard arrester units was adapted for this installation, possibly that serving the original No.4 wire. In the process of re-erection, the fifth layer of tubes, as for Stage 3, was introduced. Sea trials began on 3 November 1948 and were completed by the end of the following May. Following these trials, the installation was removed from HMS *Warrior* and returned to Farnborough.

The final installation at RAE – Stage 5 – was on Laffan's Plain, parallel to, and to the north of, Runway 07/25, a location free from wind eddies around and above buildings. The installation alongside Runway 04/22 was removed, components from both this equipment and that returned from HMS *Warrior* being used on the new site. Trials and demonstrations continued at its new location, the project culminating at Stage 6 in the one-and-only occurrence – on 12 November 1953 – of a catapulted launch and landing sequence of a Hawker Sea Hawk aircraft, with its undercarriage remaining retracted throughout the exercise.

APPENDIX 12

THE BH.5 CATAPULT (BH.V ACCELERATOR) –

ENGINEERING DETAILS

The BH.5 Catapult originally was designed to launch an aircraft of 30,000lb weight at 75kts {13,600kg, 38m/sec} or of 14,000lb at 85kts {6,400kg, 43m/sec} with a mean acceleration of the order of 3½g. The maximum/mean acceleration ratio was not to be greater than 1·2 and the rate of build-up of acceleration was not to exceed 10g/sec; all this to occur within the launch time of about two seconds' duration. In common with the greater majority of ships' catapults and accelerators, the power unit was a hydro-pneumatic jigger comprising a cylinder of 40in {1m} bore with a length of almost 15ft {4·6m} with a rope reeving of 8:1. This produced a working stroke at deck level of 119ft {36m}.

The overall length of the prototype launcher and its peripherals on Jersey Brow was 250ft {76m} with the track length itself being 142ft {43m}. The machinery pit was located with its centre line 61ft {19m} west of the track. It is not known why this distance was selected; it meant that the cabling between the drive unit and the deck shuttle would be lengthy – perhaps it represented a typical shipboard installation. The machinery pit was an underground cavern 101ft {31m} long with a width varying between 33ft and 16ft {10m & 5m}, having a clear headroom in excess of 12ft {3·7m}. (The edge of this pit was only 15ft {4·6m} from one of the NAD offices!) The head and tail ropes between the drive unit and the deck shuttle ran in culverts with fairleads in access pits. The control pit was 10ft x 8ft {3m x 2·4m} and the remote compressor room was 37ft x 18ft {11m x 5m}. Of all this machinery, the only parts to be seen above ground were the shuttle, the mechanical positioner and the collapsible wheel chocks.

The method of control of the BH.5 differed from previous installations. All early systems depended upon the flow control of the pressurising fluid onto the face of the piston. If the driving force was compressed air, the supply was through a balanced poppet valve – or sleeve valve – with the displaced fluid on the back of the piston being controlled by discharge through a variable orifice, the size of which depended upon the movement of the piston through its stroke. (With cordite operation, the pressure build-up was conditioned by the predetermined amount and composition of the cordite elements forming the cartridge.)

In the BH.5, the system needed to be refined in order that precision control of pressure and its build-up would be effective throughout its launching stroke. With the greater power input required for this installation it was felt that control of the discharge fluid side rather than the pressure air side would achieve this. Garstin (#243 p.4) describes this:

'The jigger moving sheaves are mounted on a cross-head which is connected to a piston rod to a main power piston in the power cylinder. ... One side of the piston is permanently connected to a group of air storage reservoirs, the other being subjected to fluid pressure, essentially water, which is maintained at sufficient pressure to hold the piston fully home against the air pressure and the release of which through a launching valve enables the characteristics of launching to be controlled. The launching valve consists of a large single face slide valve, which is operated hydraulically; this valve is open only during the launch.

'The air having at all times free access to the piston, it is convenient to control the characteristics of the catapult by restricting the escape of fluid from the hydraulic side of the piston. This can be done by adjusting the opening of the launching valve to generate a large pressure drop across it. Since the launching valve is operated hydraulically, the best method of control is by limiting the supply or escape of the fluid from the operating cylinder by means of a profiled rod, or "carrot", attached to one of the two operating pistons (one to open the valve, the other to close it) and moving through an orifice or choke ring.'

The function of the launching valve may better be appreciated when it is realised that the bore diameter was some 15in {380mm}. The carrots in the operation of the launching valve again were the means of providing variable orifice control dependent upon the position of the valve gate itself. Mechanical adjustment of a third carrot in the system allowed the catapult to be pre-set to the appropriate condition for launching an aircraft according to its weight. The ideal of having a single control for this weight variation was achieved by this feature. For any further variation, a new carrot profile would need to have been determined, and its performance checked by experiment.

Garstin (#244) draws attention to the differences between the Jersey Brow installation and those aboard ship. While in both prototype and production systems the major components were identical, certain operational features appeared to advantage in the shipboard systems. Generally the power available on board ship was considerably greater than that at Farnborough; this was evident in the facilities for recharging the compressed air and hydraulic pressure vessels. Also the control system at Farnborough had been subject to development procedures, contributing to a significant advantage for the subsequent shipboard installations. Due to the distance between the shipboard ram-rooms (machinery flats) and the flight deck, the equivalent weight of the moving parts of the shipboard installation was 17,500lb {8,000kg} compared with 16,400lb {7,500kg} at Jersey Brow. The duty of the jigger was to cater for this weight additional to that of the aircraft to be launched, which could fall within the range of 50–120 percent of the weight of the aircraft itself. It was this relatively high ratio of machinery/aircraft weight which encouraged the research into alternative methods of launching aircraft – leading to the BS series of steam catapults.

THE BS.4 STEAM CATAPULT – ENGINEERING DETAILS

For details of this interesting application – being outside his 'hands-on' experience – the author has been compelled to depend entirely upon the writings of others. Of these, many examples exist, both as popular articles in engineering journals and as comprehensive technical reports. The following information has been obtained from a couple of highly informative articles, published more than three decades apart. The earlier source is that which appeared in *The Engineer* of 28 January 1955 reported under the title of 'Steam Catapult for Aircraft Carriers'. The latter, with a more historic content, appeared in the *Model Engineer* of 8 May 1999 as 'The Steam Catapult, How it Works' from the pen of George Ray (#79), who was actively involved in its development.

The history of the steam catapult and its *raison d'être* is treated earlier in these pages. The system was developed from Colin C. Mitchell's 1938 patent covering a slotted tube as a prime mover for catapulting aircraft. The originator, at that time, held an engineering post with Messrs MacTaggart Scott & Co (Engineers) of Edinburgh in the design and manufacturing of flight deck machinery. His wartime service – as Engineer Commander, RNVR – embodied a peripatetic commission covering all aspects relating to flight deck machinery. Post-war he held senior positions with Messrs Brown Brothers, also of Edinburgh, in a parallel field of activity. It was in this latter post that he developed the slotted tube principle into a working example of the steam catapult which eventually became the standard equipment for all navies operating shipborne, fixed-wing aircraft.

Basically, the steam catapult was a twin-cylinder, single-acting, one-stroke, free-piston steam engine using high pressure steam from the ship's main propulsion boilers. The system comprised steam storage vessels, the slotted cylinders with their associated pistons, a retardation system for the pistons and a hydraulic jigger for manoeuvring the moving parts of the installation. The catapult was used for launching aircraft only by what for many years was known as the 'tail-down' system but is correctly named the 'bridle' method. The cylinders and associated machinery were located immediately below flight deck level

with only the towing shuttle for launching the aircraft protruding through a linear slot in the flight deck itself. The main feature of the installation was the design of the steam cylinder to permit the shuttle to run along a slot in the entire length of its wall and by a suitable sealing device to contain the pressure of the working fluid therein. How this was solved is best seen in the illustrations showing the pair of cylinders and details of the piston assemblies.

Each cylinder assembly was of 18in {450mm} bore and was formed in unit lengths of 6ft {1·8m}. The total length of the catapult was generally within the limits of 150–200ft {46–61m}. The cylinder carried a continuous slot along the top which was closed with a free-hinged clamp/cover containing a pressure seal. This seal was a continuous steel strip of 1in x ½in {25x13mm} section fitting between the cover and a lip on the cylinder. The cover functioned as a brace to retain the hoop strength containing the force tending to open out the cylinder walls during operation. There was a distinct advantage in this feature: it avoided the provision of a large number of external ribs to maintain the strength integrity of the cylinder walls (as distinct from that seen in the German V.1 slotted-tube launcher).

Each piston ran within its cylinder with a projection passing through the slot, this in turn connected to the drive unit which propelled the aircraft along the deck. To achieve this, the sealing strip had to be diverted ahead of the piston to permit its passage, and then be replaced to maintain the steam pressure behind the piston. Again, the illustrations depict the principle of operation. At the start, the sealing strip was seated between the cover and the cylinder. As the piston was propelled forward the driving key lifted the sealing strip to allow the drive to pass under the strip. The drive key behind the lifting element carried a set of dogs (crenellated projections) to engage a matching set on the shuttle assembly. Behind this was the driving iron which re-seated the pressure seal immediately downstream of the power piston and sealing block. Although the seal was not perfect, only a small amount of steam leakage actually took place.

Each piston assembly comprised some seven discrete elements all forming a unit about 15ft {4·6m} long and weighing about 2,000lb {910kg}. The two units, together with the bridging piece and towing shuttle, amounted to 5,000lb {2,270kg}; this was the total mass of moving mechanical parts. The piston assembly was formed with the power piston at the rearmost end. The back face of this was where the

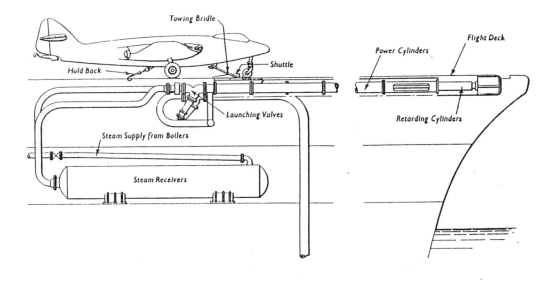

Diagrammatic arrangement showing the main elements of a steam catapult with a Hawker Sea Hawk type aircraft in the ready-to-launch position. The launching track is shown foreshortened, It would have been about five times the length of this particular aircraft.
(*The Engineer*)

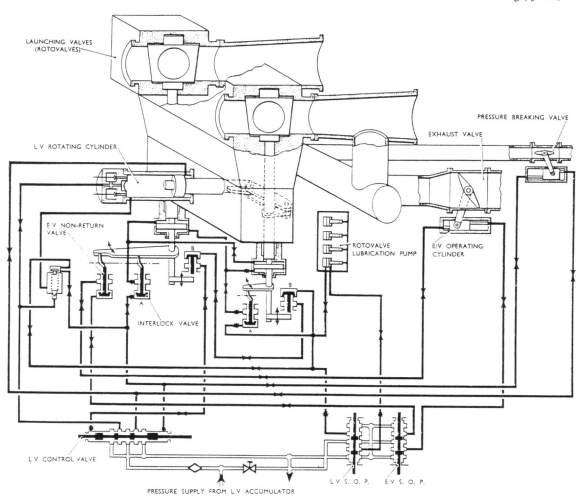

Basic diagram of the hydraulic circuitry for actuating the steam launching and exhaust valves for the steam catapult. (Institution of Mechanical Engineers)

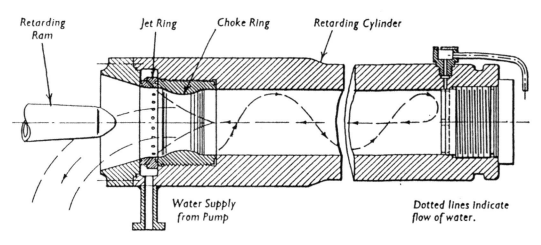

The *modus operandi* of the spike-and-barrel for braking the piston/shuttle assembly of a steam catapult is depicted showing the section through the retarding cylinder (barrel). Water entered through the jet ring, passing through the choke ring with a vortex motion to the back end of the cylinder where its rotation was stopped and the flow reversed as a slug through the axis of the vortex. The ram (spike) entered a virtually solid plug of water and was brought to rest under the extremely high pressure generated by its motion. (*The Engineer*)

An early example of the twin piston unit of the BXS.4 steam catapult showing the original form of driving key and roller-mounted shuttle. The section of cylinder surrounding the starboard arrester spike is thought to be the length immediately upstream of the arrester barrel. January 1962.

(© Crown Copyright/MOD)

steam pressure was applied and it therefore carried a representative sample of piston rings. Forward of this and separated by a spacer unit was a guide piston, to the front of which was attached the retardation spike. To the top and to the front of the intermediate spacer unit was fixed the driving key, loosely connected to the shuttle assembly, which was the unit which broke the pressure seal ahead of the piston, allowing the drive to pass beneath the strip. Immediately behind was the driving iron which returned the sealing strip, after the passage of the protrusion to the shuttle, to restore the sealing strip to its original position. Behind this was the sealing block which fully closed the seal and allowed the steam pressure behind the piston to do its work in propelling the unit forwards.

Having boosted an aircraft from rest to some 120kts {61m/sec} the moving parts of the catapult had to be brought to a standstill within the length of the arrester spike, in other words, 6ft {1·8m}. This produced a mean deceleration rate of the order of 125g but capable of peaking to a maximum of 800g. In other words, a braking force of some 1,800 tons was required to act within a fraction of a second. These figures could be doubled if a breakaway during the launch was experienced.

For bringing the piston assembly to rest, the tried and tested 'spike-and-barrel' system was adopted and adapted. Instead of retaining the water within the barrel by a frangible disc, an engineer designer from the manufacturers, J. R. C. Waterson of Messrs Brown Brothers developed a method in which a continuous stream of water flowing through the barrel was just as effective. The barrel was left open at the after end – the point of penetration by the spike – with a finely machined nozzle ring as an orifice. The bore of the orifice was a matter of only a few 'thou' {thousandths of an inch = fraction of a millimetre} larger than the maximum diameter of the spike. The inlet nozzles were inclined axially and tangentially so that the water emerging from them followed a helical path into the bore of the barrel. When this water flow reached the closed end of the barrel its rotary motion was arrested by a series of vanes on the face of the end plug and directed axially back along the axis of the barrel, spilling out through the centre of the vortex formed by the jets. By this means, although there was plenty of water sloshing around the vicinity of the unit (which could be tolerated and dealt with), the barrel could be kept full of water and ready for use without the need for constant vigil and action by the deck crew.

At the end of the launching stroke, the tip of the spike at the forward end of the piston assembly would enter the open end of the barrel and

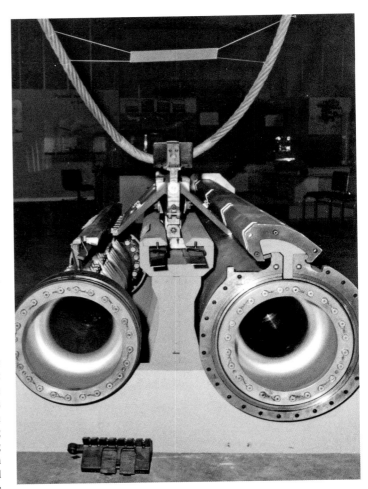

Rear view of the dual piston assembly of a BS.4 steam catapult. This was a later development of the system in which the static reaction to the launching forces on the shuttle was absorbed by the centre rail support beam between the cylinders. The right hand – starboard – piston is seen contained within a sample length of the cylinder with its swinging cover and sealing strip to the slot in the cylinder wall. OD May 1964.

(© Crown Copyright/MOD)

the displaced water was forced to pass at high speed through the decreasing annulus between the spike and the nozzle ring. Since the speed of the water through this annulus was very high, the pressure of the water in the barrel was raised accordingly and this pressure provided the retarding effect upon the piston assembly until its progress was stopped. A bonus was gained by using the kinetic energy of the water issuing through the choke ring annulus causing it to impinge upon a curved annular surface – a bucket, in section rather like that of a Pelton turbine runner – at the base of the spike. It has been calculated that the momentum exchange via this bucket accounted for some 40 percent of the braking effort. The water pressure in the barrel during an arrestment exceeded 100,000psi {>6,800bar}; in fact the compressibility of the water had to be taken into account in the design of the system. The forward structure of the ship itself was designed to accommodate the high forces experienced during the launching run and retardation of the moving components of the catapult. When under operation the allowance for linear expansion of the catapult was three inches {75mm}.

Greaves (#246) and Ray (#ibid.) show that 1,000lb of steam at 500psi {450kg at 34bar} was required for less than two seconds for each launch. With a launch rate of one every 30 seconds, the ship's boilers were required to supply steam at a rate of 120,000lb/hr {54,000kg/hr}

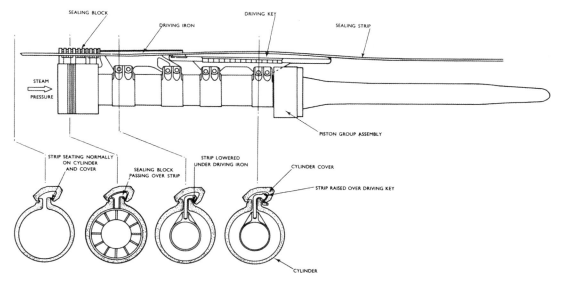

Illustrating the principle of opening and sealing the slot in the cylinder for the passage of the piston/shuttle assembly of the BS.4 steam catapult. August 1959. (© Crown Copyright/MOD)

just to supply the catapult. For prolonged operations this could have created an embarrassing demand upon the ship's engine room. The steam receivers for the catapult – acting as buffer storage between the boiler rooms and the catapult – had capacity of 2,385cuft {67m³}. It is not known the extent to which expansive working was employed in the system; probably very little as the controls – from line drawings – appear to be fairly straightforward, but in detail were more complicated. Ray again supplies information on the system:

'Effectively there was no expansive working in the catapult. However, since no steam was supplied from the ship's boilers during launch, the launch cycle could be said to be wholly expansive as far as the original dry steam storage vessels were concerned. The volume of 2,385cuft quoted is that of the dry steam installation. Wet accumulators were smaller and as a result of their method of operation were not solely expansive, since steam was generated in the accumulator during the launch cycle.'

Throughout the decade following the trials in HMS *Perseus* (1950) Greaves (*#ibid.*) records a number of relatively minor developments of the steam catapult. The original principle and the major components remained unchanged. Trials were conducted with an articulated piston assembly and attention was given to the lubrication of the piston throughout its stroke in an attempt to reduce abrasion.

In the early commissioning phases the main launch control valves became the subject of much thought and attention, involving at one stage a significant change of approach. Ray (*#ibid.*) records that the original lift valves, through the tortuous steam route through them from steam receivers to catapult cylinders, produced considerable pressure drops. Ward (*#207*) describes (1956) some of the problems encountered during the trials of the first installation pair in HMS *Ark Royal*[55]. It was necessary for the launching valves to open quickly, but under a programmed control. This was attained by use of a variable orifice device in the hydraulic control circuit dubbed, through its profile, the carrot valve which penetrated an orifice ring. Ward (*#ibid.*) states that in these trials – during which 409 dead-load launches were achieved – nine different carrots were used, each with a slightly modified profile from its fellow. A further six carrots, manufactured by Messrs Brown Bros, were used when live launches were conducted. Other problems appeared to stem from the use of dry steam as the motive power. Experiments were conducted in parallel at RAE Bedford, the outcome

being the provision of a new, tapered-plug type of launching valve and the adoption of wet steam as the power source.

The original lift valves were replaced by a pair of ROTO valves produced by Messrs Harland and Wolff of Belfast. The pressure drop across these, when fully open, was relatively small. The ROTO was a rotary plug valve giving a full bore diameter of 12in {300mm}. Its rate of opening required to be controlled yet to provide full bore within a fraction of a second of initiation, under high pressure steam conditions. The rate of opening was critical to contain the rate of increase of the accelerating forces during the launch (change of 'g') within specified limits.

Greaves (*#ibid.*), five years after Ward's paper, describes the effect of these modifications, with ongoing research being conducted into other associated features. With the steam catapult, the tensioning of the system in readiness for a launch, including the holdback unit attached to the aircraft for restraining forward motion from engine thrust, needed addressing, together with the bridle retrieval at the end of a launch. These features are discussed elsewhere in this text.

Ray (*#ibid.*), who had the responsibility for designing the BS.6 systems for the projected and cancelled carrier CVA-01 in 1966, states that with the demise of steam as the principal source of propulsion for surface ships in the Royal Navy it would appear unlikely that the steam catapult could be justified in any future RN ship. A successor to the steam catapult could be of great engineering interest. The author is indebted to George Ray for the following details of the range of steam catapults. The indication 'X' refers to experimental equipment.

BXS.1 – the prototype, fitted in HMS *Perseus*. On removal from this ship it was installed at RAE Bedford, becoming known as the BSX.4 'Raised Catapult'.

BXS.2 – the first catapult ordered for the new NAD site at Thurleigh, RAE Bedford.

BXS.3 – the trials installation at ARDE Shoeburyness. This had a short stroke of 47ft {14m} and was cordite-powered.

BS.4 – comprised the production series of catapults for the Royal Navy, French navy, Dutch navy, various Commonwealth navies and the Argentinian navy. An operational difference appeared in the type of control valves fitted. The upward load reaction during launch was absorbed via the deck covers. The designation applied whatever the length of catapult stroke. Each catapult was identified by a serial number for ready reference as to which ship had which specification.

BSX.5 – the 'flush catapult' at RAE Bedford. The number '5' indicated the major design change to accommodate the upward force from the shuttle by using a centre rail instead of channels attached to the deck covers.

BS.5 – the shipboard centre rail catapults fitted to HMS *Eagle*[51], the forward unit being 150ft long and the waist unit 200ft.

BS.6 – the projected catapult for the new carrier CVA-01, a centre rail unit with the waist catapult having a 250ft stroke.

APPENDIX 14
DIRECT ACTING (WATER SPRAY) ARRESTER GEAR (DAAG)

ENGINEERING DETAILS

Much of the content of this section is taken from Thomlinson (*#296*) who describes the equipment installed at RAE Bedford. He starts by saying that the equipment falls into a class known as 'constant run-out' gear. This means that in operation the full stroke of the gear is used regardless of the weight and speed of engagement of the aircraft. The upper limit of performance was for aircraft of 50,000lb AUW engaging at 130kts {22,700kg at 67m/sec} but it was also required to accept aircraft of 15,000lb AUW engaging at speeds of 100kts {6,800kg at 51m/sec} without exceeding a maximum retardation of 1·5g. To meet these requirements it was estimated that a pull-out of the arrester rope of 612ft {186m} would be required.

The arrangement consisted of a length of 1,280ft {390m} of about 1 inch {25mm} diameter steel wire rope reeved about two deck-edge sheaves located 250ft {76m} apart and placed symmetrically about the centre line of the runway. Each end of the rope was fitted with a simple cylindrical piston, each of which was free to slide, with a clearance of about 0·003in/0·008in {0·08mm/0·20mm}, in a steel tube of 3¾in {95mm} bore diameter. The tubes each were 510ft {155m} long, made up from section lengths of 24½ft {7m}. The rope passed out of the tube through a simple sleeve-type gland at the end adjacent to its corresponding deck-edge sheave. The tubes may point away from the sheaves in any direction although it seemed logical to lay them parallel with the runway. In one installation at Bedford they were at right angles to the runway thereby giving the minimum interference to camera sightlines.

Along the crown of each tube were some 400 ⅜in {9·5mm} diameter tapped holes into which were screwed stainless steel plugs each provided with a 0·143in {3·6mm} diameter hole down its axis, the inner end being bell-mouthed. The size and spacing of these holes were important features of the design and were referred to as the 'orifice programme'. The tubes were laid horizontally so that by admitting water to the tubes at inlets close to the gland ends, the tubes were flooded completely with water – the filling process continuing until water overflowed from the plug orifices. Water was prevented from seeping through the rope gland (under a head of a few inches of pressure) by injecting a 'shot' of grease into the gland clearance. The open end of the tube – abaft the piston, with the piston in its starting position – was curved upwards so that the lower lip of the outlet was above the crown of the tube.

When an aircraft engaged that part of the rope system between the deck sheaves (the centre span) the pistons were drawn along the tubes and the fluid between the piston and the gland was expelled through those orifices. The hydraulic pressure thus generated acted on the pistons and this resistant force was transmitted to the aircraft via the wire rope. The pistons had a nominal stroke of 500ft {150m} which, within the geometry of the system, permitted an aircraft travel of 612ft {186m}.

To the after face of each piston was attached a fibre rope (known as the tail rope) some 550ft {168m} long which passed out of the open end of the tube. This was drawn into the tube during operation and used for recovery of the system into its ready condition. This tail rope was also used to provide a tensioning device to absorb any slack in the system and a breaking link which fractured as soon as an aircraft picked up the centre span wire.

The first trials of this system, according to Lewis (*#58 p.73*), were

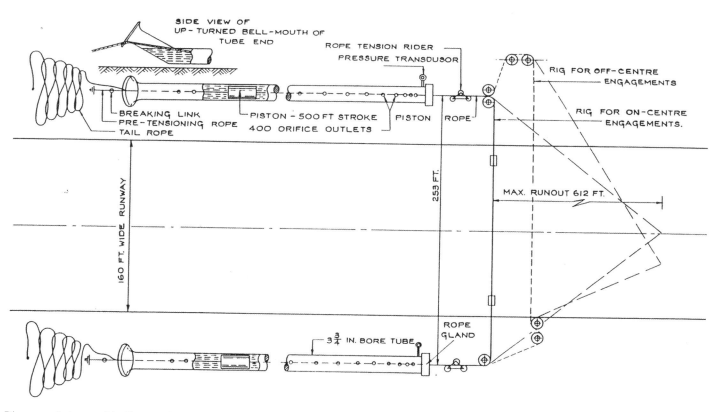

Diagrammatic layout of the first experimental Direct Acting Arrester Gear (DAAG), or water spray gear, at Thurleigh. The discharge was to atmosphere, producing an impressive pair of curtains of spray in the process. (© Crown Copyright/MOD)

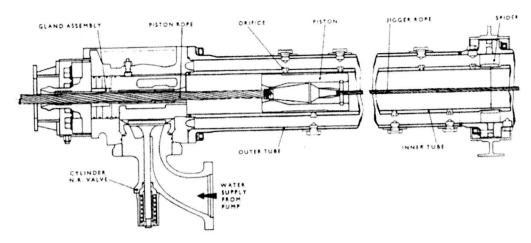

Sectional drawing of the DAAG, showing the piston at the arrestment end of its travel. (© Crown Copyright/MOD)

conducted at RAE Bedford in August 1962. The demonstration of this equipment was a fascinating experience with a multitude of water jets shooting sky-high, spraying all and sundry as the droplets fell back to the ground.

SHIPBOARD INSTALLATION

The foregoing describes a basic system which was developed in a number of ways. The first was to adapt the system for use on shipboard, to become known as the DA.1 Arresting Gear. In this case the tube lengths were 200ft {61m}, each containing 95 gallons {432 litres} of water, and the spacing of the deck-edge sheaves was reduced to 80ft {24m}. A new orifice programme (using 155 orifices of 5⁄32in {4·0mm} diameter) was designed to accept aircraft of 32,000lb AUW at an engaging speed of 105kts {15,500kg at 54m/sec} with a rate of retardation in excess of 2g. The pressure generated during an arrestment was of the order of 5,000psi {340bar}. Other refinements were introduced, with hydraulic recovery and pre-tensioning of the system. The problem of having an efficient gland through which the wire rope progressed was a difficult one to solve. An improvement was found in the use of locked coil ropes in which the outer surface was much smoother than a stranded rope but the penalty of extra stiffness within the rope itself needed to be accepted. One major feature was the encapsulation of the spray tubes within an outer tube so that the water ejected from the system could be recovered and re-used; this was important when the water contained an anti-freeze component. (A pity Her Majesty's matelots were denied the spectacle of seeing much water being squirted upwards – negative rainfall?) A prototype system, for a single arrester wire, was installed in HMS *Eagle*[51] during her 1967 refit (the other wires being linked to Mark 13/14 arrester systems). This installation was made in September 1969 during which an experimental 2:1 reeving of the wire was introduced. The intention here was the anticipated shortening of the retardation stroke, hence the arrester-tube length. The result of this was discouraging and reversion to the full length was made. The prototype installation at Thurleigh – the DAX.1 – achieved 2,000 landings in its life, while that aboard HMS *Eagle*[51] – the DAX.2 – was used for some 600 landings.

With regard to the system envisaged and developed for CVA-01, the specification envelope required aircraft within the weight range of 15,000lb to 40,000lb {6,800–18,100kg} to be arrested from speeds of engagement ranging from 110kts at the lighter weight to 125kts for the upper weight {56–64m/sec}, with maximum retardation at 4·2g. From consideration of ship and equipment the arrested run had to be contained within a pull-out of 270ft {82m}, with a deck-edge sheave spacing of 100ft {30m}. This project was awarded the title of DA.2 Arresting Gear.

For each of the deck arrester wires there were to be a pair of these units, one each side of the ship, placed below deck, along a fore-and-aft alignment. Each unit would comprise an inner tube of 5¾in bore, contained within the outer tube of 10in bore {146mm, 254mm}. Along each inner tube there would be upwards of 200 orifices of ¾in {19mm} diameter. Pressures of up to 3,000psi {204bar} were expected to be generated within the inner tubes while that in the outer tubes would be kept to 600psi {41bar} by way of relief valves. The pitch of the orifices – the orifice programme – was graduated so that as the piston speed decreased, so did the orifice area ahead of the piston with a reasonably constant pressure, and therefore the rate of retardation was maintained throughout the arrestment. Within the design parameters and a carefully designed orifice programme, the gear was capable of arresting a wide range of aircraft weights and engagement speeds without requiring additional control mechanisms. The units were pre-tensioned and reset by a hydraulic jigger system connected to a tail rope in each inner tube. This jigger was 14ft long with a 16:1 reeve. It contributed nothing to the braking effort of the system, the working fluid being vented to an accumulator during arrestment. Auxiliary equipment was required to provide a 20 seconds reset cycle with all units full in the standby situation, covering leakage, spillage and make-up.

When CVA-01 was cancelled, development of DA.2 continued and eventually was installed as the arrester deck in HMS *Ark Royal*[55] for her recommissioning in 1970. In this configuration with four arrester wires, four arrester tube units, each 228ft {70m} long, were located along the deckhead each side of the upper hangar. DA.2 was the only equipment capable of accepting Phantom aircraft.

AIRFIELD OVERSHOOT SYSTEM

For the airfield arrester system the design parameters matched those of the original installation at Thurleigh, using the 600ft {180m} maximum pull-out, operating with a 50,000lb {22,700kg} aircraft entering at 130kts {66m/sec}. The energy absorption of the system was calculated to be $37·5 \times 10^6$ft.lb {$5·2 \times 10^6$kgfm}. Two centre spans, 20ft {6m} apart, were provided with the rope sheaves at 200ft {61m} centres spanning a standard 150ft {46m} wide runway. The centre spans (supported 4½in {115mm} above the runway on split rubber discs) were connected at each extremity by beckets and purchase cables into a pair of arrester units, one laid along each side of the runway.

Each arrester unit (energy absorber) comprised a pressure tube of 4in {100mm} bore, 520ft {160m} long fitted with about 400 spray orifices in alignment atop the tube. This pressure tube was located at the bottom of an 18in {460mm} diameter jacket pipe, buried 30in {760mm} deep in the ground. For operation the jacket was half-filled with the braking fluid (water containing an anti-freeze), which filled the fully submerged pressure tube through the spray orifices. Tensioning of the system was achieved by pull-lift winches. The resetting procedure automatically re-primed the submerged pressure tube through the spray orifices on resetting the system. It is not known how resetting was accomplished, but it is conjectured to have been achieved by a tractor.

It can be seen that this system differed from the carrier-borne installation in that the energy absorbers were considerably longer and the jacket pipe much larger than the shipboard application. To my way of thinking, this system, from the maintenance aspect, must have suffered a great disadvantage in that it was buried completely in the ground.

BIBLIOGRAPHY

PUBLISHED WORKS

The following books are varied as to the content of related matter and in the standard of presentation but each has a contribution to make, however great or small.

1. Almond, P., *Aviation – The Early Years* (from The Hulton Getty Picture Collection), Könemann, 1997 (ISBN 5–89508–682–7)
2. Andrews, C. F. & Morgan, E. B., *Supermarine Aircraft Since 1914*, Putnam, 1981 (ISBN 0–370–10018–2)
3. Ashworth, C., *Military Airfields of the Central South and South-East*, 'Action Stations' Series, Patrick Stephens, 2nd edn, 1990 (ISBN 1–85260–376–3)
4. Barnes, C. H., *Handley Page Aircraft Since 1907*, Putnam, 1976 (ISBN 0–370–00030–7)
5. Baxter, R., *Farnborough Commentary*, Patrick Stephens, 1980 (ISBN 0–85059–434–0)
6. Beaver, P., *The British Aircraft Carrier*, Patrick Stephens, 3rd edn, 1987 (ISBN 0–85059–877–X)
7. Biggs, M. & Ott, F., *80 Years of the British Aircraft Carrier*, Friends of the Fleet Air Arm Museum (pb), 1994 (ISBN 0–9513139–3-2)
8. Blake, R., Hodgson, M. & Taylor, W., *The Airfields of Lincolnshire Since 1912*, Midland Counties Publications, 1984 (ISBN 0–904597–32–6)
9. Bolster, C. M., *Assisted Take-Off of Aircraft*, Norwich University, Northfield, Vermont, USA, 1950
10. Bowyer, C., *Spitfire*, Bison Books, 1994 (ISBN 0–86124–140–1)
11. Bowyer, M. J. F., *Military Airfields of the Cotswolds and the Central Midlands*, 'Action Stations' Series, Patrick Stephens, 2nd edn, 1990 (ISBN 1–85260–372–0)
12. Brett, B., *Modern Sea Power*, Hamlyn (W. H. Smith), 1986 (ISBN 0–603–03899–9)
13. Brown, D., *Aircraft Carriers – World War 2 Fact Files*, Macdonald & Jane's (pb), 1977 (ISBN 0–354–01009–3)
14. Brown, D. L., *Miles Aircraft Since 1925*, Putnam, 1970 (ISBN 0–370–00127–3)
15. Brown, E. M. *Wings on My Sleeve*, Airlife, 2nd edn, 1978 (ISBN 0–9504543–6-2)
16. Brown, E. M., *Wings of the Weird and Wonderful (Volume 1)*, Airlife, 1983 (ISBN 0–906393–30–2)
17. Brown, E. M., *Wings of the Weird and Wonderful (Volume 2)*, Airlife, 1985 (ISBN 0–906393–44–2)
18. Brown, E. M., *Wings of the Navy – Flying Allied Carrier Aircraft of World War 2*, Airlife, 1987 (ISBN 0–906393–87–6)
19. Brown, E. M., *Testing for Combat*, Airlife, 1994 (ISBN 1–85310–319–5)
20. Chapman, J. H. B., *The Development of the Aircraft Carrier*, 'Transactions of the Royal Institution of Naval Architects 1960', Vol.102
21. Chesneau, R., *Aircraft Carriers of the World, 1914 to the Present. An Illustrated Encyclopedia*, Brockhampton Press, 1998 (ISBN 1–86019–87–5-9)
22. Cooper, P. J., *Forever Farnborough – Flying the Limits 1904–1996* Hikoki (pb), 1996 (ISBN 0–9519899–3-6)
23. Cronin, R. C., *Royal Navy Shipboard Aircraft Developments 1912–1931*, Air Britain (Historians) Ltd, 1990 (ISBN 0–85130–165–7)
24. Crosley, R. M., *Up in Harm's Way – Flying with the Fleet Air Arm*, Airlife, 1995 (ISBN 1–85310555–4)
25. Darman, P., *Great Carrier Aircraft*, Blitz Editions, 1996 (ISBN 1–85605–343–1)
26. Dennis, R., *Farnborough's Caterpillars – An Account of Research Flying and Parachute Escapes at the Royal Aircraft Establishment*, Footmark Publications, 1996 (ISBN 0–9515738–8-8)
27. Dickson, B. W. A., *Aircraft – from Airship to Jet Propulsion 1908–1948*, Haldrett Press, c.1961 (Privately published for Vickers-Armstrong Ltd)
28. Dobson, M. D. (ed), *Wings over Thurleigh – An Aeronautical Research Heritage 1954 to 1994*, published privately, 2002
29. Ellis, C., *A History of Combat Aircraft*, Hamlyn, 1980 (ISBN 0–600–32112–6)
30. Ellis, W. H. B., *Hippocrates R. N. – Memoirs of a Naval Flying Doctor*, Robert Hale, 1988 (ISBN 0–7090–3269–2)
31. Ferguson, H. M., *Deck-Landing Projector Sight*, 'GEC Journal', Vol.29, No.2, 1962
32. Ferguson, J. D., *Aboard 'The Ark' – for the last time*, 'Aircraft Illustrated' (Journal), August 1978
33. Finnis, B., *The History of the Fleet Air Arm – From Kites to Carriers*, Airlife, 2000 (ISBN 1–84037–182-X)
34. Friedman, N., *British Carrier Aviation*, Conway Maritime Press, 1988 (ISBN 0–85177–488–1)
35. Garstin, D. J. I., *Hydraulics in Flight Deck Machinery*, 'Proceedings of the Institution of Mechanical Engineers', Vol.180, 1966
36. Gillcrist, P. T., *Feet Wet, Reflections of a Carrier Pilot*, Airlife, 1990 (ISBN 1–85310–191–5)
37. Gunston, B., *A Century of Flight*, Brian Trodd (pb), 1989 (ISBN 0–85361–131-X)
38. Gunston, B., *Plane Speaking –– A Personal View of Aviation History*, Patrick Stephens, 1991 (ISBN 1–85260–166–3)
39. Hadley, D., *Barracuda Pilot*, Airlife, 1992 (ISBN 1–85310–195–8)
40. Halpenny, B. B., *Military Airfields of Greater London*, 'Action Stations' Series, Patrick Stephens, 1984 (ISBN 0–85059–585–1)
41. Hare, P. R., *The Royal Aircraft Factory*, Putnam, 1990 (ISBN 0–85177–843–7)
42. Hayward, R., *The Fleet Air Arm in Camera – 1912–1996*, Sutton Publishing, 1996 (ISBN-0–7509–1254–5)
43. Hobbs, D., *Aircraft of the Royal Navy Since 1945*, Maritime Books (pb), c.1983 (ISBN 0–90771–06–8)
44. Hobbs, D., *The Fleet Air Arm in Focus (Part One)*, Maritime Books (pb), 1990 (ISBN 0–907771–45–9)
45. Hobbs, D., *Aircraft Carriers of the Royal and Commonwealth Navies*, Greenhill Books, 1996 (ISBN 1–85367–252–1)
46. Holyoak, V., *On the Wings of the Morning – RAF Bottesford 1941–1945*, privately published, 1995 (ISBN 0–9526739–0-8)
47. Hoyt, E. P., *Carrier Wars*, Robert Hale, 1990 (ISBN 0–7090–4018–0)
48. Humble, R., *The Rise and Fall of the British Navy*, Macdonald, 1986 (ISBN 0–356–72227–1)
49. Ireland, B., *The Rise and Fall of the Aircraft Carrier*, Marshall Cavendish, 1979 (ISBN 0–85685–711–4)
50. Jackson, A. J., *De Havilland Aircraft Since 1909*, Putnam, 1978 (ISBN 0–370–30022-X)
51. James, D. N., *Gloster Aircraft Since 1917*, Putnam, 1971 (ISBN 0–370–00084–6)
52. John, Rebecca, *Caspar John*, Collins, 1987 (ISBN 0–00–217136–8)
53. Jones, A. H. & Jones, M. H., *No Easy Choices – A Personal Account of Life on the Carrier HMS Illustrious 1940–1943*, Square One Publications, 1994 (ISBN 1–872017–85–1)

54. Jones, F., *Air Crash*, Robert Hale, 1985 (ISBN 0–7090–2161–5)

55. King, B., *Royal Naval Air Service 1912–1918*, Hikoki Publications, 1997 (ISBN 0–951–9899–5-2)

56. King, P., *Knights of the Air*, Constable, 1989 (ISBN 0–09–468100–7)

57. Kirby, R., *Avro Manchester – The Legend behind the Lancaster*, Midland Publishing, 1995 (ISBN 1–85780–028–1)

58. Layman, R. D., *Before the Aircraft Carrier – The Development of Aviation Vessels 1849–1922*, Conway Maritime, 1989 (ISBN 0–85177–516–0)

59. Lewis, P. M. H., *The British Fighter Since 1912*, Putnam, 1967

60. Lewis, R. G., *A History of Aircraft Carrier Flight Deck Equipment*, Monograph privately published by the author, 2000

61. Loynes, D. N., *Flight Deck Machinery*, lecture paper (details unknown), 1968

62. McDonald, S., *Historic Aircraft*, Grange Books, 1995 (ISBN 1–85627–821–2)

63. Marriott, L., *Royal Navy Aircraft Carriers – 1945–1990*, Ian Allan, 1985 (ISBN 0–7110–1561–9)

64. Mason, T., *The Secret Years – Flight Testing at Boscombe Down 1939–1945*, Hikoki, 1998 (ISBN 0–951899–9-5)

65. Middleton, D., *Tests of Character – Epic Flights by Legendary Test Pilots*, Airlife, 1995 (ISBN 1–85310–481–7)

66. Miller, D. & Peacock, L., *Carriers – the Men and the Machines*, Salamander Books, 1991 (ISBN 0–86101–561–4)

67. Mondey, D., *The Hamlyn Concise Guide to British Aircraft of World War II*, Chancellor Press, 2nd edn, 1994 (ISBN 1–85152–668–4)

68. Nichols, M. (ed), *Carriers, Air power at Sea*, Key Publishing (pb), 2001

69. Pearcy, A., *A Short Illustrated History of the Royal Aerospace Establishment, Bedford*, Airlife, 1999 (ISBN 1–85310–360–8)

70. Penrose, H., *British Aviation – The Great War and Armistice*, Putnam, 1969 (ISBN 370–00128–1)

71. Penrose, H., *British Aviation – The Ominous Skies 1935–1939*, HMSO, 1980 (ISBN 0–11–290298–7)

72. Polmar, N., *Aircraft Carriers – A Graphic History of Carrier Aviation and its Influence on World Events*, Macdonald, 1969 (ISBN 356–02805–4)

73. Poolman, K., *Illustrious*, William Kimber, 1955

74. Poolman, K., *Escort Carrier 1941–1945*, Ian Allan, 1972 (ISBN 7110–0273–8)

75. Poolman, K., *Escort Carrier – HMS* Vindex *at War*, Leo Cooper, 1983 (ISBN 0–436–37705–5)

76. Popham, H., *Into Wind – A History of British Naval Flying*, Hamish Hamilton, 1964 (ISBN 241–01771–8)

77. Powell, H. P., *Test Flight*, Allan Wingate, 1956

78. Preston, A., *Modern Air Power – Carriers*, Bison Books, 1990 (ISBN 1–85627–006–879)

79. Preston, A., *The World's Great Aircraft Carriers*, Chancellor, 1999 (ISBN 0–75370–146–4)

80. Pudney, J., *Laboratory of the Air*, HMSO, 1948 (SO Code 70–541*)

81. Rawlings, Revd J. D. R., *Pictorial History of the Fleet Air Arm*, Ian Allan, 2nd imp, 1974 (ISBN 0–7110–0436–6)

82. Ray, G., *The Steam Catapult, How it Works, 'Model Engineer'*, 8 May 1999

83. Reynolds, C. G., *The Carrier War*, Time-Life Books, 1987 (ISBN 0–8094–3304–4)

84. Rolt, L. T. C., *Isambard Kingdom Brunel – A Biography*, Longmans, Green, 1957

85. Rooke, D. S., *Almost into Wind*, Guild Bindery Press (USA), 1993 (ISBN 1–55793–040–6)

86. Roskill, S. W., *The Navy at War, 1939–1945*, Wordsworth Editions (pb), 1998 (ISBN 1–85326–697–3)

87. Salmon, P., *Catapults and Catapulting of Aeroplanes, 'Proceedings of the Royal Aeronautical Society'*, Vol. XXXVI, No. 260, August 1932

88. Smith, D. J., *Britain's Military Airfields 1939–45, 'Action Stations'* Series, Patrick Stephens, 1989 (ISBN 1–85260–038–14)

89. Smith, N. L., *HMS* Tracker, Friends of the Fleet Air Arm Museum (pb), 1987 (ISBN 0–951–3139–0-8)

90. Sturtivant, R., *Fleet Air Arm at War*, Ian Allan, 1982 (ISBN 0–7110–1084–6)

91. Sturtivant, R., *British Prototype Aircraft*, Promotional Reprint Company, 1995 (ISBN 1–85648–221–9) (Previously published as *British Research and Development Aircraft*, Haynes, 1990 [ISBN 0–85429–697–2])

92. Sturtivant, R. & Ballance, T., *The Squadrons of the Fleet Air Arm*, Air Britain, 1994 (ISBN 0–85130–223–8)

93. Sturtivant, R. & Burrow, M., *Fleet Air Arm Aircraft 1939 to 1945*, Air Britain (Historians) Ltd, 1995 (ISBN 0–85130–232–7)

94. Taylor, H. A., *Airspeed Aircraft Since 1931*, Putnam; Brassey's Ltd, 1970 (ISBN 370–001–10–9)

95. Taylor, H. A., *Fairey Aircraft Since 1915*, Putnam; Brassey's Ltd, 1974 (ISBN 0–370–0065-X)

96. Thetford, O., *Aircraft of the Royal Air Force Since 1918*, Putnam; Brassey's Ltd, 7th edn, 1979 (ISBN 0–370–30186–2)

97. Thetford, O., *British Naval Aircraft Since 1912*, Putnam; Brassey's Ltd, 5th edn, 1982 (ISBN 0–370–30021–1)

98. Thomas, D. A., *A Companion to the Royal Navy*, Harrap, 1988 (ISBN 0–245–54572–7)

99. Till, Dr G., *Air Power and the Royal Navy, 1914–1945*, Jane's Publishing Co, 1979 (ISBN 0–354–01204–5)

100. Treadwell, T. C., *Strike from beneath the Sea*, Tempus, 1999 (ISBN 0–7524–1704–5)

101. Turnill, R. & Reed, A., *Farnborough, the Story of RAE*, Robert Hale, 1980 (ISBN 0–7091–8584–7)

102. Tute, W., *The True Glory*, Centurion Press, 1989 (ISBN 1–870630–15–7)

103. Wallace, G., *Carrier Observer*, Airlife, 1993 (ISBN 1–85310–307–1)

104. Ward, Cdr N. D., *Sea Harrier over the Falklands*, Orion, 1994 (ISBN 1–85797–102–7)

105. Watts, A. J., *The Royal Navy – An Illustrated History*, Brockhampton Press, 1999 (ISBN 1–86019–9836)

106. Wellham, J. W. G., *With Naval Wings*, Spellmount, 1995 (ISBN 1–873376–33–2)

107. Williams, R., *Royal Navy Aircraft Since 1945*, Airlife, 1989 (ISBN 0–87021–996–0)

108. Anon, *The Engineer*, Editorial, 4 March 1955 (quoted in the '*Journal of the Organ Club*', issue 2/55, April 1955, Editor, G. G. J. Cooper)

109. Misc, *Gas Turbines* (IMechE, 1951), privately bound volume for G. G. J. Cooper, being a collection of 35 papers published in the '*Proceedings of the Institution of Mechanical Engineers*' (including the first James Clayton paper presented by A/Cdre F. Whittle) over the period 1945–50 on the early development of gas turbine technology.

110. Ireland, B., *Jane's Naval Airpower – The Complete History from 1914 to the Present Day*, Harper Collins, 2003 (ISBN 0–00–711152–5)

111. McCart, N., *The Illustrious & Implacable Classes of Aircraft Carrier – 1940–1969*, Fan Publications, 2000 (ISBN 1–901225–04–6)

112. Jordan, D., *Aircraft Carriers*, Silverdale Books, 2002 (ISBN 1–8560–5700–3)

113. Treadwell, T. C., *Strike from beneath the Sea, A History of Aircraft-Carrying Submarines*, Tempus, 1999 (ISBN 0–7524–1704–5)

114. Cooper, P. J., *Farnborough – 100 Years of British Aviation*, Midland, 2006 (ISBN 1 85780 239 9)

115. Marriott, L., *Catapult Aircraft*, Pen & Sword, 2006 (ISBN 1–84415– 419–X)

IN-HOUSE PUBLICATIONS AND TECHNICAL REPORTS

The following is an index of published material of RAE origin and first appeared in either a House Journal or as the subject of a Departmental Technical Report. Originally these papers would have had a limited distribution and would not have been available for public reference. Again, I have copies or have been privileged to have consulted originals in the DERA archive at Pyestock. Authors whose names also appear in Appendix 2, being colleagues of mine, obviously, were known to me personally.

201. Arnold, A. A., *Seafire XV Arresting Trials*, RAE, Catapult Test Note 110, December 1944

202. Arnold, A. A., *Spitfire XXI Deck Arresting Trials – HMS* Pretoria Castle, RAE, Catapult Test Note 113, December 1944

203. Arnold, A. A., *Report on Accelerator and Arrester Gear Proofing Trials on HMS Vengeance,* RAE, Catapult Test Note 121, February 1945

204. Arnold, A. A., *Mechanical Catapult Loading Gear*, RAE, Technical Note NA.21, October 1950

205. Baines, D. C., *The BH.5(1) Catapult has Launched its Last Aircraft*, 'RAE News', (House Magazine) Vol. XI, No.2, February 1958

206. Beard, Cdr E. C., *Catapult Aircraft Launching Bridles*, RAE, Technical Note Naval CE.2, January 1950

207. Beard, Cdr E. C., *The Development of High Speed Loading Arrangements for the Catapulting of Naval Aircraft*, RAE, Technical Note Naval CE.3, January 1950

208. Boddington, L., *Arrester Gear for Aerodromes, Proposed Scheme for Experiments*, RAE, Paper ref: D.O/1358/LB/38, December 1938

209. Boddington, L., *Assisted Take-Off for Future Naval Aircraft*, RAE, ARC Paper No.8812, July 1945

210. Boddington, L., *Landing of Future Naval Aircraft*, RAE, ARC Paper 8981, September 1945

211. Boddington, L. & Nixon, R. J., *Operation of Undercarriage-less Aircraft from H M Aircraft Carriers – Experimental Equipment for the Landing Experiments*, RAE, Technical Note Naval 71, ARC Paper 9288, January 1946

212. Boddington, L., *Deck Landing Trials of Vampire at RAE and on HMS* Ocean, RAE, Technical Note Naval Aircraft 74, 1946

213. Boddington, L. & Bartlett, J. L., *Aircraft Carriers – Safety Barriers*, RAE, Technical Note NA.82, ARC Paper 10265, January 1947

214. Boddington, L., *The Effects on Layout and Design of the Future Carrier Operating Undercarriageless Aircraft*, RAE, Technical Note NAD.14, ARC Paper 12065, January 1949

215. Boddington, L., *Catapulting with the Assistance of R.A.T.O.G.*, RAE, Technical Note NAD.27, August 1949

216. Boddington, L. & Thomlinson, Dr J., *Landing Experiments with Undercarriageless Vampire Aircraft on a Flexible Deck*, RAE, Report NA.198, February 1950

217. Boddington, L., *The Skew Deck Layout for Carrier Flight Decks*, RAE, Technical Memo NA.45, October 1951

218. Brown, D. K., *Ship Assisted Landing and Take Off – A Brief History of Flight Deck Machinery in the R N*, Admiralty, DCNA, October 1985

219. Brunton, W. G., *Aircraft Arresting Hooks – Examination of Wear of Various Types of Hook and the Effects on the Aircraft and Arrester Wires*, RAE, Report Catapult Note 43, 1942

220. Bullen, N. I., *The Influence of Rope Stretch on Tension Variations in Arresting Gears*, RAE, Technical Note NA.229, ARC Paper 14401, July 1951

221. Bullen, N. I., *The Calculation of the Performance of Drag Chain Arresting Gears,* RAE, Technical Note NA.240, February 1952

222. Caunter, C. F., *A Historical Summary of the Royal Aircraft Establishment: 1918–1948*, RAE, Report 2150A, November 1949

223. Child, S. & Caunter, C. F., *A Historical Summary of the Royal Aircraft Factory and its Antecedents: 1878–1918*, RAE, Report Aero 2150, ARC Paper 10969, March 1947

224. Chisman, S. W., *Development of a Cordite Operated Slotted Tube Catapult*, RAE, Technical Note NA.262, April 1954

225. Collier Webb, W/Cdr D. R. – Personal communication, with details from his database relating to aircraft accidents at Research Stations

226. Cooper, G. G. J., *Arresting Trials of Firebrand TF.IV Aircraft on HMS* Pretoria Castle, RAE, Catapult Technical Note 139, June 1945

227. Cooper, G. G. J., *Intensive Deck Landing Trials with Firebrand TF.III & TF.IV Aircraft on HMS* Pretoria Castle – *19th September 1945*, RAE, Naval Aircraft Test Note 153, 1945

228. Cooper, G. G. J., *Intensive Deck-Landing Trials with Seafire 45 Aircraft on HMS* Pretoria Castle – *12th July–15th October 1945*, RAE, Naval Aircraft Test Note 157, 1945

229. Cooper, G. G. J., *Observations upon the Arrester Gear in HMS* Pretoria Castle *in Connection with Recent Barrier Landings*, RAE, Naval Aircraft Test Note 158, 1945

230. Davies, J., *Naval Aircraft Arrester Hooks – Report of Trials at RNAS Arbroath, September-October 1943*, RAE, Catapult Note 43/2, December 1943

231. Davies, J., *Mosquito Aircraft – Report on Arrestation Trials at RNAS Arbroath, January 1944*, RAE, Catapult Note 54, February 1944

232. Davies, J., *Proofing Trials of Aircraft Arrester Gear Installation – Spitfire VIII Aircraft at RNAS Arbroath, April 1944*, RAE, Catapult Section Test Note 101, May 1944

233. Duddy, R. R. & Lean, D., *Automatic Control During Deck Landings*, RAE, Technical Note Aero 221, ARC Paper 14391, November 1951

234. Dyer, I. S., *An Outline of the History and Development of the Royal Aircraft Establishment, Llanbedr*, RAE, Technical Note IR.139, May 1973

235. Dyer-Smith, J. E., Thomlinson, Dr J. & Greaves, D. G., *Report on a Visit to USA to Review Work on Aircraft Carrier Flight Deck Machinery and Airfield Arrester Gears*, RAE, Technical Note Naval 45, January 1961

236. Ewans, J. R. & Warren, V. G., *Rocket-Assisted Take-Off Tests on a Fulmar*, RAE, ARC Paper 6220, August 1942

237. Fancourt, Capt H. L. StJ., *Some Aspects Concerning the Introduction of Twin-Engined Naval Aircraft*, Admiralty, ARC Paper 10563, April 1947

238. Farmer, E. R., *Report on Failure of Fuselage of Firebrand I under Side Load During Tests on "Illustrious" Type Naval Arrester Gear*, RAE, Catapult Section Test Note 109, November 1944

239. Fawell, H. D., *Report on the Development of the Direction Controlled Take-Off Scheme*, RAE, Catapult Note 42, December 1942

240. Fawell, H. D., *Note on Investigation of Arresting an Airacobra with Tricycle Undercarriage,* RAE, Catapult Section Note 51, August 1943

241. Fawell, H. D., *Portable Inertia Arrester Gear – Supplementary Note on the Further Development of the Drag Chain Arrester Gear,* RAE, Catapult Section Test Note 44/4, February 1944

242. Fawell, H. D., *Arresting an Airacobra Tricycle Undercarriage Aircraft – Report of Trials at RNAS, Arbroath,* RAE, Catapult Section Note 51/1, March 1944

243. Garstin, Lt/Cdr D. J. I., *The Performance of a British Naval Aircraft Catapult, BH.5(1),* RAE, Technical Note Naval CE.4, ARC Paper 14541, October 1951

244. Garstin, Lt/Cdr D. J. I., *Trials of BH.5 Type Catapult in HMS* Eagle, RAE, Technical Note Naval CE.5, April 1952

245. Goodall, Sir Stanley V., *Limitations on Carrier-Borne Aircraft,* RAE, ARC Paper 4860, December 1940

246. Greaves, Cdr D. G. & Buesnel, F. J., *A Survey of Steam Catapult Development,* RAE, Technical Note Naval 47, January 1961

247. Green, F. M., *Gravity Assisted Take-Off for Aircraft,* RAE, Report AD.3156, ARC Paper 4936, October 1940

248. Griffith, A. A. C., *Performance of the Mk.12 Arresting Gear,* RAE, Technical Note Naval 4, March 1955

249. Griffith, A. A. C., *Development of the Mk.13 Arresting and Safety-Barrier Gear* RAE, Technical Note Naval 10, August 1955

250. Grimes, A., *Some Theoretical Aspects of Ventral Catapulting,* RAE, Technical Note NA.215, September 1950

251. Grimes, A., *Angled Deck Trials, HMS* Triumph – *February 1952,* RAE, Technical Note NA.242, ARC Paper 14960, March 1952

252. Grimes, A., *The Factors Affecting Deck Landing Speeds on Aircraft Carriers,* RAE, Technical Note NA.251, ARC Paper 15860, February·1953

253. Grimes, A., *Rapid Launching from a Pair of Shipborne Catapults,* RAE, Technical Note NA.254, ARC Paper 16282, July 1953

254. Grimes, A., *Ventral Catapulting of Undercarriageless Aircraft,* RAE, Technical Note NA.261, ARC Paper 16833, March 1954

255. Hogg, H. & Warren, C. H. E., *Model Investigation of the Landing of an Undercarriageless Jet Fighter on a Flexible Deck,* RAE, Report Aero 2076, ARC Paper 9156, October 1945

256. Hotson, A. W. & Rouse, Cdr P. M., *A Comparison of Some Layout Schemes for Aircraft Carriers, Including the Use of Flexible Decks,* RAE, Technical Note NA.256/CE.7, June 1953

257. Hotson, A. W., *Landing Area Layout on Angled Deck Aircraft Carriers, and the Practicability of a Safety Fence to Protect the Park, Lift and Catapult Areas,* RAE, Technical Note NA.258, August 1953

258. Hotson, A. W., *Rapid Handling Trials of Vampire Aircraft on a Flexible Deck,* RAE, Technical Note NA.259, August 1953

259. Hotson, A. W., *The Shape of Arrester Hooks for Naval Aircraft,* RAE, Technical Note Naval 5, January 1955

260. Howie, A. K. M., *Preliminary Investigation of Fatigue Life of Arrester Gear Steel Wire Rope,* RAE, Technical Memorandum Naval 192, November 1964

261. Hufton, P. A., Ewans, J. R. & Warren, V. G., *Note on Various Methods of Improving the Take-Off Performance of the "Stirling",* RAE Report BA.1657, ARC Paper 5088, April 1941

262. James, C. A., *The Interference Problem between an Aircraft and Bridle During Catapult Take-Off,* RAE, Technical Memorandum Naval 211, August 1968

263. Johnston, J. B. B., *A Review of Bridle Catchers and Bridle Catching Problems,* RAE, NAD 7003, c.1965

264. Kell, C. & Plascott, R. H., *Operation of Jet-Propelled Aircraft from Aircraft Carriers – the Effect of High Temperature Gas Streams on Decks,* RAE, Technical Note Aero 1662, July 1945

265. Lean, D., *The Carrier Deck-Landing Properties of Five Jet Aircraft,* RAE, Report No. Aero 2465, June 1952

266. Lean, D., *A Review of Current Deck Landing Techniques,* RAE, Technical Note Aero 2206, ARC Paper 15545, November 1952

267. Lean, D., *Interim Note on Trials with an Optical Glide Path Indicator,* RAE, Technical Memo Aero.320, ARC Paper 15612, December 1952

268. Martin, P. R., *German Naval Catapults,* RAE, Technical Note NA.85, October 1947

269. Martin, P. R., *The Development of Slotted-Tube Catapults Using Hydrogen Peroxide, by the Walter Werke, Kiel,* RAE, Technical Note NA.182, April 1948

270. Martin, P. R. & Chisman, S. W., *The RAE Fabric-Net Safety Barrier for Jet Aircraft – Preliminary Development for Use with Sea Vampires,* RAE, Technical Note NA.208, April 1950

271. Martin, P. R., Willis, J. M. N. & Cox, S. G., *The RAE Safety Barrier Mk.I for Sea Hornet Aircraft,* RAE, Technical Note NA.186, ARC Paper 14028, December 1950

272. Martin, P. R. & Cox, S. G., *Review of Safety Barrier Engagements in British Aircraft Carriers, October 1948 – October 1950,* RAE, Technical Note NA.226, ARC Paper 14735, September 1951, ARC

273. Martin, P. R. & Chisman, S. W., *Safety-Barrier Trials of Aircraft Having Metal-Bladed Counter-Rotating Co-Axial Propellers,* RAE, Technical Note NA.222, February 1951

274. Martin, P. R. & Cox, S. G., *The RAE Fabric Net Safety Barrier for Jet Aircraft,* RAE, Technical Note NA.241, ARC Paper 15208, April 1952

275. Martin, P. R., *The Safety-Barrier Problem in Aircraft Carriers,* RAE, Technical Note NA.225, ARC Paper 16054, April 1953

276. Mollart-Rogerson, J., *The Development of an Improved Design of Aircraft Hold Back Gear for use during Tail-Down Launching,* RAE, Naval 27, March 1946

277. Nixon, R. J., *Note on Permissible Increases in Wind Conditions Down Deck and on Accelerators at the Beginning of Stroke,* RAE, Catapult Section Note 52, September 1943

278. Noble, E. F. H., *Review of Naval Aircraft and Arrester Gear Performance,* RAE, Technical Note NA.252, ARC Paper 15861, February 1953

279. Pierce, D., Bullen, N. I. & Stevens, D. R., *Arresting Experiments with a Ball and Bead Type of Drag Chain,* RAE, Technical Memorandum NA 67, February 1953

280. Pierce, D., *Experiments with Drag Chains for Arresting Aircraft – Alternatives to Anchor Chain,* RAE, Technical Memorandum NA. 71, March 1953

281. Poole, J., *Naval Air Department's Activities Reviewed, 'RAE News'* (House Magazine), Vol. 20, No.11, Nov: 1967

282. Rouse, Cdr P. M., *A Review of Catapult Development,* RAE, Technical Note Naval 8, ARC Paper 17825, March 1955

283. Salmon, P., *A Review on Launching Schemes for Aircraft,* RAE, Paper ref: D.O/1364, May 1939

284. Sharwood, T. C. & Andrews, L. W., *Deck Landing Trials of Firefly Mk.6 Aircraft Fitted with Fairey Hydraulic Damper for Arrester Hook (Mod.1210) and Raised Snap Gear on HMS* Vengeance, *February 1951,* RAE, Technical Note NA.228, September 1951

285. Sharwood, T. C., *Recording Arrangements and Procedure Used During Aircraft Deck Landing Trials on HMS* Illustrious *in 1951,* RAE, Technical Note NA.231, ARC Paper 15474, September 1952

286. Smith, C. S., *Dead Load Trolley for "Flush" Catapult at NAE,* RAE, Technical Note CE.6, July 1953

287. Sterne, L. H. G. & Kirk, F. N., *Consideration of Some Problems Affecting the Design of Future Carrier-Borne Aircraft,* RAE, Technical Note Naval 7, ARC Paper 17435, February 1955

288. Symons, P. J. R. & Raynor, D., *Uprating of the Mk.13 Arresting Gear,* RAE, Technical Note Naval 35 (Part 1) August 1959

289. Thomlinson, Dr J., *Performance Characteristics of Naval Arrester Gear*, RAE, Catapult Note No.39, 1942

290. Thomlinson, Dr J., *Carrier Borne Aircraft – New Requirements. Note on the Special Requirements for Assisted Take-Off*, RAE, Catapult Section Note 49, August 1943

291. Thomlinson, Dr J., *Flying Off Auster Aircraft from L.S.T.s*, RAE, Technical Note NA.76, April 1946

292. Thomlinson, Dr J., *Forward Airfields – Part II: Airfield Equipment*, RAE, Technical Note Aero.1920/NA.86, ARC Paper 10032, 1948

293. Thomlinson, Dr J. (compiler), *The Development of a Flexible Deck on which to Land Undercarriageless Aircraft*, RAE, Technical Note NA.87, ARC Paper 11455, January 1948

294. Thomlinson, Dr J., *Theoretical Treatment of the Aircraft Arresting Gear Problem*, RAE, Technical Note NA.194, ARC Paper 12579, May 1948

295. Thomlinson, Dr J. & Pierce, D., *Pneumatically Supported Flexible Landing Deck for Undercarriageless Aircraft*, RAE, Technical Note NA.192, ARC Paper 12485, April 1949

296. Thomlinson, Dr J., *Proposals for Arresting Gear, Take-Off and Ground Handling Aids, for Forward Airfields*, RAE, Technical Note NA.196, June 1949

297. Thomlinson, Dr J., *Design Study of a Friction Type Arresting Gear for Aircraft*, RAE, AARD Paper No.59, ARC Paper 15270, March 1950

298. Thomlinson, Dr J., *Behaviour of Ropes Under Longitudinal and Transverse Impact, with Particular Reference to Aircraft Arresting Gears*, RAE, Report NA.227, ARC Paper 14877, February 1952

299. Thomlinson, Dr J. & Pierce, D., *Development and Testing of an Aircraft Arresting Gear for 12,000lb, 120 Knots and 3½g Maximum Retardation*, RAE, Technical Note NA.238, ARC Paper 14963, March 1952

300. Thomlinson, Dr J., *A Note on the Present Position Regarding Fundamental Knowledge of the Arresting Gear Problem*, RAE, Technical Memo NA.60, October 1952

301. Thomlinson, Dr J., *Report on Two Visits to the United States Navy Bureau of Aeronautics to Provide Information on Flexdeck Aircraft*, RAE, Technical Note NA.253 (Part 2), ARC Paper 16087, April 1953

302. Thomlinson, Dr J., *A Study of the Aircraft Arresting Hook Bounce Problem*, RAE, Technical Note NA.263, ARC Paper 16985, May 1954

303. Thomlinson, Dr J., *A Linear Hydraulic Energy Absorber for Aircraft Arresting Gears*, RAE, Technical Note Naval 54, February 1964

304. Thomlinson, Dr J., *Direct Acting Gear*, 'RAE News', (House Magazine) Vol. 20, No.11, Nov: 1967

305. Tye, W. & Fagg, S. V., *Notes on Assisted Take-Off*, RAE, Report AD.3149, ARC Paper 4870, August 1940

306. Ward, Lt E. W., Follett, A. J. & George, F. G., *Report on the Installation and Trials of BXS.1 Catapult in HMS Perseus*, RAE, Technical Note Naval CE.9, May 1954

307. Ward, Lt/Cdr E. W., George, F. G., Holloway, P. G. & Stevens, J., *Trials of BS.4 Catapults in HMS Ark Royal*, RAE, Technical Note Naval 12, February 1956

308. Williams, P., *Scientist at Sea*, 'RAE News', (House Magazine) Vol. X, No. 9, September 1957

309. Willis, J. M. N., *Arrester Gear Trials of Seafire XV, with Sting Hook Installation*, RAE, Catapult Section Test Note 105, August 1944

310. Willis, J. M. N., *Proofing Trials of Aircraft Arrester Gear Installation – Firebrand I at 15,000lb Weight*, RAE, Catapult Section Test Note 106, August 1944

311. Willis, J. M. N. & Rowley, K. U., *Investigation of Performance of Safety Barrier Gear – Interim Report*, RAE, Catapult Technical Note 62, January 1945

312. Willis, J. M. N., *The Suitability of American Type Safety Barrier Gear for Use with British Aircraft and Aircraft Carriers*, RAE, Catapult Technical Note 63, March 1945

313. Willis, J. M. N., *A Survey of Requirements for Arrested Landings of Naval Aircraft*, RAE, Naval Aircraft Technical Note 64, December 1945

314. Willis, J. M. N., Chisman, S. W. & Bullen, N. I., *Measurement and Suppression of Tension Waves in Arresting Gear Rope Systems*, RAE, Technical Note NA.204, January 1950 (R&M 2981, 1957)

315. Willis, J. M. N. & Pierce, D., *Tests on Liquid Spring Rope Anchorages for Arresting Gears*, RAE, Technical Note NA.247, October 1952

316. Wilson, A. E., *Catapult Crew to Your Stations*, 'RAE News', (House Magazine) Vol. 22, No. 7, July 1969

317. (Anon – Staff of NAD), *Facilities and Equipment for Research and Development in Naval Aviation*, RAE, Technical Memo Naval 59, ARC Paper 18674, September 1956

318. (Anon), *The Naval Air Department, Royal Aircraft Establishment Bedford (An Outline of Work and Facilities)*, RAE Publicity Brochure, c.1970

319. (Anon), *Pyestock – A Celebration of the Gas Turbine Engine*, DERA Publicity Brochure, DERA, Pyestock, 1996

320. (Anon), *Extension of Plan of Dispersal Park*, RAE Drawing PDO 2853/18, dated 22.4.1944

321. (Anon), *Farnborough Aerodrome and Surrounding Buildings*, Royal Aircraft Factory Drawing No.2064X, dated June 1916. (Occurs as DERA Neg: O-946)

322. (Anon), *Gas Turbines*, a privately bound, personal collection, by G. G. J. Cooper of papers, on gas turbine technology published by the Institution of Mechanical Engineers during 1945–50.

A reference to (ARC****) indicates the serial, identification number of an Aeronautical Research Council paper within the DERA Archive, at Pyestock. An ARC reference may be additional to other Technical Note references for the same paper.

Many copies of the RAE House Journal *RAE News* carry references to, and short articles by, members of Naval Aircraft Department. These sources have been drawn upon in the foregoing pages, mostly without direct reference to them. They are to be found in the following issues:

June 1948, November 1948, August 1949, December 1949, April 1951, January 1952, December 1953, March 1954, December 1955, May 1957, September 1957, February 1958, May 1958, July 1958, November 1967 & July 1969.

GLOSSARY

AA	Anti-aircraft (artillery fire – 'Ack-Ack')
AARD	Aircraft and Airfields Research & Development (Committee)
A&AEE	Aircraft and Armament Experimental Establishment, Boscombe Down
abaft	'Behind' – naval parlance
abeam	'To one side' of a ship or aircraft – naval parlance
acre	Imperial unit of land area, approximating to 0·4ha
ADDLs	Airfield Dummy Deck Landings
ADRDN	Assistant Director Research and Development, Navy – MAP post
AERE	Atomic Energy Research Establishment, Harwell
AFC	Air Force Cross – a military honour
AFEE	Airborne Forces Experimental Establishment, Beaulieu & elsewhere
aft	'Towards the stern' of a ship or aircraft.
ahead	'In front of the ship' – naval parlance
ALD	Arrested Landing Deck – NAD site at RAE Bedford
ante	Before (textual reference)
ARC	Aeronautical Research Council
ASI	Air Speed Indicator, aircraft instrument connected to pitot head
astern	'Behind the ship' – naval parlance
ASW	Air to Surface Weapon
ATOG	Assisted Take-Off Gear, BH.III Accelerator
AUW	All-Up Weight
bar	Barometric pressure, 1 bar = 14·7psi = 101·35kN/m²
'Bats'	See DLCO
BDAG	Brake Drum Arrester Gear
bolter	Aircraft signalled to baulk its landing during flight deck recovery and to make a new approach
bow(s)	The front, 'sharp end' of a ship
BSS	British Standard Specification, issued by the British Standards Institute
c.	*Circa* – approximate date
CALE	Catapult Aircraft Loading Equipment Base – NAD site at RAE Bedford
CAM	Catapult Aircraft Merchantman (cargo ship carrying an aircraft mounted on a rocket catapult – known as a CAM-ship)
Capt	Captain (Naval Officer)
CATOG	Cordite Assisted Take-Off Gear, C.1 Accelerator
CBE	Commander of the British Empire – a civil/military honour
Cdr	Commander (Naval officer)
cg	Centre of gravity
chopper	Naval parlance for helicopter
C-in-C	Commander-in-Chief
CO	Commanding Officer
CVE	Carrier Vessel, Escort
DAAG	Direct Acting Arrester Gear – also DAG & DAX
DACR	Director of Airfields and Carrier Requirements – Admiralty post
DCAG	Drag Chain Arrester Gear
DCTO	Direction Controlled Take-Off
D-Day	6 June 1944, date of allied landings in Normandy, World War 2
DERA	Defence Evaluation & Research Agency – the all-embracing name for major Government Research Establishments including NGTE, RAE, RRE, etc
DLCO	Deck Landing Control Officer – Batsman or 'Bats'
DLPS	Deck Landing Projector Sight
DNC	Directorate of Navy Construction, Admiralty post
DPN	Diamond pyramid number, a dimensionless comparative figure indicating the intensity of surface hardness of a metal. The number increases with the hardness
DSC	Distinguished Service Cross – a military honour
DSO	Distinguished Service Order – a military honour
DTD	Directorate of Technical Development, MAP/MoS Department
E-in-C	Engineer-in-Chief's Department – Admiralty post
erk	Slang reference to a member of RAF ground crew
ETPS	Empire Test Pilots' School, Farnborough, later at Boscombe Down
FAA	Fleet Air Arm
fanzine	'Fans' Magazine' – a journal devoted to the particular interests of members of a society or club
FASTA	Farnborough Air Sciences Trust Association
ff	Pages following (textual reference)
fig.	Figure (textual reference)
Fl Lt	Flight Lieutenant (RAF Officer)
fo'c'sle	Forecastle, the forward extremity of the weather deck – naval parlance
for'ard	Forward, towards the bow – naval parlance
ft	Foot/feet, Imperial unit, approximating to 300mm
ft.lbf	Foot pounds-force, Imperial unit for energy measurement, approximating to 1·36J
'g'	A factor relating to gravitational forces, particularly in acceleration and retardation
g	Gram weight
goofer	Non-participating observer of shipboard flying activities
'Guns'	Chief Gunnery Officer in ship's crew
gutser	A belly landing of an aircraft, with its undercarriage retracted
ha	Hectare
HMAS	His/Her Majesty's Australian Ship
HMCS	His/Her Majesty's Canadian Ship
HMS	His/Her Majesty's Ship
HMSO	His/Her Majesty's Stationery Office
hp	Horsepower = 0·746kW
hr	Hour(s)
Hz	Hertz, unit of frequency – cycles per second
ibid.	'in the same place' (textual reference)
IJN	Imperial Japanese Navy
illn.	Illustration (textual reference)
in	Inch, inches, Imperial dimension approximating to 25mm
ISBN	International Standard Book Number
kg	Kilogram(s)
kgfm	Kilogram metres-force, measurement of energy
km	Kilometre(s)
kN	Kilonewton(s)
knot	Nautical mile, Imperial measure equal to 6080ft, normally used as a rate of speed approximating to 0·51m/sec
kt, kts	Knot, knots
kW/kVA	Kilowatt(s)/Kilovoltamp(s)
l	Litre(s)
lb	Pound(s) weight/force. Imperial measurement approximating to 0·45kg
LCT	Landing Craft, Tank (Sometimes, LST = Landing Ship, Tank)
L/L	Lend/Lease arrangements with the USA during World War 2

LMS	London, Midland and Scottish (Railway)
LSO	Landing Signals Officer (superseded DLCO)
Lt	Lieutenant (Naval or Royal Marines officer)
Lt Cdr	Lieutenant-Commander (Naval officer)
Lt Col	Lieutenant-Colonel (Army or Royal Marines officer)
m	Metre(s)
MAC	Merchant Aircraft Carrier (bulk cargo ship with flight deck and operational aircraft – known as a MAC-ship)
MAP	Ministry of Aircraft Production, later MoS
MDAP	Mutual Defense Assistance Program – Lend-Lease 'Mark.II' (Post-war)
ml	Millilitre(s)
mm	Millimetre(s)
MoS	Ministry of Supply
MoW	Ministry of Works, later Ministry of Public Building and Works
MP	Member of Parliament
mph	Miles per hour, Imperial unit of speed approximating to 0·45m/sec
m/sec	Metres per second
MV	Merchant Vessel
MW	Megawatt(s)
NAD	Naval Aircraft Department, later Naval Air Department
NAE	National Aeronautical Establishment, Bedford: subsequently, RAE Bedford
NARC	Naval Aircraft Research Committee
NATO	North Atlantic Treaty Organisation
NGTE	National Gas Turbine Establishment (subsequently, RAE) Pyestock, Farnborough
OD	Ordnance Survey Datum (in feet) above sea level
'oggin	The sailor's name for his watery environment
op.cit.	'in the work cited' (textual reference)
OS	Ordnance Survey
pitot	Sensing element to ASI, to indicate air speed
port	The left-hand side of a ship, or aircraft, when facing forward
PPRAeS	Past President, Royal Aeronautical Society
psi	Pounds per Square Inch, Imperial unit of pressure, approximating to 0·07kgf/cm², 6·89kN/m² or 0·07bar
psig	Pounds per Square Inch (Gauge)
PUAG	Purpose Use Arrester Gear (Arresting Gear, Rotary Hydraulic Mark.21)
qv	'which see' (textual reference)
RAE	Royal Aircraft Establishment, Farnborough, Hampshire; later Royal Aerospace Establishment; at the time of writing DERA, now QinetiQ, Farnborough
RAF	Royal Air Force (prior to 1918, Royal Aircraft Factory)
ranging	The procedure of bringing an aircraft up from the hangar deck and positioning it on the flight deck for take-off
RATOG	Rocket Assisted Take-Off Gear
RDF	Radio Direction Finding
RFC	Royal Flying Corps – precursor of the RAF
RINA	Royal Institution of Naval Architects
RMS	Royal Mail Ship
RN	Royal Navy
RNAS	Royal Naval Air Station – if followed by a name
RNAS	Royal Naval Air Squadron – if followed by a number
rpm	Revolutions per minute
RPV	Remotely Piloted Vehicle (pilotless target drone)
RRE	Royal Radar Establishment, Malvern
RSC	Rolled Steel Channel, structural steel bar, 'U'-shaped section

RSJ	Rolled Steel Joist, structural steel bar, 'I'-shaped section
S	Serviceable, a state of readiness of an aircraft
SBAC	Society of British Aerospace Constructors
SBAG	Safety Barrier & Arrester Gear base, NAD site at RAE Bedford
sec	Second(s), unit of time
Sgt	Sergeant, Army or Royal Marine rank
SI	Système International, European agency for Metrication
snoop	Bridle catcher at the release end of an accelerator, probably onomatopœic in origin
sqft	Square foot/feet, Imperial unit of area, approximating to 0·09m²
Sqn Ldr	Squadron Leader (RAF Officer)
starboard	The right-hand side of a ship, or aircraft, when facing forward
stern	The rear of a ship, the 'blunt' end
striking down	The procedure for removing aircraft from the flight deck into the hangar below
swg	Standard Wire Gauge, British Standard for thickness of metal sheet and wire
TBR	Torpedo Bomber reconnaissance
ton	Imperial unit of weight, approximating to 1·016 tonnes
UK	United Kingdom of Great Britain and Northern Ireland
UP	Unrotating Projectile – small, solid propellant, rocket motor
U/S	Unserviceable, a state of readiness of an aircraft
USA	United States of America, also US
USAAF	United States Army Air Force
USN	United States Navy
USS	United States Ship
UTS	Ultimate tensile strength. The value at which a test piece of steel or other metal breaks. Normally the figure is in tons per square inch, generally known as 'ton', e.g. 55 ton.
VE Day	9 May 1945, Victory in Europe Day, end of war in Europe
VIP(s)	Very Important Personage(s)
VJ Day	15 August 1945, Victory over Japan Day, end of World War 2
VSTOL	Vertical/Short Take-Off & Landing aircraft
WC	Water Closet
W Cdr	Wing Commander (RAF officer)
World War I	World War I, 1914/1918
World War 2	World War II, 1939/1945
yard	Imperial dimension, approximating to 0·9m
(#***)	Bibliographic reference to Appendix 15
°	Degrees, measurement of angle; also sensitivity of photographic film
°C	Degrees, measurement of temperature – Centigrade/Celsius
°F	Degrees, measurement of temperature – Fahrenheit
"	Inch, inches (see 'in' ante)
'	Foot, feet, (see 'ft' ante)
I	Roman numeral for 1
V	Roman numeral for 5
X	Roman numeral for 10
L	Roman numeral for 50
C	Roman numeral for 100
D	Roman numeral for 500
M	Roman numeral for 1,000

APPENDIX 17
CREDITS AND ACKNOWLEDGEMENTS

CREDITS

By far the greatest number of images in this record have their origin in the erstwhile RAE photographic archive housed at DERA, Pyestock. This collection of photographic plates and technical reports was amassed at Farnborough during the period of 73 years when it was known as the Royal Aircraft/Aerospace Establishment. The privilege accorded me by the archive personnel in the generous provision of making this material available to me is appreciated and herein acknowledged. Permission for the reproduction of these images was granted at the time of my research by the Intellectual Property Department of DERA. This permission was confirmed subsequently by the Defence Procurement Agency. Each illustration bearing the caption '© Crown Copyright/MOD' is reproduced with the permission of the Controller of Her Majesty's Stationery Office, to whom acknowledgement is made and the author's thanks recorded.

Additionally, a significant number of photographs – identified throughout as 'Falcon' – have been provided from the Falcon Aviation Publications collection, the imprint of Peter J. Cooper (my son), whose permission for their reproduction is gratefully acknowledged here.

Images carrying the reference 'Fleet Air Arm Museum' are reproduced by permission from the curator of the museum collection, RNAS Yeovilton. Extracts from some of the texts held by the FAAM are identified and reproduced with the curator's permission. (Jerry Shore, and latterly Catherine Cooper [no relation] have been the contacts here.) Useful contributions also have been made by the Society of Friends of the Fleet Air Arm Museum.

Two images are included from the RAE departmental files held at the National Archives, Kew. These are identified accordingly, with their appropriate reference numbers included.

Acknowledgement is made to John M. Dibbs and the Plane Picture Company who own the copyright for the three superb, high-quality air-to-air colour photographs of Hawker naval aircraft.

My thanks are due to the United States Navy, via NAVAIR Public Affairs (Renee Hatcher), for provision of, and permission to publish, three of their photographs.

I am indebted to the United Kingdom Atomic Energy Authority (UKAEA), Harwell, for permission to reproduce their photograph of recent excavations on the site of the erstwhile RAE Mark III Catapult at that ex-RAF Station. Particular mention is made of Paul Atyeo, Malcolm Crook and Nick Hance.

The presentation of photographs by Rolls-Royce Heritage Trust of the company's 'Kestrel' series of aero-engines (from which the composite picture of the power plant for the RAE Mark III Catapult was made) is acknowledged with gratitude. Particular thanks are due to Richard Haigh.

A number of images from my own collection are indicated as such. Some of these are detailed photographs of exhibits in the Fleet Air Arm Museum, taken by me, and are reproduced here by permission of the curator.

The Admiral Arrives. The author is welcomed aboard HMS *Pretoria Castle* by her captain, G. C. Dickens RN. Unbeknown to him (and suffering from *mal-de-l'air*) he was involuntary stand-in at a rehearsal for a visit by an admiral the following day. The aircraft is Grumman Avenger I (JZ181) flown from 778 Trials Squadron at RNAS Ford. October 1945. (Author)

In addition to the above, I have obtained valuable information, sometimes accompanied by images, willingly supplied by companies, professional bodies and individuals. The first group includes Messrs Brown Bros & Co; Messrs MacTaggart Scott & Co (both of Edinburgh); Marconi Plc, Chelmsford, (successors to the General Electric Company of Wembley); Weir Pumps Limited Manchester, (successors to Mather & Platt of Manchester). Professional bodies include the Institution of Mechanical Engineers (Amanda Spiers); The Royal Aeronautical Society (Brian Riddle, Librarian); The Royal Institution of Naval Architects (John Date); and publishers of the journals *Aircraft Illustrated*, *The Engineer*, *Engineering* and *Model Engineer*. My appreciation of, and grateful thanks for, each of these is hereby acknowledged.

Quotations appearing in the text from published works are reproduced by permission of the relevant publishers and/or authors.

I must apologise for any occurrence in which a courtesy of acknowledgement unfortunately may have been overlooked. Such oversight must be put down to ignorance or my inability to trace the origin of the material which has been reproduced in this work.

ACKNOWLEDGEMENTS

In preparing this treatise I have been indebted to a greater or lesser extent to many, both ex-colleagues and new acquaintances, who, through my manifold enquiries, have become friends.

Firstly, thanks for, and appreciation of, the assistance and support given by my own family must be given priority. My wife, Reneanne, undoubtedly takes pole position for her patience, toleration and good humour over a period of years when I appeared, perhaps single-mindedly, to be surrounded by a seemingly disordered array of papers, books and photographs; not always confined to the spare bedroom. My four sons, Peter, Paul, Timothy and Andrew, in turn, have made their appropriate contribution to this book. Peter must be given prime position as, in the guise of Falcon Aviation Publications, he has provided a large measure of textual information as well as being the supplier of images in proliferation.

For his unending encouragement I must acknowledge the contributions by Captain Eric 'Winkle' Brown RN who provided the Foreword and to whom reference is already made in the text.

For much of the RAE information I am indebted to the staff of its successor, DERA, (now QinetiQ) Farnborough, particularly to: Mrs E. (Liz) Peace, Company Secretary; Anthony Bowdery, Intellectual Property Department; Maurice Shakespeare, representing the Royal Aircraft Establishment Heritage Group and the Farnborough Air Sciences Trust Association (FASTA); Helen Gristwood – Archive; Richard Snell – Archive; Tony Knight – Airfield Manager; Chris K. D. Mitchell – Archive; and Brian Kervell – past-Custodian of the RAE Museum. David Frith was responsible for the computer artistry in producing good images from poor or damaged originals. To these must be added, from DERA, Bedford: Gordon Ingle, Derek Lewis and Anthony J. Smith; and from DERA, Malvern: Ron Henry.

The contributions from aviation authors and historians unconnected with the foregoing are acknowledged these include: Bill Baguley, Walter C. Carter, Dr Robert B. Kirby, Mike Hodgson (Thorpe Camp Preservation Group, Woodhall Spa), Ralph Lewis, Peter Parnham (Lincolnshire Aviation Heritage Centre, East Kirkby), George Ray, Brian Walker, Wing Commander D. R. Collier Webb (RAF Retd).

Being within the privileged, life-long camaraderie of Farnborough's ex-apprentices, some of these ex-lads have shared with me an involvement in the preparation of this book. As the 'grand old man' of them all – my own entry was in 1936 – I exercise my patriarchal privilege over those youngsters who followed me and who merit special attention. They are: Stanley W. Chisman (1937); Donald Pierce (1937); Richard Dennis (1942); Sqn Ldr Clive Ellam (1948); George Ray (1956); Maurice Shakespeare (1960); and Peter J. Cooper (1963).

Finally, this book would not have been possible without the valued contribution throughout the publication process made by Nick Grant and Matthew Wharmby of Ian Allan Publishing together with Alexander Stilwell, their advisory editor. To all, my most sincere thanks.

THE AUTHOR

Geoff Cooper made his lusty appearance in this scene on a snowy, Sunday morning, 14th March 1920. He was born at his paternal grandparents' home on the Broadlands Estate, Romsey in Hampshire; later, the home of Earl Mountbatten (who married Lady Edwina, the daughter of the big house). Grandmother rented a smallholding of some three acres at Ashfield Lodge while grandfather and father managed their engineering business in a large warehouse type building located in the railway station yard at Romsey. While the business was a general practice in engineering, it majored on the equipping of flour and grist mills, powered by wind, water or steam. Due to the general industrial slump in the late 1920s, and the closure of small mills by the giants, Hovis, Rank, Spiller *et al* trade began to slacken. The business was closed in 1931 and father moved his family to Farnborough – my maternal grandparents' home – where he had taken an appointment in charge of operating and maintenance of machinery in the council main drainage depot.

I attended Farnborough Secondary School, leaving with good grades in the London General School Certificate, and was accepted as an Engineering Apprentice at the RAE, starting there in September 1936. My weekly wage was seven shillings and twopence (36p) of which, on father's insistence, I had to contribute a half-crown (13p) towards board and lodging. (My parents, father having taken on a managerial post, had moved away during my last days at school.) As an apprentice, most of my final year was spent in the Catapult Section of the RAE Main Design Office. On completion of my training I was offered a post as Laboratory Assistant within the Section, with promotion to Technical Assistant Grade III after a year. It was here that I acquired the knowledge and expertise of supporting the Fleet Air Arm in the provision of improved tools and equipment for its operational duties.

Elsewhere in this text I refer to my experience of flying trials at sea covering a period of four years 1942-1945. This involved the conducting of experimental trials in a total of thirty-one different aircraft carriers (listed in Appendix 4), amounting to forty-six visits, and amassing a total of 134 days at sea. These included visits to Clydeside, Rosyth, Scapa Flow, Belfast Lough, Liverpool Docks, and Portsmouth as points of departure. On occasion I was directly involved as passenger/observer in the catapult launching and arrested landing procedures.

Although my association with Naval Aircraft Department (successor to the Catapult Section) was severed at the end of December 1945, in my new engineering post with the Ministry of Works, it was only a matter of months before I was involved again with RAE. I was recruited into the Heavy Research Plant at my new London workplace and before long I was engaged in the provision of a variety of new and interesting engineering research facilities of the government research establishments at Westcott, Malvern, Pyestock and Farnborough. My new projects at RAE were the Generator Test House [P.1], High Speed Wind Tunnel [R133A], Laboratory Buildings 134 [Q134] & 177 [R177], conversion of F1E into Guided Weapons Workshop [P67] and the conversion of the Hillside Convent building into the new RAE Technical College & Apprentices' Workshop [R.1] with the Ball Hill Development [X & Y Areas] following on. In the 1960s I was involved in the design, construction and commissioning of the Cell 4 complex (for development of the Concorde engines and intakes in supersonic flight) at Pyestock, followed by a spell supervising the experimental use of the man-carrying centrifuge at the RAF Institute of Aviation Medicine – a lodger unit at RAE. While at Pyestock I operated a contract with Vickers Supermarine involving Joe Smith – of Mitchell and 'Spitfire' fame – in the design of some heavy engineering equipment; a far cry from his expertise on monocoque fuselages of Spitfire *et al*.

In 1968 I moved from Farnborough to take up a senior engineering appointment with the Building Research Establishment at Watford as manager of its Equipment Design Office and Manufacturing Workshops. Having spent a lifetime in close association with engineering research facilities, it is in my retirement that I involved myself in the nostalgic exercise which mushroomed – through the encouragement of others – into something far greater than originally anticipated. The foregoing is the result, which I hope will prove to be of interest to those who care to read.

During the compiling of this treatise I gave a twenty-five minutes off-the-cuff dissertation on the activities of Naval Aircraft Department to Onyx, a video production company for the ITV Meridian set-up. A section of this was screened by Southern TV in their six-part series 'Farnborough Above and Beyond' - now available on VHS video.